CONTINUUM VECTOR DERIVATIVES

W9-BLA-775

Cartesian Coordinates

$$\nabla \cdot \mathbf{A} = \frac{\partial A_x}{\partial x} + \frac{\partial A_y}{\partial y} + \frac{\partial A_z}{\partial z}$$

$$\nabla \times \mathbf{A} = \left(\frac{\partial A_z}{\partial y} - \frac{\partial A_y}{\partial z} \right)\hat{\mathbf{x}} + \left(\frac{\partial A_x}{\partial z} - \frac{\partial A_z}{\partial x} \right)\hat{\mathbf{y}} + \left(\frac{\partial A_y}{\partial_x} - \frac{\partial A_x}{\partial_y} \right)\hat{\mathbf{z}}$$

$$\nabla V = \frac{\partial V}{\partial x}\hat{\mathbf{x}} + \frac{\partial V}{\partial y}\hat{\mathbf{y}} + \frac{\partial V}{\partial z}\hat{\mathbf{z}}$$

$$\nabla^2 V = \frac{\partial^2 V}{\partial x^2} + \frac{\partial^2 V}{\partial y^2} + \frac{\partial^2 V}{\partial z^2}$$

Cylindrical Coordinates

$$\nabla \cdot \mathbf{A} = \frac{1}{r}\frac{\partial}{\partial r}(rA_r) + \frac{1}{r}\frac{\partial A_\phi}{\partial \phi} + \frac{\partial A_z}{\partial z}$$

$$\nabla \times \mathbf{A} = \left(\frac{1}{r}\frac{\partial A_z}{\partial \phi} - \frac{\partial A_\phi}{\partial z} \right)\hat{\mathbf{r}} + \left(\frac{\partial A_r}{\partial z} - \frac{\partial A_z}{\partial r} \right)\hat{\boldsymbol{\phi}} + \frac{1}{r}\left[\frac{\partial}{\partial r}(rA_\phi) - \frac{\partial A_r}{\partial \phi} \right]\hat{\mathbf{z}}$$

$$\nabla V = \frac{\partial V}{\partial r}\hat{\mathbf{r}} + \frac{1}{r}\frac{\partial V}{\partial \phi}\hat{\boldsymbol{\phi}} + \frac{\partial V}{\partial z}\hat{\mathbf{z}}$$

$$\nabla^2 V = \frac{1}{r}\frac{\partial}{\partial r}\left(r\frac{\partial V}{\partial r} \right) + \frac{1}{r^2}\frac{\partial^2 V}{\partial \phi^2} + \frac{\partial^2 V}{\partial z^2}$$

Spherical Coordinates

$$\nabla \cdot \mathbf{A} = \frac{1}{r^2}\frac{\partial}{\partial r}\left(r^2 A_r \right) + \frac{1}{r \sin \theta}\frac{\partial}{\partial \theta}(A_\theta \sin \theta) + \frac{1}{r \sin \theta}\frac{\partial A_\phi}{\partial \phi}$$

$$\nabla \times \mathbf{A} = \frac{1}{r \sin \theta}\left[\frac{\partial}{\partial \theta}(A_\phi \sin \theta) - \frac{\partial A_\theta}{\partial \phi} \right]\hat{\mathbf{r}} + \frac{1}{r}\left[\frac{1}{\sin \theta}\frac{\partial A_r}{\partial \theta} - \frac{\partial}{\partial r}(rA_\phi) \right]\hat{\boldsymbol{\theta}}$$

$$+ \frac{1}{r}\left[\frac{\partial}{\partial r}(rA_\theta) - \frac{\partial A_r}{\partial \theta} \right]\hat{\boldsymbol{\phi}}$$

$$\nabla V = \frac{\partial V}{\partial r}\hat{\mathbf{r}} + \frac{1}{r}\frac{\partial V}{\partial \theta}\hat{\boldsymbol{\theta}} + \frac{1}{r \sin \theta}\frac{\partial V}{\partial \phi}\hat{\boldsymbol{\phi}}$$

$$\nabla^2 V = \frac{1}{r^2}\frac{\partial}{\partial r}\left(r^2 \frac{\partial V}{\partial r} \right) + \frac{1}{r^2 \sin \theta}\frac{\partial}{\partial \theta}\left(\sin \theta \frac{\partial V}{\partial \theta} \right) + \frac{1}{r^2 \sin^2 \theta}\frac{\partial^2 V}{\partial \phi^2}$$

PRO DEO ET VERITATE

duPont-Ball Library
STETSON
UNIVERSITY
DeLand, Florida

QC631 .V48 1988

Fields and electrodynamics

3 4369 00027130 6

QC
631
.V48
1988

FIELDS AND ELECTRODYNAMICS

A COMPUTER-COMPATIBLE INTRODUCTION

DATE DUE

MAY 12 1995

MAY 12 1995 S

WITHDRAWN

FIELDS AND ELECTRODYNAMICS

A COMPUTER-COMPATIBLE INTRODUCTION

PIETER B. VISSCHER

University of Alabama

JOHN WILEY & SONS

New York · Chichester · Brisbane · Toronto · Singapore

To Helga

Copyright © 1988, by John Wiley & Sons, Inc.

All rights reserved. Published simultaneously in Canada.

Reproduction or translation of any part of
this work beyond that permitted by Sections
107 and 108 of the 1976 United States Copyright
Act without the permission of the copyright
owner is unlawful. Requests for permission
or further information should be addressed to
the Permissions Department, John Wiley & Sons.

Library of Congress Cataloging in Publication Data:

Visscher, Pieter B.
  Fields of electrodynamics. Pieter B. Visscher
    p.   cm.
  Bibliography: p.
  ISBN 0-471-82884-X
  1. Electrodynamics.   I. Title.
QC631.V48   1988   87-31873   537.6–dc19

Printed in the United States of America

10 9 8 7 6 5 4 3 2

PREFACE

The first thing that will become apparent to a reader perusing this text is that it takes a very different approach to the theory of electrodynamics than that found in other texts. The general approach taken by electrodynamics texts has hardly changed at all in the last 50 (and not very much in the last 100) years. During that time, however, the way in which calculations are actually done has changed drastically, because of the introduction of computer techniques. Retention of the traditional pedagogical approach was justified during the first two or three decades of the computer age by arguments that held computer techniques were useful only to advanced students who had access to expensive special equipment, and that such techniques were too difficult for beginning students to understand. The first argument has, needless to say, become untrue. The second argument is unfortunately true of most books on computer methods in electrodynamic calculations, which assume a thorough understanding of the continuum theory of electrodynamics and regard discrete theories as approximations to this continuum theory.

The main point of the present text is that this order (continuum, then discrete) is both pedagogically and logically inferior to the reverse order. I introduce a very simple discrete version of the equations of electrodynamics (Maxwell's equations), from which the continuum equations follow in a natural way. As I hope the reader will see, this has four important advantages over the conventional order: it (1) drastically simplifies the minimum mathematical preparation necessary for understanding electrodynamics: (2) it allows a more logical approach to the subject; (3) it develops intuition into electrodynamic phenomena much more quickly through experience with simulations; and (4) it lays the groundwork both for the use of modern calculational tools in electrodynamics and for modern discrete theories in other fields of physics and engineering.

The basic phenomena of electrodynamics (electromagnetic waves and radiation as well as electrostatics and magnetostatics) can be understood and quantitatively calculated by using only elementary algebra. All of the essential features of electrodynamics are present in the discrete theory: Gauss', Ampère's, Faraday's, and Coulomb's laws and Poynting's theorem are all exactly true. Even some fairly sophisticated concepts of topology (such as oriented manifolds and boundaries), which are necessary for a complete understanding of Ampère's law, can be introduced without calculus. The theory of partial differential equations (or for that matter calculus itself) is needed only if one wishes to consider the continuum limit; a complete physical and intuitive understanding of the field can be achieved without them. [It should not be inferred that I am against learning about the theory of partial differential equations. But even a student who understands them thoroughly will, I believe, understand electrodynamics better if his or her experience with it is not completely abstract.]

Everyone who has taught or studied electrodynamics is aware of a very confusing inconsistency in the conventional approach. The electric field is introduced through Coulomb's law as a manifestation of electric charge, and then a few chapters later the field is riding on an electromagnetic wave with no charges in sight. Somewhere in between, the hapless student is forced to completely change his

276953

notion of what an electric field is. There is a compelling pedagogical reason for tolerating this inconsistency when the conventional approach is used, of course: introducing the fundamental notion of an electric field initially as the solution to a partial differential equation is extremely hard on the student whose understanding of partial differential equations is hazy. This inconsistency is unnecessary in this text's discrete approach, as can be seen in Chapter 9.

Intuition is developed in the conventional approach by generalization from a very limited set of exactly soluble problems, involving idealizations such as infinitely long wires, infinite charged planes, and point charges. These solutions are very useful, and are treated in detail in this text. However, it is invaluable to be able to display the behavior of electric and magnetic fields in more realistic situations. The simulation programs that accompany this text allow a student to do this for a discretization of almost any problem he or she can think of.

For a student who intends to go on in a field in which computer techniques for solving partial differential equations are important (which includes at least all fields of engineering and the physical sciences, and to an increasing extent other disciplines as well), the present approach has the obvious advantage that he or she will end up with a basic understanding of how such calculations can be done. In a practical application some other method may be more efficient than the very simple ones used in this book, but the latter will at least provide a place to start. In addition, as computers become faster (and humans don't become smarter) the advantages of simplicity over computer-time efficiency can only continue to increase. For a student who intends to go on in physics, the methods learned here will be very useful in learning about a number of recent developments in physics in which discrete systems play a central role, such as lattice gauge theories of elementary particles and renormalization-group theories of condensed matter.

I thank Jane Boyd for typing the initial draft of Part C and helping in the revision of Part A, and Joseph E. Cates for writing the Apple Pascal versions of the simulation programs and revising the Apple Basic versions. I also acknowledge the hospitality of Los Alamos National Laboratory, where parts of the manuscript were written.

Pieter B. Visscher

CONTENTS

*Asterisks mark sections that can be skipped without serious loss of continuity.

Dynamical Systems

CHAPTER ONE

MECHANICAL DYNAMICAL SYSTEMS

In this book we will study the interaction of charges and currents with changing ("dynamic") electric and magnetic fields. This interaction is a special case of what is called in general a "dynamical system." In earlier physics courses you have encountered simple mechanical dynamical systems, such as weights and pulleys, masses on springs, colliding objects, rotating objects, and the like. Dynamical systems are also very important in engineering (e.g., feedback control systems such as thermostats), biology (e.g., population models, which we will discuss later), and many other fields.

Compared with these other dynamical systems, the electromagnetic system is rather abstract (you can't see electric fields, as you can weights and pulleys). It is most easily understood by someone with a clear understanding of the related concrete dynamical systems. It is the author's experience that the difficulty most commonly encountered by students in understanding electromagnetic systems is not the abstractness of the variables (e.g., electric fields) but simply the lack of experience with dynamical systems of any sort, at least in a context in which their relation to electromagnetic systems is clear.

Accordingly, we will begin with a series of examples of dynamical systems involving simple objects such as masses, springs, and rabbits. The student can then concentrate on learning the notation, mathematical description, and calculational techniques that will be needed for electromagnetic systems. In particular, the very important concept of a field will be presented from a non-electromagnetic point of view.

SECTION 1.1   A MASS ON A SPRING

Consider a mass m on a horizontal frictionless surface, connected to a spring whose spring constant is k (Figure 1.1). We know the motion is governed by Newton's second law ($F = ma$), which we can write as

$$m(dv/dt) = -kx \tag{1.1}$$

and that the velocity v is related to the displacement x (measured from the equilibrium position) by

$$\frac{dx}{dt} = v \tag{1.2}$$

3

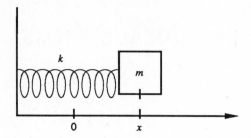

Figure 1.1 A mass on a spring.

These two first-order ordinary differential equations for two variables x and v can be turned into a single second-order one $(m\,d^2x/dt^2 = -kx)$ for x alone, but it turns out that for actual calculation the first-order ones are almost always preferable.

There are two ways to approach the study of differential equations such as Eqs. 1.1 and 1.2. One is to look for an analytic solution, that is, a solution that can be expressed as a formula involving simple functions such as exponentials, sines, or logarithms. The numerical value of the displacement x can then be calculated from the formula. The other way is to try to compute the answer numerically directly from the equations. In the present case, the first of these approaches has a lot to be said for it because we happen to have learned in mechanics how to do it. Our system is just the simple harmonic oscillator whose solution is known to be

$$x(t) = A \sin(\omega t + \phi) \tag{1.3}$$

where $\omega^2 = k/m$ and A, and ϕ are arbitrary constants (the amplitude and phase). We can learn a lot about the motion (for example how its period is related to k/m) from our knowledge about the sine function, which would be harder to deduce from a numerical solution. However, the sad fact is that in the real world (or even the world of upper-division physics) problems that are exactly soluble like this (often called "textbook problems") are few and far between. There are, of course, many "textbook problems" even in electricity and magnetism and some of them are quite useful in understanding the general principles that govern more complicated problems; we will analyze a number of them later. However, it is generally true that the more complicated equations become, the harder it becomes to find analytic solutions to them, and the less useful the analytic solutions are in understanding general principles. So to develop techniques we will need later in studying more complicated problems, we will forget that we have an analytic solution to this one and treat it numerically.

So let us suppose we know the displacement x and velocity v of the mass initially, say $x(t = 0) = 0$ and $v(0) = 1$ m/s, and look at the problem of updating these variables every Δt seconds, where Δt is some small time interval. For definiteness let us choose $\Delta t = 0.1$ s, and assume $k/m = 1.0$ s^{-2}. The meaning of the derivative in Eq. 1.1 is, of course, that the finite-difference equation

$$m\frac{v\left(t + \tfrac{1}{2}\Delta t\right) - v\left(t - \tfrac{1}{2}\Delta t\right)}{\Delta t} = -kx(t) \tag{1.4}$$

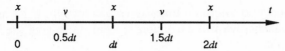

Figure 1.2   Time line, showing when the position (x) and the velocity (v) are defined.

can be made arbitrarily nearly correct by making Δt small enough. Similarly, Eq. 1.2 means the same is true of

$$\frac{x\left(t + \frac{1}{2}\Delta t\right) - x\left(t - \frac{1}{2}\Delta t\right)}{\Delta t} = v(t) \tag{1.5}$$

[You may be used to defining the derivative in terms of the "forward difference" $v(t + \Delta t) - v(t)$ instead of the "central difference" in Eq. 1.4. These are the same in the limit $\Delta t \to 0$, and Eq. 1.4 is much more accurate for nonzero Δt.]

> *Notation:*   Above, we have been careful to distinguish between the infinitesimal dt and the finite difference Δt. However, if we don't want to worry about mathematical niceties, they amount to the same thing, and for simplicity we will use the notation dt for both. Mathematicians will have no difficulty telling which is meant from the context.

Equations 1.4 and 1.5 can be regarded as equations for $v(t + \frac{1}{2} dt)$ and $x(t + \frac{1}{2} dt)$, respectively, and used to update them repeatedly. For example, we may use our value for $x(0)$ in Eq. 1.5 if $t = \frac{1}{2} dt$; it then gives us the next value $x(dt)$. However, to do so requires knowing $v(\frac{1}{2} dt)$; because the initial condition we know is for $v(0)$, we are in trouble. But suppose we knew instead $v(\frac{1}{2} dt) = 1$ m/s (after all, dt is small so they're almost the same.) Then we could get a value for $x(dt)$:

$$x(dt) = x(0) + dt\left[v\left(\tfrac{1}{2} dt\right)\right] = 0.1 \text{ m} \tag{1.6}$$

To go on to get $x(2dt)$ similarly, we'd need $v(1.5\, dt)$. The equation for this is

$$v(1.5\, dt) = v(0.5\, dt) - dt\left(\frac{k}{m}\right)x(dt) = (1.0 - 0.01) \text{ m/s} \tag{1.7}$$

which we've been able to evaluate because all variables on the right-hand side are known. It's not hard to see that we can go on like this forever, calculating x only at integer multiples of dt and v at half-integer multiples, as shown in Figure 1.2. When we finish calculating the variable at time t, we know those at all earlier times and can therefore use Eq. 1.4 or 1.5 to calculate the variable at $t + \frac{1}{2} dt$. Doing this numerically gives the following table of displacements and velocities, as the reader should verify:

Table 1.1

t/dt	0	0.5	1.0	1.5	2.0	2.5
x (in m)	0		0.10		0.20	
v (in m/s)		1.00		0.99		0.97

The idea of defining x and v at alternate times is very common in the numerical treatment of physical problems; we will encounter it again in solving the Maxwell equations of electrodynamics.

We could make a graph of x or v as a function of t from these data; it would of course look like a sine function if we used enough t's.

▶   **Problem 1.1**

    a. Complete the table out to $t = 4\,dt$, by solving Eqs. 1.4 and 1.5 by hand (in the manner of Eqs. 1.6 and 1.7) with $dt = 0.1$ s, $k/m = 1.0$ s^{-2}.

    b. Find the exact solution to the differential equations Eqs. 1.1 and 1.2 from Eqs. 1.3 and 1.2, using initial conditions $x(0) = 0$ and $v(0) = 1$ m/s. (What are the values of A and ϕ?) Check the accuracy of the $x(0.4)$ you found in part **a**.

    c. Show that the motion described by the discrete-time Eqs. 1.4 and 1.5 for $dt = 1.0$ is periodic, using the initial conditions of Table 1.1. (In this discrete context, "periodic" means the x's and v's repeat themselves after a time T, an integer multiple of dt, called the **period**.) Compute T. How far is this from the continuous-time ($dt \to 0$) period? Does $dt = 2$ give sensible results, or is it too big?   ◀

The reason numerical calculation has become a practical alternative to analytic solution in recent years is, of course, the advent of the computer. In programming a computer to simulate the motion of the mass in Figure 1.1, it is just as easy to solve Eqs. 1.4 and 1.5 numerically at each time as it is to evaluate the formula in Eq. 1.3. (In fact, if the computer designer hadn't done a lot of the work for you by writing a subprogram to compute the sine, the latter would be **much** harder.) A computer program (called MASS) has been written to solve Eqs. 1.4 and 1.5 repeatedly on a microcomputer, indicating the resulting displacements $x(t)$ by displaying the moving mass on a screen. It is described in the program guide at the end of the text.

▶   **Problem 1.1**

    d. Eqs. 1.4 and 1.5 become more nearly the same as the original differential equations (Eqs. 1.1 and 1.2) as dt approaches zero, so the solution $x(t)$ should approach the exact one in this limit. Using MASS, check the convergence of this finite-difference method to the exact value (from the analytic form in part **b**) of $x(5.0)$ by using $dt = 1.0, 0.1$, and 0.01.   ◀

This type of program is called a simulation program. It simulates the motion of the mass (i.e., it mathematically calculates the displacement of the mass at various times) without the actual motion having to take place. To write such a program, we must first specify the time interval dt between successive calculated values of the displacement. This refers not to the time it takes to calculate the values, but to the interval (measured on a real clock) between the times when the real system has these displacements. This dt is therefore called the "real time" interval. The time required by the computer to do this calculation will be called the computer time interval dt_c. Usually dt_c is not equal to the real time dt; in such cases the simulation is known as a "false time" simulation. False-time simulations are necessary for phenomena whose real rate is too fast for computers to follow (e.g., molecular vibrations) or so slow that we don't want to wait for them to finish before looking at the simulation results (e.g., the rotation of the spiral arms of the Milky Way galaxy). But for phenomena of intermediate speeds, it is possible to arrange things so $dt_c = dt$; this is a real-time simulation. The computer then calculates each displacement exactly when it is occurring in the real system. Real-time simulations are often useful, especially if we intend to use the simulation results to control the motion itself. For example, the computer in a lunar landing craft must calculate the trajectory of the craft in real time (or even a little ahead of real time; this is still loosely referred to as "real time" calculation) to control the rocket motors and turn them off when it hits the surface.

Most of the simulations in this course, however, will necessarily be false-time ones, because typical electrodynamic phenomena are extremely fast.

In a simulation, almost any value can be chosen for dt, but there is less control over the computer time dt_c required for one updating step: this is uniquely determined by the speed of the computer and the program we are using. For example, in using one version of MASS to simulate the motion of the mass on an Apple II microcomputer, it takes 0.035 s to update (using Eqs. 1.4 and 1.5) the displacement and velocity once (this time can be measured by timing 1000 updates with a watch.) Thus $dt_c = 0.035$ s.

■ **EXAMPLE 1.1** What parameters should be used in MASS to simulate a pendulum of length $l = 1.5$ m on an Apple II microcomputer in real time ? Solution: As you will recall from mechanics, the equation of motion of a pendulum is the same as that of a mass on a spring, with k/m replaced by g/l. So we use $k/m = (9.8 \text{ m/s}^2)/1.5 \text{ m} = 6.53 \text{ s}^{-2}$, with $dt = dt_c = 0.035$ s, in Eqs. 1.4 and 1.5. [To get k/m for a real spring, it is easiest to measure the period T and use $k/m = (2\pi/T)^2$.]

SECTION 1.2 TWO MASSES CONNECTED BY SPRINGS

The above problem with a single mass may not seem to have much to do with electricity and magnetism. One respect in which it is too simple to be a good analogy is that its variables refer to a single object, which is (at a given time) at a single point in space. To obtain wave propagation, which is essential in electromagnetism, we need to have variables at many (sometimes even all) points in space. As a step in this direction, let us consider a system consisting of **two** masses, called mass 1 and mass 2. The motion will be trivial (the two masses will independently undergo the harmonic oscillator motion we just described) unless the two masses **interact**. So let us connect a spring between them, as well as connecting each of them to a rigid support, as shown in Figure 1.3. At each time t, we will have to keep track of the displacements of **both** masses: we will call them $x(1, t)$ and $x(2, t)$. We use this notation rather than $x_1(t)$ and $x_2(t)$, which may be more familiar, for the following reasons. First, this is the way a subscripted array is expressed in many computer languages, including FORTRAN, BASIC, and Pascal. Second, we eventually want to think of the first argument as a position variable, taking on many (eventually a continuum of) values just as t does, so we treat it as we do t. Third, $x(1, t)$ is a lot easier to type.

The corresponding notations for the velocities of the two masses are $v(1, t)$ and $v(2, t)$. The forces on mass 1 are $-k[x(1, t)]$ (from the wall at its left) and

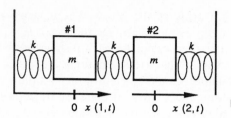

Figure 1.3 Two masses connected by springs.

$+k[x(2, t) - x(1, t)]$ (from the other mass at its right). Putting the net force into Newton's second law yields, for mass 1,

$$m\left[\frac{dv(1, t)}{dt}\right] = k[x(2, t) - 2x(1, t)] \tag{1.8}$$

We can turn this into a finite-difference equation just as we did for one mass

$$v\left(1, t + \tfrac{1}{2}dt\right) - v\left(1, t - \tfrac{1}{2}dt\right) = dt(k/m)[x(2, t) - 2x(1, t)] \tag{1.9}$$

and the corresponding equation for the second mass is

$$v\left(2, t + \tfrac{1}{2}dt\right) - v\left(2, t - \tfrac{1}{2}dt\right) = dt(k/m)[x(1, t) - 2x(2, t)] \tag{1.10}$$

These two equations are analogs of Eq. 1.4. The equations for updating the displacements are identical (except for notation) to Eq. 1.5:

$$x\left(1, t + \tfrac{1}{2}dt\right) - x\left(1, t - \tfrac{1}{2}dt\right) = dt[v(1, t)] \tag{1.11}$$

$$x\left(2, t + \tfrac{1}{2}dt\right) - x\left(2, t - \tfrac{1}{2}dt\right) = dt[v(2, t)] \tag{1.12}$$

Solving these equations by hand becomes more and more difficult, but let's try this one for the case $dt = 0.1$ s, $k/m = 1$ s^{-2} and initial conditions $x(r, 0) = 0$, $v(1, \tfrac{1}{2}dt) = 1$ m/s, $v(2, \tfrac{1}{2}dt) = 0$. (When we want to write an equation that can refer to either mass, we will replace the label 1 or 2 with the variable r, the location label, which takes values 1 and 2.) Physically, these initial conditions correspond to placing both masses in their equilibrium positions (zero displacements) and banging mass 1 with a hammer; you expect it eventually to start mass 2 moving as well. You can verify from Eqs. 1.9–1.12 that the results are as they appear in Table 1.2:

Table 1.2

t/dt	$x(1, t)$	$v(1, t)$	$x(2, t)$	$v(2, t)$
0.0	0.0		0.0	
0.5		1.0		0.0
1.0	0.1		0.0	
1.5		0.98		0.01
2.0	0.198		0.001	

▶ **Problem 1.2**

Complete Table 1.2 by hand, out to $t = 4\, dt$.

SECTION 1.3   A CHAIN OF MASSES CONNECTED BY SPRINGS

Let us continue to increase the number of variables in our problem, by using N masses instead of just 2, as shown in Figure 1.4. Then one of the masses chosen arbitrarily (say the rth; the label r can take on the values $1, 2 ..., N$) has forces $-k[x(r, t) - x(r - 1, t)]$ (exerted by the spring on its left) and $+k[x(r + 1, t) - x(r, t)]$ (exerted from the right). Its (differential) equation of motion is

$$m \frac{dv(r, t)}{dt} = k[x(r - 1, t) - x(r, t)] + k[x(r + 1, t) - x(r, t)] \quad (1.13)$$

We have to be careful if $r = 1$; then there is no leftward position ($r - 1 = 0$). We may pretend there is, but the object there is an immovable wall rather than another mass m, so that

$$x(0, t) = 0 \tag{1.14}$$

This is called a **boundary condition** for our problem. If we assume an immovable-wall boundary condition at the other boundary as well, the condition

$$x(N + 1, t) = 0 \tag{1.15}$$

gives the correct equation for $v(N, t)$. To solve Eq. 1.13, we need both initial conditions (on x and v) and the boundary condition; given these, there is a unique solution.

The finite-difference version of Eq. 1.13 is

$$v\left(r, t + \tfrac{1}{2} dt\right) - v\left(r, t - \tfrac{1}{2} dt\right) = dt \left(\frac{k}{m}\right)[x(r + 1, t) - 2x(r, t) + x(r - 1, t)]$$

$$(1.16)$$

The finite-difference equation for updating x is the same as Eq. 1.11 or 1.12:

$$x\left(r, t + \tfrac{1}{2} dt\right) - x\left(r, t - \tfrac{1}{2} dt\right) = dt\, v(r, t) \tag{1.17}$$

If we put $N = 2$ into these equations they are equivalent to the ones derived specifically for $N = 2$ (Eqs. 1.9–1.12), and will therefore give the results we found in Table 1.2. We repeat this table here as Table 1.3, using a slightly different format

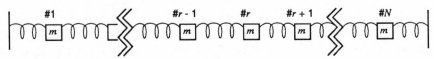

Figure 1.4   A chain of N masses connected by springs.

showing the boundary conditions and indicating by arrows which data are involved in calculating $v(1, 0.15)$ and $x(2, 0.2)$.

Table 1.3

r	0	1	2	3
$x(r,0)$	0.0	0.000	0.000	0.0
$v(r,0.05)$	0.0	1.000	0.000	0.0
$x(r,0.1)$	0.0	0.100	0.000	0.0
$v(r,0.15)$	0.0	0.980	0.010	0.0
$x(r,0.2)$	0.0	0.198	0.001	0.0

The pattern of arrows is the same for any other $v(r, t)$ or $x(r, t)$. The arrows contain the same information as the position and time indices in Eqs. 1.16–1.17, and are considerably easier to visualize. This pattern lends itself well to calculation by a computer program. The program that solves Eqs. 1.16–1.17 is called MASSES; it displays longitudinal (along-the-chain) oscillations for an N-mass system just as MASS did for the one-mass system, and is described in the program guide.

> Note that the program names in this text are generic, referring to a class of programs that perform a particular function, one on each version of the program diskette, such as Applesoft Basic, IBM-PC Turbo Pascal, Apple Pascal, and so on. To determine which program on your disk performs this function, check the Program Guide for that disk.

▶ **Problem 1.3**

Verify that the finite-difference equations Eqs. 1.16 and 1.17, with Eqs. 1.14 and 1.15 as boundary conditions reduce to the equations of Section 1.2 when $N = 2$ and to Section 1.1 (except for a factor of 2) when $N = 1$. Explain the factor of 2. ◀

▶ **Problem 1.4**

a. Check Table 1.2 or Table 1.3 using MASSES.
b. Determine (within ± 0.01 s) how long it takes mass 1 to return to $x = 0$. Hint: Increase the total time t until you see it return. Use the printed final displacement and velocity to improve your estimate of the return time. Repeat the process using smaller dt's. ◀

We can solve these finite-difference equations for $N > 2$ with the same initial conditions as in the $N = 2$ case (i.e., where mass 1 was initially banged with a hammer; we generalize to N masses by giving the others zero displacement and velocity). This is tedious to do by hand but interesting to watch in a simulation using MASSES; mass 1 moves to the right, giving a push to 2, but later bounces back to the left. As time goes on the motion involves masses successively further down the chain. The motion gets very complicated; we will see in Section 2.2 that the motion is much easier to understand for certain special initial conditions.

When there are several masses, as in Figure 1.4, a display of longitudinal motions (on paper or on a computer screen) can get quite confusing. For example, if the initial velocity is large, one mass can move past its neighbor, so it isn't clear

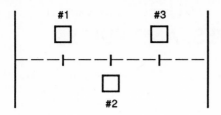

Figure 1.5 Transverse displacements of a chain of masses.

anymore which equilibrium position corresponds to which mass. If the velocity is small, the pixel moves only occasionally and the display is jerky. To avoid these problems, from now on we will display the results as **transverse** displacements (i.e., displacements perpendicular to the chain). In Figure 1.5, for example, these are vertical displacements because the chain is horizontal; each mass remains directly above or below its equilibrium position.

There are two ways to think of such a display: you can think of it as a purely schematic way of displaying motions that are really longitudinal, or as an actual picture of a chain of masses undergoing real transverse motions. We discussed only longitudinal motions in Sections 1.1 and 1.2 because it is easier to analyze the forces in this case (see Figures 1.3 and 1.4). It turns out, however, that if we stretch the springs to a tension T (by moving the walls further apart) and the displacements are small, the equation governing transverse motion has exactly the same form as Eq. 1.13, with k replaced by an effective spring constant k', which depends on the tension. (In general $k' < k$; transverse oscillations are slower than longitudinal ones.)

▶   **Problem 1.5**

Derive this analog of Eq. 1.13. Give the value of k' in terms of the tension T and the distance d between the masses.   ◀

The program that displays transverse motions of a chain of masses is called TVS, and is described in the program guide. Keep in mind, however, that it solves exactly the same equations as MASSES; only the display is different. Using the initial conditions we used above for MASSES (banging mass 1 with a hammer, this time transversely) mass 1 moves upward. This pulls mass 2 upward, and mass 1 eventually starts back down; the subsequent motion is complicated, just as before.

CHAPTER 2

FIELDS, WAVES, AND PULSES

SECTION 2.1 FIELDS

The state at any time t of the chain of masses described in Section 1.3 is uniquely determined by the list of displacements $x(1, t), x(2, t), \ldots, x(N, t)$ and the list of velocities $v(1, t), \ldots, v(N, t)$. These two lists of numbers are referred to as the **displacement field** and the **velocity field**. The term "field" is used to describe any collection of numbers, each of which corresponds to a position in space. The concept of a field arises very frequently in physics, and we will discuss several other examples in this book, such as the electric field, the magnetic field, and the density field. For the displacement fields we have been discussing, the position is specified by the index $r = 1, 2, \ldots, N$, which labels the mass.

> *Digression on terminology:* A mathematician might say that what we are calling a "field" is a **function** whose domain is the set $\{1, 2, \ldots, N\}$ of position labels r and whose range is the set of real numbers. We will use the word "function" in a narrower sense, however, requiring it to be defined at **all** space points r (in a one-dimensional space such as the chain we are considering, these are all real numbers) by some rule such as $f(r) = r^2$ or $f(r) = \sin(r)$.

In one way, the displacement field is more difficult to conceptualize than other types of fields because both the field's argument r and its value x refer to position-type variables, and mixing them up can be disastrous. It is probably better to think about a transverse field than a longitudinal one, because r and x are measured in different directions. We will try to avoid this confusion by always using the word "position" to mean r, the discrete label specifying which equilibrium position the mass occupies, and "displacement" to mean x, the continuous variable describing how far it has moved.

We can specify a field by giving a table of its values. For example, the initial condition for the velocity field in the example at the end of Section 1.3 (where only the first mass had a nonzero velocity) is given in the following table for $N = 5$.

Table 2.1

r	1	2	3	4	5
$v(r, \frac{1}{2} dt)$	1 m/s	0	0	0	0

The displacement and velocity fields change with time; for example (using $k/m =$

1 s^{-2}, $dt = 0.1$ s) the displacement field at $t = 5$ s and the velocity field at $t = 5.05$ s are

Table 2.2

r	1	2	3	4	5
$x(r, 5 \text{ s})$	0.102	-0.174	-0.021	0.504	0.383
$v(r, 5.05 \text{ s})$	-0.048	0.292	-0.518	-0.269	0.180

as you can verify using TVS or MASSES.

One often defines a particular field from a function, such as $f(r) = r^2$ or $f(r) = \sin r$; we will do this in the next section. The purpose of the present section is to make it clear that a field need not be related to any simple mathematical function—for example, you would be hard-pressed to find a simple formula for the field x or v defined in the above table.

SECTION 2.2   SINUSOIDAL WAVES

We can define a particular kind of displacement field (called "sinusoidal") from the sine function. For any two constants A and Q,

$$x(r, 0) = A \sin(Qr) \tag{2.1}$$

defines a displacement field for our N masses. That is, we can construct a table giving the N values of the field by substituting $r = 1, 2, \ldots, N$ into this equation. Taking $A = 1$, $Q = 1$, and $N = 7$, for example, we get $x(1.0) = \sin(1) = 0.841$ (note that we use radian measure for angles in this text, as is almost universally done in physics). We can determine the other values similarly in Table 2.3.

Table 2.3

r	1	2	3	4	5	6	7
$x(r, 0)$	0.841	0.909	0.141	-0.757	-0.959	-0.279	0.657

This field is depicted by squares in Figure 2.1; the sine function we obtained it from is included as a dotted line. We could use this as an initial condition (with zero initial velocities) for the masses on springs in program TVS; the evolution we would see would be rather complicated. For certain other choices of Q, however, this sinusoidal field has a very nice property: it is given by Eq. 2.1 at **all** times, with varying amplitude $A(t)$. These choices of Q amount to requiring that Eq. 2.1 vanish for $r = N + 1$ as suggested by the right-hand boundary condition (Eq. 1.15). (The field depicted in Figure 2.1 obviously doesn't satisfy this; it's a perfectly good field but it doesn't stay sinusoidal. You can see this using TVS, as described below.) The requirement on Q is

$$\sin[(N + 1)Q] = 0 \tag{2.2}$$

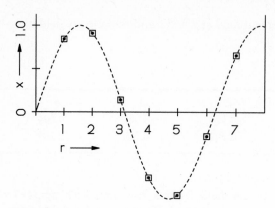

Figure 2.1 Transverse displacements of a chain, corresponding to the displacement field in Table 2.3.

so that

$$Q = \frac{M\pi}{(N + 1)} \tag{2.3}$$

for some integer M; M turns out to be the number of half-wavelengths of the wave that fit on the chain. The case of $M = 3$ half-wavelengths is shown in Figure 2.2, for $N = 7$ and $A = 1$.

▶ **Problem 2.1**

Construct a table giving numerical values for the displacement field determined by the function $x(r) = \sin[3\pi(r/8)]$, for a chain with $N = 7$. (Remember to use radian measure!) Is the table consistent with Figure 2.2? ◀

The proof that the displacement field defined by Eq. 2.1 remains sinusoidal at all times is most easily accomplished if the complex exponential form is used for the

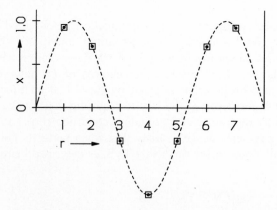

Figure 2.2 Sinusoidal displacement field with three half-wavelengths.

sine function. Recall that Euler's formula $e^{i\theta} = \cos\theta + i\sin\theta$ (where $i = \sqrt{-1}$) gives

$$\cos\theta = (e^{i\theta} + e^{-i\theta})/2 \tag{2.4}$$

$$\sin\theta = (e^{i\theta} - e^{-i\theta})/2i \tag{2.5}$$

[Those unfamiliar with complex numbers can skip this proof and still follow the rest of this section. But an understanding of complex numbers will be necessary before we get to sinusoidal electromagnetic waves and optics in Chapter 24.] Thus a sinusoidal displacement (Eq. 2.1) with varying amplitude $A(t)$ can be written

$$x(r, t) = A(t)\frac{e^{iQr} - e^{-iQr}}{2i} \tag{2.6}$$

Let us guess that the velocity, which is "initially" (i.e., at $t = \frac{1}{2}dt$) zero, is also sinusoidal with a different amplitude $B(t)$:

$$v(r, t) = B(t)\frac{e^{iQr} - e^{-iQr}}{2i} \tag{2.7}$$

The finite-difference equations (Eqs. 1.16 and 1.17) are then (substituting the above for x and v):

$$\left[B\left(t + \tfrac{1}{2}dt\right) - B\left(t - \tfrac{1}{2}dt\right)\right]\frac{e^{iQr} - e^{-iQr}}{2i}$$

$$= \left[dt\left(\frac{k}{m}\right)A(t)\right]\frac{e^{iQ(r+1)} - e^{-iQ(r+1)} - 2(e^{iQr} - e^{-iQr}) + e^{iQ(r-1)} - e^{-iQ(r-1)}}{2i}$$

$$\tag{2.8}$$

$$\left[A\left(t + \tfrac{1}{2}dt\right) - A\left(t - \tfrac{1}{2}dt\right)\right]\frac{e^{iQr} - e^{-iQr}}{2i} = dt\, B(t)\frac{e^{iQr} - e^{-iQr}}{2i} \tag{2.9}$$

[In writing Eq. 2.8 for $r = N$ we have used our special condition (Eq. 2.2) on Q by assuming Eq. 2.1 holds at $r = N + 1$.] Each equation has terms proportional to e^{iQr} and terms proportional to e^{-iQr}; if we could make the coefficients of e^{iQr} equal on the two sides and similarly for e^{-iQr}, the equations would, of course, be satisfied. The coefficients of $e^{iQr}/2i$ are equal if

$$B\left(t + \tfrac{1}{2}dt\right) - B\left(t - \tfrac{1}{2}dt\right) = dt\left(\frac{k}{m}\right)A(t)(e^{iQ} - 2 - e^{-iQ})$$

$$\tag{2.10}$$

$$= 4\, dt\left(\frac{k}{m}\right)\sin^2\left(\frac{Q}{2}\right)A(t)$$

$$\left[A\left(t + \tfrac{1}{2}dt\right) - A\left(t - \tfrac{1}{2}dt\right)\right] = dt\, B(t) \tag{2.11}$$

and the coefficients of e^{-iQr} turn out to give identical equations.

But these last two equations are exactly the ones we solved in Section 1.1 for a **single** oscillating mass; the ratio k/m has simply been replaced by $4(k/m)\sin^2(Q/2)$. Given initial conditions for $A(t = 0)$ and $B(t = \frac{1}{2} dt)$, we can certainly calculate $A(t)$ and $B(t)$ to satisfy them, just as in Section 1.1. The displacement amplitude $A(t)$ follows a simple harmonic motion with an effective spring constant

$$k_{\text{eff}} = 4k \sin^2(Q/2) = 4k \sin^2[M\pi/2(N + 1)] \tag{2.12}$$

Thus we have a "standing wave" on the chain of masses. The frequency of the oscillation is

$$\omega = (k_{\text{eff}}/m)^{1/2} = 2(k/m)^{1/2} \sin[M\pi/2(N + 1)] \tag{2.13}$$

which can be much slower than that of a single mass on a spring k [i.e., $(k/m)^{1/2}$] for long waves ($M \ll N$).

The program TVS will set up a sinusoidal initial displacement, so these standing waves can be displayed.

▶ **Problem 2.2**

    a. If $k/m = 1 \text{ s}^{-2}$ and $N = 18$ compute the frequency and period of a standing wave with 3 half-wavelengths (Figure 2.2).

    b. Simulate this wave using the program TVS with $dt = 1$, and verify your result for the period by counting the real time needed for one oscillation. (This just means counting the number of updates, i.e., the number of times each mass moves, because $dt = 1$.)

    c. Check whether $dt = 1$ is adequately small by using $dt = 2.0, 1.0, 0.5, 0.2$, and 0.1, and calculating $x(9, 12)$. [Warning: strange things sometimes happen for very large dt.] Can you calculate the $dt \to 0$ limit of $x(9, 12)$ analytically? [Hint: Think about Eq. 1.3 and Eq. 2.13.]   ◀

The interval $dt = 1$ may seem large (indeed, it would be too large for accuracy in the simulations of the previous sections). However, because the standing wave motion is slow when the wavelength is long (Eq. 2.13), it turns out to be quite accurate. In any problem, it is not important whether dt is small in an absolute sense, but whether it is small in comparison to the important time scale in the problem. This time scale (the period of the oscillation in our case) is much larger than $dt = 1$.

▶ **Problem 2.3**

The case $M = (N + 1)/2$ is particularly easy to analyze. Draw the initial displacement for this case. Because every second mass is undisplaced (and remains so) each other mass oscillates as though its neighbors were fixed. Find the frequency of simple harmonic oscillation for this equivalent fixed-neighbor system, and check your answer against Eq. 2.13 for the original N-mass system.   ◀

▶ **Problem 2.4**

What happens when $M = 2N$? (Try it using TVS, if you like.) Show in general from Eq. 2.1 that for any M, M and $2N + 2 - M$ give essentially the same wave.   ◀

SECTION 2.3   THE PRINCIPLE OF SUPERPOSITION

An initial displacement field need not involve the entire chain of masses, as sinusoidal ones do. We can, for example, start by displacing only one mass, say 1. The results are similar to those we found in Section 1.3 by giving one mass an initial velocity; the motion is rather complicated. However, things become much simpler if we introduce a smooth pulse such as that in Figure 2.3.

If we use zero initial velocities and let this evolve using program TVS, we find that it splits into two smaller pulses moving in opposite directions, as shown in Figure 2.4. We will see later precisely why it does this. At the moment, we just want to observe a remarkable fact about these pulses: after reflection from the walls (during which they reverse amplitude) they hit each other again and each passes on as though the other weren't there. This violates our intuitive expectation of what happens when two objects, such as automobiles, hit each other: when they collide, they are normally damaged somehow by the collision (Figure 2.5). They don't just pass through each other like ghosts. We can look at this phenomenon more carefully, using TVS to set up the two traveling pulses just before they collide (Figure 2.6, top left). (The way in which TVS does this will be explained in Section 2.6.) We can also set up one pulse and not the other (Figure 2.6, top right). We find that the shape of each pulse after they pass through each other is exactly the same whether the other was present or not.

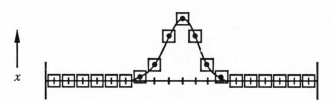

Figure 2.3   Displacement field of a smooth pulse.

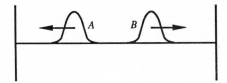

Figure 2.4   Schematic picture of the system of Figure 2.3 at a later time.

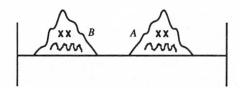

Figure 2.5   Expected appearance after collision.

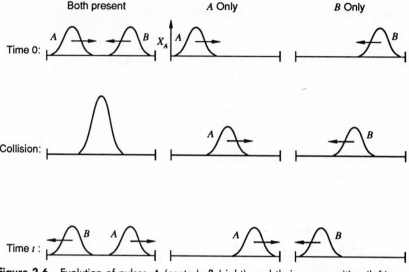

Figure 2.6 Evolution of pulses *A* (center), *B* (right), and their superposition (left).

Let us try to state this conclusion quantitatively. Denote the displacement and velocity of pulse A by $x_A(r, t)$ (top center graph of Figure 2.6) and $v_A(r, t)$. These are specified at $t = 0$ (or $t = \frac{1}{2} dt$ for v) by an initial condition, and computed at later times from the equations of motion (Eqs. 1.16 and 1.17). We can define similarly $x_B(r, t)$ (top right graph) and $v_B(r, t)$. Each field is defined for all masses r, but $x_A(r, 0)$ happens to be zero for $r > N/2$ (i.e., on the right half of the chain), so we see it only on the left half of the top center graph. Similarly pulse B is nonzero only on the right half (top right graph). The initial condition with both pulses present (top left) can be regarded as the sum (or "superposition") of the initial conditions for pulses A and B separately:

$$x(r, 0) = x_A(r, 0) + x_B(r, 0) \qquad (2.14)$$

$$v(r, 0) = v_A(r, 0) + v_B(r, 0) \qquad (2.15)$$

And what we have observed at a time t after the "collision" is that the displacement $x(r, t)$ (bottom left) again looks like a superposition, this time of the later displacements of the individual pulses, $x_A(r, t)$ and $x_B(r, t)$ (bottom right graphs):

$$x(r, t) = x_A(r, t) + x_B(r, t) \qquad (2.16)$$

$$v(r, t) = v_A(r, t) + v_B(r, t) \qquad (2.17)$$

This is a general statement of the principle of superposition: if x_A, v_A and x_B, v_B are any two solutions to the equations of motion, then their sum (Eqs. 2.16 and 2.17) is also a solution.

What do we need to do to prove this? We are assuming x_A, v_A and x_B, v_B satisfy our equations of motion (Eqs. 1.16 and 1.17). Saying they satisfy Eq. 1.17

means

$$x_A\left(r, t + \tfrac{1}{2} dt\right) - x_A\left(r, t - \tfrac{1}{2} dt\right) = dt\, v_A(r, t) \tag{2.18}$$

$$x_B\left(r, t + \tfrac{1}{2} dt\right) - x_B\left(r, t - \tfrac{1}{2} dt\right) = dt\, v_B(r, t) \tag{2.19}$$

We must show that the total x and v fields given in Eqs. 2.16 and 2.17 satisfy the same equations of motion. Plugging them into Eq. 1.17 gives

$$\begin{aligned}
x_A\left(r, t + \tfrac{1}{2} dt\right) + x_B\left(r, t + \tfrac{1}{2} dt\right) - x_A\left(r, t - \tfrac{1}{2} dt\right) - x_B\left(r, t - \tfrac{1}{2} dt\right) \\
= dt\left[v_A(r, t) + v_B(r, t)\right]
\end{aligned} \tag{2.20}$$

This is clearly just the sum of Eqs. 2.18 and 2.19, which we are assuming to be satisfied. Thus we have proved the superposition satisfies Eq. 1.17. The reader should check that the other equation of motion (Eq. 1.16, for updating v) is also satisfied by the superposition of the A and B fields; again, this involves writing Eq. 1.16 for the A fields and then for the B fields and adding the resulting equations. The result should be the desired equation of motion (Eq. 1.16) for v.

Let us make one other observation, which we will regard as being part of the superposition principle. You can see from Eqs. 1.16 and 1.17 that if we take $x(r, t)$ and $v(r, t)$ corresponding to a solution, and multiply them both by a constant C, they still are a solution. That is, the equation remains true when it is multiplied through by C everywhere. Thus instead of using only a sum ($x = x_A + x_B$) in Eqs. 2.16 and 2.17 we could have written an arbitrary linear combination of solutions A and B:

$$x(r, t) = C_A x_A(r, t) + C_B x_B(r, t) \tag{2.21}$$

$$v(r, t) = C_A v_A(r, t) + C_B v_B(r, t) \tag{2.22}$$

The principle we have just proved is abstract enough that it may be hard to remember what it really says. It may help to state the principle of superposition in words, which can be done something like this: "Superposed fields evolve independently." That is, if a field is constructed at $t = 0$ by superposing several fields, its value at a later time can be obtained by evolving each of its constituent fields by itself and adding the results. We get the same result we would have gotten by evolving the total field. (By "evolving" a field, we mean calculating its change with time using equations of motion, in this case Eqs. 1.16 and 1.17. It is probably more grammatical to say "the field evolves" than "we evolve the field," but both expressions are often used.)

▶   **Problem 2.5**

Consider a chain of $N = 4$ masses with $k/m = 1$. Choose any two initial conditions (each consisting of a displacement and a velocity field), and call them x_A, v_A and x_B, v_B. That is, pick numerical values for $x_A(1, 0), v_A(1, \tfrac{1}{2} dt), x_A(2, 0), \ldots$ (16 in all). Most of them can be zero, but make sure at least one of the A's and one of the B's is nonzero. Sketch the A and B displacements (i.e., make two sketches like Figure 2.3). Use TVS to calculate $x_i(r, t = 5.0)$ and $v_i(r, t = 5.05)$ for $i = A$ and B, using $dt = 0.1$.

Use it again with the initial conditions $x(r,0) = x_A(r,0) + x_B(r,0)$ and $v(r,0) = v_A(r,0) + v_B(r,0)$, and verify that the principle of superposition holds in this case. ◀

Our proof of the principle of superposition may seem so trivial that you may wonder if it isn't always true: are there situations in which superposed fields don't evolve independently? The answer is very definitely yes; we will encounter some in Section 4.2, in discussing population models. The key difference is that the equations of motion we are dealing with now are **linear**, and the population models are nonlinear. If dx/dt had been given by Eq. 1.2 as v^2 instead of v, for example, the equation of motion for the superposed fields would have had on its right-hand side $(v_A + v_B)^2 = v_A^2 + 2v_Av_B + v_B^2$, which is not the sum of the right-hand sides of the equations for x_A and x_B. There is an extra "cross term" $2v_Av_B$, which we can think of as causing an interaction between the two waves. The above proof doesn't work because of the cross term. In this case a pulse emerging from a collision with another pulse would be very definitely different from what it would have been if the other pulse had been absent. On the other hand, the equation of motion of the chain-of-masses system we have discussed is only one of many equations of motion for which the superposition principle holds. It holds, for example, for Maxwell's equations of motion that govern electromagnetic fields. The principle is valid for any **linear** system. A linear system is one with a linear equation of motion, i.e., all of whose terms involve the first power of one of the fields. Not only constants and higher powers, but any other functions of the fields such as $\cos(x)$ are forbidden. As a convenient mnemonic for all this, one can restate the principle of superposition as follows: "Superposed fields evolve independently under linear equations of motion."

SECTION 2.4 WAVES ON A NECKLACE: PERIODIC BOUNDARY CONDITIONS

It is awkward to have to deal with waves on a finite chain of N masses with fixed ends, because the reflections from the ends tend to obscure the phenomena we are studying (for example, pulse propagation). One solution is to use an infinite chain. However, we obviously can't really do any calculations for an infinite chain, even by computer. Many of these difficulties can be solved by arranging the masses into a ring (or necklace) so that mass 1 and mass N are connected to each other by a spring, rather than to fixed walls; there are then no reflections. Such an arrangement is shown in Figure 2.7.

The program that computes the behavior of this necklace is called RING. It is difficult to display on the computer screen the actual ring as in the figure, so the

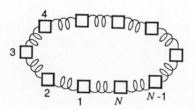

Figure 2.7 Ring of masses and springs.

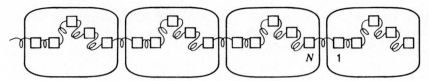

Figure 2.8   Ring of masses viewed as a periodic system.

masses are displayed in a straight line just as in TVS. To visualize how they are connected, you can imagine the screen rolled up so the leftmost mass (1) is next to the rightmost one (N). Or if you prefer, you may imagine that you are jammed together at a long table with many other students, each looking at the same simulation on his own screen, as in Figure 2.8. Your Nth mass is connected to your right-hand neighbor's 1st mass; this has the same effect as connecting it to **your** 1st mass because his displacements and velocities are the same as yours. You can think of the motion shown in Figure 2.8 as occurring on an **infinite** chain of masses, yours being labeled $1, 2, \ldots, N$, your right-hand neighbor's being $N + 1, N + 2, \ldots, 2N$, your left-hand neighbor's being $1 - N, 2 - N, \ldots, 0$, and so on. Thinking of it this way, you must assume the displacements to satisfy

$$x(r + N, t) = x(r, t) \tag{2.23}$$

for all r and t. That is, x is a **periodic** function of r, with period N. (Note that we're talking here about **spatial** periodicity; the motion may be periodic in time as well, but that's a different question.) The velocity must also be periodic:

$$v(r + N, t) = v(r, t) \tag{2.24}$$

Equations 2.23 and 2.24 play the role of boundary conditions, and are called **periodic boundary conditions**.

We can calculate the motion of this necklace of masses using the same equations (Eqs. 1.16 and 1.17) we used for the fixed-end chain, except at the ends where we must use mass 1 instead of $N + 1$ as the neighbor of mass N. Thus Eq. 1.16 reads

$$v\left(N, t + \tfrac{1}{2} dt\right) - v\left(N, t - \tfrac{1}{2} dt\right) = dt \left(\frac{k}{m}\right) [x(1, t) - 2x(N, t) + x(N - 1, t)]$$

$$\tag{2.25}$$

Such special formulas need not be used in a computer program such as RING: we use the infinite-periodic-chain picture (Figure 2.8), and store extra displacements $x(0, t)$ and $x(N + 1, t)$ that are simply set equal to $x(N, t)$ and $x(1, t)$, respectively, after each update, as demanded by Eq. 2.23.

The advantage of periodic boundary conditions is that reflections are eliminated; a wave that reaches the right-hand side of the screen simply disappears there and appears on the left. In the infinite-periodic-chain picture, you think of it as having jumped off of your screen and onto your right-hand neighbor's, while simultaneously your left-hand neighbor's wave jumped onto your screen.

Periodic boundary conditions are introduced at this point because doing so makes it possible to construct simple solutions from which all others can be obtained by superposition, as shown in Section 2.6.

■ **EXAMPLE 2.1**   Let us look at a ring of four masses with the same initial conditions as in Table 1.3; these are shown in Table 2.4.

Table 2.4

r	1	2	3	4
$x(r,0)$	0.0	0.0	0.0	0.0
$v(r,0.05)$	1.000	0.0	0.0	0.0
$x(r,0.1)$	0.100	0.0	0.0	0.0
$v(r,0.15)$	0.980	0.010	0.0	0.010
$x(r,0.2)$	0.198	0.001	0.0	0.001

At $t \leq 0.1$ the new boundary condition makes no difference. But note the difference in how the disturbance spreads out after that: instead of being stopped by a boundary at $r = 0$, the disturbance wraps around to the other side of the system ($r = 4$).

▶   **Problem 2.6**

a. Complete Table 2.4 by hand out to $t = 0.4$.

b. Verify that program RING gives the same result, and use it to compute $x(r, 1.0)$.   ◀

SECTION 2.5   STRAIN IN A CHAIN OF MASSES

We have written the equations of motion for a chain of masses (Eqs. 1.16 and 1.17) in terms of the displacement $x(r, t)$ and velocity $v(r, t)$ of each mass. For some purposes it is better to use different variables, the strains (or elongations) in the springs, instead of the displacements. This gives a more symmetrical set of equations, which allows easier treatment of conservation of energy and is mathematically identical to the Maxwell equations of electrodynamics in one dimension.

The elongation of the spring connecting masses r and $r + 1$ is its total length (Figure 2.9) minus the equilibrium (unstretched) length a. The strain in this spring is defined as the elongation divided by the equilibrium length:

$$s\left(r + \tfrac{1}{2}, t\right) = \left[x(r + 1, t) - x(r, t)\right]/a \qquad (2.26)$$

We will label this spring by $r + \tfrac{1}{2}$ (an odd half-integer). The tension is the spring constant times the elongation, or $kas(r + \tfrac{1}{2}, t)$, which is the rightward force on

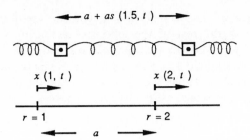

Figure 2.9   Definition of strain s in a chain of masses.

mass r. The leftward force is $kas(r - \frac{1}{2}, t)$, so

$$m \frac{dv\,(r, t)}{dt} = ka\left[s\left(r + \tfrac{1}{2}, t\right) - s\left(r - \tfrac{1}{2}, t\right)\right] \tag{2.27}$$

is the equation for updating the velocities (Eq. 1.13). The discrete-time version of this equation is

$$v\left(r, t + \tfrac{1}{2}\,dt\right) - v\left(r, t - \tfrac{1}{2}\,dt\right) = dt\left(\frac{ka}{m}\right)\left[s\left(r + \tfrac{1}{2}, t\right) - s\left(r - \tfrac{1}{2}, t\right)\right] \tag{2.28}$$

Note that this is simpler than the corresponding equation in terms of the displacement (Eq. 1.16). The discrete equation for updating the strain is obtained from Eq. 1.17 and Eq. 2.26:

$$s\left(r', t' + \tfrac{1}{2}\,dt\right) - s\left(r', t' - \tfrac{1}{2}\,dt\right) = dt\left[v\left(r' + \tfrac{1}{2}, t'\right) - v\left(r' - \tfrac{1}{2}, t'\right)\right] \tag{2.29}$$

We will use t' to denote a half-integer multiple of dt, and r' to denote a half-integer $(r + \frac{1}{2}$, in the notation of Eq. 2.26). We have written Eq. 2.29 in terms of r' to emphasize its similarity to Eq. 2.28; there is a symmetry between s and v in this formulation of the problem that was absent in the x, v formulation.

■ **EXAMPLE 2.2**   We will consider the initial conditions of Example 2.1 (in which only mass 1 had an initial velocity $v = 1$). The initial strains are also zero because the initial displacements were assumed zero, and the initial conditions are as shown in the first lines of Table 2.5.

Table 2.5

r	0.5	1.0	1.5	2.0	2.5	3.0	3.5	4.0
$s(r, 0.00)$	0.0		0.0		0.0		0.0	
$v(r, 0.05)$		1.0		0.0		0.0		0.0
$s(r, 0.10)$	0.1		−0.1		0.0		0.0	
$v(r, 0.15)$		0.980		0.010		0.0		0.010

Using Eqs. 2.28 and 2.29 with $dt = 0.1$ and $a = k/m = 1$, you can verify that the subsequent evolution is as given in the rest of the table (we are using periodic boundary conditions with $N = 4$ masses). Note that the velocity field at $t = 0.15$ is the same as in Table 2.4, and the strain field $s(r, t)$ at each time is exactly what Eq. 2.26 would give you from the $x(r, t)$ of Table 2.4. This is as it should be: we are describing exactly the same physical motions, using different variables.

▶   **Problem 2.7**

    a. Extend Table 2.5 to $t = 0.4$ using Eqs. 2.28 and 2.29. Verify that $s(r, 0.2)$ is consistent with Table 2.4.   ◀

The disadvantage of this strain formulation is that while one can compute the strain field easily from the displacement field, one cannot do the reverse. The displacement field is not even uniquely determined by the strain (how would you decide whether the chain was undisplaced or uniformly displaced, given zero strains?) If we need to know the displacement, we may as well use Eq. 1.17 and compute the strains after we are done. The strain formulation also limits our choice of boundary conditions: we can't assume fixed ends, but we can use periodic boundary conditions. The compensating advantage of the strain formulation is that its equations are equivalent to Maxwell's equations and that it makes it easier to decompose a wave into pulses, which we will do in Section 2.6.

    The simulation program that updates the strain and velocity using Eqs. 2.28 and 2.29 is called MAX1 (because these equations are equivalent to the Maxwell equations in one dimension) and is described in the program guide.

▶   **Problem 2.7**

    b. Use MAX1 to verify Table 2.5 and the results of Problem 2.7a. Compute $s(r, 1.0)$.   ◀

You can use MAX1 to look at the evolution of a smooth strain pulse (with initial condition described by a graph like Figure 2.3, but with s instead of x as the vertical coordinate). You will find behavior similar to Figure 2.4 (the pulse separates into two), although the velocity fields are quite different.

SECTION 2.6   DECOMPOSITION OF WAVES: DELTA PULSES

The principle of superposition is extremely useful in analyzing the behavior of systems with linear equations of motion. It has the effect of drastically reducing the number of different initial fields for which we need to solve the equations in order to be able to predict how any initial field evolves. Once we have solved the equations for some set of initial fields, we can superpose these to obtain solutions for a much larger set of initial fields.

■ **EXAMPLE 2.3 One Mass on a Spring**   The initial "fields" in this case are just the two numbers $x(0)$ and $v(\frac{1}{2} dt)$. Let us label A the initial conditions we used in Section 1.1:

$$x_A(0) = 0, \qquad v_A(\tfrac{1}{2} dt) = 1$$

Another possible initial condition is the one with zero initial velocity:

$$x_B(0) = 1, \qquad v_B(\tfrac{1}{2} \, dt) = 0$$

From the discrete-time equations of motion (Eqs. 1.16 and 1.17) we can get the displacements $x_A(t)$, $x_B(t)$ and velocities $v_A(t)$, $v_B(t)$ at later times. We can then superpose these to obtain many other solutions to the equations of motion. Let us take a linear combination (Eqs. 2.21 and 2.22) with coefficients C_A and C_B, which the principle of superposition tells us is also a solution. Its initial values are

$$x(0) = C_A x_A(0) + C_B x_B(0) = C_B$$

$$v(\tfrac{1}{2} \, dt) = C_A v_A(\tfrac{1}{2} \, dt) + C_B v_B(\tfrac{1}{2} \, dt) = C_A$$

What set of initial conditions can be written this way? All of them, in fact. Given **any** $x(0)$, $v(\tfrac{1}{2} \, dt)$, we need merely choose $C_A = v(\tfrac{1}{2} \, dt)$, $C_B = x(0)$ for these initial conditions to be satisfied. Thus the two initial conditions A and B form a "complete set," meaning that any other one can be written as a linear combination. If we know how these two initial states evolve, we know **everything** about the system.

We can generalize these ideas and apply them to more complicated systems. For the N-mass system, we can define $2N$ initial states, in each of which **one** displacement or one velocity is equal to 1.0, and all the others are zero. Each is represented by a table with a single nonzero entry such as (for $N = 5$)

Table 2.6

r		1	2	3	4	5	
$x(r,0)$		0	0	1	0	0	
$v(r, dt/2)$		0	0	0	0	0	

Clearly we can construct ten tables like this, one for each initial state. Just as in the previous example, we can represent any arbitrary initial state as a linear combination of these ten states. However, the initial states we have defined do not evolve in any simple way; they become complicated at later times. What we would like to do is find a complete set of initial states (ten in the case $N = 5$) which evolve in a simple way. It turns out to be easiest to do this in the strain formulation of Section 2.5. If we do some experimentation with program MAX1, we will find that there are some states that seem to evolve simply, in a system with periodic boundary conditions (i.e., a necklace). Let's try an initial condition in which only two of the springs are stretched and all the velocities are zero. This initial configuration is shown in Figure 2.10. It looks qualitatively like the smooth strain pulse mentioned in Section 2.5, which splits into two pulses, going to the right and to the left, respectively. Qualitatively, this will also happen for the two-mass pulse. Because they are not very smooth, the leftward and rightward moving pulses do not retain their form exactly as time passes, at least for most choices of dt. However, there is a special choice of dt for which the evolution is very simple: $dt = 1$. Using this value, the pulses moving to the right and left look the same at every time and are as simple

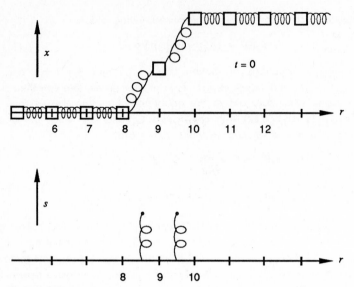

Figure 2.10   Initial displacements (top) and strains (bottom) for two-mass pulse.

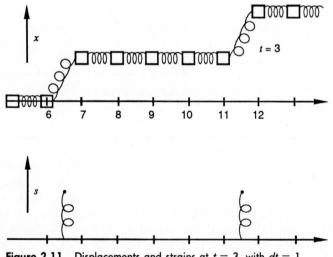

Figure 2.11   Displacements and strains at $t = 3$, with $dt = 1$.

as they could possibly be: only one spring is stretched at each time (see Figure 2.11), and always by the same amount. The only change from one time to the next is that this displacement moves along the chain by one mass.

▶   **Problem 2.8**

You may argue here that $dt = 1$ is not small compared to the characteristic time during which changes occur in this pulse: the displacement goes from zero to its maximum value during just one time interval dt. Compute the actual error this causes in the evolution of a sharp pulse (Figure 2.10) on a long necklace ($N = 18$) by using MAX1 to calculate the strains at $t = 3$ for $dt = 1$ and then for $dt = 0.1$. What is the maximum difference, as a fraction of the initial strain? (Input by hand approximately the maximum value allowed by the program for the initially-nonzero strains, and zero initial velocities.) Our justification for using $dt = 1$ here is that we will eventually use these pulses to construct much wider pulses that change on a much longer time scale. Compute the error in the evolution of a Gaussian strain pulse of width 4 (use the maximum or default height) with zero initial velocities, by using MAX1 (again compare $dt = 1$ to $dt = 0.1$). Is the error smaller for the smoother pulse?   ◄

It is not obvious why the pulse of Figure 2.10 evolves the way we have said, so we should solve the equation of motion by hand for this initial condition. The result is shown in the table below. To avoid running into boundaries we have started the pulse on springs 8.5 and 9.5 (for $N = 18$ this is about the middle of the chain). The table includes only those masses (7 through 11) that are displaced in the first 3 seconds; the blanks represent zeros.

Table 2.7

r	7.0	7.5	8.0	8.5	9.0	9.5	10.0	10.5	11.0
$s(r,0)$				1		1			
$v(r,0.5)$					0				
$s(r,1)$				1		1			
$v(r,1.5)$			1		0		-1		
$s(r,2)$		1		0		0		1	
$v(r,2.5)$	1		0		0		0		-1

Check to make sure that these are really the numbers that come out of the equations of motion (Eqs. 2.28 and 2.29). Equation 2.29 gives for example

$$s(7.5, 2) = s(7.5, 1) + [v(8, 1.5) - v(7, 1.5)] = 0 + [1 - 0] = 1$$

and Eq. 2.28 gives

$$v(8, 2.5) = v(8, 1.5) + [s(8.5, 2) - s(7.5, 2)] = 1 + [0 - 1] = 0$$

(we omit factors of k/m and dt, because both are 1.)

Except at the beginning, the two pulses in Table 2.7 clearly evolve independently of each other. We can pick out one pulse alone by using for our initial condition the displacement and velocity after the pulses have clearly separated, say at times 2.0 and 2.5. This gives Table 2.8 (we have displaced all the times by 2, in effect).

Table 2.8

r	9.5	10.0	10.5	11.0	11.5	12.0	12.5
$s(r,0)$	0		1				
$v(r,0.5)$		0		-1			
$s(r,1)$	0		0		1		
$v(r,1.5)$		0		0		-1	

This table evidently represents a pulse that starts on spring 10.5 and moves to the right. At each time, only one spring has a nonzero strain; during the following time interval, the mass to its right has a negative velocity in order to relieve the strain. The displacement associated with this pulse depends on our choice of x-origin; if we assume that the masses in front of (i.e., to the right of) the pulse are undisplaced, then the displacement of each mass behind the pulse is negative. Each mass is stationary except during the single time interval when the pulse passes it.

▶   **Problem 2.9**

Construct a table like Table 2.8 for a leftward-moving pulse (see Table 2.7), which begins on spring 7.5 at $t = 0$.   ◀

We could call the pulse shown in Table 2.8 pulse A (with strain s_A and velocity v_A), the one which starts on spring 11.5 pulse B, and so on. However, we would clearly run out of letters before we ran out of pulses on a long chain, so each pulse is labeled according to the spring it starts on (as a subscript) and a superscript is added to indicate which way it is going ($+$ for right, $-$ for left). Then the strain of the pulse in the table is denoted $s_{10.5}^+(r, t)$, and its velocity by $v_{10.5}^+(r, t)$. The strain of the pulse that starts at an arbitrary spring, say r', is denoted $s_{r'}^+(r, t)$. The initial value (at $t = 0$) of this field is 1 if $r = r'$ and zero otherwise. This function of r and r' is commonly called the Kronecker delta function $\delta_{r'r}$, so we can write $s_{r'}^+(r, 0) = \delta_{r'r}$. For this reason we will call these pulses "delta pulses." At later times t the pulse has moved over by t/dt masses, so in general

$$s_{r'}^+(r, t) = \delta_{r'+(t/dt), r} \tag{2.30}$$

Similarly,

$$s_{r'}^-(r, t) = \delta_{r'-(t/dt), r} \tag{2.31}$$

The corresponding formulas for the velocity fields (using the result of Problem 2.9) are

$$v_{r'}^+(r, t) = -\delta_{r'+(t/dt), r} \tag{2.32}$$

and

$$v_{r'}^-(r, t) = \delta_{r'-(t/dt), r} \tag{2.33}$$

▶   **Problem 2.10**

Suppose we had **not** assumed $k/m = 1$. Show that we can still produce delta pulses that evolve in a simple way, by choosing $dt = (k/m)^{-1/2}$. That is, find initial conditions similar to those in Table 2.8 that evolve in a simple way. Use the equations of motion to construct a table such as Table 2.8 for this case. What is the velocity v_p of these pulses? You can also verify this using MAX1 or MASSES.   ◀

We now have a large collection of pulses, two for each spring r', with displacements $s_{r'}^+$ and $s_{r'}^-$. Our objective was to see whether we could build up an **arbitrary** pulse as a linear combination of them, such as

$$s(r, t) = \sum_{r'} C_{r'}^+ s_{r'}^+ (r, t) + \sum_{r'} C_{r'}^- s_{r'}^- (r, t) \tag{2.34}$$

$$v(r, t) = \sum_{r'} C_{r'}^+ v_{r'}^+ (r, t) + \sum_{r'} C_{r'}^- v_{r'}^- (r, t) \tag{2.35}$$

Here $C_{r'}^+$ and $C_{r'}^-$ are any constant coefficients. We will prove a "pulse decomposition theorem" in Section 2.7 that states that this is essentially true; there is only one initial state that cannot be written this way.

■ **EXAMPLE 2.4**   There are two special waves whose strains and velocities don't change with time. One has uniform velocity (the same for all masses) and zero strain, and the other has uniform strain and zero velocity. Express the uniform-strain wave as a linear combination of delta pulses. Solution: You found in Problem 2.9 that the leftward delta pulse has positive velocities, whereas the rightward one (Table 2.8) has negative ones. It is reasonable to assume, therefore, that by superimposing a rightward and a leftward pulse at each spring, the velocities will cancel. We can show this formally by using the Kronecker delta-function representation of the pulses (Eqs. 2.30 through 2.33). Setting $C_r^+ = C_r^- = s_0$ (some constant), Eq. 2.35 gives

$$v(r, t) = \sum_{r'} s_0 (-\delta_{r' - (t/dt), r}) + \sum s_0 \delta_{r' + (t/dt), r}$$

where the sums are over all the springs r' in the system. But the Kronecker delta function has the property $\sum \delta_{r' - b, r} = 1$ for any b (the summand is zero unless $r' = r + b$, in which case it is 1; in a finite periodic system we define $\delta_{r', r}$ as 1 if r' and r differ by a multiple of N.) Thus $v(r, t) = -s_0 + s_0 = 0$. Equation 2.34 yields in the same way $s(r, t) = s_0 + s_0 = 2s_0$, a constant.

▶   **Problem 2.11**
Find the linear combination of delta pulses that yields uniform velocity and zero strain.
                                                                                                          ◀

Let us look now at the consequences of the pulse-decomposition theorem. The first sum on the right side of Eq. 2.34, which we will denote by $s^+(r, t)$, is a superposition of pulses that are all moving to the right at the same rate. The velocity field associated with this superposition is the first sum on the right side of Eq. 2.35, which we denote by $v^+(r, t)$. However the wave s^+, v^+ appears initially is exactly the way it appears at all later times, it will simply be displaced to the right further and further. That is, given the initial fields $s^+(\cdot, 0)$ and $v^+(r, \frac{1}{2})$, we can compute them at any later time t by moving them a distance t. Mathematically,

$$s^+(r, t) = s^+(r - t, 0) \tag{2.36}$$

$$v^+(r, t + \tfrac{1}{2}) = v^+(r - t, \tfrac{1}{2}) \tag{2.37}$$

because a pulse that is at r at time t must have started at $r - t$. [In general, this

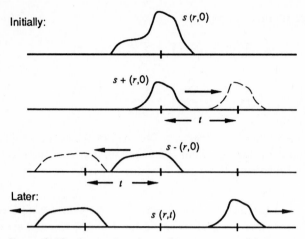

Figure 2.12 Separation of an arbitrary wave into leftward- and rightward-moving parts.

would be $r - v_p t$, where v_p is the speed of the pulse. However, because $k/m = 1$, $v_p = 1$ (Problem 2.10).] The same thing can be done with the second sums in Eqs. 2.34 and 2.35, which we call $s^-(r, t)$ and $v^-(r, t)$; they are given at any time from their initial conditions by

$$\cdot \; s^-(r, t) = s^-(r + t, 0) \tag{2.38}$$

$$v^-\left(r, t + \tfrac{1}{2}\right) = v^-\left(r + t, \tfrac{1}{2}\right) \tag{2.39}$$

These ideas are illustrated in Figure 2.12. The initial strain field $s(r, 0)$ (shown at top of figure) is separated into its rightward-moving part $s^+(r, 0)$ and its leftward-moving part, shown in the two graphs below it. (We do not show the velocity field, which must be known in order to decompose the fields in this way.) During a time interval t, these parts move rightward and leftward respectively by a distance t. The entire strain field can then be obtained by superposing these two displaced parts, as shown at the bottom. (This separation into rightward- and leftward-moving waves can be illustrated using program MAX1, but one must be careful not to inadvertently include the wave we will discuss in the next section, which can't be represented this way.)

It is now possible to explain how MAX1 generates left- and right-moving smooth pulses. Given any smooth initial strain field $s(r, 0)$ we can construct it from rightward-moving delta pulses by setting $C_{r'}^+ = s(r', 0)$ and $C_{r'}^- = 0$ in Eq. 2.34. (The sum over r' then has only one term, with $r' = r$, because $s_{r'}^+(r, 0) = 0$ unless $r = r'$.) The initial condition for the velocity is then given by Eq. 2.35, which the program computes.

▶ **Problem 2.12**

Construct in the above way a rightward-moving pulse with the initial strains shown in Figure 2.10 or Table 2.7 (it must have different initial velocities). Use Eqs. 2.28 and 2.29 to make a table like Table 2.7. Does the pulse move without changing shape, as expected? As another check, you can input this initial s and v by hand to MAX1. ◀

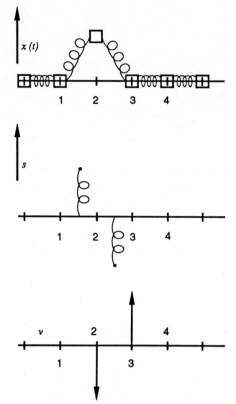

Figure 2.13   A displacement delta-pulse.

We can create a rightward-moving displacement delta-pulse (with only one displaced mass) by superposing two rightward-moving strain delta pulses (Table 2.8) of opposite signs on neighboring springs (Figure 2.13 and Table 2.9).

Table 2.9

r	1.0	1.5	2.0	2.5	3.0	3.5	4.0
$x(r,0)$	0		1		0		0
$s(r,0)$		1		-1		0	
$v(r,\frac{1}{2})$	0		-1		1		0

Given an arbitrary initial displacement $x(r,0)$, we can construct it by superposing translated copies of this delta pulse, just as we did above for an arbitrary initial strain. Such pulses are used in the TVS and RING programs to generate right-moving pulses for given initial displacements.

▶   **Problem 2.13**

Draw a picture with the same displacements as in Figure 2.13 but for a leftward-moving pulse. Give x, s, and v in a table like Table 2.9, for $t = 0, 0.5, 1$, and 1.5.   ◀

SECTION 2.7 PROOF OF THE PULSE-DECOMPOSITION THEOREM

The central idea of this theorem is that any wave can be decomposed into a simple set of waves, the delta pulses. Unfortunately this is not quite true; there is a special "corrugation" wave, which we will describe below, that cannot be so decomposed. However, the existence of this wave doesn't really interfere with our original objective: to find a complete set of waves (from which anything else can be made as a linear combination) each of which evolves in a simple way. We only need to add this special wave to our set of delta pulses.

The initial conditions for the corrugation wave are $s_{corr}(r, 0) = 0$, $v_{corr}(r, \frac{1}{2}) = (-1)^r$. It is shown in Table 2.10 for a ring (note that N must be even for v to satisfy the periodic boundary conditions):

Table 2.10

r	...	0.0	0.5	1.0	1.5	2.0	2.5	3.0	...
$s_{corr}(r, 0.0)$	...		0		0		0		...
$v_{corr}(r, 0.5)$	...	1		-1		1		-1	...
$s_{corr}(r, 1.0)$	...		-2		2		-2		...
$v_{corr}(r, 1.5)$	...	-3		3		-3		3	...
$s_{corr}(r, 2.0)$	...		4		-4		4		...
$v_{corr}(r, 2.5)$	...	5		-5		5		-5	...

At times $t > 0$ the chain has a corrugated appearance. You will notice a pattern of alternating signs in the table, and that the absolute values of s and v increase the same amount during each interval (i.e., they are proportional to t.) This pattern suggests trying formulas such as

$$s_{corr}(r, t) = 2(-1)^{r+t-1/2}t \tag{2.40}$$

$$v_{corr}(r, t) = 2(-1)^{r+t-1/2}t \tag{2.41}$$

▶ **Problem 2.14**

Show that Eqs. 2.40 and 2.41 correctly give the evolution of the initial condition $v_{corr}(r, \frac{1}{2}) = (-1)^r$ for $dt = 1$ and $k/m = 1$, by showing algebraically that they satisfy the equations of motion (Eqs. 2.28 and 2.29) for the chain. ◀

The linear growth of this wave with time is completely unphysical. The actual motion of such a wave (which would be obtained if we used a small dt) is that of a standing wave with $Q = \pi$ (see Section 2.2). The amplitude would oscillate between $+1$ and -1. The linear growth we obtain here is entirely a consequence of using a large value for dt. This sort of thing happens very frequently in solving finite-difference equations. In the field of numerical analysis this value of dt would be called "marginally stable" (the word "unstable" being reserved for values that cause exponential, rather than just linear, growth with t; it turns out that any $dt > 1$ is unstable.) A competent numerical analyst would never use such a dt. However, the lack of stability turns out not to cause any insurmountable problems, and is a small price to pay for the beautifully simple evolution of our delta pulses, which only occurs for $dt = 1$. For waves that are smoother (for example sinusoidal waves with $Q \ll \pi$), $dt = 1$ gives very good accuracy.

We will now prove the **Pulse-decomposition theorem**:

Any wave on a chain of masses with periodic boundary conditions can be written as a linear combination of delta pulses and our special corrugation wave with $v(r, \tfrac{1}{2}dt) = (-1)^r$. That is, given the initial stress s and velocity v of an arbitrary wave (satisfying Eqs. 2.28 and 2.29, with $dt = 1$, $k/m = 1$ for simplicity), we can find coefficients $C_{r'}^{+}$, $C_{r'}^{-}$, and C_{corr} so that

$$s(r, t) = \sum C_{r'}^{+} s_{r'}^{+}(r, t) + \sum C_{r'}^{-} s_{r'}^{-}(r, t) + C_{corr} s_{corr}(r, t) \qquad (2.42)$$

$$v(r, t) = \sum C_{r'}^{+} v_{r'}^{+}(r, t) + \sum C_{r'}^{-} v_{r'}^{-}(r, t) + C_{corr} v_{corr}(r, t) \qquad (2.43)$$

Here s_{corr}, v_{corr} are given by Eqs. 2.40 and 2.41 or Table 2.10.

To prove this theorem, we only need to verify Eqs. 2.42 and 2.43 for $t = 0$ and $t = \tfrac{1}{2}$ respectively; the principle of superposition then implies that they are true at all t, because all the constituent waves are solutions of the equations of motion of the chain. The most straightforward way to compute the coefficients is simply to write Eqs. 2.42 and 2.43 for $t = 0$ and $t = \tfrac{1}{2}$, insert the actual values of s and v for the arbitrary wave, and solve for the C's. There are $2N + 1$ variables and only $2N$ equations, so we ought to be able to solve them; we should even be able to pick one coefficient arbitrarily. However, to prove this we would have to show that the matrix involved in these simultaneous equations had a nonzero determinant, which is a somewhat messy task.

We can go about it in a slightly different way, so that it will be clearer what is going on physically. We will compute the C's one at a time, each time subtracting the corresponding term of Eqs. 2.42 and 2.43 from $s(r, 0)$ and $v(r, \tfrac{1}{2})$. Let us first calculate C_{corr}. This is related to the amount of corrugation in our wave, a sort of "corrugation amplitude," which we will denote by A_{corr}. We can estimate the corrugation amplitude by adding the s's and v's with alternating signs, defining:

$$A_{corr} = \sum_{r} (-1)^{r+1/2} s(r, 0) + \sum_{r} (-1)^r v(r, \tfrac{1}{2}) \qquad (2.44)$$

This expression was chosen so that A_{corr} vanishes for all the delta pulses, as the reader can verify. For the corrugation wave (Eq. 2.41) it is exactly N (the number of masses on the chain.) Thus, for a general wave (Eqs. 2.42 and 2.43) which is a superposition of these, it is

$$A_{corr} = \sum C_{r'}^{+} \cdot 0 + \sum C_{r'}^{-} \cdot 0 + C_{corr} \cdot N \qquad (2.45)$$

which gives us C_{corr}:

$$C_{corr} = \frac{A_{corr}}{N} = N^{-1} \left[\sum (-1)^{r+1/2} s(r, 0) + \sum (-1)^r v(r, \tfrac{1}{2}) \right] \qquad (2.46)$$

Once we know the coefficient of the corrugation term, we can subtract it from our given wave to get a corrugation-free remainder, expressible in terms of delta pulses alone. (This operation is sometimes referred to as "projecting out" the corrugation wave.)

There is now a straightforward procedure for decomposing this remainder into delta pulses. Consider the following initial condition for $N = 4$ masses on a ring (the numbers are randomly chosen integers multiplied by 4 to facilitate dividing by N):

Table 2.11

r	0.5	1.0	1.5	2.0	2.5	3.0	3.5	4.0
$s(r,0.0)$	8		4		12		4	
$v(r,0.5)$		16		8		12		4

We first want to subtract out the corrugation term in Eqs. 2.42 and 2.43 whose amplitude is given by Eq. 2.46 as $C_{corr} = 4^{-1}[-8 + 4 - 12 + 4 - 16 + 8 - 12 + 4] = -7$. So we subtract the corresponding term in Eq. 2.43, namely -7 times the corrugation wave with velocities $-1, 1, -1, 1$, to get

Table 2.12

r	0.5	1.0	1.5	2.0	2.5	3.0	3.5	4.0
$s(r,0.0)$	8		4		12		4	
$v(r,0.5)$		9		15		5		11

You can check that this has zero corrugation amplitude, using Eq. 2.44. We want to express this remaining wave as a superposition of delta pulses. We do this by subtracting out delta pulses to eliminate the s's and v's one at a time, until there is nothing left. We can eliminate $s(0.5, 0)$ by subtracting a multiple of $s_{0.5}^{+}(r, t)$, the rightward delta pulse starting at $r = 0.5$, shown in Table 2.13:

Table 2.13 Delta pulse $s_{0.5}$

r	0.5	1.0	1.5	2.0	2.5	3.0	3.5	4.0
$s(r,0.0)$	1		0		0		0	
$v(r,0.5)$		-1		0		0		0

Evidently the appropriate multiple is 8, leaving

Table 2.14

r	0.5	1.0	1.5	2.0	2.5	3.0	3.5	4.0
$s(r,0.0)$	0		4		12		4	
$v(r,0.5)$		17		15		5		11

We can now eliminate $v(1, \frac{1}{2})$ by subtracting out 17 times the leftward-moving delta pulse starting at $r = 1$; this gives

Table 2.15

r	0.5	1.0	1.5	2.0	2.5	3.0	3.5	4.0
$s(r, 0.0)$	0		-13		12		4	
$v(r, 0.5)$		0		15		5		11

Obviously we can continue this process until we get to the rightmost entry in the table; you can check that after we eliminate $v(3, 0.5)$ we have

Table 2.16

r	0.5	1.0	1.5	2.0	2.5	3.0	3.5	4.0
$s(r, 0.0)$	0		0		0		-11	
$v(r, 0.5)$		0		0		0		11

which is exactly -11 times the rightward pulse starting at $r = 3.5$. Subtracting this, we are left with exactly zero; we have decomposed the original wave into delta pulses, as promised.

Can we be sure that we will always end up with zero for $v(N, 0.5)$ after we eliminate $s(N - \frac{1}{2}, 0)$, as happened in this case? Fortunately, the answer is yes. Since $v(N, 0.5)$ is then the only nonzero term in the corrugation amplitude (Eq. 2.44), we have $A_{\text{corr}} = (-1)^N v(N, 0.5)$. But we made $A_{\text{corr}} = 0$ at the beginning, and subtracting delta pulses never changed it (they have zero A_{corr}); thus it is still zero,

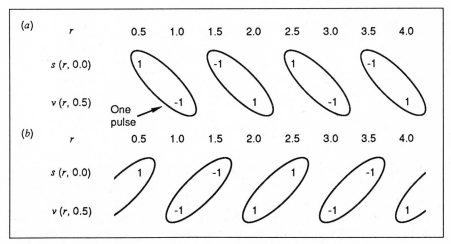

Figure 2.14   Wave made up of delta pulses moving right *(a)* or left *(b)*.

and so is $v(N, 0.5)$. Therefore, our procedure will work for arbitrary even N and arbitrary initial conditions; we have proved the pulse-decomposition theorem.

You may have noticed that we did not use the last pulse available to us, the leftward delta pulse starting at $r = 0.5$. This circumstance means that our result is not unique—we could have started anywhere along the chain, making the coefficient of any desired delta pulse vanish. Thus, different sets of coefficients may give the same wave. Subtracting two such sets of coefficients would yield a nonzero set of coefficients giving a wave whose stress and velocity were identically zero—that is, our delta pulses are not linearly independent. It is easy enough to display such a set of coefficients. Construct a train of delta pulses of alternating sign moving to the right, as shown in Figure 2.14a, by using $C_r^+ = (-1)^{r-1/2}$. But this same wave can equally easily be thought of as a train of leftward pulses, as indicated in Figure 2.14b. If you watched it move you would see the signs reverse, but you couldn't tell whether it was going to the right or to the left. The difference of these trains has $C_r^+ = (-1)^{r-1/2}$, $C_r^- = (-1)^{r+1/2}$, and its s and v fields are identically zero, proving that the delta pulses are not linearly independent.

SECTION 2.8   ENERGY AND POWER IN A CHAIN OF MASSES

In this section we will write down the energy of a periodic chain of N masses, and look at how it changes with time. This will be useful later when we discuss energy in electromagnetic fields, because the mathematics is essentially the same. It is easiest to do in the strain formulation, because the potential energy of a spring depends only on the strain and not on the displacements of the ends individually. The potential energy of a spring with strain s is $\frac{1}{2}k(as)^2$, and the kinetic energy of a mass is of course $\frac{1}{2}mv^2$ (we assume the springs to be massless, so they have no kinetic energy). Although we expect the total energy of the chain to be conserved, we expect energy to be transferred from masses to springs and vice versa, so that the kinetic and potential energies of individual masses and springs will change with time. We can get the change in $\frac{1}{2}mv^2$ from $t - \frac{1}{2}dt$ to $t + \frac{1}{2}dt$ by multiplying Eq. 2.28 by $v(r, t + \frac{1}{2}dt) + v(r, t - \frac{1}{2}dt)$ and by $m/2$:

$$\frac{m}{2}v^2\left(r, t + \tfrac{1}{2}dt\right) - \frac{m}{2}v^2\left(r, t - \tfrac{1}{2}dt\right)$$

$$= \frac{dt}{2}kas\left(r + \tfrac{1}{2}, t\right)v\left(r, t - \tfrac{1}{2}dt\right) - \frac{dt}{2}kas\left(r - \tfrac{1}{2}, t\right)v\left(r, t - \tfrac{1}{2}dt\right) \quad (2.47)$$

$$+ \frac{dt}{2}kas\left(r + \tfrac{1}{2}, t\right)v\left(r, t + \tfrac{1}{2}dt\right) - \frac{dt}{2}kas\left(r - \tfrac{1}{2}, t\right)v\left(r, t + \tfrac{1}{2}dt\right)$$

This expression looks rather inelegant, but it does have a simple physical interpretation. Note that $kas(r + \frac{1}{2}, t)$ is the force exerted by a spring $r + \frac{1}{2}$ on the mass to its left, at r (see Figure 2.15). The product $kasv$ of the force on an object and the velocity of the object is the power transferred by the force (the rate at which work is done on the object), so we can interpret $kas(r + \frac{1}{2}, t)v(r, t - \frac{1}{2}dt)$ as the power

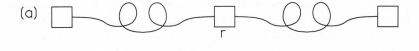

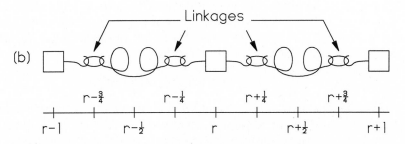

Figure 2.15 (a) Part of a chain of masses. (b) Detail showing linkages.

transferred to the left across the linkage (Figure 2.15b) between spring $r + \frac{1}{2}$ and mass r. The corresponding term in Eq. 2.47 was $(\frac{1}{2}dt)$ times this power, which is the energy transferred over a time interval of length $\frac{1}{2}dt$. The most sensible such interval to associate it with is that from $t - \frac{1}{2}dt$ to t because these are the times at which s and v were evaluated in this term. We will label this linkage $r + \frac{1}{4}$ as in Figure 2.15b, and this time interval by its midpoint $t - \frac{1}{4}dt$. We can then define a **power field** by

$$P\left(r + \tfrac{1}{4}, t - \tfrac{1}{4}dt\right) = -kas\left(r + \tfrac{1}{2}, t\right)v\left(r, t - \tfrac{1}{2}dt\right) \tag{2.48}$$

◀

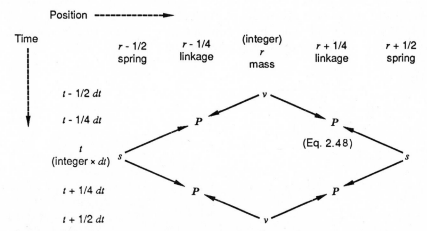

Figure 2.16 Schematic diagram showing which velocity and strain are used to compute each value of the power field.

The minus sign is inserted so that P is the power transferred to the right, which is the usual sign convention.

Note now that each of the other three terms in the kinetic energy change (Eq. 2.47) during the interval between time $t - \frac{1}{2} dt$ and $t + \frac{1}{2} dt$ can also be interpreted as a power transferred to mass r from the right or the left, during either the first or the second half of this interval. The easiest way to understand the relationships among the indices r and t is by making a table (Figure 2.16) of position vs time showing where each variable (s, v, or P) is defined. The arrows into each P indicate which s and v to multiply to yield that P; they are easy to remember because they are the closest s and v to P in the table. Formally we can define P for any odd multiple r of $\frac{1}{4}$ and any odd multiple t of $\frac{1}{4} dt$ by

$$P(r, t) = -kas(\text{closest}) v(\text{closest}) \tag{2.49}$$

where s(closest) is evaluated at $r \pm \frac{1}{4}$ and $t \pm \frac{1}{4} dt$ (the choice of signs is dictated by the fact that s is defined for only one combination). Because each P represents power flowing to the right, the two P's on the left side of Figure 2.16 represent power flowing into mass r. Thus, they should appear with a plus sign in the change of kinetic energy. The two P's on the right represent power flowing out and appear with a minus sign. Thus

$$\left(\frac{m}{2}\right)v^2\left(r, t + \frac{1}{2} dt\right) - \left(\frac{m}{2}\right)v^2\left(r, t - \frac{1}{2} dt\right)$$

$$= P\left(r - \frac{1}{4}, t - \frac{1}{4} dt\right)\left(\frac{dt}{2}\right) + P\left(r - \frac{1}{4}, t + \frac{1}{4} dt\right)\left(\frac{dt}{2}\right) \tag{2.50}$$

$$- P\left(r + \frac{1}{4}, t - \frac{1}{4} dt\right)\left(\frac{dt}{2}\right) - P\left(r + \frac{1}{4}, t + \frac{1}{4} dt\right)\left(\frac{dt}{2}\right)$$

Using the definition (Eq. 2.49) of P, you should verify that this is the same as Eq. 2.47.

As an example, let us look at the power transferred during the passage of a strain delta pulse, using $k = 1$, $m = 1$, $a = 1$, and $dt = 1$. We begin by tabulating the evolution of s and v, as we did in Table 2.8. We can then calculate a value of P for each neighboring s and v by multiplying them, as indicated in Figure 2.16. We show only the three nonzero values of P in Table 2.17:

Table 2.17

r	0	0.25	0.5	0.75	1.0	1.25	1.5	1.75	2.00
$s(r,0)$			1				0		
$P(r,0.25)$				$P = +1$					
$v(r,0.5)$	0				-1				0
$P(r,0.75)$						$P = +1$			
$s(r,1.0)$			0				1		
$P(r,1.25)$								$P = +1$	
$v(v,1.5)$	0				0				-1

As we would expect, power is transferred to the right by this pulse (i.e., $P > 0$). To check the kinetic energy balance (Eq. 2.50), convert this to a table of kinetic and potential energies and energy transfers $P\,dt/2$:

Table 2.18

r	0	0.25	0.5	0.75	1.0	1.25	1.5	1.75	2.00
$(\frac{1}{2}ka^2)s^2(r,0)$			$\frac{1}{2}$				0		
$(\frac{1}{2}dt)P(r,0.25)$				$+\frac{1}{2}$					
$(\frac{1}{2}m)v^2(r,0.5)$	0				$\frac{1}{2}$				0
$(\frac{1}{2}dt)P(r,0.75)$				0		$+\frac{1}{2}$			
$(\frac{1}{2}ka^2)s^2(r,1.0)$			0				$\frac{1}{2}$		
$(\frac{1}{2}dt)P(r,1.25)$				0				$+\frac{1}{2}$	
$(\frac{1}{2}m)v^2(r,1.5)$	0				0 ← 0				$\frac{1}{2}$

For $r = 1$ and $t = 1.0$, for example, Eq. 2.50 involves the four P's indicated by arrows. The kinetic energy of mass $r = 1$ decreases by $\frac{1}{2}$ because the energy $P(1.25, 0.75)\,dt/2 = \frac{1}{2}$ is transferred away to spring $r = 1.5$. This energy shows up as an increase in the potential energy $(ka^2/2)s^2$ of spring 1.5. The equation for the change in potential energy can be derived analogously to Eq. 2.50. It is

$$\left(\frac{ka^2}{2}\right)s^2\left(r,\, t + \tfrac{1}{2}dt\right) - \left(\frac{ka^2}{2}\right)s^2\left(r,\, t - \tfrac{1}{2}dt\right)$$

$$= P\left(r - \tfrac{1}{4},\, t - \tfrac{1}{4}dt\right)\left(\frac{dt}{2}\right) + P\left(r - \tfrac{1}{4},\, t + \tfrac{1}{4}dt\right)\left(\frac{dt}{2}\right)$$

$$- P\left(r + \tfrac{1}{4},\, t - \tfrac{1}{4}dt\right)\left(\frac{dt}{2}\right) - P\left(r + \tfrac{1}{4},\, t + \tfrac{1}{4}dt\right)\left(\frac{dt}{2}\right) \qquad (2.51)$$

The right-hand side is formally identical to that of Eq. 2.50, although the allowed values of r and t are different.

▶   **Problem 2.15**

Derive Eq. 2.51 for the potential energy balance from Eqs. 2.29 and 49.   ◀

▶   **Problem 2.16**

a. Construct tables similar to Tables 2.17 and 2.18 for the case $k = m = a = 1$, $dt = \frac{1}{2}$. That is, tabulate s, P, and v in the first table, and $s^2/2$, $P\frac{1}{2}dt = P/4$, and $\frac{1}{2}v^2$ in the second. Use the same initial $s(0)$, $V(t = \frac{1}{2}dt)$. Give all nonzero values for $t \leq 1.5\,dt = 0.75$. Write the kinetic energy balance equation during the interval from 0.25 to 0.75 (of length dt) for mass $r = 1$, and verify it numerically. (You should also check this for masses 0 and 2, but you need not write these down.)

b. Do the same for the potential energy balance for spring $\frac{1}{2}$, between $t = 0$ and 0.5.   ◀

We promised at the beginning of this section to write down the total energy of our chain of masses, and establish that it is independent of time. Not only haven't we done this, but we can't: the potential energy and the kinetic energy are never defined at the same time, but at alternate multiples of $dt/2$. In Table 2.18, each row representing a multiple of $dt/2$ had $\frac{1}{2}ka^2s^2$ or $\frac{1}{2}mv^2$, but not both. Equation 2.50 gives us the change in kinetic energy (KE) from $t - \frac{1}{2}dt$ to $t + \frac{1}{2}dt$, but doesn't mention the KE after half this interval, at time t. However, note that the change in KE has two terms that involve the first half of the interval (that is, in which the time argument of P is $t - \frac{1}{4}dt$, and two that refer to the second half ($t + \frac{1}{4}dt$). It would make sense to regard the former two terms as describing that part of the change that occurs over the first half of the interval, that is, to define a "kinetic energy" at the time when it wasn't defined before by

$$\mathrm{KE}(r, t) \equiv \mathrm{KE}\left(r, t - \tfrac{1}{2}dt\right) + P\left(r - \tfrac{1}{4}, t - \tfrac{1}{4}dt\right)\tfrac{1}{2}dt - P\left(r + \tfrac{1}{4}, t - \tfrac{1}{4}dt\right)\tfrac{1}{2}dt \tag{2.52}$$

where $\mathrm{KE}(r, t - \frac{1}{2}dt)$ means $(m/2)v^2(r, t - \frac{1}{2}dt)$, which was well-defined all along. The definition (Eq. 2.52) applies only when t is an even multiple of dt (i.e., a time when v is not defined). However, from Eqs. 2.52 and 2.50 you can easily show that Eq. 2.52 is correct (as an equation, not a definition) for the other t's as well.

From Eq. 2.51, we can guess a corresponding definition for the PE at times for which s is undefined:

$$\mathrm{PE}(r, t) \equiv \mathrm{PE}\left(r, t - \tfrac{1}{2}dt\right) + P\left(r - \tfrac{1}{4}, t - \tfrac{1}{4}dt\right)\tfrac{1}{2}dt - P\left(r + \tfrac{1}{4}, t - \tfrac{1}{4}dt\right)\tfrac{1}{2}dt \tag{2.53}$$

The definition (Eq. 2.52) has a simple interpretation in terms of Table 2.18; instead of adding or subtracting all four P's indicated by arrows to $\mathrm{KE}(t = 0.5) = \frac{1}{2}$, to get $\mathrm{KE}(t = 1.5)$, we can just add or subtract the top (earlier) two to define $\mathrm{KE}(t = 1.0)$, which turns out to be 0. This allows us to fill in both kinds of energies at *all* multiples of $\frac{1}{2}dt$:

Table 2.19

r	0	0.25	0.5	0.75	1.0	1.25	1.5	1.75	2.00
Energy ($t = 0$)	(0)		$\frac{1}{2}$		(0)		0		(0)
($\frac{1}{2}dt$)$P(t = 0.25)$				$+\frac{1}{2}$		0			
Energy ($t = 0.5$)	0		(0)		$\frac{1}{2}$		(0)		0
($\frac{1}{2}dt$)$P(t = 0.75)$				0		$+\frac{1}{2}$			
Energy ($t = 1$)	(0)		0		(0)		$\frac{1}{2}$		(0)
($\frac{1}{2}dt$)$P(t = 1.25)$								$+\frac{1}{2}$	
Energy ($t = 1.5$)	0		(0)		0		(0)		$\frac{1}{2}$

where the newly defined energies are in parentheses. The kinetic energy at $t = 0$ must be obtained by using Eq. 2.52 in reverse (i.e., solving it for the earlier KE), as indicated by the upward arrows in the table.

▶ **Problem 2.16**

c. Table 2.19 lacks excitement because all the new energies we defined were zero. Show that this is not true for $dt = \frac{1}{2}$ by filling in the missing energies in the table in Problem 2.16a. [In fact, we get more excitement than we bargained for: there is a **negative** energy $PE(\frac{1}{2}, 1.5\, dt) = -1/32$, which is bothersome because the old PE's were obtained by squaring s, and must therefore be positive. There is no reason why the newly defined energies should necessarily be, however. Physically, we want the energies to be positive in the continuous-time ($dt \to 0$) limit, but in this limit each newly defined energy at a time t is very close to a (positive) old one at $t - \frac{1}{2}\, dt$, and will therefore be positive or zero.] ◀

Finally, we can check that energy is conserved. This is actually just a consequence of our definitions (Eqs. 2.52 and 2.53): the newly defined KE of a mass included exactly the energy lost by the neighboring springs, so the total energy is unchanged. We can see this formally by writing the total energy of the chain at time t:

$$E(t) = \sum_r KE(r, t) + \sum_{r'} PE(r', t) \qquad (2.54)$$

where $r = 1, 2 \ldots, N$ labels the masses and r' labels the springs. From Eqs. 2.52 and 2.53 we get the change in total energy:

$$E(t) - E\left(t - \tfrac{1}{2}\, dt\right) = \sum_r P\left(r - \tfrac{1}{4}, t - \tfrac{1}{4}\, dt\right) - \sum_r P\left(r + \tfrac{1}{4}, t - \tfrac{1}{4}\, dt\right)$$

$$+ \sum_{r'} P\left(r' - \tfrac{1}{4}, t - \tfrac{1}{4}\, dt\right) - \sum_{r'} P\left(r' + \tfrac{1}{4}, t - \tfrac{1}{4}\, dt\right) \quad (2.55)$$

The first sum involves P at positions of the form (integer $- \frac{1}{4}$), as does the fourth (because $r' = $ integer $- \frac{1}{2}$), so they cancel. The second and third terms involve positions (integer $+ \frac{1}{4}$) and also cancel. The energy change is therefore zero.

▶ **Problem 2.16**

d. Table 2.19 is consistent with conservation of energy because the sum of each row of energies is $\frac{1}{2}$. Verify conservation of energy in this way for the table you made in Problem 2.16c. ◀

▶ **Problem 2.17**

a. Another way to define the kinetic energy $K = \frac{1}{2}mv^2$ at a time when v is undefined would be to use the mean of $v(r, t - \frac{1}{2}\, dt)$ and $v(r, t + \frac{1}{2}\, dt)$. Using the **arithmetic** mean gives a result different from and less convenient than Eq. 2.52. Show, however, that using the **geometric** mean, that is, defining

$$KE(r, t) \equiv \tfrac{1}{2}mv\left(r, t - \tfrac{1}{2}\, dt\right)v\left(r, t + \tfrac{1}{2}\, dt\right)$$

gives the same result as Eq. 2.52. (b) Write the corresponding formula for $PE(r, t)$, and show that it is equivalent to Eq. 2.53. These geometric-mean forms are more convenient for numerical calculations in which the energy but not the power is desired, because they do not require calculating the power. ◀

CHAPTER THREE

ABSTRACT FIELDS AND THE CONTINUUM LIMIT

SECTION 3.1 VERTEX AND EDGE FIELDS: THE GRADIENT

In this section we will introduce a different set of terminology for the displacement and strain fields on a chain of masses that we have been discussing. This terminology is completely unnecessary for calculations involving one-dimensional chains, but we will find it useful later in understanding Dirac delta functions and vector calculus.

The first concept we will introduce is that of a **one-dimensional lattice**, which is simply a more abstract way of talking about the places in space where our displacement and strain fields are defined. Part of a one-dimensional lattice is shown in Figure 3.1. (The use of the term "lattice" is most appropriate in two dimensions, but we will use it in one and three dimensions as well.) It is a collection of points called **vertices** and line segments called **edges**, such that each vertex touches exactly two edges and each edge touches exactly two vertices. We will use the letter v to denote a vertex label; do not confuse it with the velocity. In the context of a chain of masses, you should think of the vertices as the equilibrium positions of the masses, and the edges as the locations of the springs. A lattice may be finite or infinite; a finite lattice would describe a ring of masses such as those described in Section 2.4. Given such a lattice, a **discrete vertex field** (which we will usually denote by w) is a specific way of assigning real numbers to the vertices. An example of a discrete vertex field is shown in Figure 3.2a. The number assigned to a particular vertex v is denoted $w(v)$. Thus, the displacement field describing a chain of masses is a vertex field; the number $w(v)$ assigned to each vertex gives the displacement (denoted $x(r)$ in the previous sections) of the mass whose equilibrium position is at that vertex. Another example of a vertex field may be obtained by thinking of Figure 3.2a as part of a road map and the numbers as the altitudes (in thousands of meters) of four towns along a road. We can similarly define a **discrete edge field** as an assignment of real numbers $s(e)$ to the edges e, such as the assignment depicted in Figure 3.2b. An example of an edge field is then the strain field in a chain of masses; another (based again on interpreting the lattice as a road map) might be the number of hours required to drive along each edge.

The strain field of a chain of masses is uniquely determined by the displacement field through Eq. 2.26, which in the present notation reads

$$s(e) = \frac{w(v_{\text{right}}) - w(v_{\text{left}})}{dr} \tag{3.1}$$

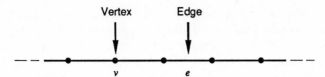

Figure 3.1 Part of a 1D lattice.

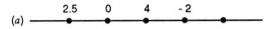

Figure 3.2 (a) A discrete vertex field. (b) A discrete edge field.

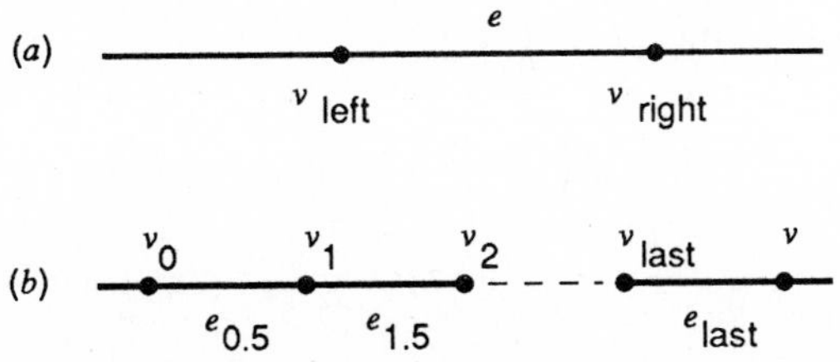

Figure 3.3 (a) Vertex labels used in defining the gradient. (b) Vertex labels used in defining the discrete integral.

Here v_{right} and v_{left} are the vertices to the right and left of the edge e, as shown in Figure 3.3a, and dr is the length of each edge, formerly denoted by a. (Note that dr, which has dimensions of length, is unrelated to the dimensionless integer r we previously used to label the vertices.) There is no reason why Eq. 3.1 should only be applied to displacement fields on chains of masses; applied to the altitude field along a road it gives you the average slope (or "grade") of the road along edge e. It is referred to in general as the **discrete gradient** of the vertex field w, denoted "grad w" and defined by

$$[\text{grad } w](e) \equiv \frac{w(v_{\text{right}}) - w(v_{\text{left}})}{dr} \tag{3.2}$$

Thus the gradient operation gives an edge field for any vertex field. When we take the continuum limit ($dr \to 0$), it will turn into the operation of differentiation.

▶   **Problem 3.1**

Calculate the discrete gradient of the vertex field in Figure 3.2a (let $dr = 1$). Give the result in a figure such as Figure 3.2b.   ◀

It is interesting to ask whether this operation can be inverted. That is, given an edge field s, can one find the vertex field w of which it is the gradient? Evidently this vertex field is not unique; one can obviously add a constant to all the values of w and get the same gradient. One can therefore choose **one** of the values of w arbitrarily, at a "reference vertex," which we will denote by v_0 (see Figure 3.3b). We can obtain a formula for w at any other desired vertex v by adding up Eq. 3.1 over all of the intervening edges:

$$
\begin{aligned}
s(e_{0.5}) &+ s(e_{1.5}) + \cdots + s(e_{last}) \\
&= \frac{w(v_1) - w(v_0) + w(v_2) - w(v_1) + \cdots + w(v) - w(v_{last})}{dr} \\
&= \frac{w(v) - w(v_0)}{dr}
\end{aligned}
\tag{3.3}
$$

where the w's at all the intervening vertices canceled out on the right-hand side. Thus

$$
w(v) = w(v_0) + \int_{v_0}^{v} s(e) \, dr
\tag{3.4}
$$

where we have defined

$$
\int_{v_0}^{v} s(e) \, dr \equiv s(e_{0.5}) \, dr + s(e_{1.5}) \, dr + \cdots + s(e_{last}) \, dr
\tag{3.5}
$$

the **discrete integral** of the edge field s from v_0 to v. For any given edge field s, Eq. 3.4 gives a vertex field w those gradient is s [in fact many such vertex fields, because $w(v_0)$ can be chosen arbitrarily.] It thus inverts the gradient operation.

▶   **Problem 3.2**

    a. Compute the discrete indefinite integral of the edge field shown in Figure 3.2b. That is, find a discrete vertex field whose gradient is that edge field. Use the leftmost vertex as v_0, and assume $w(v_0) = 0$. Give your result in a figure similar to Figure 3.2a.

    b. Do the same using the **rightmost** vertex as v_0. [Hint: You can't use Eq. 3.4 directly]. ◀

SECTION 3.2   CONTINUUM FIELDS

In Section 2.2, we used formulas such as $\sin(Qr)$ to specify a displacement field. Such a formula defines a function $f(r)$ of a real variable; that is, it is defined at all points of a continuum, not just at the vertices of a lattice. We only used the formula at a few points, however, so we cannot claim to have used a continuum field; it was just a convenient formula for determining a discrete field. In a physical system that is intrinsically discrete, such as a chain of masses, the values of f between vertices are irrelevant. However, there are systems for which a truly continuum description is useful. One such system is the elastic string depicted in Figure 3.4a. We can use the

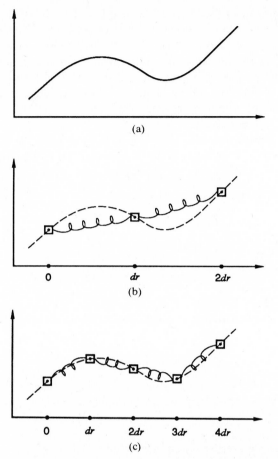

Figure 3.4   (*a*) An elastic string. (*b*) Model of string in the previous figure. (*c*) A smaller *dr*.

methods we have developed by modeling it as a chain of masses (Figure 3.4*b*), each representing a length *dr* of the string. We will describe the behavior of this model in detail in Section 3.3; at present we just want to discuss what kind of field we should use to describe a particular configuration of the string, such as the one in the figure. Clearly, if we had a rapidly spatially varying wave (say a sinusoidal wave whose wavelength was of order *dr* or smaller) specifying the displacement at intervals of *dr* would not be enough; we would need to use a smaller *dr* (Figure 3.4*c*) to describe it correctly and compute its behavior. The behavior would change as we decreased *dr*; hopefully it would stop changing when *dr* became very small (much less than the wavelength) and would approach a limit (the "continuum limit") that would correctly describe the observed behavior of the string. To calculate this limit we must define what we mean by a "continuum field," which we will do in this section.

   We will define two types of continuum fields, continuum vertex fields and continuum edge fields, as well as their gradients (derivatives). In most applications these are equivalent to derivatives of ordinary functions that you presumably already know how to compute, and the distinction may seem academic. If you are not enthusiastic about mathematical niceties of this sort, you can probably skip the rest of this chapter and still understand most of the rest of the book. However, you may find that you want to come back and read it later; we will have occasion to use certain concepts, such as the Dirac delta function, which can be conveniently introduced through the concepts that we define in this section. If it's any consolation, our method of introducing the Dirac delta function is considerably less abstract than the usual one[1] in which it is defined as a linear functional on an abstract space of infinitely differentiable test functions.

We will define a **continuum vertex field**, denoted by a letter with a subscript c, such as w_c, as a consistent **rule** that gives a discrete vertex field (which we will denote by w_{dr}) for any lattice with spacing dr. Examples are:

$$\text{Rule } w_c: \quad w_{dr}(v) = 2v^3 \tag{3.6}$$

$$\text{Rule } x_c: \quad x_{dr}(v) = 4\sin(2v) \tag{3.7}$$

$$\text{Rule } u_c: \quad u_{dr}(v) = 0 \quad \text{if } v < 0$$
$$1 \quad \text{if } v > 0 \quad \text{(a step function)} \tag{3.8}$$

(We are using v here to denote a vertex of our 1D lattice, which is also a real number.) By "consistent" we mean that if a vertex v is contained in two different lattices (with different dr) it gives the same value for w_{dr}. For example,

$$w_{2\,dr}(v) = w_{dr}(v) \tag{3.9}$$

if v is a vertex of both lattices. The examples above are trivially consistent because they don't mention dr; the consistency requirement is meant to exclude rules such as $w_{dr}(v) = dr$.

> At this point I can hear an alert reader saying "Now wait a minute. What you have defined as a "continuum vertex field" is nothing but an ordinary real-valued function of a real variable: a rule assigning a number w to each number v. Why introduce all this extra terminology?" The reader is quite correct: A continuum vertex field is essentially equivalent to an ordinary function. However, we need it in order to define continuum **edge** fields, which are **not** all equivalent to ordinary functions—in fact they include the Dirac delta function, as we will soon see.

We will define a **continuum edge field** in exactly the same way as a continuum vertex field, as a consistent rule giving a discrete edge field s_{dr} for any lattice. However, we can't use quite the same definition of "consistent" as we did for a vertex field (that it gives the same value for the same edge with different dr's) because we will never find the same edge on two lattices with different dr's (see Figure 3.5)—the edges necessarily have different lengths. But note that Figure 3.5 does show two edges e_1 and e_2 of the lattice of spacing dr that, when put together, add up to an edge e of the $2\,dr$ lattice. A reasonable consistency condition would be that the values of s also add up, in the sense that

$$s_{dr}(e_1)\,dr + s_{dr}(e_2)\,dr = s_{2\,dr}(e)\,2\,dr \tag{3.10}$$

[1]Schwartz, L., *Theorie des distributions*, Tome I et II, Actualities scientifiques et industrielles 1091 and 1122, Hermann & Cie, Paris 1950, 1951.

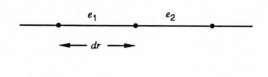

Figure 3.5 Lattices with spacings dr and $2dr$.

If you think of s as a strain field, this condition is clearly required for physical reasonableness: it says that the total elongation of the edge e is the sum of the elongations of its two pieces e_1 and e_2.

▶ **Problem 3.3**

Show that the consistency condition (Eq. 3.10) is equivalent to requiring that for each pair of vertices v and v' that are on both the dr and $2\,dr$ lattices,

$$\int_v^{v'} s_{dr}\, dr = \int_v^{v'} s_{2\,dr}\, dr \tag{3.11}$$

That is, the discrete integral (Eq. 3.5) is the same for both lattices. ◀

We can now define the gradient operation on continuum vertex fields. This is in fact almost trivial, albeit a bit abstract. The gradient of a continuum vertex field w_c should clearly be a continuum edge field, which we may call s_c. To define it, we must specify a rule giving a **discrete** edge field for each lattice. The rule w_c determines a discrete **vertex** field w_{dr} for this lattice, and we know how to find its gradient (Eq. 3.2). This is our required discrete edge field.

Let's now look at some examples of continuum edge fields. The above definition of the gradient gives us a way to generate these.

■ **EXAMPLE 3.1** Find the gradient of the continuum vertex field w_c (Eq. 3.6). The gradient $s_c = \operatorname{grad} w_c$ is given by the rule

$$s_{dr}(e) \equiv [\operatorname{grad} w_{dr}](e) = \frac{w_{dr}(e + \tfrac{1}{2}dr) - w_{dr}(e - \tfrac{1}{2}dr)}{dr}$$

$$\tag{3.12}$$

$$= \frac{2(e + \tfrac{1}{2}dr)^3 - 2(e - \tfrac{1}{2}dr)^3}{dr} = 6e^2 + \tfrac{1}{2}dr^2$$

where $e \pm \tfrac{1}{2}dr$ is a vertex to the right/left of the edge e.

■ **EXAMPLE 3.2** The gradient of the continuum vertex field u_c (a step function, Eq. 3.8) is given by the rule

$$s_{dr}(e) \equiv [\operatorname{grad} u_{dr}](e) = \frac{u_{dr}(e + \tfrac{1}{2}dr) - u_{dr}(e - \tfrac{1}{2}dr)}{dr}$$

$$\tag{3.13}$$

$$= \begin{cases} 1/dr & \text{if the edge } e \text{ contains the origin} \\ 0 & \text{if it does not} \end{cases}$$

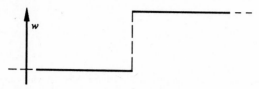

Figure 3.6   A string with a step-function displacement.

because both values of u_{dr} are the same (both 0 or both 1) if the two ends of the edge are on the same side of the origin. This continuum edge field s_{dr} is known as the Dirac delta function; it is the derivative of the step function. In the context of an elastic string, a step function displacement has a sudden discontinuity in the vertical displacement (Figure 3.6), which could of course only be achieved approximately in a real string. The strain is then a delta function, zero everywhere except at the discontinuity and undefined (very large, in a real string) there.

▶   **Problem 3.4**

Compute the gradient of the continuum vertex field x_c (Eq. 3.7). That is, give a formula for $s_{dr}(e)$ for arbitrary dr and e.   ◀

▶   **Problem 3.5**

Any function w_f can be used to define a continuum vertex field w_c, which has a gradient in the sense in which we have defined it. (The subscript f stands for "function.") This includes some rather strange functions which, like the step function, do not have derivatives in the ordinary sense. Define a function $w_f(r)$, to be 1 when r is rational and 0 when r is irrational. Find the gradient $s_c = \text{grad } w_c$ for this case (i.e., give the rule determining the discrete edge field s_{dr} at any arbitrary edge e, extending from $e - \frac{1}{2}dr$ to $e + \frac{1}{2}dr$, in terms of the rationality or irrationality of the endpoints.)   ◀

We are used to getting a function when we compute a gradient (i.e., a derivative.) In what sense can a continuum edge field be thought of as a function? Let us consider a continuum edge field s_c, and try to define a corresponding function, say s_f. We must determine the value of s_f at each point y, from the rule s_c that gives a discrete edge field $s_{dr}(e)$ for each dr. For a given dr, we can find an edge that is within $dr/2$ of y. If we let dr approach zero, these edges approach y arbitrarily closely, which suggests defining $s_f(y)$ as their limit (if this exists):

$$s_f(y) = \lim_{dr \to 0} s_{dr}(e) \tag{3.14}$$

wherein e is understood to be the closest edge to y, for each dr. In the case of Example 3.1 (the gradient of $2v^3$), this gives

$$s_f(y) = \lim\left(6e^2 + \tfrac{1}{2}dr^2\right) = \lim 6e^2 = 6y^2 \tag{3.15}$$

which is exactly the usual derivative. It is not hard to prove that this is true in general. If a continuum vertex field (thought of as a function w_f) is differentiable, this limit s_f will exist for $s_c = $ grad w_c, and s_f will be exactly the derivative of w_f in the usual sense. However, w_c may not be differentiable, in which case the limit does not exist for all y, and we cannot define a function s_f associated with the continuum edge field s_c. One such non-differentiable function is the step function (Example 3.2). In this case the limit in Eq. 3.14 gives zero for $y \neq 0$, because for $dr < y$ the closest edge to y cannot include the origin. At $y = 0$, the closest edge **always** includes the origin, so we have $\lim 1/dr$, which of course doesn't exist as $dr \to 0$. Although the gradient in this case is a perfectly well-defined continuum edge field (the Dirac delta "function"), it cannot be associated with any ordinary function. Objects of this sort are often called "generalized functions"; they can be defined in a number of different ways, most of which are either much more abstract or much less rigorous than our present definition in terms of continuum edge fields.

▶   **Problem 3.6**

Compute the limit in Eq. 3.14 for the continuum edge field you found in Problem 3.4. Also compute the ordinary derivative of the function $w(v) = 4\sin(2v)$. Are they the same?   ◀

▶   **Problem 3.7**

Does the limit in Eq. 3.14 exist for the continuum edge field you found in Problem 3.5?
                                                                                                                         ◀

▶   **Problem 3.8**

Prove that if a continuum vertex field w comes from a differentiable function, then the derivative we defined is exactly the continuum derivative. For simplicity assume $y = 0$, and consider only lattices such that y is a face center (so $\pm \frac{1}{2} dr$ are vertices).   ◀

So far we have found a procedure for finding a function s_f associated with a given continuum edge field s_c. Can we do the reverse, and define a continuum edge field for a given function? We might try to do this the way we do it for a vertex field, by simply evaluating the function at the position of the edge: $s_{dr}(e) = s_f(e)$. This does not work, as you can see from Example 3.1, where it gives $s_{dr}(e) = 6e^2$, but the actual continuum edge field is $s_f(e) = 6e^2 + \frac{1}{2} dr^2$.

▶   **Problem 3.9**

Show that this only works for linear functions. That is, show that $s_{dr}(e) = s_f(e)$ defines a consistent continuum edge field s_c if and only if $s_f(y) = ay + b$ for some constants a and b.   ◀

However, we can make use of the fact that we know how to associate functions with continuum **vertex** fields. All we need to do is integrate the function s_f, find the associated vertex field w_c, and then differentiate it (i.e., take its gradient). This gives

$$s_{dr}(e) = \frac{w_{dr}\left(e + \frac{1}{2}dr\right) - w_{dr}\left(e - \frac{1}{2}dr\right)}{dr}$$

$$= dr^{-1} \int_{e-dr/2}^{e+dr/2} s_f(y)\, dy$$

(3.16)

That is, we can construct an edge field by integrating the function over each edge. Thus any function that is integrable in the conventional sense (and in particular any continuous function) determines a continuum edge field using Eq. 3.16.

> We are assuming here that the reader knows what we mean by an integral of an ordinary function. We could avoid this assumption, in fact, by **defining** the integral of a function as the limit of our discrete integral. That is, for any function $s_f(y)$, we can **approximate** $s_{dr}(e)$ by $s_f(e)$, take the discrete integral, and then take the limit $dr \to 0$. It is not hard to see that this is equivalent to the usual Riemann integral.

▶ **Problem 3.10**

> Show that the continuum edge field defined by Eq. 3.16 is "consistent" in the sense of Eq. 3.10.   ◀

We will sometimes be interested in an **approximate** value of the discrete edge field $s_{dr}(e)$ defined by Eq. 3.16. If s_f doesn't vary much over the interval of integration, we can replace the integrand by its value at the center of the interval, that is, $s_f(e)$. This gives

$$s_{dr}(e) \cong s_f(e) \tag{3.17}$$

In Example 3.1 this approximation is wrong by $\frac{1}{2}dr^2$.

▶ **Problem 3.11**

> Show that if s_f is an analytic function (i.e., it has a power series expansion around each point e) the error in the approximate Eq. 3.17 is always of order dr^2 (i.e., it may contain higher powers, but no lower ones).   ◀

In cases in which we intend to take the $dr \to 0$ limit anyway, the approximation is therefore adequate, and we will use it frequently.

We have obtained a number of important results in this section that are worth summarizing:

1. Every function determines a continuum vertex field, and *vice versa*; the two concepts are essentially equivalent.
2. Every continuous function determines a continuum edge field, but some important edge fields are **not** obtainable from functions.
3. Every continuum vertex field (in effect, every function) has a gradient, which is a continuum edge field. **If** the function is differentiable in the usual sense, this continuum edge field is obtainable from a function, namely the usual derivative.

SECTION 3.3   THE DYNAMICS OF A STRING: PARTIAL DIFFERENTIAL EQUATIONS

In the previous section we described how one could model a system such as an elastic string or bar by chains of masses, which describe the string better and better as one approaches the continuum limit. We will now work out this model in detail.

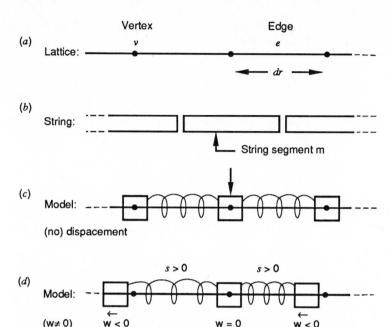

Figure 3.7 A chain-of-masses model for a longitudinally strained string.

Imagine the string to be divided into short segments of length dr, as shown in Figure 3.7b. The masses all have the same magnitude $m = \lambda\,dr$, where λ is the mass per unit length of the string or bar. Figure 3.7d shows a displaced configuration of the string (unlike the figures of the previous section, this shows a **longitudinal** displacement and strain field.) There is a restoring force on the segment to the left of an edge e (due to tension) if the centers of it and the segment to the right of the edge are pulled apart. This force is proportional to the strain at that edge:

$$F = Cs \qquad (3.18)$$

where C is a constant. (For a bar, C is Young's modulus times the cross-sectional area. It is the same for a string if s is positive, which we can ensure by first subjecting the string to a uniform positive strain, larger in magnitude than the largest negative strain in our wave, and then superposing our wave.) The force given by Eq. 3.18 is the same as that which would be produced by a perfectly elastic spring with spring constant $k = C/dr$. Thus, all the results we have obtained for a chain of masses carry over to the elastic string if we use these values for m and k.

▶ **Problem 3.12**

Show that Eq. 3.18 is correct for **transverse** motions as well, with a different value of C. Assume that the string has a uniform (longitudinal) tension in it, and its motions and nonuniform strains are small and entirely transverse. What is the value of C? ◀

We can then copy the discrete equations of motion for the string directly from Section 2.5, switching to our present notation (v for vertices, a for dr, C/dr for k, and $\lambda \, dr$ for m). We will need to use u for the velocity field to avoid confusion with the vertex label v. Eq. 2.29 becomes

$$s\left(e, t + \tfrac{1}{2}dt\right) - s\left(e, t - \tfrac{1}{2}dt\right) = dt \, \frac{u\left(e + \tfrac{1}{2}dr, t\right) - u\left(e - \tfrac{1}{2}dr, t\right)}{dr} \qquad (3.19)$$

Note that the last factor on the right hand side is just the discrete gradient (Eq. 3.2), so that this can be written more concisely as

$$\frac{s\left(e, t + \tfrac{1}{2}dt\right) - s\left(e, t - \tfrac{1}{2}dt\right)}{dt} = [\text{grad } u](e, t) \qquad (3.20)$$

The left-hand side also has the form of a discrete derivative (with respect to time rather than space), for which we may as well also establish a more concise notation

$$\frac{ds(e, t)}{dt} \equiv \frac{s\left(e, t + \tfrac{1}{2}dt\right) - s\left(e, t - \tfrac{1}{2}dt\right)}{dt} \qquad (3.21)$$

We will refer to this as the "discrete time derivative." We will use the same notation for the continuum time derivative, but it will always be clear from the context which is meant. Then the equation of motion for s becomes

$$\frac{ds(e, t)}{dt} = [\text{grad } u](e, t) \qquad (3.22)$$

The equation of motion for the velocity field u (Eq. 2.28) can also be written in terms of a discrete time derivative:

$$\frac{du(v, t)}{dt} = \frac{C}{\lambda} \frac{s\left(v + \tfrac{1}{2}dr, t\right) - s\left(v - \tfrac{1}{2}dr, t\right)}{dr} \qquad (3.23)$$

The right-hand side looks a lot like the discrete gradient we defined in Section 3.1, except for the fact that it acts on an edge field s instead of a vertex field. However, we can turn a vertex field into an edge field (and vice versa) by shifting our lattice over by $\tfrac{1}{2}dr$ so that all the vertices turn into edges and edges into vertices. The resulting lattice (Figure 3.8) is called the **dual lattice**. Note that by shifting to the dual lattice, taking the gradient, and switching back, we can in effect define the gradient of an edge field. In less abstract terms, this amounts to defining the rightmost factor in Eq. 3.23 as the discrete gradient of s, so that

$$\frac{du(v, t)}{dt} = \frac{C}{\lambda}[\text{grad } s](v, t) \qquad (3.24)$$

(As in the case with many of the concepts we have introduced in one dimension, the dual lattice may seem an unnecessary construct here, but we will find it very useful in understanding the relationships among the various vector derivatives in two and three dimensions.) We now have two concise discrete equations (Eqs. 3.22 and 3.24) for updating the strain and velocity fields of a chain of masses. They are no different from the ones we used in Section 2.5, except for notation.

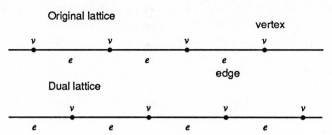

Figure 3.8 The original and dual lattices.

The purpose of the present section is to try to use these equations in the continuum limit to describe the motions of a string. Ideally, given initial continuum stress and velocity fields s and u [determined by formulas like $s(y) = \cos(Qy)$, for example], we would like a prescription for determining a formula for $s(y, t)$ and $u(y, t)$ at later times t, just as the discrete updating equations (Eqs. 3.22 and 3.24) give the later discrete fields. Unfortunately, there is no such general prescription; this is an important difference between discrete and continuum problems which makes continuum problems much harder. The best we can do is to guess at formulas for $s(y, t)$ and $u(y, t)$ and then test whether they're "right" or not. We want them to describe a real string, which obeys the discrete equations for small dr and dt; thus by "right" we mean that $s(y, t)$ and $u(y, t)$ obey the discrete equations in the continuum limit $dr \to 0$, $dt \to 0$. Assuming for the moment that s and u are ordinary differentiable functions (not other continuum fields such as delta functions or step functions), the discrete equations (Eqs. 3.22 and 3.24) turn in this limit into continuum equations involving ordinary derivatives, namely

$$\frac{ds(y, t)}{dt} = \frac{du(y, t)}{dy} \tag{3.25}$$

and

$$\frac{du(y, t)}{dt} = \frac{C}{\lambda} \frac{ds(y, t)}{dy} \tag{3.26}$$

These equations are referred to as **partial differential equations** or PDE's, because a derivative of a function of more than one variable is called a "partial derivative." In writing partial derivatives, d is sometimes replaced by the symbol ∂ to remind the reader that it refers to a partial derivative. Almost all of the derivatives in this book are partial, so we will not use ∂ because it would convey no extra information.

▶ **Problem 3.13**

Eliminate s from Eqs. 3.25 and 3.26 to obtain a (second-order) partial differential equation for u alone. This is referred to as the one-dimensional **wave equation**. ◀

The evolution of a continuum string is described by the two simultaneous PDEs (Eqs. 3.25 and 3.26), but only (as we mentioned above) in the sense that we can use them to check whether a guessed solution is correct. There are useful tricks and search strategies for guessing these solutions, which you can learn about in books on partial differential equations, but there is no guarantee that you can find a solution for any given PDE's and initial conditions. For this particular pair of PDE's, however, it turns out that you can. We are fortunate enough to have found the most general solution of its discrete analog in Section 2.6. This is a linear combination of two waves, one moving to the right and the other to the left, with wave velocities $\pm c$, where $c^2 = dr^2 k/m$ (see Problem 2.10). In terms of our present parameters, this is

$$c^2 = C/\lambda \tag{3.27}$$

The general solution in the discrete case was

$$s(e, t) = s^+(e - ct, 0) + s^-(e + ct, 0) \tag{3.28}$$

$$u\left(v, t + \tfrac{1}{2}dt\right) = u^+\left(v - ct, \tfrac{1}{2}dt\right) + u^-\left(v + ct, \tfrac{1}{2}dt\right) \tag{3.29}$$

where $s^+(e, 0)$ and $s^-(e, 0)$ are arbitrary initial conditions for the stress and

$$u^+\left(e + \tfrac{1}{2}dr, \tfrac{1}{2}dt\right) \equiv -cs^+(e, 0) \tag{3.30}$$

(this can be seen from Table 2.8 for the case $c = 1$, or from Problem 2.10) and

$$u^-\left(e - \tfrac{1}{2}dr, \tfrac{1}{2}dt\right) \equiv cs^-(e, 0) \tag{3.31}$$

Suppose now that we choose differentiable functions $s_f^\pm(y, 0)$, and obtain the discrete $s^\pm$ from Eq. 3.16. We could then define the discrete $u^\pm$ (Eqs. 3.30 and 3.31). The resulting s and u (Eqs. 3.28 and 3.29) would comprise an exact solution of the discrete equations for each choice of dr (with $dt = dr/c$). Taking the continuum limit $dr, dt \to 0$ of these equations gives the continuum equations (Eqs. 3.25 and 3.26) for the functions

$$s_f(y, t) = s_f^+(y - ct, 0) + s_f^-(y + ct, 0) \tag{3.32}$$

$$u_f(y, t) = u_f^+(y - ct, 0) + u_f^-(y + ct, 0) \tag{3.33}$$

where

$$u_f^+(y, 0) \equiv -cs_f^+(y, 0) \tag{3.34}$$

$$u_f^-(y, 0) \equiv cs_f^-(y, 0) \tag{3.35}$$

We have thus shown that for any two differentiable functions $s^+(y, 0)$ and $s^-(y, 0)$ (we will omit the subscript f when it is clear we mean functions rather than discrete fields) Eqs. 3.32 and 3.33 give exact solutions of the continuum equations. These are completely general solutions in that every solution can be written in this form (see Problem 3.16).

You may have wondered what happened to the other solution of the discrete equations we found in Section 2.7, the corrugation wave. This does not have a continuum analog (you can check that its limit as $dr \to 0$ does not exist), and we can never get it from a continuous function s_f.

▶   **Problem 3.14**

Verify using the usual rules of differentiation that Eqs. 3.32 and 3.33 give solutions of the PDEs (Eqs. 3.25 and 3.26), for any (differentiable) choice of $s^{\pm}(y, 0)$.   ◀

▶   **Problem 3.15**

a. For the continuum rightward-moving wave with initial strain $s(y, 0) = \cos(y)$, find formulas for $u(y, t)$ and $s(y, t)$ for general y and t. Verify by direct substitution that these satisfy Eqs. 3.25 and 3.26.

b. Find $u(y, t)$, $s(y, t)$ for a wave with initial strain field $s(y, 0) = \cos(y)$ and zero initial velocity field, and check that they satisfy the PDE's.   ◀

▶   **Problem 3.16**

Consider **arbitrary** differentiable initial strain and velocity fields $s(y, 0)$ and $u(y, 0)$. Find formulas for $u(y, t)$ and $s(y, t)$ for arbitrary y and t, in terms of $s(y, 0)$ and $u(y, 0)$. Verify that they satisfy the PDE's.

CHAPTER FOUR

POPULATION MODELS

SECTION 4.1 SINGLE-VARIABLE POPULATION GROWTH MODEL

We have been dealing so far with physical systems whose equations of motion are well known from mechanics. We are now going to discuss less well-known systems; we will simply assume equations of motion for them and return later to the question of how one would go about determining the correct equations. Let us first consider a simple population problem with one variable, specifically the growth of a colony of hares. Denote the number of hares at time t by $H(t)$. The change in the hare population between time t and time $t + dt$ is of course related to the number of the female hares who give birth during that time interval, and we may as well take this number to be proportional to the total number of hares. So the equation of motion for $H(t)$ is

$$H(t + dt) - H(t) = R\, dt\, H(t) \tag{4.1}$$

We have written the proportionality constant as $R\, dt$ to make this look like earlier equations of motion (Chapter 2). The constant R can be interpreted as a net growth rate (birth rate minus death rate) per unit time, giving the fractional increase in H per unit time. We will assume $dt = 1$ to simplify the notation below.

To use Eq. 4.1 to update $H(t)$, it is easiest to solve it for $H(t + 1)$:

$$H(t + 1) = (1 + R)H(t) \tag{4.2}$$

▶ **Problem 4.1**

Compute by hand $H(t)$ for $t = 1, 2$, and 3 using the initial condition $H(0) = 100$ and $R = 0.2$. ◀

▶ **Problem 4.2**

Write a computer program to solve Eq. 4.2, using $dt = 1$. It should solicit input values for R, $H(0)$, and the last time to be computed, t_{max}. It should update $H(t)$ for each $t \le t_{max}$, printing t and $H(t)$ on one line for each t. Give the results for $R = 0.2$, $H(0) = 100$, and $t_{max} = 10$. (Note: one point will be deducted if you use an array when a single variable would suffice.) Alternatively, you may use a commercial spreadsheet program to do the same calculation. ◀

▶ **Problem 4.3**

Consider a colony of hares whose total population $H(t)$ is measured every day for four days. The results are ($dt = 1$ day)

Table 4.1

t	$H(t)$
0	100
1	110
2	121
3	133

a. Assuming the equation governing the population growth has the form of Eq. 4.1, find the growth rate R.

b. Compute $H(4)$ and $H(5)$.   ◀

Starting with any initial condition $H(0)$, we can compute $H(1)$ by substituting $t = 0$ into Eq. 4.2: $H(1) = (1 + R)H(0)$. Successively substituting $t = 1, 2, \ldots$ gives us the following table:

Table 4.2

t	$H(t)$
0	$H(0)$
1	$(1 + R)H(0)$
2	$(1 + R)^2 H(0)$

We might guess from this table that for any t,

$$H(t) = (1 + R)^t H(0) \tag{4.3}$$

You should show that this is a solution of Eq. 4.2. This is an example of exponential growth (i.e., growth in which t appears in the exponent). For arbitrary dt, the factor in front of $H(0)$ is $(1 + R\, dt)^{t/dt}$, which can be written $[(1 + R\, dt)^{1/R\, dt}]^{Rt}$. The expression in square brackets has a limit as $dt \rightarrow 0$, namely e, the base of the natural logarithms. So the $dt \rightarrow 0$ limit (the "continuous time" limit) of Eq. 4.3 is simply

$$H(t) = e^{Rt} H(0) \tag{4.4}$$

which is the solution of the differential equation form of Eq. 4.1,

$$dH(t)/dt = RH(t) \tag{4.5}$$

▶   **Problem 4.4**

Compute the continuous-time solution (Eq. 4.4) for the system described in Problem 4.1. What is the maximum discrepancy from the $dt = 1$ results, for $0 \leq t \leq 3$?   ◀

SECTION 4.2   INTERACTING POPULATION MODELS

Suppose our hares live in a forest that is also occupied by wolves. The likely interactions under these circumstances are that the wolves will eat the hares; thus the wolf population $W(t)$ will affect the hare population. We can represent this in our mathematical model by an additional term in the equation of motion for $H(t)$ (Eq. 4.1):

$$H(t + 1) - H(t) = R_H H(t) - V_W W(t) \qquad (4.6)$$

(We are assuming $dt = 1$ for convenience again.) Here R_H is the hare growth rate (called R above) and V_W represents the voraciousness of the wolves. Quantitatively, V_W is the number of hares each wolf eats during the time interval dt. We must also have an equation for updating the wolf population. This will involve the net growth rate R_W of the wolf population, which will be negative: in the absence of hares they die out. We will also want a term describing the effects of the hares on the wolves. This effect is nutrition, and is described quantitatively by N_H, the nutritiousness of the hares (specifically, the extra number of cubs a wolf is enabled to have and support by killing one hare.) The equation for updating the wolf population is then

$$W(t + 1) - W(t) = R_W W(t) + N_H H(t) \qquad (4.7)$$

Equations 4.6 and 4.7 are easy to solve by computer; a program called POP that solves them is described in the program guide. A number of interesting effects can be investigated, such as the effects of overpopulation. If the populations are reasonably balanced, both will grow approximately exponentially (Figure 4.1a). However, if the wolf population starts out high and the hare population low, the wolves will eat the hares into extinction and thereafter become extinct themselves because of the lack of food (Figure 4.1b).

Equations 4.6 and 4.7 involve a serious oversimplification, in that they are linear in H and W. One consequence of this is that when the hares become extinct owing to the voraciousness term $-V_W W(t)$, the hare population actually becomes negative. This term assumes the wolves continue to eat hares even when there are no more hares. Clearly, when there are few hares, it is more reasonable to assume the number eaten would be proportional to the number of hares available to be eaten, as well as the number of wolves stalking them. The same is true of the nutritiousness term; eating hares can't do the wolves any good when there are no wolves, so the

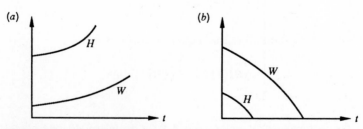

Figure 4.1   (a) Growing hare and wolf populations. (b) Extinction.

benefit should be proportional to $H(t)W(t)$ instead of just $H(t)$. This gives us a pair of nonlinear equations

$$H(t+1) - H(t) = R_H H(t) - V_W W(t) H(t) \tag{4.8}$$

$$W(t+1) - W(t) = R_W W(t) + N_H H(t) W(t) \tag{4.9}$$

(The meanings of V_W and N_H are different here from those in Eqs. 4.6 and 4.7.) A program called NONLIN solves equations such as these. However, nonlinear systems are much more complicated than linear systems. The electromagnetic systems we will mainly be concerned with are linear, so we won't dwell on nonlinear ones here.

SECTION 4.3   POPULATION FIELDS

We would like to generalize our hare population model to give information about the way hares are distributed in space. Suppose the forest they live in is very long, and is crossed by several fences as in Figure 4.2. We can then count the number of hares in each region bounded by fences, which we will call a "cell." The cells are labeled $c_1, c_2, \ldots$ as shown in the figure. We will denote an arbitrary one of these cells as c, and the population of this cell at the time t (a multiple of some time interval dt) by $H(c, t)$. This is a "field" in the sense defined in Section 1.2, because it gives a number for each of several locations in space (labeled earlier by r, and here by c.) We will call $H(c, t)$ the population field at time t. It can be specified in a table. For example the initial population field $H(c, 0)$ might be as follows:

Table 4.3

Cell label, c	c_1	c_2	c_3	c_4
$H(c, 0)$	13	24	56	19

meaning that there are 13 hares in cell c_1, 24 in c_2, and so on.

The population field $H(c, t)$ can change with the passage of time not only because hares are born or die within cell c, but because they may cross the fences.

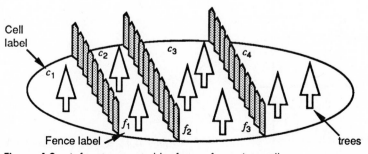

Figure 4.2   A forest separated by fences $f_1, \ldots$ into cells $c_1, c_2, \ldots$

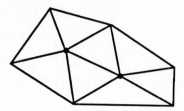

Figure 4.3   A 2D network of fences.

We will label the fences by $f_1, f_2, \ldots,$ and refer to an unspecified one simply as f. Let us denote the net number of hares crossing the fence f between time t and time $t + dt$ by $M(f, t + \frac{1}{2}dt)$. We label the time of the migration count M by $t + \frac{1}{2}dt$, because it refers to the time interval whose midpoint is $t + \frac{1}{2}dt$. Note that this definition of M is ambiguous: does it mean the net number crossing from left to right, or vice versa? We will resolve this ambiguity by letting the fence label f specify both the position of the fence and the direction of migration we will count as positive. We will refer to this combination of position and direction as an "oriented fence." The label f will always refer to an oriented fence. When we label an oriented fence, as in Figure 4.4 below, we will draw an arrow to indicate the positive direction of migration. Thus, $M(f)$ is the net number crossing to the right across the fence labeled f in the figure (i.e., the total number crossing to the right minus the total number crossing to the left). But $M(f')$ is the net number crossing to the **left** across the fence labeled f', because this fence has an arrow pointing left; its orientation is leftward. If it seems pedantic to you to be so explicit (you may wonder why the signs aren't chosen in the obvious way, so M is the number going to the right) try to assign orientations to the 15 fences of the two-dimensional network in Figure 4.3 in the "obvious" way, and try to explain why a different assignment is less obvious.

For simplicity, to write equations of motion for $H(c, t)$, assume no hares are born or die, so that the population of each cell changes only through migration. This is a sort of "law of conservation of hares." (The results we get from this assumption will be applicable to electric charge, which is also subject to a conservation law.) Then the equation for updating the hare population $H(c, t)$ of the cell c shown in Figure 4.4 is clearly

$$H(c, t + dt) - H(c, t) = M\left(f, t + \tfrac{1}{2}dt\right) + M\left(f', t + \tfrac{1}{2}dt\right) \qquad (4.10)$$

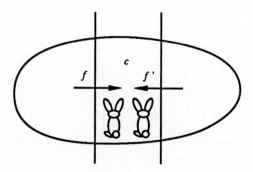

Figure 4.4   A cell c bounded by two fences f and f'.

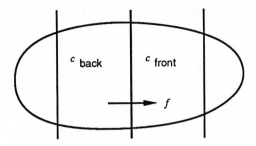

Figure 4.5   A fence f and the two cells used in estimating the migration $M(f)$.

where f and f' are the two inwardly oriented fences bounding c. The right side of Eq. 4.10 is just the total number of hares entering the cell c, which must be equal to the change in the population of that cell (the left-hand side). This equation is called the "equation of continuity" in physics, and it is valid in any situation in which some quantity (hares here, electric charge later) is conserved.

We could now use the equation of continuity to update the initial population field of Table 4.3, if we knew the values $M(f, \frac{1}{2}dt)$ taken on by the migration field for the interval from 0 to dt. How do we determine $M(f, \frac{1}{2}dt)$? Recall that for the chain of masses in Chapter 1, we chose arbitrarily two initial fields, x at $t = 0$ and v at $t = \frac{1}{2}dt$. The analog of this for the population system, which is choosing $H(c, 0)$ and $M(f, \frac{1}{2}dt)$ arbitrarily, does not make sense; the population at $t = 0$ largely determines the migrations across the fences, so they cannot be chosen independently. Consider the fence f in Figure 4.5. The hares that cross f to the right must have originated in the cell c_{back} in back of f (meaning in back of you as you face in the direction of the arrow.) Clearly the more hares there are in the cell c_{back} at some time t, the more are likely to cross f between t and $t + dt$; we might expect the number to be proportional to $H(c_{\text{back}}, t)$. Similarly the number crossing leftward across f from the cell c_{front} in front of f might be proportional to $H(c_{\text{front}}, t)$. In general the proportionality coefficient (call it k) would depend on which f and c are involved (some fences might be lower, for example), so we will denote it by $k(c, f)$. Thus, the number of hares migrating across f can be written

$$M\left(f, t + \tfrac{1}{2}dt\right) = \text{Number crossing forward} - \text{number crossing backward}$$

$$= k(c_{\text{back}}, f)H(c_{\text{back}}, t) - k(c_{\text{front}}, f)H(c_{\text{front}}, t) \qquad (4.11)$$

We will return to the general question of how to determine the correct equation in Section 4.6; for now we will just assume Eq. 4.11.

Now the continuity equation (Eq. 4.10) and the migration equation (Eq. 4.11) can be used to calculate the evolution of any initial population field, such as that in Table 4.3. However, if the cells are all different, and the migration constant k depends on where we are, this is clearly a complicated problem. Usually in physics we deal instead with **uniform** systems (empty space, in electricity and magnetism) that have the same properties everywhere. Clearly, then, we should choose the cells to be the same size, so we can use the same equations for all of them, as shown in Figure 4.6. Here we have a regular array of cells, with fences separated by some fixed distance, which we can call dr. This allows us to label the cells by their

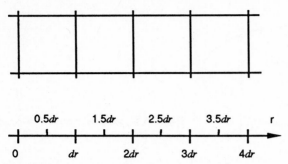

Figure 4.6 A uniform system of cells separated by fences.

distance r along the horizontal coordinate axis shown; more specifically, we label each cell by the r-coordinate of its center. Thus, instead of having to invent arbitrary labels $c_1, c_2, \ldots$, as was necessary in the nonuniform system of Figure 4.2, we can use $0.5dr, 1.5dr, \ldots$ for the labels c in Eqs. 4.10 and 4.11. We can similarly use positions to label the fences: the ones in Figure 4.6 are 0, dr, $2dr$, $3dr$, and $4dr$. However, remember that we have to specify a direction for the fence also, or else the migration $M(f, t)$ is ambiguous. Unlike Figure 4.3, this array is a regular one, so we can choose the direction to be that of increasing coordinate r, that is, to the right. When we use a position r to label an oriented fence, it is understood that we mean the **rightward** oriented fence at position r.

You may notice a similarity between Figure 4.6 and the lattices we defined in Chapter 3. They are in fact exactly equivalent; if you change "vertex" to "fence" and "edge" to "cell," all of the previous definitions are applicable here. We are using the "cell" terminology here only to emphasize that we are thinking of our one-dimensional lattice as a section through a three-dimensional system, rather than as an intrinsically one-dimensional object like a string.

Let us now write the equation of continuity (Eq. 4.10) for this regular array of cells. If the cell c in Figure 4.4 is labeled as shown in Figure 4.7a, by its position r (a half-integer multiple of dr) then the oriented fence to its left (f in the earlier figure) is labeled $r - \frac{1}{2}dr$. The oriented fence to its right (labeled $r + \frac{1}{2}dr$) is at the

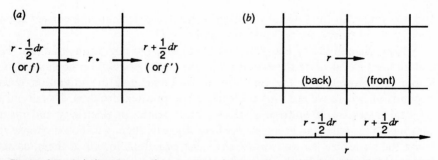

Figure 4.7 Labeling (in a uniform system) of (a) a cell r and its bounding fences and (b) a fence and its adjacent cells.

same position as that labeled f' in the earlier figure, but it is **not** the same oriented fence because its arrow points right, whereas that of f' points left. If we denote by $\bar{f}'$ the fence with the same position as f' but the opposite orientation, then clearly the net migrations M just differ by a sign:

$$M(\bar{f}') = -M(f') \qquad (4.12)$$

In this case the oriented fence $\bar{f}'$ is $r + \frac{1}{2}dr$, as shown in Figure 4.7a. So we can substitute $r + \frac{1}{2}dr$ for $\bar{f}'$ in Eq. 4.12, and Eq. 4.10 gives an equation of continuity for our regular array:

$$H(r, t + dt) - H(r, t) = M\left(r - \tfrac{1}{2}dr, t + \tfrac{1}{2}dt\right) - M\left(r + \tfrac{1}{2}dr, t + \tfrac{1}{2}dt\right) \quad (4.13)$$

The equation for M (Eq. 4.11) also simplifies for our regular array of cells (Figure 4.6). This is a uniform system (one whose properties are the same everywhere), so all the fences are the same height. In such a system k will be independent of position, and Eq. 4.11 becomes

$$M\left(f, t + \tfrac{1}{2}dt\right) = k\left[H(c_{\text{back}}, t) - H(c_{\text{front}}, t)\right]$$

where c_{back} and c_{front} are as shown in Figure 4.5. For an arbitrary fence f at position r ($r = 0, dr, 2dr, \ldots$ as in Figure 4.6), clearly c_{back} is at $r - \frac{1}{2}dr$ and c_{front} is at $r + \frac{1}{2}dr$, as shown in Figure 4.7b. Thus M is given by

$$M\left(r, t + \tfrac{1}{2}dt\right) = k\left[H\left(r - \tfrac{1}{2}dr, t\right) - H\left(r + \tfrac{1}{2}dr, t\right)\right] \qquad (4.14)$$

Now we have a set of equations (Eqs. 4.13 and 4.14) for computing the evolution of the population and migration fields in a uniform system. Let's look at a specific numerical example. Suppose we wanted to determine the evolution of a system of $N = 4$ cells and three fences (the system shown in Figure 4.6), for an initial condition in which there are 10 hares in each of the two central cells and none in the outer two. The initial population field is shown in the top line of Table 4.4. We must choose a value for the constant k (k determines how often hares jump fences), say $k = 0.2$. Using Eq. 4.14 for $t = 0$, we would get, for example (using $r = 1.0dr$),

$$M\left(dr, \tfrac{1}{2}dt\right) = k\left[H(0.5dr, 0) - H(1.5dr, 0)\right] = 0.2[0 - 10] = -2$$

Collecting the results into a table gives

Table 4.4

r	$0.5dr$	$1.0dr$	$1.5dr$	$2.0dr$	$2.5dr$	$3.0dr$	$3.5dr$
$H(r,0)$	0		10		10		0
$M(r, 0.5dt)$		-2		0		2	
$H(r, dt)$	2		8		8		2
$M(r, 1.5dt)$		-1.2		0		1.2	
$H(r, d2t)$	3.2		6.8		6.8		3.2

The values of $H(r, dt)$ were obtained from Eq. 4.13 using $t = 0$: for example

$$H(1.5dr, dt) = H(1.5dr, 0) + M(1.0dr, dt/2) - M(2.0dr, dt/2)$$

$$= 10 + (-2) - 0 = 8.$$

As we did for the chain of masses in Chapter 1, we must treat the region near the boundaries differently from the rest. We have used impermeable-barrier boundary conditions. That is, we have assumed the hares cannot cross the fences at $r = 0$ and $r = N\,dr$:

$$M(0, t) = M(N\,dr, t) = 0 \qquad\qquad (4.15)$$

▶   **Problem 4.5**

a. Construct a table like Table 4.4 for a chain of $N = 4$ cells, using initial populations 0, 20, 0, 0 and $k = 0.2$, $dt = 1$, up to a time $t_{max} = 4$. What populations would you expect after a very long time?   ◀

The equations of motion (Eqs. 4.13 and 4.14) for H and M in a regular chain of cells can be easily updated by computer; the simulation program that does so is called MIG1.

▶   **Problem 4.5**

b. Use MIG1 to check your answers to Problem 4.5a.
(You may have to rescale the populations to get it to display nicely.) How close are the populations at $t = 15$ to what you expect after an infinite time?   ◀

SECTION 4.4   UNIFORM SYSTEMS: DENSITY AND CURRENT

The system we discussed in the previous section was uniform in that all the cells were the same except at the boundary, but nonuniform in the sense that its properties near the fences were different from its properties away from the fences. In particular, the fences impeded the motions of the hares, which were free to move around between fences. However, the mathematical description we have developed could be used just as well if the fences were purely imaginary and served only to delineate the cells for counting purposes. We could still count the number of hares crossing each imaginary fence in each time interval, and the change in each cell's population would still be given exactly by the continuity equation, Eq. 4.10. It is not obvious now that the migration equation (Eq. 4.14) is still true. But that wasn't obvious before either, and we have promised to discuss it later, so we will just assume it for now. So all of our formulas (and the program MIG1) from Section 4.3 work just as well when the fences are imaginary.

If there are no real fences, then we are free to put the imaginary ones where we like; in particular we may choose dr as we like. Two choices, for a particular distribution of hares, are shown in Figure 4.8a. The two sets of numbers shown for H represent the same system, but are quite different because the lower ones represent cells of a greater width dr. The conventional method of representing the

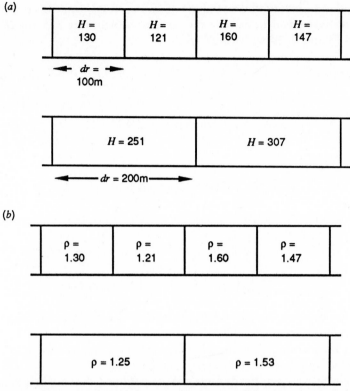

Figure 4.8   (*a*) Two population fields describing the same system. (*b*) The corresponding density fields.

distribution of hares in a way that is less sensitive to the choice of cell size is, of course, to define a **density**

$$\rho(c) = H(c)/dr \tag{4.16}$$

so that the same data look like Figure 4.8*b* and the density is about the same no matter what *dr* we choose. Because we have been treating our long forest of hares as a one-dimensional system (the vertical position has no significance, only the horizontal position does) we have defined a **linear** density, that is, a number of hares per unit length of the cell. It is a more direct measure of the "crowdedness" of the hares than the population was.

Now that we've converted the population field into a density field to make its value less dependent on *dr*, let's turn our attention to the migration field $M(f, t)$, which gives the number of hares crossing fence *f* during the time interval of length *dt* centered at *t*. Clearly it will increase with *dt* in much the same way that the population *H* increased with *dr*. We can eliminate this dependence by defining the **current**

$$j(f, t) \equiv M(f, t)/dt \tag{4.17}$$

which is the number of hares per unit time crossing fence *f*, during the interval centered at t. We can rewrite the equations of motion for H and M (Eqs. 4.13 and 4.14) in terms of the equivalent density and current: substituting $\rho\, dr$ for H and $j\, dt$ for M in Eq. 4.13 and dividing by $dr\, dt$ gives

$$\frac{\rho\left(r, t + \tfrac{1}{2}dt\right) - \rho\left(r, t - \tfrac{1}{2}dt\right)}{dt} = -\frac{j\left(r + \tfrac{1}{2}dr, t\right) - j\left(r - \tfrac{1}{2}dr, t\right)}{dr} \qquad (4.18)$$

(Note that the t here differs by $\tfrac{1}{2}dt$ from that in Eq. 4.13.) This can be written more concisely in terms of the discrete time derivative defined by Eq. 3.21 and the discrete gradient (Eq. 3.2), as

$$d\rho(r, t)/dt = [\text{grad } j](r, t) \qquad (4.19)$$

Making the same substitutions in the migration equation (Eq. 4.14) gives

$$\begin{aligned}
j\left(r, t + \tfrac{1}{2}dt\right) &= (k\, dr/dt)\left[\rho\left(r - \tfrac{1}{2}dr, t\right) - \rho\left(r + \tfrac{1}{2}dr, t\right)\right] \\
&= K\left[\rho\left(r - \tfrac{1}{2}dr, t\right) - \rho\left(r + \tfrac{1}{2}dr, t\right)\right]
\end{aligned} \qquad (4.20)$$

where we can use the constant $K \equiv k\, dr/dt$ instead of k to describe the migration rate. Equations 4.19 and 4.20 are the evolution equations for the density and current; except for rescaling the variables they are the same as Eqs. 4.13 and 4.14 and can be solved by the same computer programs.

The relationship between the discrete density field and the total population of a system provides a good example of the concept of a discrete integral, which we will encounter frequently in electromagnetism. The total population H_{tot} of the system is, of course, the sum over the cells c in the system of the population field $H(c)$; using Eq. 4.16 to write $H(c)$ as $\rho(c)\, dr$, we obtain

$$H_{\text{tot}} = \sum_{c} \rho(c)\, dr \qquad (4.21)$$

Using the correspondence we set up in Section 4.3 between 1D cell fields and edge fields, we see that the sum in Eq. 4.21 is exactly what we defined in Section 3.1 as a

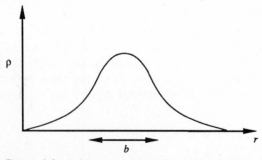

Figure 4.9 Schematic graph of the density field in a 1D system.

Figure 4.10   A 1D distribution of ants.

discrete integral. Thus the total population of the system is exactly the discrete integral of the density field ρ.

What is the best choice of dr to use in defining the density? Suppose the density is as shown in Figure 4.9 (we have smoothed it out, since it's only defined at intervals of dr). An actual distribution of hare positions corresponding to the density we drew in Figure 4.9 is shown in Figure 4.10. The positions are spread out along a 10-centimeter line. Each short line represents the position of one hare. Perhaps we should say that the hares are ants here, given the small scale of the picture. At the ends of the r-axis, the density is low enough so that the ants can be seen individually; toward the middle they are just smeared together. There are a total of 2000 ants, and their positions were chosen randomly by a computer program in such a way that the probability of choosing a position r is proportional to the function graphed in Figure 4.9 (more precisely, a bell-shaped curve whose formula is $\exp[-(r-5)^2/4)]$). Figure 4.11a is the density field $\rho(r)$ for $dr = 2.5$ cm, represented as a histogram. Evidently this does not give much information about the shape of the density curve (Figure 4.9). The problem is that $dr = 2.5$ cm is

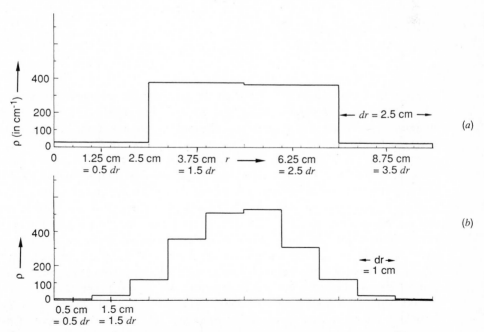

Figure 4.11   Density fields corresponding to the ant distribution in the previous figure, for various dr.

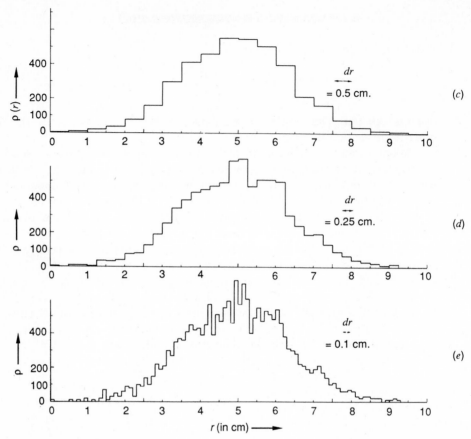

Figure 4.11 Continued.

as large as the distances of interest in the problem, for example the distance over which the density changes significantly. We have indicated this distance by the line marked *b* in Figure 4.9. It isn't a quantitatively exact notion, but the *b* we have shown is clearly not 10 times too large or 10 times too small; it is of the right order of magnitude. In Figure 4.10, *b* seems to be about 2.5 cm or so, and we should not have chosen $dr \cong b$. So we need a smaller *dr*; the next figure (Figure 4.11*b*) shows the density field for *dr* = 1 cm. This clearly shows more detail of the bell-shaped curve, and *dr* = 0.5 cm (Figure 4.11*c*) shows even more. It is tempting to just keep decreasing *dr*, but at *dr* = 0.25 cm we begin to see some odd bumps. These correspond to random fluctuations in the populations of the cells, which are not very important if the cells are large so there are very many ants in each one (this is called the "law of large numbers" by statisticians) but which become increasingly noticeable as the cells are made smaller. At *dr* = 0.1 cm (Figure 4.11*e*) the fluctuations make quite a mess of the curve, and by *dr* = 0.0125 cm, the situation is hopeless. These cells are so small that there are only a few ants in most of the cells. You can even tell in Figure 4.11*g* which cells have one ant; these have $\rho(r)$ = 1/0.0125 = 80. Each of the other cells has some integer multiple of this density.

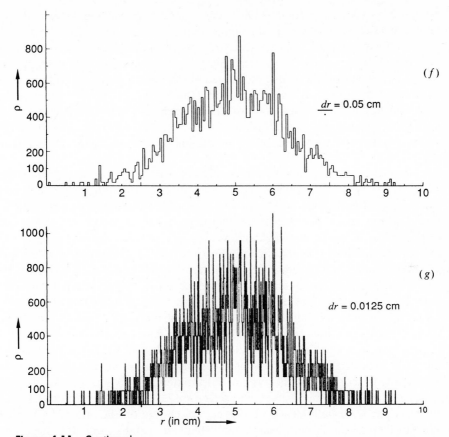

Figure 4.11   Continued.

Clearly these problems can be avoided by making sure that dr is much larger than the average spacing between ants, so there will be many ants in each cell. Let's call this average spacing a; in this case $a = 10$ cm$/2000 = 0.005$ cm, and we require $dr \gg a$.

Let us summarize the criteria for a reasonable value of dr. It must be large compared to the spacing between ants but small compared to the distance over which the density changes: $a \ll dr \ll b$. In the case we have looked at, the best value is about $dr = 0.5$ cm, which satisfies this criterion ($0.005 \ll 0.5 \ll 2.5$) reasonably well.

SECTION 4.5   DESCRIBING REAL DIFFUSIVE SYSTEMS: THE CONTINUUM LIMIT

Up to this point we have written equations describing the evolution of a density field without much regard for how accurately they might describe a real physical system. In this and the next section we will address this question, and see that in a certain limit (the "continuum limit") the equations of Section 4.4 actually provide an exact description of a real diffusive system.

A **diffusive system** is one in which a population of objects (which could be hares or molecules) move at random. The migrating hares provide one example; another would be the spreading of dye molecules throughout a perfectly still glass of water in which a drop of dye has been placed. (In theory anyway—in practice the only way to ensure it remains still is to use Jello instead of water. In water the dye will be carried by convection, not diffusion.) By putting the water in a capillary tube, we can even make this system one-dimensional so we can apply the equations of Section 4.3.

One situation in which we can clearly **not** expect the simple migration equation (Eq. 4.20) to be correct is when the density varies significantly over the distance dr. In terms of the distance b (Figure 4.9) over which the density varies significantly, this condition is $dr \geq b$. Clearly the migration across a "fence" (we call it a "face" of the cell in nonpastoral contexts) depends on how many molecules are close to the face. If the density varies significantly over the cell, the density close to the face may not be the same as the average density ρ that appears in Eq. 4.20. Evidently, to have any hope of an accurate description we need $dr \ll b$, so the density varies very little within a cell, or from one cell to the next. This is the same as one of the criteria we found in the previous section for the density to make sense. Another requirement (to avoid large fluctuations in ρ) was that there be many particles in each cell. Taken together, these criteria require that the total number of objects be very large. Even the 2000 ants in Figure 4.10 are barely enough for the density field to provide a reasonable description of the system. Fortunately, in physics we usually deal with systems in which the number of objects is even larger: in electrostatics experiments, for example, we usually deal with charges containing on the order of 10^{12} electrons. We can then get away with using a very small dr—that is, dividing the system into very many small cells (many more than the 800 in Figure 4.11g)—without encountering problems with fluctuations. The scale b on which the density changes can be much greater than dr, and a graph of the density looks almost like a continuous function (see Figure 4.12). We will refer to such a discrete field having $dr \ll b$ as "slowly varying" (referring to space variation, not time variation) or just as "smooth." In physics we deal mostly with smooth fields.

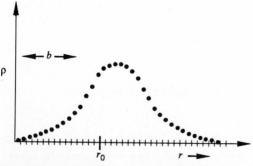

Figure 4.12 A possible smooth density field in a system with about 40 cells; the density at each cell is represented by a dot.

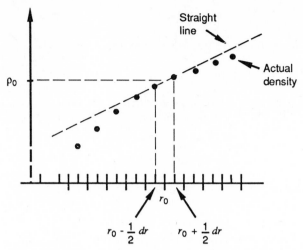

Figure 4.13 Detail of Figure 4.12 near $r = r_0$.

The limit of smooth fields ($dr \ll b$, or $dr/b \to 0$) is sometimes referred to as the "continuum limit" because of the smallness of dr (r becomes almost continuously variable, rather than discrete). Whether we call it the "continuum" or "smooth-field" limit will depend on whether we wish to emphasize the smallness of dr (relative to b) or the largeness of b (relative to dr). It is important to realize that they mean exactly the same thing. In a real system, the latter is usually more appropriate, because we cannot let dr get arbitrarily small (for a given type of system with a certain average interparticle spacing) but we can in principle let b be arbitrarily large, by doing experiments on large systems.

Let us now ask what discrete equation will describe a diffusive system correctly in the continuum limit. This equation will not necessarily be unique; it turns out, in fact, that a rather large class of equations predict the **same** evolution in the continuum limit. What is it about an equation for the current j that determines how a smooth density evolves? Let us apply the particular equation we have been using (Eq. 4.20) to the density field of Figure 4.12, for a particular face at position r_0. An enlargement of the part of the field near r_0 is shown in Figure 4.13. A small piece of any smooth curve looks almost straight; a straight-line approximation to the density near r_0 is shown in the figure. The equation of the line is

$$\rho_{\text{straight}}(r) = \rho_0 + g(r - r_0) \tag{4.22}$$

Here the constants ρ_0 and g are the value of ρ at r_0 and the slope or (continuum) gradient at r_0, respectively. This approximation gets bad when $r - r_0$ is of the same order of magnitude as the distance over which ρ curves significantly; this is essentially the distance we have been calling b. So in the continuum ($dr \ll b$) limit, Eq. 4.22 is a good approximation when $r - r_0$ is of the order of dr. This includes the r's of the cells whose densities enter Eq. 4.20, namely $r_0 + \frac{1}{2}dr$ and $r_0 - \frac{1}{2}dr$. So we may use the straight-line approximation (Eq. 4.22) in Eq. 4.20 in this limit, giving

$$j\left(r_0, t + \tfrac{1}{2}dt\right) = K\left[\rho_0 + g\left(-\tfrac{1}{2}dr\right)\right] - K\left[\rho_0 + g\left(+\tfrac{1}{2}dr\right)\right]$$
$$= -K\,dr\,g \tag{4.23}$$

Thus the current depends only on the density gradient g for smooth densities. The proportionality constant is called the **diffusivity**. This is denoted by D and defined by the fact that

$$j = -Dg \tag{4.24}$$

in the limit of smooth density fields, so that in this case $D = K\,dr$.

▶ **Problem 4.6**

Show explicitly that the error involved in Eq. 4.23 vanishes in the continuum limit $dr \ll b$, as follows: Suppose that the density in Figure 4.13 were given exactly by

$$\rho(r) = \rho_0 + g(r - r_0) + \rho_2(r - r_0)^2 + \rho_3(r - r_0)^3$$

for some constants ρ_0, g, ρ_2, and ρ_3 (i.e., the beginning of a Taylor expansion about r_0 of an arbitrary analytic function of r). Evaluate $j(r_0)$ (Eq. 4.20). Show that the discrepancy with Eq. 4.23 is proportional to $D\,dr^2$, and find the proportionality constant. [A possible definition of b in this case would be such that this proportionality constant is $b^{-2}g$, so the relative error is $(dr/b)^2$.]   ◀

What if we had used a different equation for the current? For example, the current at r might depend on the populations of the next-nearest-neighbor cells $r - 1.5dr$ and $r + 1.5dr$ shown in Figure 4.14 as well as on the nearest-neighbor cells. Denoting the coefficient describing this dependence by K', we would have

$$j(r, t) = K\rho(r - 0.5dr) - K\rho(r + 0.5dr)$$
$$+ K'\rho(r - 1.5dr) - K'\rho(r + 1.5dr) \tag{4.25}$$

We can use the straight-line approximation (Eq. 4.22) for ρ here also, giving

$$j(r, t) = Kg(-0.5dr - 0.5dr) + K'g(-1.5dr - 1.5dr)$$
$$= -(K + 3K')\,dr\,g \tag{4.26}$$

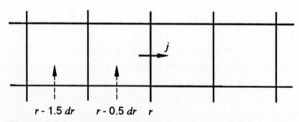

$r - 1.5\,dr \qquad r - 0.5\,dr \quad r$

Figure 4.14 The nearest and next-nearest neighbors of the face at r.

This has the same form ($j = -Dg$) as the current predicted by the simpler Eq. 4.20 (which is the same as Eq. 4.25 without the K' terms), and the diffusivity is

$$D = (K + 3K')\, dr \qquad (4.27)$$

So the currents predicted by the two different discrete equations (Eqs. 4.20 and 4.25) are exactly the same in the continuum limit, as long as the coefficients K and K' in each equation are chosen so the diffusivities D are the same. A modified version of the MIG1 program, called MIG2, has been written to demonstrate this: it updates a density field using Eq. 4.25 instead of using Eq. 4.20 as MIG1 does. They give quite different results for nonsmooth initial densities.

▶ **Problem 4.7**

    a. Use the nearest-neighbor current equation (Eq. 4.20) and the continuity equation (Eq. 4.18) to update by hand the density field in this table (we use $dr = 2$ so cell and face positions are all integers)

r	1	3	5	7	9	11	13
$\rho(r)$	0	0	0	100	0	0	0

using $K = 0.2$, $dt = 1$, up to $t = 2$. Use impermeable boundary conditions $[j(0, t) = j(14, t) = 0]$. What is the diffusivity D?

    b. Do the same thing for the next-nearest neighbor current equation (Eq. 4.25) with the same D, using $K' = 0.02$, $dt = 1$, $dr = 2$. Choose K so that D is the same as in part a. Does the result resemble that of part a? What is the maximum discrepancy between the ρ's for parts a and b, at $t = 2$?

    c. Check parts a and b using MIG2. ◀

However, as the density becomes smoother the results become more similar. For example, a bell-shaped initial condition with width W becomes smoother as W increases. For $W \gg dr$ (i.e., as we approach the continuum limit) the results of the two equations should be more and more nearly the same.

▶ **Problem 4.8**

    a. Use MIG2 to evolve a bell-shaped initial density with width $W = 12$ (use $N = 18$ cells) using Eq. 4.20 (i.e., $K' = 0$) and $K = 0.2$, and again using Eq. 4.25 with $K' = 0.02$, $K = 0.14$. Write down and compare the resulting ρ's or populations after $t = 10$. What is the maximum discrepancy?

    b. Repeat, using $W = 6$ instead. Does the maximum discrepancy change with W in the way you expected? ◀

The smoothest possible density is the straight-line density $\rho = \rho_0 + gr$; it is clear from Eqs. 4.23 and 4.26 that the two equations will give exactly the same results in this case.

If we added still more terms to our current equation (Eq. 4.25) to take still further cells into account, we would still have a current of the form $j = -Dg$ in the continuum limit, with more terms in D. So there is a very large set of discrete equations which will work equally well. However, there are limits to how much we can change the discrete equation (Eq. 4.20) for the current. Notice that the changes we have made all preserve the symmetry of the system. In particular the system has a left–right symmetry at each face: from the point of view of the face at r where we are trying to determine the current, the cells at $r - \frac{1}{2}dr$ and $r + \frac{1}{2}dr$ look the same. Thus, the influence of the density in cell $r - \frac{1}{2}dr$ on the current out of it at r must be the same as the influence of the density in cell $r + \frac{1}{2}dr$ on the current out of it at r: this is why we used the same value K for the coefficients in Eq. 4.20 of the densities at both of these cells. The sign difference between the coefficients in Eq. 4.20 is because the net current out of $r - \frac{1}{2}dr$ is j, whereas that out of $r + \frac{1}{2}dr$ is $-j$; this is simply a consequence of the way we arbitrarily assigned an orientation to the face at r (Section 4.3).

To see how important this symmetry property is, let's write an equation similar to Eq. 4.20 but lacking the symmetry. An easy way to do this is to change the sign of the first term:

$$j(r, t) = +K\rho\left(r + \tfrac{1}{2}dr, t\right) + K\rho\left(r - \tfrac{1}{2}dr, t\right) \tag{4.28}$$

Table 4.5 shows the consequences of using this wrong-symmetry equation of motion (together with the continuity equation Eq. 4.18) to update a simple initial condition (essentially that of Problem 4.5a) using $dt = 1$ and $K = 0.2$.

Table 4.5

r	$0.5\,dr$	$1.0\,dr$	$1.5\,dr$	$2.0\,dr$	$2.5\,dr$	$3.0\,dr$	$3.5\,dr$	$4.0\,dr$	$4.5\,dr$
ρ	0		0		100		0		0
$j(r, 0.5\,dt)$		0		20		20		0	
$\rho(r, dt)$	0		-20		100		20		0
$j(r, 1.5\,dt)$		-4		16		24		4	
$\rho(r, 2\,dt)$	4		-40		92		40		4

Clearly this is giving physically impossible results for the density; we have a negative number of hares in cell $1.5\,dr$ at time dt, for example. This is because we had a positive current out of it at time $0.5\,dt$ even though it had no hares; it happened because of the wrong sign in Eq. 4.28. We are most interested in the behavior of smooth densities, and the density in Table 4.5 is not smooth. However, a little thought will convince you that Eq. 4.28 gives impossible results for smooth densities as well. For example, if ρ is the same in every cell (uniform density of hares) there will of course be no net current [and the correct-symmetry equation (Eq. 4.20) gives this] but the wrong-symmetry equation (Eq. 4.28) gives a uniform nonzero current for this case.

The final result of this section is that in the continuum ($r \ll b$) limit, any of a large class of equations for determining the current (all of which have the correct symmetry) will give the same result. This result has been found to agree very well with the results of experiments on diffusive systems. Our only reason for using the particular equation (Eq. 4.20) involving only nearest-neighbor cells was that it is the simplest; any other would have done as well.

The idea we have introduced above (that large classes of systems whose small-scale descriptions are very different may all have the same large-scale properties) is the basis of a very large field of present-day physics research. It is associated with terms such as "universality" and "renormalization group," and has led to important advances in our understanding of various physical phenomena, particularly **critical phenomena** that occur in materials that undergo phase transformations, but also in such varied areas as the aggregation of dust particles and the flow of oil through porous rocks.

SECTION 4.6   THE EXPERIMENTAL DETERMINATION OF PHYSICAL LAWS

Many physics textbooks contain descriptions of the "scientific method" whereby physical laws are derived from experiments. These descriptions are always oversimplified; few discoveries are made in completely logical ways. But such descriptions may nonetheless help one to understand why we believe certain physical laws, even if the don't accurately describe how they were discovered. Another version of the scientific method, which admittedly is just as oversimplified as anyone else's will be presented here. This version does, however, provide a useful framework for thinking about why we believe in such things as the diffusion equation and Maxwell's equations.

Suppose we were given a diffusive system to watch (say, ants diffusing aimlessly in one dimension, such as along the clothesline shown in Figure 4.15), and asked to deduce equations governing its motion. We would begin by making some observations of the system; suppose we chose to use the discrete density $\rho(r, t)$ and the current $j(r, t)$ to describe it. If we measured these by actually watching the ants, of course, we could guess immediately that it was a diffusive system, and write down the right equations. So let's make the problem harder: suppose we can't see the ants, but each cell and each face has an antmeter mounted on it, as in Figure 4.16. All we can observe are the readings on the antmeters, which are ρ's for the ρ-antmeters in the cells and j's for the j-antmeters on the faces. We don't know, however, that ρ is a density and j a current. Suppose we can do experiments in which we fix the initial condition $\rho(r, 0)$ and observe $j(r, \frac{1}{2}dt)$ and $\rho(r, dt)$ on the antmeters. Let's guess that the equation is linear, so that the j field we see at $t = \frac{1}{2}dt$ is given by an equation of the form

$$j\left(r, \tfrac{1}{2}dt\right) = K_1\rho\left(r - 0.5dr, 0\right) + K_2\rho\left(r + 0.5dr, 0\right)$$

$$+ K_3\rho\left(r - 1.5dr, 0\right) + K_4\rho\left(r + 1.5dr, 0\right) + \cdots$$

(4.29)

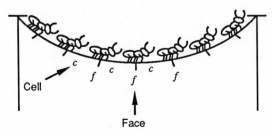

Figure 4.15   Ants diffusing along a clothesline.

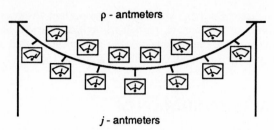

ρ - antmeters

j - antmeters

Figure 4.16 Clothesline with antmeters.

with a possible dependence on many ρ's. To isolate the effects of $\rho(r_0, 0)$ at a particular cell r_0 from those of all the other cells, it would be best to start with an initial condition in which $\rho(r, 0) = 0$ for all the other cells ($r \neq r_0$) and $\rho(r_0, 0)$ is nonzero, say 1.0 in some units. Then only one term in Eq. 4.29 is nonzero, and each value we measure for j determines one of the K's. For example, if we observed the following table of j's (in which $r_0 = 2.5 dr$)

Table 4.6

r	0.0	0.5 dr	1.0 dr	1.5 dr	2.0 dr	2.5 dr	3.0 dr	3.5 dr	4.0 dr	4.5 dr
$\rho(r, 0)$	0		0		1.0		0		0	
$j(r, 0.5 dt)$		−0.01		−0.1		0.1		0.01	
$\rho(r, dt)$	0.01		0.1		0.8		0.1		0.01	

we could deduce by substituting $r = 3.0 dr$ into Eq. 4.29 that

$$K_1 = j\left(3.0 dr, \tfrac{1}{2} dt\right) / \rho(2.5 dr, 0) = 0.1 \tag{4.30}$$

and similarly $K_2 = -0.1$ (using $r = 2.0 dr$), $K_3 = 0.01$, and $K_4 = -0.01$. This determines everything about Eq. 4.29 for j.

The next thing we have to determine is the equation for ρ (which will turn out to be the continuity equation). This equation is somewhat different because $\rho(r, dt)$ depends on $\rho(r, 0)$ as well as on $j(r, \tfrac{1}{2} dt)$. We could determine this fact experimentally by constructing a system with uniform ρ, that is, $\rho(r, 0)$ independent of r, and noting that $\rho(r, dt)$ is still nonzero. In such a case we can still hope that the **change** in ρ depends only on $j(r, \tfrac{1}{2} dt)$. To find the coefficients, we could set up a current field with only one nonzero current, and analyze the resulting change $\Delta\rho$. Alternatively (because it might be experimentally difficult to set up a prescribed current field) we could use the data in Table 4.6. The change $\rho(3.5 dr, dt) - \rho(3.5 dr, 0) = 0.1$ should depend mostly on the **nearest** current, $j(3 dr, \tfrac{1}{2} dt) = 0.1$; we could then deduce that the coefficient must be $j/\Delta\rho = 1$, and end up with the continuity equation.

We can now observe that the unknown system has the same symmetry properties as the diffusive systems of Section 4.5 (i.e., $K_1 = -K_2$, $K_3 = -K_4$, as revealed by the left–right symmetry of j in Table 4.6). Thus the behavior of a smooth density is determined entirely by the diffusivity $D = -j/g$, which is given in this case by Eq. 4.27 as $D = (K_1 + 3K_3)\, dr = 0.13dr$. For practical purposes such as predicting when the ant density will rise to an unacceptable level at the position of a pair of pants hanging on the clothesline, this is all we need to know about the system.

We have been making a rather artificial assumption, that we can set up any initial condition we choose. In fact, it may not be easy to do this precisely, especially for a nonsmooth initial condition such as we used in Table 4.6. However, there is an alternative way to deduce the diffusivity D that determines the response to smooth densities, without worrying about the coefficients K that determine the response to nonsmooth densities. This is simply to set up a uniform density gradient, that is, a density is linear in r:

$$\rho = \rho_0 + gr \tag{4.31}$$

and measure the resulting uniform current j. Then D is given by $-j/g$ (Eq. 4.24). Such a ρ is much easier to set up experimentally than a nonsmooth initial condition such as that in Table 4.6, because any density field (even a very nonsmooth one) will become more smooth with the passage of time. An initial condition that is only approximately linear will become more and more linear as time passes; this can be demonstrated with the MIG1 program, using, for example, a zig-zag shaped initial density.

The fact that the diffusivity D can be measured most easily using smooth initial conditions does not entirely eliminate the usefulness of studying the response to nonsmooth initial conditions, even if we can only do it imprecisely. In particular, in order to know that the diffusivity was the right thing to measure we had to know that system had the symmetry properties of a diffusive system. The symmetry is something that is quite easy to deduce from the response to a localized (nonsmooth) initial condition like Table 4.6. For example, if the system had obeyed the "wrong-symmetry" Eq. 4.28 the signs of all the j's would have been positive in the table. We have only to observe that a nonzero $\rho(r_0, 0)$ in cell r_0 causes outward j's at both faces to know that the symmetry is that of a diffusive system.

Let us then write a list of steps (outlining a "scientific method") that we might actually take to experimentally determine the equations for our system, given only the antmeter readings:

1. Determine the symmetry (and conservation laws, if any) of the equations. This is most easily done by observing the response to localized initial conditions; the experiment need not be very accurate because we are after qualitative, not quantitative, information.

2. Measure the coefficient(s) that determine the evolution of a smooth initial field. For a diffusive system, this is the diffusivity, and it can be determined from the behavior of a system with a uniform density gradient (Eq. 4.24).

It may not be easy to determine accurately the individual coefficients K in a discrete difference equation, but as long as we're only interested in the evolution of smooth densities, only the diffusivity D matters. We can use **any** discrete equation having that diffusivity, so we may as well use the simplest. This is the one in which each density variable influences only the current closest to it in space, namely Eq. 4.20.

We could use an analogous procedure to determine experimentally the equations of motion (supposing that we didn't know them already) for a chain of masses (or its continuum analog, the elastic string). To do this, we would set up initial conditions in which a single stress is nonzero. Assuming a linear dependence of the velocity at time $\frac{1}{2}dt$ on nearby stresses, the fact that only the nearest-neighbor velocities are noticeably nonzero would lead us to the gradient equation for updating the velocities (Eq. 3.23). A similar treatment of the stress would give the stress equation (Eq. 3.21).

In Chapter 9, we will use a similar procedure to determine the Maxwell equations of electrodynamics.

SECTION 4.7   CONTINUUM DENSITY FIELDS: PDEs FOR DIFFUSION

In Section 3.2 we defined the notions of "continuum vertex field" and "continuum edge field." These notions can be carried over into the context of one-dimensional diffusion using the correspondence described in Section 4.3 (vertex → face, edge → cell). The results are a "continuum face field" and a "continuum cell field," and the definitions are exactly the same as before: each is a consistent rule for determining a discrete field for each choice of lattice. As before, face fields are essentially equivalent to continuous functions, but cell fields need not be.

In this section we suppose that we have a system whose discrete density and current fields are approximately given by continuum fields. These might be given by ordinary functions (most of this section is devoted to that case) but they might not. Consider a very special distribution of objects, in which there is exactly one object, and it is located at the origin (see Figure 4.17). There is no way to describe the density of this system by an ordinary function. But we can calculate the discrete density for any choice of lattice, and we find $\rho(c) = 1/dr$ for the cell c containing the origin, and $\rho = 0$ for any other. This is a rule giving the discrete density for each choice of lattice, that is, a continuum cell field. It is, in fact, exactly the Dirac delta function (Example 3.2).

If the continuum fields that approximately describe our real density and current fields are given by ordinary differentiable functions, then these must satisfy the discrete updating equations (Eq. 4.19 and Eq. 4.20) in the continuum limit. Just as in the case of the equations for a string (Section 3.3), the discrete time derivative

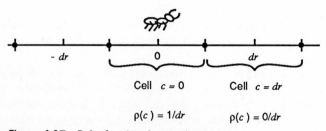

Figure 4.17 Delta-function density distribution.

and discrete gradient in these equations become continuum derivatives. Equation 4.19 (the continuity equation) becomes in this limit the partial differential equation (PDE)

$$d\rho(y, t)/dt = -[\text{grad } j](y, t) \tag{4.32}$$

the one-dimensional continuum continuity equation. ("Continuum continuity equation" sounds redundant but isn't: "continuity equation" is just an undescriptive name that has come to be applied, for complicated historical reasons, to an equation that is true of discrete as well as continuum variables.) The discrete equation for the current becomes (in terms of the diffusivity D defined in Section 4.5) the PDE

$$j(y, t) = -D[\text{grad } \rho](y, t) \tag{4.33}$$

(sometimes called Fick's law of diffusion).

When we wrote the PDEs for an elastic string (Section 3.3), it was possible to find a formula (Problem 3.16) giving a solution for **any** initial conditions. In the diffusion case there is no such simple formula. Most functions you might write down for the initial condition do not have solutions in terms of simple functions; soluble problems are the exception rather than the rule. In fact, the attention paid to continuum PDEs in electrodynamics textbooks like this one is far out of proportion to their direct applicability in the real world. There are several reasons for this. One is that a few of the problems that **can** be solved exactly are very important, for example the electrostatic field of a point charge, which we will calculate later. A second reason is that there are exact solutions that, although they do not describe any real system precisely, can be used to illustrate properties characteristic of many real systems: for example sinusoidal electromagnetic waves, which we will also discuss later. A third reason, which was more relevant before computers became widely available, is that it is time-consuming to solve discrete problems numerically by hand, but easy to make up continuum textbook problems that are exactly soluble.

"Solving" the PDEs (Eqs. 4.32 and 4.33) amounts, therefore, to guessing a solution and then verifying that it satisfies the equations.

▶ **Problem 4.9**

Show that $\rho(y, t) = ay^2 + 2aDt$ provides a solution to Eqs. 4.32 and 4.33 (a is a constant). What is $j(y, t)$?   ◀

▶ **Problem 4.10**

a. Show that $\rho(y, t) = At^{-1/2} \exp(-y^2/4Dt)$ is a solution of these PDEs. Write $j(y, t)$ explicitly.

b. This expression does not make sense at $t = 0$, of course; as $t \to 0$ it approaches zero if $y \neq 0$, and diverges for $y = 0$. Show that the corresponding continuum cell field **does** have a limit as $t \to 0$, and calculate it. [A sequence of continuum fields $\rho(t)$ is said to have a limit ρ' as $t \to 0$ if the corresponding discrete fields $\rho_{dr}(t)$ approach ρ'_{dr} for each dr. In this case, ρ_{dr} must be calculated by integrating ρ over a cell rather than evaluating it at the center, because ρ varies rapidly within the cell.]

c. Use program MIG1 to set up the $dr = 2$ discretization of this initial field, using $N = 18$, $D = 0.8$ (i.e., $k = 0.2$) and a suitable value for A. Put the origin $y = 0$ near the center of the screen. Compare the population ρ_{dr} at $t = 20$ with the formula in part (a) (you may evaluate it at the cell center rather than integrating over the cell) for $y = 0$, $y = 3dr = 6$, and $y = 6dr$. What is the largest fractional error? (If you can't calculate the initial population quantitatively from A, just verify that the ratios of these populations are right.) ◀

▶ **Problem 4.11**

By eliminating j, write a single (second-order) PDE involving $\rho(r, t)$ alone. (This is called the "diffusion equation" or the "heat equation.")

CHAPTER FIVE

SOURCES AND SINKS; STATIC FIELDS

SECTION 5.1 SOURCES

In discussing density and current in Chapter 4, we assumed that the diffusing objects (hares or ants) did not enter or leave the system. In physics we frequently need to allow for such a possibility. Suppose, for example, that someone is bringing hares to the cell labeled $1.5dr$ in Figure 5.1. If he brings $B(r, t)$ hares to cell r ($= 1.5dr$, here) during the time interval centered at t, this modifies the continuity equation (Eq. 4.13) for the change in population

$$H\left(r, t + \tfrac{1}{2}dt\right) - H\left(r, t - \tfrac{1}{2}dt\right)$$

by a term $B(r, t)$. When we divide this by $dr\, dt$ to get an equation like Eq. 4.19 for the rate of change of density, this term becomes $B(r, t)/dr\, dt$, the rate of introduction of hares per unit length per unit time. We will refer to this as the **source density** $s(r, t)$ (not to be confused with the strain of Chapter 2):

$$s(r, t) = B(r, t)/dr\, dt \tag{5.1}$$

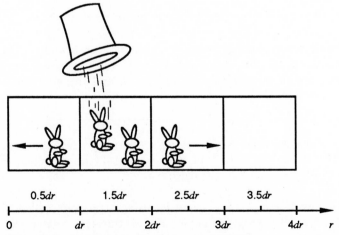

Figure 5.1   A diffusive system with a source of hares.

This gives a modified discrete continuity equation

$$d\rho(r, t)/dt = -[\text{grad } j](r, t) + s(r, t) \tag{5.2}$$

[The current equation (Eq. 4.20 or Eq. 4.33) is unchanged.] To update ρ and j from these equations, we need to solve them for $\rho(r, t + \frac{1}{2}dt)$ and $j(r, t + \frac{1}{2}dt)$ respectively. Using the definitions of the discrete derivatives from Chapter 3, this gives

$$\rho(r, t + \tfrac{1}{2}dt) = \rho(r, t - \tfrac{1}{2}dt)$$
$$+ (dt/dr)[j(r - \tfrac{1}{2}dr, st) - j(r + \tfrac{1}{2}dr, st)] + dt\, s(r, t) \tag{5.3}$$

$$j(r, t + \tfrac{1}{2}dt) = (D/dr)[\rho(r - \tfrac{1}{2}dr, t) - \rho(r + \tfrac{1}{2}dr, t)] \tag{5.4}$$

Let us apply these new equations to the example of Figure 5.1. Here the source density is nonzero only in cell 1.5dr; suppose it is a steady source (independent of time), say 80 hares per hour per meter. Assuming zero initial densities, a diffusivity $D = 1.0 \text{ m}^2/\text{hr}$, $dr = 2$ m, and $dt = 1$ hr, the evolution of the density is as in Table 5.1:

Table 5.1

r (in m)	1		3		5		7
$\rho(r, 0)$ (in hares/m)	0		0		0		0
$j(r, 0.5)$ (in hares/hr)		0	[$s = 80$]	0		0	
$\rho(r, 1)$	0		80		0		0
$j(r, 1.5)$		-40	[$s = 80$]	40		0	
$\rho(r, 2)$	20		120		20		0
$j(r, 2.5)$		-50	[$s = 80$]	50		10	
$\rho(r, 3)$	45		150		40		5

The arrows indicate which data go into the calculation of ρ and j. The density becomes asymmetric about the source cell because we have used impermeable-barrier boundary conditions, $j(0, t) = j(8 \text{ m}, t) = 0$, and the boundaries are not symmetric about $r = 3$ m.

▶ **Problem 5.1**

Compute the evolution of this system up to $t = 3$ hr using a **transient** source, that is, $s(1.5dr, dt/2) = 80$ hares hr^{-1} m^{-1} but $s(r, t) = 0$ at all other times (and other cells). Use impermeable boundaries at $r = 0$ and $r = 8$. ◀

We can follow the evolution of this system by using a program called SOURCE, which is described in the program guide. For a transient source, the density eventually becomes uniform (independent of r) and steady (independent of t) because no hares are being introduced and those that were initially introduced

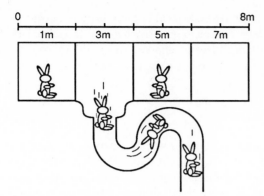

Figure 5.2   A diffusive system with a sink.

spread throughout the system. With a steady source, however, no such steady state is achieved; the density just continues to increase.

▶   **Problem 5.2**

a. Use SOURCE to check Table 5.1; what is $\rho(r, 10)$?

b. Check your result from Problem 5.1 and give $\rho(r, 10)$.   ◀

SECTION 5.2   SINKS

A sink is simply a negative source; if the number $s(r, t)$ of hares added to cell r per unit time per unit length is negative, its magnitude is the number of hares **removed** from cell r per unit time. A system with a sink in cell $r = 3$ meters is shown in Figure 5.2. We describe the sink by a negative s, say $s(3\ \mathrm{m}, t) = -50$ hares m^{-1} hr^{-1}. Note that this is a slightly idealized sink; the rate at which real materials (water, for example) go down a real sink depends on how much there is to start with, that is, on the density $\rho(r, t)$. In particular, if none is there, none can go down. For our sink, $s = -50$ independently of ρ; to make sure we can remove these 50 hares, we'd better have some hares initially. Let's start with an initially uniform density $\rho(r, 0) = 100$ hares m^{-1}.

Then Eqs. 5.3 and 5.4 give Table 5.2 (using again $dr = 2$ m, $dt = 1$ hr, $D = 1.0$ m^2/hr)

Table 5.2

r (in m)	1		3		5		7
$\rho(r, 0)$ (in hares/m)	100		100		100		100
$j(r, 0.5)$ (in hares/hr)		0	[$s = -50$]	0		0	
$\rho(r, 1)$	100		50		100		100
$j(r, 1.5)$		25	[$s = -50$]	-25		0	
$\rho(r, 2)$	88		25		88		100
$j(r, 2.5)$		31	[$s = -50$]	-31		-6	
$\rho(r, 3)$	72		6		75		97

Note that after 3 hours the hare density is about to become negative; this is inevitable in any situation in which we keep removing hares at the same rate without adding any. In situations involving electric charge, this will not be a problem; negative charge is not physically impossible, as negative hares are.

SECTION 5.3   STATIC FIELDS

We want to investigate the circumstances under which a diffusive system with steady sources and sinks can approach a steady state, in which the density and current are **static** (independent of time, from the Greek word for "standing" or "stationary"). That could not happen in the cases considered above because the total population $P(t)$ of the system [the sum of the populations $\rho(r, t)\, dr$ over all the cells r] steadily increased or decreased. The rate of change of the total population is evidently the sum of the source rates

$$\frac{dP(t)}{dt} = \sum_r s(r, t)\, dr \tag{5.5}$$

Remember that the symbol on the left-hand side is a discrete derivative

$$\frac{P\left(t + \tfrac{1}{2}dt\right) - P\left(t - \tfrac{1}{2}dt\right)}{dt}$$

▶   **Problem 5.3**

Prove Eq. 5.5 from Eq. 5.2, in a system with impermeable boundaries (i.e., $j = 0$ at the ends).   ◀

We can hope to approach a static field only if the net source rate (Eq. 5.5) vanishes. We can accomplish this by having both positive and negative sources. Consider a system with a positive source of strength A at position r_+ and an equal and opposite negative one at r_-, so that the source density is

$$s(r, t) = A\delta(r - r_+) - A\delta(r - r_-) \tag{5.6}$$

where δ is the Dirac delta function defined in Section 4.7. This system is shown in Figure 5.3a; it has impermeable boundaries at which $j = 0$. If we start it at $t = 0$ with zero density and let it evolve, we will find some positive density appearing near the source and negative density near the sink, as shown in Figure 5.3b. (You can think of the sink as producing antiparticles, if you like, which annihilate with the particles produced by the source when they diffuse to the same point.) The maximum density is always at the source, but the particles that diffuse leftward from the source are blocked by the boundary and have no sink to fall into, so the density in that region rises. Similarly, the density near the right-hand boundary becomes negative. If you wait a long time, you will see it approach the density field shown in Figure 5.4a.

It turns out that we can calculate the static density field in this system exactly, both for nonzero dr and in the continuum limit. The discrete equations that

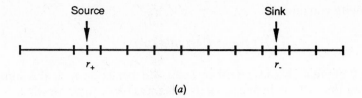

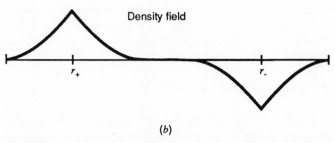

(b)

Figure 5.3   (a) A diffusive system with a source and a sink. (b) The (continuum) density field after a short time.

determine the evolution are Eq. 5.2 for $d\rho/dt$ and the continuity equation, Eq. 4.33. If the field becomes static after a long time, $d\rho/dt$ approaches zero and drops out, leaving

$$0 = -[\text{grad } j](r) + s(r) \tag{5.7}$$

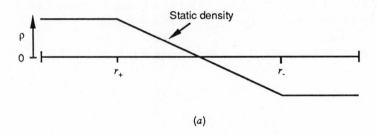

(a)

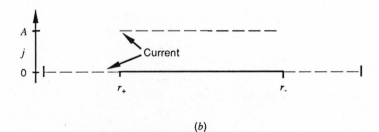

(b)

Figure 5.4   (a) The static density field. (b) The continuum static current.

and the continuity equation

$$j(r) = -D[\text{grad } \rho](r) \qquad (5.8)$$

(We omit the time index because neither j nor ρ depends on it in this limit.) We begin by solving the first of these equations for j. One way to do this is to observe that its gradient s is the difference of two delta functions. Therefore, because the integral of a delta function is a step function, j is the difference of two step functions. We will do it in a more direct way, however. Except at r_+ and r_-, Eq. 5.7 says that grad $j = 0$, that is, j is constant. This means that $j = 0$ at every face between the left boundary and the cell r_+ because $j = 0$ at the left boundary. For cell r_+, Eq. 5.7 becomes

$$j\left(r_+ + \tfrac{1}{2}dr\right) - j\left(r_+ - \tfrac{1}{2}dr\right) = dr\, s(r_+) = dr\ A/dr = A \qquad (5.9)$$

so that $j = A$ at the face $r_+ + \tfrac{1}{2}dr$ just to the right of this cell. Between r_+ and r_-, j is again constant at the value A. As we cross r_-, the analog of Eq. 5.9 has $-A$ instead of A, so j decreases to zero again, as it must to satisfy the right-hand boundary condition. Thus the final result for the current is

$$j(r, t) = 0 \quad \text{if } r < r_+$$

$$= A \quad \text{if } r_+ < r < r_- \qquad (5.10)$$

$$= 0 \quad \text{if } r > r_-$$

This current is shown (in the continuum limit) in Figure 5.4b. We must now calculate the static density. Suppose it starts at the left boundary at a value ρ_l, which we will calculate later (Problem 5.4). Because $j = 0$ between there and r_+, Eq. 5.8 says that ρ is flat in this region, as we anticipated in Figure 5.4a. However, when we evaluate Eq. 5.8 at the face $r_+ + \tfrac{1}{2}dr$, we find that

$$\rho(r_+ + dr) - \rho(r_+) = -dr\, j\left(r_+ + \tfrac{1}{2}dr\right)/D = -dr\, A/D \qquad (5.11)$$

so that

$$\rho(r_+ + dr) = \rho_l - \frac{dr\, A}{D} \qquad (5.12)$$

The same equation at the next face gives another decrease of $dr\, A/D$:

$$\rho(r_+ + 2dr) = \rho_l - 2\frac{dr\, A}{D} \qquad (5.13)$$

and it is clear that for any multiple a of dr,

$$\rho(r_+ + a) = \rho_l - \frac{aA}{D} \qquad (5.14)$$

In particular, if $a = r_- - r_+$,

$$\rho(r_-) = \rho_l - (A/D)(r_- - r_+) \tag{5.15}$$

To the right of r_-, ρ no longer changes, so our final result for ρ is that it is ρ_l for $r < r_+$, is given by Eq. 5.14 between r_+ and r_-, and by Eq. 5.15 for $r > r_-$, in agreement with Figure 5.4.

It is possible to solve for ρ in the continuum limit, regarding Eq. 5.8 as a differential equation for ρ. It is useful to go through this calculation because it is identical to (and easier to picture intuitively than) one we will do when we come to electrostatics, the calculation of the potential inside a capacitor. The current is discontinuous at the source and sink, so we cannot solve the differential equation for the whole system at once. We must confine ourselves to the three regions into which these points divide the system, and within each of which all the functions are continuous. Between the source and the sink, j is constant ($= A$) so we can integrate Eq. 5.8 with respect to r, giving

$$\rho(r) = -(A/D)r + C \tag{5.16}$$

where C is a constant. To the left of the source, this integration just gives $\rho(r) = a$ constant, the left-hand boundary value that we called ρ_l before. To calculate the constant C, we must find some relationship between the functions describing ρ on the two sides of the discontinuity. This is a very simple version of a problem that will occupy us a great deal in electrostatics, that of establishing boundary conditions at interfaces. The relationship we need in this case is that ρ is **continuous** across the boundary (in spite of the fact that j is not). The justification for this involves how singular we allow the source to be; we have allowed it to have a delta function but nothing more singular. (In fact, we haven't even encountered anything more singular. However, such things do exist—the derivative of a delta function, for example). This implies that its integral, the current, can have step discontinuities but nothing more singular than that (such as delta functions), and in turn that the density can have discontinuous slope, but nothing more singular (such as an actual discontinuity). The continuity of ρ at r_+ means that its value as we approach this point from the left (namely, ρ_l) must equal the value as we approach it from the right, namely, $-(A/D)r_+ + C$, from Eq. 5.16. This determines C. Similar considerations at r_- determine the (constant) value of ρ to the right of the sink, and we end up with the same formulas for ρ in the continuum limit that we had in the discrete case, namely Eqs. 5.14 and 5.15.

▶   **Problem 5.4**

Calculate the value of ρ_l in terms of A, D, r_+, r_-, and the total length L of the system, assuming that the initial density was zero (and therefore the total population remains zero, by Eq. 5.5). You may use either a discrete or a continuum method.   ◀

▶   **Problem 5.5**

a. For the special case that $L = 18 dr$, $dr = 2$, r_+ and r_- are $3.5 dr$ from the left and right boundaries respectively, and $D = 1.0$, calculate ρ_l in terms of A.

b. Use the SOURCE program to simulate this case for a convenient value of A, and check your answer to part a.   ◀

▶   **Problem 5.6**

The above argument that a steady-state static field is impossible with just positive sources holds only in a **finite** system. (If the added population is spread throughout a system whose size approaches infinity, the rate of population increase in each cell may approach zero, which is not inconsistent with a static field.) We will see later that in an infinite two-dimensional system, a single point source with zero initial density **does** in fact produce a static field. It **almost** does so in one dimension, in the following sense:

a. Consider a delta-function source $s(r, t) = A\delta(r)$, where A is a constant. Show that $\rho(r) = C - (A/2D)|r|$ is a static solution of the discrete diffusion equations (Eqs. 5.3 and 5.4), for any constant C. Sketch $\rho(r)$ and $j(r)$ for $C = 10A\,dr/D$. The only drawback of this solution is that for any C, it goes negative for $|r| > 2CD/A$; because diffusion from a positive source can never produce a negative density, it cannot be the correct long-time limiting field.

b. Use SOURCE with a delta-function source at the center of the screen to see what the distribution really looks like after a long time. (Use many cells, to approximate an infinite system.) Can you make any connection with part a? Does $\rho(r)$ approach a limit as $t \to \infty$? Does $j(r)$?   ◀

CHAPTER SIX

TWO- AND THREE-DIMENSIONAL SYSTEMS

SECTION 6.1 BEDSPRINGS AND TWO-DIMENSIONAL GRADIENTS

The two-dimensional (2D) analog of the one-dimensional (1D) masses-and-springs system of Chapter 1 is a "bedspring" consisting of masses connected by springs of length dr as shown in Figure 6.1. To label the masses and springs, we will draw a lattice of $dr \times dr$ squares (Figure 6.1b). The vertices (corners of the squares) will be labeled by their position vectors $r = (x, y)$ (that at $x = dr$, $y = dr$ is labeled in the figure), and the edges will be labeled by the position vectors e of their centers. Each vertex r is the equilibrium position of one mass, and will also serve to label that mass. Similarly, each edge e is the equilibrium position of a spring, whose label will be e. We will consider transverse motions only, so the springs move only in and out of the paper; call the displacement of the mass at vertex r, $w(r)$ ($w > 0$ means out of the paper, $w < 0$ into it). According to our definition (Chapter 2) of a field as a rule for assigning a number to each of several positions in space, w is a field; we will call it a **two-dimensional vertex field**. (Note that we are now using the symbol x to denote a cartesian component, whereas in the one-dimensional systems of Chapters 1 and 2 it denoted the displacement we now call w.)

To compute the force exerted by each spring, we will need to know the transverse elongation of each spring, or alternatively the (transverse) strain. The strain $s(e)$ in spring e can be defined by

$$s(e) = \left[w(r_{\text{front}}) - w(r_{\text{back}}) \right]/dr \tag{6.1}$$

which is unambiguous only if we associate an orientation with the edge e, so we can tell which vertex is the front and which is the back. We indicate this orientation with an arrow, as in Figure 6.2. When we deal with a coordinatized grid like that of Figure 6.2, we will usually orient edges in the direction of increasing coordinate x or y, as we have in Figure 6.2. We will call s a 2D **edge field**.

Equation 6.1 defines a procedure for calculating an edge field from any vertex field, a generalization of the gradient we defined in one dimension. The gradient for any dimensionality is defined by

$$[\text{grad } w](e) \equiv \left[w(r_{\text{front}}) - w(r_{\text{back}}) \right]/dr \tag{6.2}$$

The easiest way to specify the numerical values of two-dimensional fields is to simply write them next to the corresponding vertices or edges on a diagram of the grid, as in Figure 6.3a. Here we have specified a vertex field w, and computed its

89

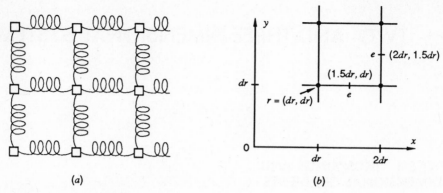

(a)   (b)

Figure 6.1   (a) A 2D "bedspring" of masses and springs. (b) Lattice whose edges and vertices label the springs and masses.

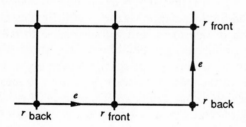

Figure 6.2   Vertex labeling used in defining the gradient.

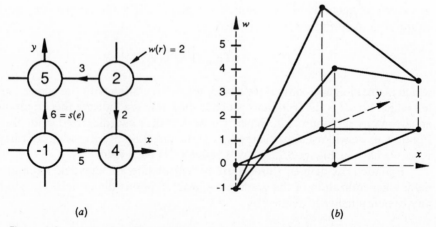

(a)   (b)

Figure 6.3   (a) A vertex field and its gradient. (b) A 3D picture of the corresponding altitude field.

gradient (using $dr = 1$). On the bottom edge, for example, Eq. 6.2 gives $[4 - (-1)]$ $= 5$. We can orient the edges however we want; given any choice of oriented edges, such as that in Figure 6.3a, an edge field is specified by giving a number $s(e)$ for each edge e. We will refer to the edge having the same position as e but the opposite orientation as "$-e$." (Cautionary note: When we label e by the coordinates (x, y) of its center, it is possible to misinterpret "$-e$" as the edge at $(-x, -y)$; we will **not** use $-e$ in that sense.) It is clear from the definition (Eq. 6.2) of the gradient that this just reverses the roles of r_{front} and r_{back}, reversing the sign of the gradient:

$$[\text{grad } w](-e) = -[\text{grad } w](e) \tag{6.3}$$

We will incorporate this feature into the definition of an edge field: if it is given at e, we can define it at $-e$ by

$$s(-e) \equiv -s(e) \tag{6.4}$$

We have exploited this fact in Figure 6.3a to choose the edge orientations so $s(e)$ is always nonnegative; this has the advantage that the arrows in Figure 6.3a indicate the direction of the gradient, as well as the orientation direction of the edge. Each arrow points in what is the "uphill" direction, if we think of w as an altitude field and the vertices as towns, as in Section 3.1 (see Figure 6.3b).

▶   **Problem 6.1**

Calculate the gradient of the vertex field shown in Figure 6.4 (use $dr = 2$). Indicate the result directly on a copy of the figure, rather than giving a list of edge coordinates or labels.   ◀

We can also specify a vertex field by giving a formula, that is, a function of the components x and y of its position vector, for computing it at any vertex $r = (x, y)$. As an example, suppose

$$w(r) = 2r^2 - 3 = 2x^2 + 2y^2 - 3 \tag{6.5}$$

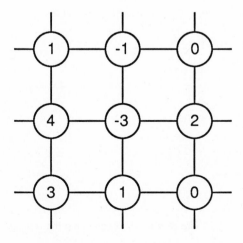

Figure 6.4   A vertex field.

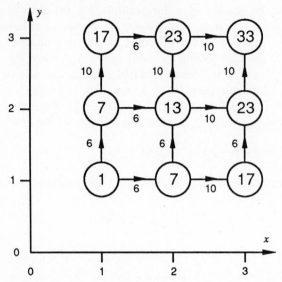

Figure 6.5   The vertex field defined by Eq. 6.5, and its gradient.

Then the values of w, for $dr = 1$, are as shown in Figure 6.5. We have also shown the gradient of w, obtained by subtracting values of w in front and in back of each edge. Given a formula for w, we can get a formula for grad w at a horizontal edge (parallel to the x-axis, as in Figure 6.6) whose center has coordinates x, y:

$$dr\,[\text{grad } w](e) = w(r_{\text{front}}) - w(r_{\text{back}}) = w\left(x + \tfrac{1}{2}dr, y\right) - w\left(x - \tfrac{1}{2}dr, y\right)$$

$$= 2\left(x + \tfrac{1}{2}dr\right)^2 + 2y^2 - 3 - \left[2\left(x - \tfrac{1}{2}dr\right)^2 + 2y^2 - 3\right] = 4x\,dr$$

so

$$[\text{grad } w](e) = 4x \qquad\qquad\qquad (6.6)$$

The formula for a vertical edge is different; the same argument gives

$$[\text{grad } w](e) = 4y \qquad\qquad\qquad (6.7)$$

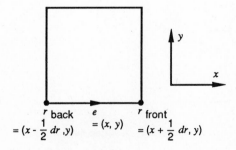

Figure 6.6   An edge along the x-axis.

[Eventually we will think of Eqs. 6.6 and 6.7 as the x and y components of a vector, but we can't really do that until we take the continuum limit. Here they are not defined at the same place (the same edge e).] You can verify that Eqs. 6.6 and 6.7 give the same numbers that were tabulated in Figure 6.5. When we write formulas like Eqs. 6.6 and 6.7, it is assumed that all edge orientations are in the direction of increasing coordinate x or y (i.e., the arrows point to the right or up).

▶ **Problem 6.2**

    a. Compute the values of the vertex field defined on a lattice with $dr = 1$ by $w(r) = 3x + y^2$, in the rectangle $0 \le x \le 3$, $0 \le y \le 3$, and show them in a diagram like Figure 6.5.

    b. Show grad w on all edges in this region.

    c. Give formulas for grad w on horizontal and vertical edges. Do the formulas depend on dr? Verify that they agree with b. ◀

▶ **Problem 6.3**

Same as Problem 6.2a, b, and c but for $w(r) = x^3 + y$. ◀

SECTION 6.2   INVERTING THE GRADIENT: PATH INTEGRAL AND CURL

In one dimension (see Section 3.2), if we knew the edge field that was the discrete gradient of a particular vertex field w, we could get w back (in terms of its value at a reference point) by adding up values of grad w. That procedure works in higher-dimensional lattices as well. Figure 6.7 shows a two-dimensional lattice with a vertex field w, which we would like to calculate at the vertex r, from its gradient $s = \text{grad } w$. We assume that we know the value $w(r_0)$ at a reference vertex r_0. We have drawn a **path** P (a set of oriented edges e) that leads from r_0 to r; the edges are assumed to be oriented in the forward direction along the path P from r_0 to r. If we denote the edges and vertices visited by $e_{0.5}, r_1, e_{1.5}, r_2, \ldots$, we can write

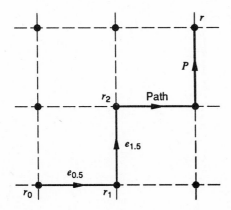

Figure 6.7 A path on a lattice.

from Eq. 6.2

$$w(r_1) = w(r_0) + s(e_{0.5}) \, dr$$

$$w(r_2) = w(r_1) + s(e_{1.5}) \, dr = w(r_0) + s(e_{0.5}) \, dr + s(e_{1.5}) \, dr$$

$$\cdots$$

$$w(r) = w(r_0) + \sum_{e \in P} s(e) \, dr \tag{6.8}$$

Here $\in$ is the set-inclusion symbol: $e \in P$ means "e is in P," so the sum is over all the edges making up the path P; it is called a (discrete) **path integral** over the path. (We will use the symbols $\sum$ and $\int$ more or less interchangeably when referring to discrete integrals.)

■ **EXAMPLE 6.1**   Integrate the edge field s shown in Figure 6.5 along the circuitous path from $(1, 1)$ to $(2, 1)$ shown in Figure 6.8. We have copied the field values onto the path, being careful to reverse the sign of s on the last two edges because the orientation of the edge is opposite that in Figure 6.5. The path integral is

$$\sum s(e) \, dr = 6 + 6 + 10 - 6 - 10 = 6. \tag{6.9}$$

Inserting this into Eq. 6.8 gives $w(r) = 1 + 6 = 7$ at the vertex $r = (2, 1)$, in agreement with Figure 6.5.

Equation 6.8 is a formula for inverting the gradient operation, that is, for calculating a vertex field w from its gradient s. It has a peculiarity, however, which was not present in one dimension. In one dimension the path from r_0 to r was uniquely

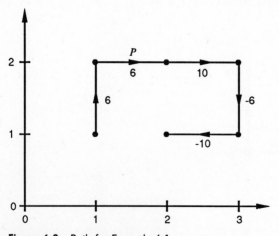

Figure 6.8   Path for Example 6.1.

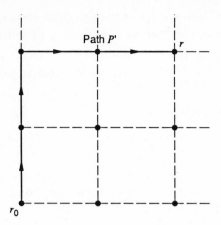

Figure 6.9 An alternate path from r_0 to r.

determined: 90° turns are not possible on a straight line. In two dimensions, there are obviously many possible paths from r_0 to r, of which we drew only one in Figure 6.7; a second one (call it P') is indicated in Figure 6.9. The argument above works for any path, so the path integral in Eq. 6.8 gives the same value for each possible path. It is not hard to see that this cannot be true for any edge field s, but depends on the special assumption we made about s, that it is a gradient. For example, suppose s were nonzero only at the edge denoted $e_{0.5}$ in Figure 6.7. Then the path integral would be nonzero for path P, but zero for path P'.

▶ **Problem 6.4**

    a. Figure 6.10 defines an edge field s which happens to be the gradient of a vertex field w. Compute the path integral of s over path P and also that over P' (Figures 6.7 and 6.9), assuming $dr = 1$. Are they the same? *Hint:* Begin by redrawing the figure with the edge orientations of Figure 6.7, changing the sign of s where necessary.

    b. Assuming $w(r_0) = -3$ at the lower left, compute $w(r)$ at all vertices shown. Do you need to compute a path integral from r_0 to r for each vertex r? Check your answer by computing the gradient and comparing to s. ◀

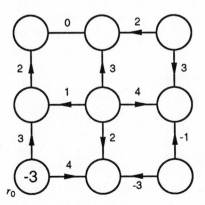

Figure 6.10 Edge field for Problem 6.4.

Let us now consider an **arbitrary** edge field s, not necessarily obtained as a gradient. We would like to find a way of determining whether it is a gradient, that is, whether it satisfies:

Condition 1: s is a gradient. (That is, there is a vertex field w such that $s = \text{grad } w$.)

We established above that Condition 1 implies

Condition 2: The path integral of s depends only on the endpoints r_0 and r, and not on the path.

Actually these two conditions are equivalent, that is, Condition 2 implies Condition 1 as well. To see this, consider an edge field s satisfying Condition 2 and define a vertex field $w(r)$ by Eq. 6.8, choosing some arbitrary value for $w(r_0)$. In general this definition is ambiguous, because the path integral may depend on the path. But we assumed it didn't (Condition 2), so $w(r)$ is well-defined. To compute its gradient at an edge e (Figure 6.11), choose any path P from r_0 to r_{back}, the vertex in back of e. We can denote by $P + e$ the path from r_0 to r_{front} (the vertex in front of e), which consists of P extended by e. The gradient is defined (Eq. 6.2) by

$$[\text{grad } w](e)\, dr = w(r_{\text{front}}) - w(r_{\text{back}})$$

$$= w(r_0) + \sum_{e' \in P+e} s(e')\, dr - \left[w(r_0) + \sum_{e' \in P} s(e')\, dr \right] \quad (6.10)$$

The second sum is over all the edges e' in P, and the first sum is over $P + e$, that is, it has exactly the same terms plus an extra term $s(e)\, dr$. So everything cancels except $s(e)\, dr$, and we get

$$[\text{grad } w](e) = s(e)$$

Thus s is a gradient (Condition 1), which is what we set out to prove.

A third condition equivalent to the above two is:

Condition 3: The path integral of s over a closed path is zero.

A **closed** path is one that has no beginning and no end, as in Figure 6.12. It is sometimes said to begin and end at the same vertex, say the vertex r_0 in the figure, but clearly there is nothing unique about this "beginning"; any other vertex on the

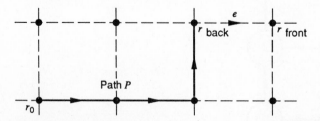

Figure 6.11 A path P from r_0 to the vertex in back of an edge e.

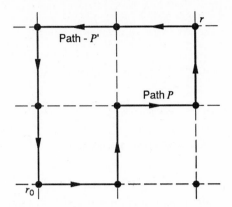

Figure 6.12 A closed path.

path would do as well. Let us first show that Condition 3 implies Condition 2 by taking two arbitrary paths P and P' from r_0 to r, and trying to prove that the path integrals are the same. Suppose P and P' are as shown in Figures 6.7 and 6.9. Note that we can get a closed path (in fact the one depicted in Figure 6.12) by putting P and P' together. However, to make the orientations consistent (our definition of a path doesn't allow two edges to meet head-to-head) we must reverse the orientations of all the edges e of one of the paths, say P'. The reversed path consists of the reversed edges $-e$, so we will refer to it as $-P'$. The path integral over $-P'$ is the same as that over P', with each edge e replaced by $-e$. This just reverses the sign of each s because $s(-e) = -s(e)$, and hence of the entire path integral. We will refer to the combination of P and $-P'$ as $P - P'$. So $P - P'$ is a closed path, and (because we're assuming Condition 3) the path integral over it is zero:

$$0 = \sum_{P-P'} s(e)\, dr = \sum_{P} s(e)\, dr + \sum_{-P'} s(e)\, dr$$

$$= \sum_{P} s(e)\, dr - \sum_{P'} s(e)\, dr \qquad (6.11)$$

However, this simply states that the path integrals over P and P' are the same, which is what we wanted. [We have omitted the dummy variable e and the $\in$ under the summations for brevity here; it is implied by the context.] What's left is to show that Condition 2 implies Condition 3, that the path integral around a closed path is zero. We can break any closed path at two points r_0 and r to get two paths P and $-P'$, such that P and P' each leads from r_0 to r. Thus we are assuming their integrals to be the same (Condition 2). But then the **right**-hand side of Eq. 6.11 is zero, and we can follow the equation backwards to conclude that the integral over the original closed path $P - P'$ was zero.

▶ **Problem 6.5**

Compute the path integral over the $2\,dr \times 2\,dr$ path (counterclockwise) around the perimeter of Figure 6.10. ◀

▶ **Problem 6.6**

Compute the path integrals over all four $dr \times dr$ square paths in Figure 6.10. ◀

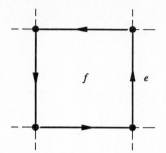

Figure 6.13 A one-face path around a face *f*.

Unfortunately, none of our three equivalent conditions on *s* is very easy to check in practice. No one would want to compute the path integrals over all possible paths to all possible points, or around all possible closed paths. However, it turns out we don't have to check **all** closed paths; it is enough to check the simplest ones, namely, those enclosing exactly one square of the grid, as in Figure 6.13. We label the squares by the letter *f* (for face) because in a three-dimensional lattice they are faces of cubes. We will show that these one-face paths are sufficient because **any** closed path, such as the one in Figure 6.12, can be related to a combination of them, shown in Figure 6.14.

The path integral of *s* around a single face *f* (per unit area of the face) is called the **curl** of *s*; because it is defined for every face *f*, it is a **face field**:

$$[\text{curl } s](f) \equiv \frac{\displaystyle\sum_{e \text{ of } f} s(e)\, dr}{dr^2} \tag{6.12}$$

Here the sum is over the four edges *e* of the face *f*, with the counterclockwise orientations shown in Figure 6.13.

The sum of the one-face path integrals around all the faces in Figure 6.14 can then be written

$$\sum_{f \in S} \sum_{e \text{ of } f} s(e)\, dr = \sum_{f \in S} [\text{curl } s](f)\, dr^2 \tag{6.13}$$

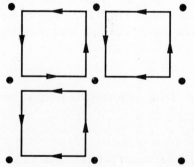

Figure 6.14 Three one-face paths, one for each face enclosed by the path in Figure 6.12.

where a "surface" S is a set of faces, in this case the three shown in Figure 6.14. We will refer to the counterclockwise closed path around S as the **boundary** of S; the standard mathematical notation for this boundary is ∂S. The inner sum is over the four edges of each face. All twelve edges in this double sum are shown in Figure 6.14; note that the two internal edges, each shared by two faces, appear twice, with opposite orientations. The corresponding terms in the sum are $s(e)$ and $s(-e)$, which cancel by Eq. 6.4. The only terms that are left are the **external** edges, that is, those of the boundary path ∂S. Thus, Eq. 6.13 becomes

$$\sum_{e \in \partial S} s(e)\, dr = \sum_{f \in S} [\text{curl } s](f)\, dr^2 \qquad (6.14)$$

This equation is called **Stokes's theorem**. The argument above proves that it is true for any 2D edge field $s(e)$ and any discrete surface (set of faces) S. We will refer to the sum on the right-hand side as the **discrete surface integral** of the face field [curl s].

As a consequence of Stokes's theorem we can write a fourth equivalent condition for s to be a gradient:

Condition 4: curl $s = 0$

This implies Condition 3 by Stokes' theorem, because the right-hand side of Eq. 6.14 is zero. Conversely, Condition 3 implies [curl s] $= 0$ because the curl **is** a path integral around a closed path.

■ **EXAMPLE 6.2**   Compute the curl of the edge field in Figure 6.10. Consider first the upper left face. The $s(e)$'s that appear in the sum (Eq. 6.12) defining the curl are 2, 1, 3, and 0. However, we must be careful to get the signs right when adding them up. The easiest way to do this is to put your finger on one of the vertices (say the upper left one) and move counterclockwise around the square. Count each $s(e)$ with a $+$ sign if your finger is moving along the orientation arrow, and $-$ if it is moving against the arrow. The result should be $-2 - 1 + 3 + 0 = 0$. It has to be zero by Condition 4 above, because s is a gradient. You should verify that the value of the curl at the other three faces is also zero.

▶ **Problem 6.7**

Compute the curl of the edge field of Figure 6.3a.   ◀

▶ **Problem 6.8**

a. Compute the curl of the edge field shown in Figure 6.15 at each of the six faces shown. (Use $dr = 1$.)

b. Verify Stokes's theorem for the shaded surface S (consisting of four faces). That is, calculate both the surface integral and the path integral.   ◀

▶ **Problem 6.9**

a. Indicate on a diagram the values of the edge field defined by $s(e) = x + 2y$ (at edges pointing along the positive x-axis as in Figure 6.6) and $s(e) = x$ (at y-pointing edges). Use $dr = 1$, and let the origin be a vertex. Show the square region $1 \le x \le 3$, $1 \le y \le 3$.

b. Compute the curl of this field, for the four faces having $1 < x < 3, 1 < y < 3$.

c. Verify Stokes's theorem for the surface S consisting of these four faces.   ◀

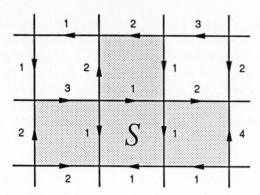

Figure 6.15 Edge field and surface S for Problem 6.8.

SECTION 6.3 GRADIENT AND CURL IN THREE DIMENSIONS

All of the results of the previous section can be carried over to three (or more) dimensions. We can calculate a vertex field from its gradient by Eq. 6.8. The four conditions for an edge field $s(e)$ to be a gradient are still equivalent, by exactly the same arguments. However, when we define the curl (Eq. 6.12, Figure 6.13) we must define [curl s](f) for faces f that are not in the plane of the paper (Figure 6.16a.) It is not enough to say that the four edges of the face are to be traversed "counter-clockwise"; if we look at the path from the other side of the face it seems clockwise. When we specify a face f, in addition to specifying where it is (by giving the coordinates of its center, say) we must specify which side of it we should look at it from in defining the curl; this is the "orientation" of the face and will be specified in pictures by drawing an arrow toward the observer. This can be thought of as distinguishing between the two sides of the face, one becoming the "front" or "$+$" side, and the other the "back" or "$-$" side. The curl is then defined with the edges oriented as in Figure 6.16a. A good way to remember the relation of the path to the

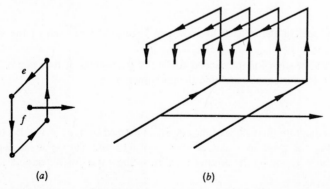

(a)   (b)

Figure 6.16 (a) Standard choice of edges used in defining the curl of a face field. (b) A discretized right hand.

face orientation is the **right-hand rule**: if your right thumb points along the orientation arrow of the face, your curled fingers (Figure 6.16b) indicate the orientation of the boundary path. In analogy with the notation used for edges, we will refer to a face at the same location as f but with opposite orientation as $-f$. It is easy to see from Eq. 6.12 that the curl at $-f$ is the same as that at f except for the orientation of the path, which reverses the sign of the curl. We will assume this as part of the definition of a face field: a face field j is specified by giving its values at faces f with specific orientations, and is defined at the reversed faces by

$$j(-f) = -j(f) \tag{6.15}$$

We would also like to be able to prove Stokes's theorem (Eq. 6.14) in 3D. It relates a discrete surface integral over a surface S (Figure 6.17) to a path integral over its boundary $P = \partial S$. One difference between two and three dimensions is that in 3D a path P does not uniquely determine a surface S of which it is the boundary. The two surfaces S and S' in Figure 6.17 have the same boundary path P; Stokes's theorem is true of either one. We will regard a surface S as a set of oriented faces. In principle they could have any orientations. However, we usually want to be able to identify a " $-$ " side and a " $+$ " side of the whole surface (for example, if it were a closed surface such as that of a cube or a sphere, these would be the "inside" and "outside," respectively.) This requires that the faces be oriented consistently. That is, they must not meet in such a way that the " $+$ " side of one face joins the " $-$ " side of the other (Figure 6.18a). A surface whose faces are oriented consistently is called an **oriented** surface, and one whose faces can be reoriented consistently is an **orientable** surface. The surfaces in Figure 6.18a are all orientable. An example of an unorientable surface is a Möbius strip (Figure 6.18b). If we start at the indicated face and move to the right while drawing arrows on faces, each consistent with the previous one, we will inevitably find an inconsistency when we have gone all the way around. In Stokes's theorem we need to know the direction of the boundary path P, which is unambiguous only if the surface S is an **oriented** surface; the orientation of the boundary is determined by the right-hand rule, as shown in Figure 6.16.

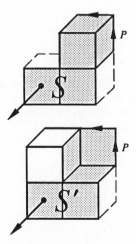

Figure 6.17 Two surfaces with the same boundary.

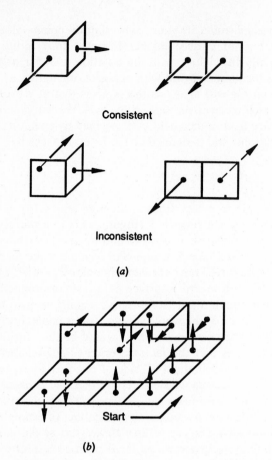

Consistent

Inconsistent

(a)

Start ─

(b)

Figure 6.18 (a) Examples of surfaces with consistently and inconsistently oriented faces. (b) A Möbius strip.

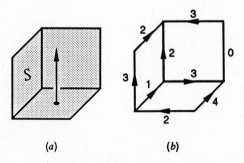

(a)   (b)

Figure 6.19 A 3D edge field. (Think of this as the corner where the floor and the wall of a room meet; the arrow indicates the inside of the room to be the "+" side of the surface.)

We can now state Stokes's theorem for a general dimensionality: if S is a discrete oriented surface, P is its boundary, and s is any edge vector field, then the surface integral of [curl s] over S is equal to the path integral of s over P (Eq. 6.14). The proof is identical to that found in Section 6.2.

▶ **Problem 6.10**

Consider a simple surface composed of three faces, as in Figure 6.19a. Compute the curl of the edge field shown in Figure 6.19b, and indicate it in a similar diagram. Verify Stokes's theorem by evaluating both sides explicitly. ◀

SECTION 6.4   FORCE FIELDS

Force fields are perhaps the most familiar examples of edge fields. Suppose we have a lattice of vertices connected by edges (Figure 6.20) and an object that moves along the edges. (In the continuum limit, it will be able to move anywhere.) The object is subject to a force that could be gravitational, electrical, elastic, or something else. A discrete **force field** is an edge field, whose value $F(e)$ at any edge e is defined as the work dW done per unit length by the force when the object moves along e from the vertex in back of e to the vertex in front:

$$F(e) = dW/dr \qquad (6.16)$$

If we think of the force as varying along e, then $F(e)$ is the average over e of the force component along e. The total work done when an object moves from a vertex r_0 to r along a path P (a set of edges) is the sum of $dW = F(e)\,dr$ over the edges, which is exactly the path integral defined in Section 6.2:

$$W(P) = \sum_P F(e)\,dr \qquad (6.17)$$

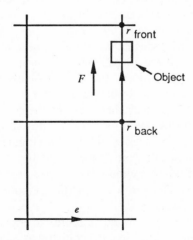

Figure 6.20 Lattice with a force field.

A force field is called **conservative** if the work it does over any closed path vanishes. This is exactly the third of the four equivalent conditions discussed in Section 6.2. Applied to a force field, these are four equivalent conditions that must be met in order for a force field (F) to be conservative:

1. $-F$ is the gradient of a vertex field V (called the **potential energy**).
2.   The work done in going from r_0 to r is independent of the path taken.
3.   No (net) work is done in traversing a closed path.
4.   curl $F = 0$.

These conditions are equivalent in any dimensionality, although in one dimension they are trivial: **all** edge fields are conservative. This is easiest to see from Condition 1, because any discrete field can be integrated in one dimension. Condition 2 is trivial, because there is essentially only one path in one dimension. Condition 4 is true vacuously, because there are no square faces in one dimension.

▶   **Problem 6.11**

A three-dimensional edge, field is defined at each edge labeled by its center position $e = (x, y, z)$ of a cubic lattice by the formula

$$F(e) = -mg$$

for edges parallel to the y-axis (mg is a constant), and by $F(e) = 0$ for x and z edges. Determine whether this field is conservative, and if so, find a formula for the potential energy $V(r)$ at any vertex $r = (x, y, z)$. Can you suggest a physical interpretation for this force field?   ◀

SECTION 6.5   DIFFUSION IN TWO AND THREE DIMENSIONS: DIVERGENCE

In this section we will generalize the equations for evolving a diffusive system (discussed in Chapter 4) to two and three dimensions. In three dimensions, the analog of our one-dimensional lattice is a cubic lattice (Figure 6.21). If we choose a coordinate system whose origin is at a vertex, the vertices are the points $r = (x, y, z)$ for which x, y, and z are all multiples of some specified lattice spacing dr. The cells of this lattice are the cubes of volume dr^3 into which this divides space. We label them by the coordinates of their centers, which are all half-integer multiples of dr. The edges are lines connecting neighboring vertices; they are labeled by the coordinates of their centers, which are characterized by having exactly one coordinate a half-integer instead of an integer multiple of dr. The faces of the cubes have two half-integer coordinates. We can think of the one-dimensional diffusive systems of Chapter 4 as linear sections through cell centers in a 3D system (Figure 6.22); this was the reason for calling the points along the line "faces" and the segments between them "cells." We will think of 2D diffusive systems also as cross-sections of 3D ones (Figure 6.23a) through cell centers. Thus, in a 2D picture (Figure 6.23b) the squares are called "cells" and the lines "faces."

In describing one-dimensional populations, we found it useful to replace the population of a cell by a density (population per unit length), which was much less

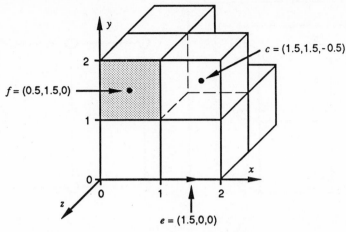

Figure 6.21   A 3D lattice (distances in units of dr).

dependent on dr in a fairly uniform system with many particles per cell. We can do exactly the same thing in three (or two) dimensions, dividing by the volume (area) of the cell respectively. (From now on we will, for the most part, deal with the 3D case, sometimes putting the 2D version in parentheses.) Thus, we will define the density in three dimensions by

$$\rho(c) \equiv \frac{\text{population of cell } c}{d\tau(c)} \tag{6.18}$$

where $d\tau(c)$ is the volume dr^3 of cell c.

In one dimension, the current $j(f, t)$ was defined as the number crossing the face f per unit time, during the time interval centered at t. This is sometimes also called the **flux** (a Latin word meaning "flow"). In higher dimensions, we can also talk about the flux (which is still the number crossing per unit time), but this is now strongly dependent on the area $da = dr^2$ of the face (or in 2D, on its length dr). So it is better to define a **current density** by

$$j(f) = \frac{\text{number crossing } f}{da\,dt} \tag{6.19}$$

which is more independent of dr, in a fairly uniform system.

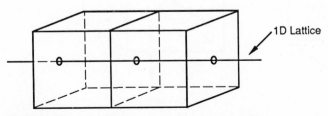

Figure 6.22   A 1D lattice as a section through a 3D one.

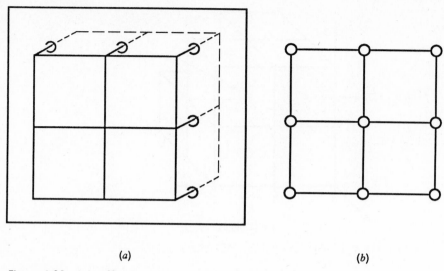

(a)   (b)

Figure 6.23   (a) A 2D cross-section through a 3D lattice. (b) 2D lattice alone.

Having generalized the density and the current fields, we can now try to generalize the discrete equations (Eqs. 4.19 and 4.20) that describe their time evolution. Let us begin with the continuity equation. This arose from the fact that particles (hares, in Chapter 4) are conserved; the change in the number in each cell is exactly the net number that enter through the faces. In three dimensions this says

$$\rho\left(c, t + \tfrac{1}{2}dt\right)d\tau - \rho\left(c, t - \tfrac{1}{2}dt\right)d\tau = \sum_{f} j(f, t)\, da\, dt \qquad (6.20)$$

where the sum is over the six inward-oriented faces of c (Figure 6.24). We can write

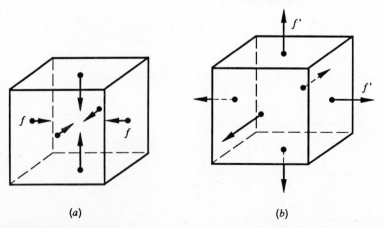

(a)   (b)

Figure 6.24   (a) A cell c and its six inward-oriented faces. (b) The six outward-oriented faces of c.

this as an equation for the discrete time derivative of ρ by dividing by dt and $d\tau$:

$$d\rho(c, t)/dt = d\tau^{-1} \sum_f j(f, t)\, da \qquad (6.21)$$

The expression on the right-hand side is one we will encounter frequently in electrodynamics. It is the **discrete divergence** of j, except that the divergence is conventionally defined in terms of **outward**-oriented faces $f' = -f$, as shown in Figure 6.24b. Given an arbitrary face field j, its discrete divergence is a cell field denoted by div j and defined by

$$[\operatorname{div} j](c) = d\tau^{-1} \sum_{f'} j(f')\, da \qquad (6.22)$$

where the sum is over the six outward-oriented faces of the cell c. Since $j(f') = j(-f) = -j(f)$ (Eq. 6.15), div j is the negative of the right-hand side of Eq. 6.21. Thus the updating equation for ρ becomes

$$d\rho(c, t)/dt = -[\operatorname{div} j](c, t) \qquad (6.23)$$

Although we derived this equation in three dimensions, it can be derived equally easily in two or one dimensions. The definition of the divergence must be altered appropriately: in 2D the sum is over the four faces of each cell (shown as lines in Figure 6.23b), and in 1D it is over two faces (which show up as points in a one-dimensional picture). We should also change the factor $d\tau^{-1} da = dr^2/dr^3$ to dr/dr^2 and $1/dr$ respectively, but of course these are all the same. If you compare Eq. 6.22 in 1D with the corresponding 1D equation we wrote down before (Eq. 4.19), you will notice that we wrote grad j instead of div j on the right-hand side. A little thought will convince you that the concepts of divergence and gradient are in fact equivalent in 1D; we used the latter in Chapter 4 only because we had not yet defined the divergence. To avoid confusion, from now on we will use the symbol "div" in the continuity equation in any dimensionality. (Another interesting equivalence occurs in 2D, where the **curl** and divergence are essentially equivalent.)

In 1D we defined a **discrete integral** of a cell field (or of an edge field; the two are equivalent in 1D), as the sum of $\rho(c)\, dr$ over a set of cells. We have defined integrals of edge fields in 2D and 3D (path integrals, Section 6.2) that generalize 1D edge-field integrals. We have also defined integrals of face fields in 2D and 3D (surface integrals, Section 6.2), which have no 1D analog. The only type of integral that remains to be defined is that of a 3D cell field, such as a density field. In 1D a cell field is essentially equivalent to an edge field, and its integral was defined in the same way. In 3D they are quite different, and the integral of a cell field is defined quite differently from that of an edge field. We will define it in a way that preserves a useful fact we found in 1D: the integral of the density over any region gives the population of that region. We will define a discrete **three-dimensional** region as any set of cells in a 3D lattice, such as that shown in Figure 6.25. We can then define the integral of a cell field ρ over R as the sum of the populations $\rho(c)\, d\tau(c)$ of the cells in R:

$$\text{integral of } \rho \text{ over } R \equiv \sum_{c \in R} \rho(c)\, d\tau(c) \qquad (6.24)$$

Figure 6.25 A discrete 3D region R and its surface S.

Then the integral of the continuity equation (Eq. 6.23) gives the discrete time derivative of the population of R:

$$\frac{d}{dt} \sum_{c \in R} \rho(c, t) \, d\tau = - \sum_{c \in R} [\operatorname{div} j](c, t) \, d\tau(c) \tag{6.25}$$

▶ **Problem 6.12**

Show that the discrete time derivative and cell-field integration operations commute:

$$\frac{d}{dt} \sum \rho(c, t) \, d\tau = \sum \frac{d\rho(c, t)}{dt} \, d\tau \tag{6.26}$$

by writing out both sides explicitly.   ◀

If particles are neither created nor destroyed, the population of R can only change due to particles flowing in or out through the surface of R (labeled S in Figure 6.25; we will sometimes refer to it as ∂R or the "boundary of R"). The rate of change of the population should therefore be the total inward flux through S. This can be expressed as a discrete surface integral $-\sum j(f) \, da$, the sum of the fluxes through the individual faces f that make up S (the minus sign is because a boundary is conventionally oriented outwards, whereas we want the inward flux). This is consistent with Eq. 6.25 only if (omitting the arguments c, f, and t)

$$\sum_R [\operatorname{div} j] \, d\tau = \sum_{\partial R} j \, da \tag{6.27}$$

This is called the **divergence theorem**, and is true of **any** face field j and any region R with boundary ∂R. To prove this, rewrite the left-hand side using the definition of the divergence:

$$\sum_{c \in R} \sum_{f \text{ of } c} j(f) \, da(f) = \sum_{f \in S} j(f) \, da(f) \tag{6.28}$$

The sum on the left side is over all six outward faces of each of the cells in R. Evidently, each internal face f (a face which is not part of the boundary S) is counted twice, as shown in Figure 6.26: once as a face of the cell behind f [giving a term $+j(f) \, da$, because f is an outward face of that cell], and once as a face of the

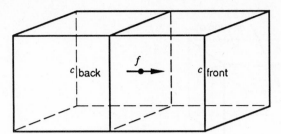

Figure 6.26 An internal face bounds two cells.

cell in front [giving $-j(f)\,da$, because f is an **inward** face of **that** cell]. Thus, the internal faces cancel out, leaving the external faces $f \in S$. These are exactly the terms that appear on the right-hand side, so we have proved the divergence theorem.

▶ **Problem 6.13**

    a. Calculate the divergence of the face field in Figure 6.27. Assume $j(f) = 0$ where it is not indicated, including all faces parallel to the plane of the paper. Show the result on a sketch.

    b. Verify the divergence theorem by evaluating both sides explicitly, when R is the three-cell region shown. ◀

To determine the time evolution of a 3D density field, we need a 3D version of Fick's law (Eq. 4.33), which is

$$j(f, t) = -D[\text{grad}\,\rho](f, t) \qquad (6.29)$$

To use this equation in 3D, we must define what we mean by the gradient of a cell field (we have so far defined it only for vertex fields). However, we can turn cells into vertices, and vice versa, by shifting our lattice over by $(dr/2, dr/2, dr/2)$ as

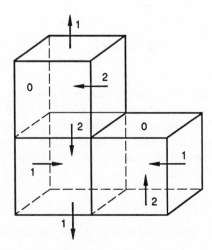

Figure 6.27 The face field $j(f)$ in Problem 6.13.

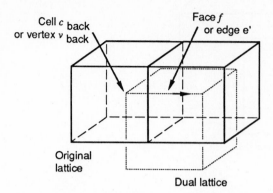

Figure 6.28 Original (solid line) and dual (dotted line) lattices.

shown in Figure 6.28. This shifted lattice is called the **dual lattice**. The density field ρ, which is a cell field when we are looking at the original lattice, becomes a vertex field if we look at the dual lattice. Stated more formally, we can define a vertex field ρ' on the dual lattice by $\rho'(v') \equiv \rho(c)$ where v' is the vertex of the dual lattice having the same position (x, y, z) as the cell c. We can then take its gradient, which is an edge field on the dual lattice, defined at edges such as the dotted line e' in Figure 6.28 by Eq. 6.2:

$$[\text{grad } \rho'](e') = [\rho'(v'_{\text{front}}) - \rho'(v'_{\text{back}})]/dr \qquad (6.30)$$

Writing the same equation in terms of ρ defines the gradient of ρ on the original lattice, at the face f with the same position and orientation direction as the dual edge e':

$$[\text{grad } \rho](f) = [\rho(c_{\text{front}}) - \rho(c_{\text{back}})]/dr \qquad (6.31)$$

This definition makes Eq. 6.29 meaningful; it is the 3D version of Fick's law, which is found experimentally to describe 3D diffusive systems in the continuum limit.

SECTION 6.6 CONTINUUM LIMITS IN THREE DIMENSIONS: VECTOR FIELDS

We defined in the previous section discrete vertex, edge, face, and cell fields in 3D. We can define continuum versions of these fields exactly as we did in Chapter 3: a continuum vertex field A is a rule for determining a consistent discrete vertex field A_{dr} for each dr, a continuum cell field ρ is a rule for determining a consistent discrete cell field ρ_{dr} for each dr, etcetera. The meaning of "consistent" is also analogous: for a cell field in 1D, it was essentially

$$\rho_{2\,dr}(C)2\,dr = \rho_{dr}(c_1)\,dr + \rho_{dr}(c_2)\,dr \qquad (6.32)$$

where C is a cell of length $2\,dr$, which comprises two cells c_1 and c_2 in the lattice of

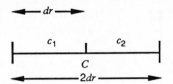

Figure 6.29 A 1D cell C, composed of two smaller cells.

spacing dr (Figure 6.29). This means physically that the population of the large cell C is the same no matter which lattice you use to compute it. In 3D, the analogous consistency condition is

$$\rho_{2\,dr}(C)(2\,dr)^3 = \sum_{c \in C} \rho_{dr}(c)\,dr^3 \qquad (6.33)$$

where the sum is over the eight small cells c in the large cell C (Figure 6.30). For a face field, the corresponding formula has a sum over four faces f making up a $2\,dr \times 2\,dr$ face F, and for an edge field it has a sum over two edges e making up an edge E of length $2\,dr$. For a vertex field, consistency means (as in 1D) that $A_{2\,dr}(v) = A_{dr}(v)$ if v is a vertex of both lattices.

   Given these definitions, we can set up correspondences between continuum fields and ordinary functions [by which we now mean functions of three variables (x, y, z)]. Just as in 1D, a continuum vertex field is essentially equivalent to an ordinary function: given a function A_f, we define the discrete field A_{dr} at a vertex $v = (x, y, z)$ by

$$A_{dr}(v) = A_f(x, y, z) \qquad (6.34)$$

Conversely, given a rule for determining A_{dr} for each dr, we can define a function at a point (x, y, z) by choosing a lattice with a vertex v there, and using $A_f(x, y, z) = A_{dr}(v)$ (it is independent of which lattice we choose by the consistency condition above).

   Determining edge and face fields from functions is a bit trickier; these will turn out to correspond to **vector** functions. This is easiest to see by thinking of a particular kind of edge field, the **force** field F discussed in Section 6.4. Recall that

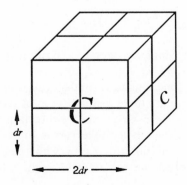

Figure 6.30 A cell C of length $2\,dr$, composed of 8 cells c.

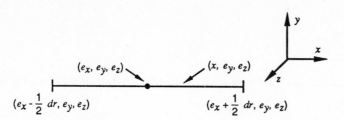

Figure 6.31 Region of integration for Eq. 6.35.

$F(e)$ for an edge e in the x-direction was related to the average of the x-component of the actual force vector **F**. If we try to determine $F(e)$ from a continuous vector function $\mathbf{F}_f(x, y, z)$, this average is an integral over the edge (Figure 6.31)

$$F(e) = dr^{-1} \int_{e_x - dr/2}^{e_x + dr/2} F_{fx}(x, e_y, e_z)\, dx \qquad (6.35)$$

where e_x, e_y, and e_z are the coordinates of the center of edge e, and F_{fx} is the x-component of the vector field $\mathbf{F}_f$. This formula looks very much like the one (Eq. 3.16) that we used in 1D for a general edge field. So we will adopt this as our general procedure for getting 3D continuum edge fields from functions: Given any vector function $\mathbf{F}_f(x, y, z)$, the associated continuum edge field [a rule yielding $F_{dr}(e)$] yields the integral in Eq. 6.35 for an edge e along the x-direction,

$$F_{dr}(e) = dr^{-1} \int_{e_y - dr/2}^{e_y + dr/2} F_{fy}(e_x, y, e_z)\, dy \qquad (6.36)$$

for one along the y-direction, and similarly for z. As in 1D, we can determine F_{dr} **approximately** by evaluating the function at the center of the edge. For an x-edge e, for example,

$$F_{dr}(e) \cong F_{fx}(e_x, e_y, e_z) \qquad (6.37)$$

The reverse procedure, getting a function from a continuum edge field, is also exactly as it is in 1D:

$$F_{fx}(x, y, z) = \lim_{dr \to 0} F_{dr}(e) \qquad (6.38)$$

where e is the x-edge closest to (x, y, z). Similarly, F_{fy} is a limit involving a sequence of y-edges.

We can relate continuum face fields to functions in a similar way. There, too, we need three different functions because there are three distinct types of faces near each point of space: those whose normal is in the x-, y-, and z-direction respectively. We can again regard these three functions as a single **vector** function, say $\mathbf{j}_f(x, y, z)$. A face f (not to be confused with the subscript f on $\mathbf{j}$, which just stands for "function") normal to the x-direction (which we will refer to as an x-face for

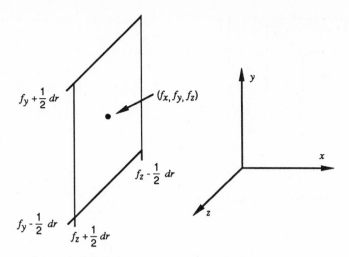

Figure 6.32   Region of integration for Eq. 6.39.

brevity) has zero extent in the x-direction but extends from $f_y - \frac{1}{2}dr$ to $f_y + \frac{1}{2}dr$ in the y-direction and from $f_z - \frac{1}{2}dr$ to $f_z + \frac{1}{2}dr$ in the z-direction (see Figure 6.32). Thus, after averaging it in the y-direction as in Eq. 6.36, we must average it in the z-direction as well. Doing so yields a repeated integral (an integral of an integral, to be discussed further in Section 6.8)

$$j_{dr}(f) = dr^{-2} \int_{f_z - dr/2}^{f_z + dr/2} dz \int_{f_y - dr/2}^{f_y + dr/2} dy \, j_{fx}(f_x, y, z) \qquad (6.39)$$

By cyclically permuting the coordinate axes (that is, changing x to y, y to z, and z to x) we can get formulas for a y-face and a z-face.

▶   **Problem 6.14**

Show that the continuum face field defined by Eq. 6.39 is consistent, in the sense defined above.   ◀

As we did for edge fields, we can approximate this double integral by the value of the function at the center of the face:

$$j_{dr}(f) \cong j_{fx}(f_x, f_y, f_z) \qquad (6.40)$$

If we want to get the vector function component $j_{fx}(x, y, z)$ back from the continuum field, we can take the limit of $j_{dr}(f)$ where f is the closest x-face to (x, y, z).

Defining a continuum cell field from a function is simpler, because a cell does not have multiple orientation directions. Thus we need only one function $\rho(x, y, z)$ to describe it (i.e., it is a **scalar** rather than a vector field). But we must average it over all three coordinates, as shown in Figure 6.33:

$$\rho_{dr}(c) = dr^{-2} \int_{c_z - dr/2}^{c_z + dr/2} dz \int_{c_y - dr/2}^{c_y + dr/2} dy \int_{c_x - dr/2}^{c_x + dr/2} dx \, \rho_f(x, y, z) \qquad (6.41)$$

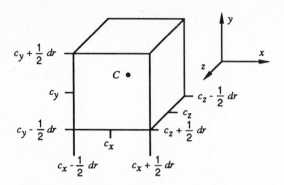

Figure 6.33   Integration region for Eq. 6.41.

We can again get the function $\rho_f(x, y, z)$ back from the continuum field by taking the limit of $\rho_{dr}(c)$ at the closest cell to (x, y, z).

Our purpose in defining a continuum field (as a distinct concept from a function) is that just as in 1D, there are some very important fields that are not obtainable from functions. An example is the 3D Dirac delta function. To define this continuum cell field, we must give a rule defining a discrete cell field δ_{dr} for each choice of lattice spacing dr. By analogy with the 1D case, this rule is

$$\rho_{dr}(c) = dr^{-3} \qquad \text{if the cell } c \text{ contains the origin}$$
$$= 0 \qquad \text{if not} \tag{6.42}$$

(The factor dr^{-3} is chosen so that the population of the cell containing the origin is 1, as in the 1D case. This is the total population as well, of course, because $\rho = 0$ everywhere else.) Just as the 1D delta function was obtainable as the divergence (or gradient) of a step function, the 3D delta function is also obtainable as the divergence of a simple function, an inverse-square vector field, as we will see in Chapter 7.

▶   **Problem 6.15**

Show that the 3D delta function can be expressed as a product of 1D delta functions:

$$\delta^{3D}(x, y, z) = \delta^{1D}(x)\,\delta^{1D}(y)\,\delta^{1D}(z) \tag{6.43}$$

using the definition of the 1D delta function (Eq. 3.13). Equation 6.43 is a relation between continuum fields, which should be interpreted as meaning that for each dr, the corresponding discrete fields are related by

$$\delta^{3D}_{dr}(c) = \delta^{1D}_{dr}(x)\,\delta^{1D}_{dr}(y)\,\delta^{1D}_{dr}(z) \tag{6.44}$$

Here c is the cell centered at x, y, z.   ◀

We have drawn pictures of discrete vector fields by putting numbers by the faces or edges, sometimes with longer or shorter arrows to indicate the magnitude of the field

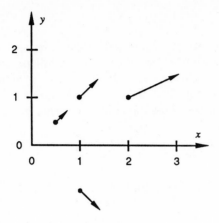

Figure 6.34 A vector field.

(as in Figure 6.35). Clearly this won't work for a continuum field—it would get more and more cluttered as dr decreased. So we usually will depict a continuum vector field obtained from a vector function $\mathbf{j}_f(x, y, z)$ by drawing the vector $\mathbf{j}_f$ itself at a few selected points (x, y, z). We will do this for both edge and face fields (in fact, in the continuum limit these are essentially equivalent). An example involving a face field follows.

■ **EXAMPLE 6.3** The field determined by the vector function $\mathbf{j}_f(\mathbf{r}) = \frac{1}{2}\mathbf{r}$ (i.e., $j_{fx}(x, y) = \frac{1}{2}x$, $j_{fy}(x, y) = \frac{1}{2}y$) is shown in Figure 6.34 (you can think of this as the $z = 0$ cross-section of a 3D system). The tail of the vector is placed at (x, y) so the head is at $(x, y) + \mathbf{j}_f$. At $(x, y) = (2, 1)$, for example, $j_{fx} = (1, \frac{1}{2})$. The corresponding discrete field with $dr = 1$ is shown in Figure 6.35.

▶ **Problem 6.16**

Consider another vector function $j_{fx}(x, y) = y/2$, $j_{fy} = x/2$. Show the continuum field defined by this function in a sketch such as Figure 6.34, and the corresponding discrete field in a sketch such as Figure 6.35. ◀

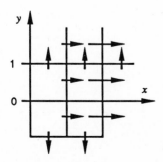

Figure 6.35 The discrete field corresponding to Figure 6.34.

SECTION 6.7   VECTOR CALCULUS

In this section we will consider continuum vector fields that are specified by differentiable functions. We will derive simple formulas for the gradient, curl, and divergence of such fields. Consider first the gradient. If $A(x, y, z)$ is a function, it (or more precisely the continuum vertex field to which it is equivalent) determines a discrete vertex field for any lattice, whose value at any vertex v at position (x, y, z) is

$$A_{dr}(v) = A(x, y, z) \tag{6.45}$$

(In this section we will not use f subscripts on functions.) We have formally defined the gradient of the continuum vertex field A, as a continuum edge field (a rule for determining a discrete edge field for any lattice): it is the particular rule that gives $\mathrm{grad}(A_{dr})$, the discrete gradient of the discrete vertex field A_{dr}. We would like to associate this continuum edge field with a vector function, which we will denote by $\mathrm{grad}\, A$. Its x-component can be expressed (see Eq. 6.38) as a limit of the discrete gradients:

$$[\mathrm{grad}\, A]_x(x, y, z) = \lim_{dr \to 0} \mathrm{grad}(A_{dr})(e) \tag{6.46}$$

where the e's are x-edges whose locations approach (x, y, z). The simplest such sequence of edges consists of edges centered at (x, y, z) with lengths dr that approach zero; one of them is shown in Figure 6.36. For these, the limit becomes

$$\lim_{dr \to 0} \frac{A_{dr}(v_{\mathrm{front}}) - A_{dr}(v_{\mathrm{back}})}{dr} = \lim_{dr \to 0} \frac{A(x + dr/2, y, z) - A(x - dr/2, y, z)}{dr}$$

$$= \frac{dA(x, y, z)}{dx} \tag{6.47}$$

by the definition of the (partial) derivative. We can determine a formula for $[\mathrm{grad}\, A]_y$ by considering a sequence of y-edges giving similarly

$$[\mathrm{grad}\, A]_y(x, y, z) = dA(x, y, z)/dy \tag{6.48}$$

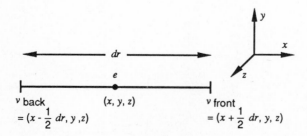

Figure 6.36   The edge e with its front and back vertices.

and similarly for z. Thus the vector function grad A can be written

$$[\text{grad } A](x, y, z) = (dA/dx, dA/dy, dA/dz) \tag{6.49}$$

This is sometimes written formally as

$$[\text{grad } A](x, y, z) = (d/dx, d/dy, d/dz)A = \nabla A \tag{6.50}$$

where the symbol ∇ (pronounced "del") is formally defined as the vector

$$\nabla \equiv (d/dx, d/dy, d/dz) \tag{6.51}$$

although this expression really has no meaning outside of Eq. 6.50 or similar equations we will derive below.

> Traditionally the gradient of a vector function is defined directly by Eq. 6.49, instead of as a limit of discrete gradients; the discrete gradients we defined earlier are regarded as approximations to the continuum ones. This is a rather roundabout procedure: one defines the derivative as a limit, and then turns around and moves away from the limit again to define discrete approximations. The two approaches yield equivalent calculational results, of course; ours has the advantage that it allows us to prove theorems such as Stokes's rigorously for discrete fields, and carry them over to fields determined by continuous functions, for which rigor is much harder to come by. It also allows us to deal with functions such as step functions for which the derivatives in Eq. 6.49 do not exist.

▶ **Problem 6.17**

Use the derivative formula (Eq. 6.49) to compute the vector function that is the gradient of the scalar function $A(\mathbf{r}) = 2r^2 - 3$ [here $\mathbf{r} = (x, y, z)$, so $A(x, y, z) = 2x^2 + 2y^2 + 2z^2 - 3$]. The discrete gradient of the 2D version of this function was given in Eqs. 6.6 and 6.7. ◀

Next we will find a formula for the **divergence** of a function. The discrete divergence turns face fields (which are related to vector functions) into cell fields (related to scalar functions), so the divergence of a vector function $\mathbf{j}(x, y, z)$ will be a scalar function, denoted div $\mathbf{j}$. The continuum cell field associated with the function div $\mathbf{j}$ has been defined formally as the rule assigning the discrete field div(j_{dr}) to each lattice. From this we can obtain the function div $\mathbf{j}$ as a limit:

$$[\text{div } \mathbf{j}](x, y, z) = \lim_{dr \to 0} [\text{div } j_{dr}](c) \tag{6.52}$$

where the cell c is the one closest to (x, y, z) in each lattice. The simplest sequence of such cells to use in this limit is a sequence centered at (x, y, z) with side lengths dr approaching zero. For each such cell, the discrete divergence in Eq. 6.52 is

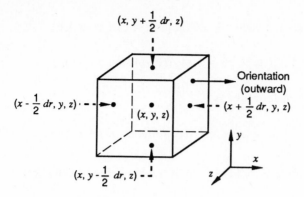

Figure 6.37   The 6 faces of the cell centered at (x, y, z); 4 are labeled by coordinates.

defined by Eq. 6.22 as a sum of $j_{dr}(f)$ over its six outward-oriented faces f (Figure 6.37). We can approximate these (with an error that vanishes as dr approaches zero) by the values of the function $\mathbf{j}$ at the face centers (Eq. 6.40). For the rightmost face, for example, this value is $j_x(x + \frac{1}{2}dr, y, z)$ (the x-component is used because this face is normal to the x-axis). For the leftmost face, it is $-j_x(x - \frac{1}{2}dr, y, z)$ (the $-$ sign is required because the outward-oriented face in Figure 6.37 is oriented oppositely to the conventional rightward orientation used in Eq. 6.40). Adding the terms for all six faces, we get

$$[\operatorname{div}\mathbf{j}](x, y, z) = \lim_{dr \to 0} dr^{-1}\Big[j_x\big(x + \tfrac{1}{2}dr, y, z\big) - j_x\big(x - \tfrac{1}{2}dr, y, z\big)$$

$$+ j_y\big(x, y + \tfrac{1}{2}dr, z\big) - j_y\big(x, y - \tfrac{1}{2}dr, z\big)$$

$$+ j_z\big(x, y + \tfrac{1}{2}dr, z\big) - j_z\big(x, y, z - \tfrac{1}{2}dr\big)\Big] \qquad (6.53)$$

$$= dj_x(x, y, z)/dx + dj_y(x, y, z)/dy + dj_z(x, y, z)/dz$$

by the definition of the derivative. We can express this formally in terms of the "del" symbol (Eq. 6.51) as a dot product

$$\operatorname{div}\mathbf{j} = (d/dx, d/dy, d/dz) \cdot \mathbf{j} = \nabla \cdot \mathbf{j} \qquad (6.54)$$

[Recall that the dot product is defined by

$$\mathbf{h} \cdot \mathbf{j} \equiv h_x j_x + h_y j_y + h_z j_z \qquad (6.55)$$

for any two vectors $\mathbf{h}$ and $\mathbf{j}$.]

▶   **Problem 6.18**

  a. Use the derivative formula (Eq. 6.53) to compute the divergence of the 3D vector function $\mathbf{j}(\mathbf{r}) = \frac{1}{2}\mathbf{r}$ (a 2D cross-section of this is sketched in Figure 6.34).

  b. Do the same for the function defined by $j_x(x, y, z) = \frac{1}{2}y$, $j_y(x, y, z) = \frac{1}{2}x$, and $j_z(x, y, z) = 0$.   ◀

Lastly, consider the curl. It turns edge fields into face fields, hence the version for differentiable functions will turn vector functions into vector functions. Any vector function $\mathbf{F}(x, y, z)$ corresponds to a continuum edge vector field as described in Section 6.6, whose curl is a continuum face field. We want to find a formula for the function (curl $\mathbf{F}$) corresponding to that continuum face field. As before, we do this by taking the limit of the discrete fields specified by this continuum face field, which are just the curls of the discrete edge fields F_{dr} determined by $\mathbf{F}(x, y, z)$. That is, the x-component of curl $\mathbf{F}$ is

$$[\mathrm{curl}\,\mathbf{F}]_x(x, y, z) \equiv \lim_{dr \to 0} [\mathrm{curl}\,F_{dr}](f) \qquad (6.56)$$

where f is the x-face closest to (x, y, z) for each dr. As in the case of the divergence, we choose a sequence of x-faces centered on (x, y, z), one of which is shown in Figure 6.38. The discrete curl is defined by Eq. 6.12 as the sum of $F_{dr}(e)/dr$ over the four edges e of the face f, oriented according to the right-hand rule (counterclockwise in this case). Approximating these by their values at the edge centers (Eq. 6.37), the top edge contributes $F_z(x, y + \frac{1}{2}dr, z)/dr$. The bottom edge yields $-F_z(x, y - \frac{1}{2}dr, z)/dr$ (the minus sign appears because the orientation of the bottom edge is opposite to the direction of increasing z assumed in Eq. 6.37 to be the edge orientation). The curl becomes

$$[\mathrm{curl}\,\mathbf{F}]_x(x, y, z) = \lim_{dr \to 0} \Big[F_z\big(x, y + \tfrac{1}{2}dr, z\big) - F_z\big(x, y - \tfrac{1}{2}dr, z\big)$$

$$+ F_y\big(x, y, z - \tfrac{1}{2}dr\big) - F_y\big(x, y, z + \tfrac{1}{2}dr\big)\Big]/dr \qquad (6.57)$$

$$= \frac{dF_z}{dy} - \frac{dF_y}{dz}$$

Similar formulas for the y and z components of curl $\mathbf{F}$ can be obtained by drawing pictures of y and z faces (or more easily, by cyclically permuting the indices in Eq.

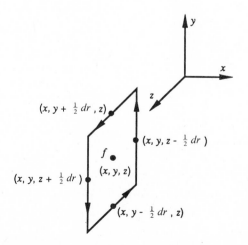

Figure 6.38 An x-face f and its four counter-clockwise edges e.

6.57). The results are

$$[\text{curl}\,\mathbf{F}]_y(x, y, z) = \frac{dF_x}{dz} - \frac{dF_z}{dx} \tag{6.58}$$

$$[\text{curl}\,\mathbf{F}]_z(x, y, z) = \frac{dF_y}{dx} - \frac{dF_x}{dy} \tag{6.59}$$

The curl can be expressed compactly in terms of the ∇ notation by noting that Eqs. 6.57 through 6.59 are the x, y, and z components of the formal vector cross product $(d/dx, d/dy, d/dz) \times \mathbf{F}$, so that

$$\text{curl}\,\mathbf{F} = \nabla \times \mathbf{F} \tag{6.60}$$

[Recall that the x-component of a cross product $\mathbf{G} \times \mathbf{F}$ is defined as $G_y F_z - G_z F_y$, and cyclic permutations give the y and z components].

■ **EXAMPLE 6.4** Compute the curl of the vector function $F_x(x,y,z) = -\frac{1}{2}y, F_y(x,y,z) = \frac{1}{2}x$, $F_z(x,y,z) = 0$ (Figure 6.39). Most of the derivatives in Eqs. 6.57 through 6.59 vanish in this case; only the z-component (Eq. 6.59) is nonzero. It is

$$[\text{curl}\,\mathbf{F}]_z = \frac{d(\frac{1}{2}x)}{dx} - \frac{d(-\frac{1}{2}y)}{dy} = \frac{1}{2} + \frac{1}{2} = 1$$

in accordance with the fact that the arrows in Eq. 6.49 seem to "curl" around the z-axis.

▶ **Problem 6.19**

Consider the vector function $F_x = 0$, $F_y(x, y, z) = cx$, $F_z(x, y, z) = ax^2 + by^2$.

a. Compute the vector function curl $\mathbf{F}$.

b. Compute the discrete edge vector field $F(e)$ defined by $\mathbf{F}$, at the twelve edges of the cube $0 \le x \le dr, 0 \le y \le dr, 0 \le z \le dr$. Use Eq. 6.37, that is, approximate $F(e)$ by evaluating the function at the center of the edge. Compute the discrete curl of this edge field.

c. Use Eq. 6.40 to approximate the face vector field defined by the function curl $\mathbf{F}$ at the six faces of this cube. Need this agree with b in general? Does it in this case?   ◀

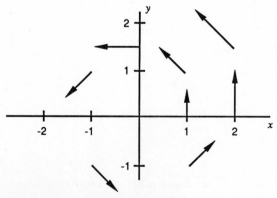

Figure 6.39 The vector function F.

▶ **Problem 6.20**

Compute the exact edge and face fields for the previous problem by integrating **F** over edges and curl **F** over faces (Eqs. 6.36 and 6.39). Which of these differ from the approximations of the previous problem? Will the discrete curl of this exact edge field always agree with the exact face field, for any **F**?   ◄

SECTION 6.8   INTEGRALS OF CONTINUUM FIELDS

We defined various integrals of discrete fields in Section 6.2 and Section 6.5. From these, we can define integrals of continuum fields formally in a straightforward way. Consider first a continuum cell field ρ, which we would like to integrate over a continuum region R. We will refer to the result as a continuum volume integral. Figure 6.40 shows such a region for the 2D analog, an area integral. By definition, ρ is a rule that determines a discrete field ρ_{dr} for any lattice. Each such ρ_{dr} can be integrated, if we can decide what region R_{dr} (a set of cells in the lattice of spacing dr) to integrate it over; the volume integral of ρ can then be defined as a limit

$$\int_R \rho\, d\tau \equiv \lim_{dr \to 0} \sum_{c \in R_{dr}} \rho_{dr}(c)\, d\tau(c) \tag{6.61}$$

where the sum on the right is a discrete integral. The choice of R_{dr} is somewhat arbitrary; a reasonable choice is to include each cell whose center is in the continuum region R, as we have done in Figure 6.40. For reasonable functions ρ and regions R, many possible rules for choosing R_{dr} would give the same answer in the continuum limit. The question is usually academic, because we will almost always choose a lattice in which the cell boundaries fall exactly on the boundary of R, so we can choose $R_{dr} = R$ exactly.

When the continuum field ρ is specified by a function $\rho(x, y, z)$ (as opposed to a Dirac delta function, for example) we may replace $\rho_{dr}(c)$ by the value of $\rho(x, y, z)$ at the center, with an error that vanishes in the limit. This amounts to a 3D analog of the familiar Riemann integral.

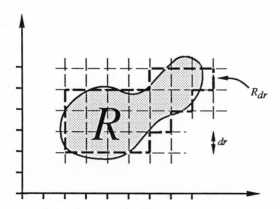

Figure 6.40   A continuum region R in 2D, with an approximation R_{dr}.

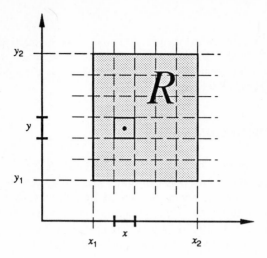

Figure 6.41 A rectangular region in 2D, showing cell (x, y).

For the special case of a **rectangular** region, an important method for evaluation of volume integrals is to transform them to multiple integrals. Consider the 2D integral over the region in Figure 6.41, bounded by the lines $x = x_1$, $x = x_2$, $y = y_1$, and $y = y_2$. If we denote the cell centers by (x, y), then the x's are centers of intervals (1D cells) on the x-axis, and the y's are centers of 1D cells along the y-axis. The sum over 2D cells is a sum over pairs (x, y), which can be expressed as a double sum over x and y; Eq. 6.61 becomes (note that $d\tau = dr^2$ in 2D)

$$\int_R \rho \, d\tau = \lim_{dr \to 0} \sum_{(x, y)} \rho(x, y) \, dr^2$$

$$= \lim_{dr \to 0} \sum_x dr \left[\sum_y \rho(x, y) \, dr \right] \tag{6.62}$$

In the limit, the object in square brackets becomes a 1D integral over y, and the sum over x becomes an integral over x, so

$$\int_R \rho(x, y) \, d\tau = \int_{x_1}^{x_2} dx \left[\int_{y_1}^{y_2} dy \, \rho(x, y) \right] \tag{6.63}$$

The 3D analog for a rectangular solid is

$$\int_R \rho(x, y, z) \, d\tau = \int_{x_1}^{x_2} dx \left\{ \left[\int_{y_1}^{y_2} dy \left[\int_{z_1}^{z_2} dz \, \rho(x, y, z) \right] \right] \right\} \tag{6.64}$$

■ **EXAMPLE 6.5** Compute the 2D integral of $\rho(x, y) = x^2 + y$ over the square region $0 \le x \le a$, $0 \le y \le a$. First calculate the integral over y:

$$\int_0^a dy \, \rho(x, y) = \int_0^a dy (x^2 + y) = x^2 a + \tfrac{1}{2} a^2 \tag{6.65}$$

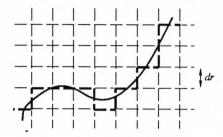

Figure 6.42 A 2D continuum path and a discrete approximation.

Then integrate over x:

$$\int_R \rho(x,y)\,d\tau = \int_0^a dx\,[x^2 a + \tfrac{1}{2}a^2] = \tfrac{1}{3}a^4 + \tfrac{1}{2}a^3 \qquad (6.66)$$

▶ **Problem 6.21**

Compute the area integral of $\rho(x,y) = a + (y^2/a) + (x^2/a)$ over the rectangular region $0 \le x \le a, 0 \le y \le 2a$. ◀

▶ **Problem 6.22**

Compute the area integral of $\rho(x,y) = \sin(x)\sin(y)$ over $0 \le x \le \pi, 0 \le y \le \pi$. ◀

▶ **Problem 6.23**

Compute the area integral of $\rho(x,y) = a + (y^2/a) + x$ over the triangular region $0 \le x \le y, 0 \le y \le a$. *Hint*: express this as a double integral with the inner integral over x; its limits can then depend on y. ◀

We can also define a continuum path integral by approximating an arbitrary path by edges of a lattice (Figure 6.42) and taking the limit of the discrete path integrals; a surface integral can be defined similarly. Then the various theorems we have proved (Stokes's theorem and the divergence theorem) turn into theorems about continuum integrals when we take the $dr \to 0$ limit.

SECTION 6.9 DYNAMICS OF BEDSPRINGS

We now return to the 2D system of springs and masses introduced in Section 6.1. We defined the displacement field $w(r)$, a 2D vertex field giving the transverse (out of the paper in Figure 6.1) displacement of the mass labeled by r, whose equilibrium position is at vertex r. We also defined its discrete gradient, the strain field $s(e)$. To write equations of motion analogous to the 1D ones of Chapter 2, we must of course introduce the velocity field $v(r)$, giving the (transverse) velocity of mass r.

To derive the discrete equations of motion for this system, consider a spring at an edge e pointing outward from a vertex r. Arguments like those of Chapter 1 show that if the entire bedspring is under longitudinal tension (i.e., the springs are initially uniformly stretched), the transverse component of the force exerted by this spring on the mass at r is proportional to the strain $s(e)$; defining a constant K so

Figure 6.43　The four edges touching a vertex r of a 2D lattice; they label the four springs which exert forces on the mass at r.

that the proportionality coefficient is Km/dr,

$$F(\text{of } e \text{ on } r) = (Km/dr)s(e) \tag{6.67}$$

Then we can write the total force $m\,dv/dt$ on the mass at r as

$$m\,dv\,(r, t)/dt = \sum_e F(\text{of } e \text{ on } r) + F_{\text{ext}}(r, t)$$

$$= (Km/dr)\sum_e s(e, t) + F_{\text{ext}}(r, t) \tag{6.68}$$

where the sum is over the four edges diverging outward from r (Figure 6.43), and we have assumed a possible external force F_{ext} (we may be pushing on mass r with a finger, for example). Note that F_{ext} is defined at the same discrete times as w and s, usually integer multiples of dt. The sum over edges resembles the sum (Eq. 6.22) defining the discrete divergence, except that s is defined at edges instead of faces. However, the edges of the original lattice are equivalent to faces of the dual lattice

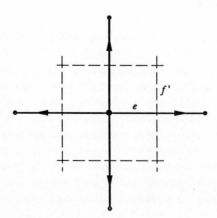

Figure 6.44　Four faces f' of the 2D dual lattice (dotted lines).

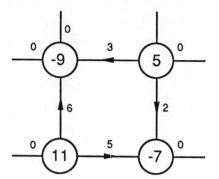

Figure 6.45 Divergence of the edge field of Figure 6.3.

(shown in 3D in Figure 6.28 and in 2D cross-section in Figure 6.44). Note that faces and edges look the same in 2D (both are line segments). We call f' a face only because we imagine it to be a cross-section of a 3D face. This provides a natural way to define the divergence of an edge field; expressed in terms of edges, the divergence (Eq. 6.22) becomes

$$[\operatorname{div} s](r) = dr^{-1} \sum_e s(e) \tag{6.69}$$

where the sum is over the four (six in 3D) outward edges from the vertex r.

As an example, the divergence of the edge field $s(e)$ computed in Figure 6.3, assuming edges at which no value is given have $s(e) = 0$, is shown in Figure 6.45.

▶ **Problem 6.24**

Compute the divergence of the field shown in Figure 6.46, at the four vertices shown. Use $dr = 1$. ◀

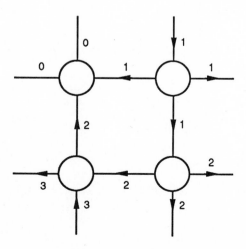

Figure 6.46 Edge field for Problem 6.24.

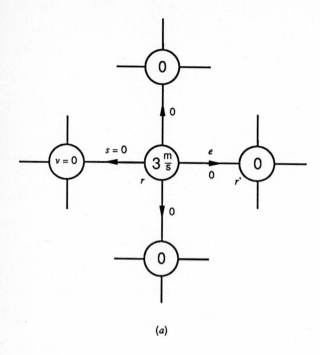

(a)

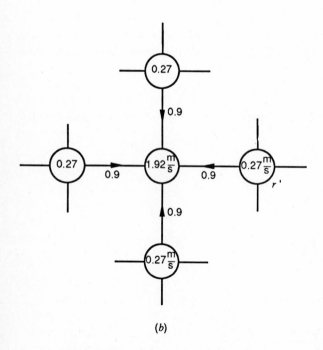

(b)

Figure 6.47   (a) Stress $s(t = 0)$ and velocity $v(t = \frac{1}{2}dt)$. (b) $s(t = dt)$ and $v(t = 1.5\,dt)$.

Thus, we can write the equation of motion for updating the stresses and velocities in a bedspring in terms of the divergence and the gradient:

$$dv/dt = K \operatorname{div} s + m^{-1}F_{\text{ext}} \tag{6.70}$$

$$ds/dt = \operatorname{grad} v \tag{6.71}$$

(The equation for ds/dt is the same as in 1D.) These equations will describe a real system of Hooke's-law springs, in the limit $dt \to 0$.

■ **EXAMPLE 6.6**   Consider an initially unstressed bedspring $[s(e, t = 0) = 0, v(r, t = -dt/2)$ $= 0]$ in which we apply an external force $F_{\text{ext}} = F_0$ to one mass r (the others have $F_{\text{ext}} = 0$), during the first time interval (centered at $t = 0$). Let us use $K = 1$ m^2 s^{-2}, $dt = 0.3$ s, $dr = 1$m, $F_0 = 10$ N, $m = 1$ kg. Then applying Eq. 6.70 at $t = 0$ yields

$$v(r, \tfrac{1}{2}dt) = v(r, -\tfrac{1}{2}dt) + K\,dt\,\operatorname{divs} + m^{-1}F_0\,dt$$

$$= 0 + K\,dt \cdot 0 + 3.0 = 3.0 \text{ m/s}$$

for the central mass r. All the other masses have zero velocity; the velocity field at $t = \tfrac{1}{2}dt$ (and the stress field at $t = 0$) are shown in Figure 6.47a. We can now update the stress s using Eq. 6.71. At each outward edge such as that labeled e in Figure 6.47a, we get

$$s(e, dt) = s(e, 0) + dt\,[0 - 3.0 \text{ m/s}]/dr = -0.9$$

This is shown in Figure 6.47b; note that the edge orientation shown there is $-e$, in order that the number written $[s(-e) = -s(e)]$ can be positive. Also shown on this figure is the velocity field at time 1.5dt; at the central vertex this has changed by $dv = K\,dt\,\operatorname{divs} = 0.3\Sigma s(e) = 0.3(-0.9 - 0.9 - 0.9 - 0.9) = -1.08$ m/s, leaving $v(r, 1.5\,dt) = v(r, 0.5\,dt) + dv = 3.0 - 1.08 = 1.92$ m/s. At the four outer vertices (labeled r' in Figure 6.47b) we must be careful about signs: div$s$ is a sum over the **outward** edges at this vertex, only one of which has nonzero s. This outward edge is $-e$ in the notation of Figure 6.47, and has $s(-e) = +0.9$; thus $v(1.5, r') = 0 + 0.9\,dt = 0.27$ m/s.

▶   **Problem 6.25**

Compute all nonzero values of $s(t = 2)$ and $v(t = 2.5)$ for this system, and indicate the results on a figure similar to Figure 6.47b.   ◀

We can consider the continuum limit of the 2D bedspring problem just as we did in Section 3.3 for a 1D chain of masses and springs. As dr gets smaller, the masses get closer together and each line of masses in the bedspring becomes more like an elastic string; the 2D web of elastic springs then resembles a trampoline. We can describe the transverse displacement and velocity by continuum vertex fields in two dimensions just as we did in one dimension. The strain becomes a 2D continuum edge field. If we also let $dt \to 0$, and all of the fields are described by (differentiable) functions, Eqs. 6.70 and 6.71 become partial differential equations for a vector function $\mathbf{s}(\mathbf{r})$ and a scalar function $v(\mathbf{r})$. We will discuss solutions of these equations in Chapter 8.

CHAPTER SEVEN

CURVILINEAR COORDINATES

SECTION 7.1 LATTICES AND FIELDS IN CURVILINEAR COORDINATES

The concepts of discrete fields we have introduced are not restricted to lattices of cubes. Space can be divided into cells in many ways, as shown in Figure 7.1. In a diffusive system, for example, one can then define a discrete density field in terms of the number of objects in each cell, and a discrete current in terms of the number that cross each face. However, we usually prefer to be more systematic and define the cells in terms of a consistent set of coordinates. We can think of our cubical cells as being defined in terms of cartesian coordinates x, y, z: the cell with one corner at the point (x_0, y_0, z_0) consists of all points (x, y, z) having $x_0 \leq x \leq x_0 + dx$, $y_0 \leq y \leq y_0 + dy$, $z_0 \leq z \leq z_0 + dz$ (Figure 7.2). We will use dx, dy, and dz to refer to the linear dimensions of the cells; for our cubical cells these were all equal to dr.

This way of describing a cell generalizes to other coordinate systems. In cylindrical coordinates, each point in space is described by coordinates r, ϕ, and z (Figure 7.3). Here r is the distance to the z-axis. (**Not** to the origin! We will define r as the distance to the origin in **spherical** coordinates.) The angle ϕ is measured about the z-axis, starting at the x-axis and increasing toward the y-axis, and z has the same meaning as in cartesian coordinates. We will define a cell as the set of points with $r_0 \leq r \leq r_0 + dr$, $\phi_0 \leq \phi \leq \phi_0 + d\phi$, $z_0 \leq z \leq z_0 + dz$, as shown in Figure 7.4. Four of the faces of this cell are flat, and two are curved cylindrical surfaces, of radii r_0 and $r_0 + dr$ respectively. Of the 12 edges, 8 are straight and 4 circular. We will often label a cell by the coordinates of its "center," defined as $(r_0 + \frac{1}{2}dr, \phi_0 + \frac{1}{2}d\phi, z_0 + \frac{1}{2}dz)$. In cartesian coordinates this coincides with the center of mass, but this is not true in general.

We will usually assume dr, $d\phi$, and dz are the same for each cell, and start r_0, ϕ_0, and z_0 off at zero, so the vertices have $r_0 = 0, dr, 2\,dr, \ldots$, $\phi_0 = 0, d\phi, 2\,d\phi, \ldots$,

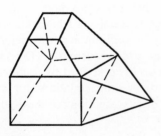

Figure 7.1 A cell in a general lattice.

128

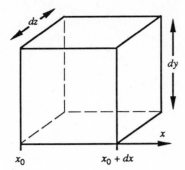

Figure 7.2   A cell in a Cartesian lattice.

$z_0 = 0, dz, 2\,dz, \ldots$. As long as 2π is an integer multiple of $d\phi$, any choice of dr, $d\phi$, and dz then defines a unique subdivision of space into cells (Figure 7.5), just as any choice of dr did for cubical cells.

We can define cells in spherical coordinates (Figure 7.6) in a similar way. The position of a point is specified by giving r, which now denotes its distance from the origin, and two angles describing the direction of the straight line connecting it to the origin. The angle the line makes with the z-axis is the co-latitude, denoted by θ. The plane formed by this line and the z-axis is at an angle ϕ (the azimuthal angle) from the x-axis; this coincides with the meaning of ϕ in cylindrical coordinates. If we choose cell dimensions dr, $d\theta$, and $d\phi$ so that 2π is a multiple of $d\phi$ and π an integer multiple of $d\theta$ (note that θ ranges from 0 to π) this defines a subdivision of space into cells such as that in Figure 7.7.

We can define discrete fields in curvilinear coordinates just as in cartesian coordinates. A discrete cell field, for example, assigns a value $\rho(c)$ to each cell c. If

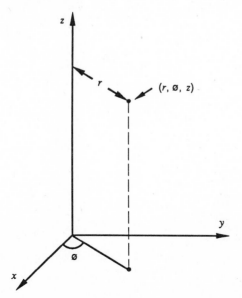

Figure 7.3   Cylindrical coordinates.

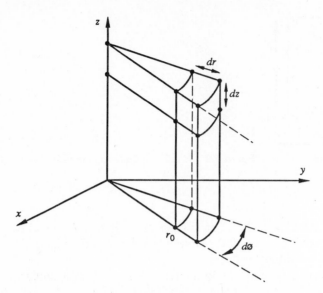

Figure 7.4   A cell in cylindrical coordinates.

ρ is a density field, $\rho(c)$ can again be defined as the number of objects in the cell divided by its volume. In curvilinear coordinates, however, the volume is not the same for each cell (and equal to dr^3) as it was for cubical cells. We will denote the volume of a cell in general by $d\tau(c)$. Thus, for cubical cells $d\tau(c) = dr^3$ for all c; in cylindrical coordinates $d\tau = \frac{1}{2}d\phi[(r + \frac{1}{2}dr)^2 - (r - \frac{1}{2}dr)^2)]\,dz$ for a cell whose center is at (r, ϕ, z). To derive this, note that $2\pi/d\phi$ cells make up an annular cylinder whose volume is dz times the area of an annulus (Figure 7.8), $\pi[(r + \frac{1}{2}dr)^2 - (r - \frac{1}{2}dr)^2]$.

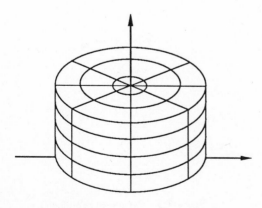

Figure 7.5   Cells in cylindrical coordinates, for the case $\phi = \pi/4$.

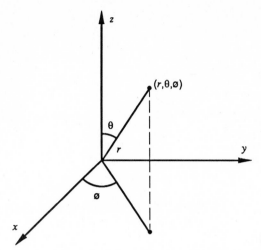

Figure 7.6   Spherical coordinates.

The discrete integral $\sum\rho(c)\,d\tau(c)$ we defined in cartesian coordinates generalizes immediately to an arbitrary coordinate system: it gives the total population of the region (a set of cells) that is summed over.

Evidently, we can also generalize the notions of vertex, edge, and face fields. A discrete vertex field for an arbitrary lattice is just a function W assigning a value $W(v)$ to each vertex v. An edge field F assigns a value $F(e)$ to each oriented edge e, and a face field assigns a value to each oriented face. As an example, we would like

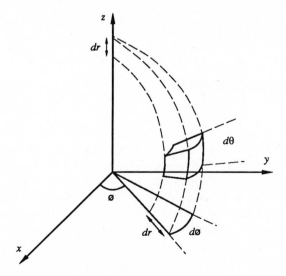

Figure 7.7   A cell in spherical coordinates.

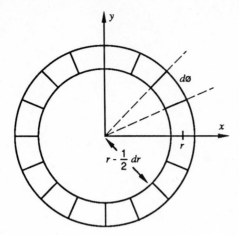

Figure 7.8 Annulus of $2\pi/d\phi$ faces (or top view of annular cylinder of cells).

to be able to define a current field $j(f)$ in terms of the number of objects that cross that face during a time interval of length dt. If we denote the area of the (possibly curved) face by $da(f)$, we want

$$j(f) \equiv (\text{no. of objects crossing})/dt\, da(f) \qquad (7.1)$$

Then the number of objects per unit time crossing a surface S composed of many faces is the discrete surface integral

$$\sum_{f \in S} j(f)\, da(f) \qquad (7.2)$$

In the next section we will define operations that relate these vertex, edge, face, and cell fields in curvilinear coordinates.

SECTION 7.2 DISCRETE GRADIENT, DIVERGENCE, AND CURL

Let us first generalize the discrete gradient from cartesian to curvilinear coordinates. This operation gives an edge field for any vertex field W. In cartesian coordinates, the gradient was defined for each directed edge e (Figure 7.9), as the difference $W(v_{\text{front}}) - W(v_{\text{back}})$ divided by the distance between these vertices. Things will work out more neatly in curvilinear coordinates if we measure this distance **along** the edge rather than in a straight line (it won't matter in the continuum limit because the edges become arbitrarily nearly straight). So we define

$$dl(e) = \text{length of the edge } e \qquad (7.3)$$

which is an analog of the cell volume $d\tau(c)$.

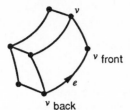

Figure 7.9 Part of a lattice, illustrating the definition of the gradient.

The reason for defining dl **along** the edge is that in a finer cell lattice (Figure 7.10a) in which there is an additional vertex v_{middle} and the edge e is divided into two edges e_1 and e_2, we have $dl(e) = dl(e_1) + dl(e_2)$. This would be false if dl were measured in a straight line (Figure 7.10b).

We can then define the gradient in an arbitrary coordinate system by

$$[\text{grad}\,W](e) = \big[W(v_{\text{front}}) - W(v_{\text{back}})\big]/dl(e) \tag{7.4}$$

If we want to know the change in W between two distant vertices v_0 and v, we need only add the changes $[\text{grad}\,W](e)\,dl(e)$ along a path P (a string of connected edges) joining them, obtaining the discrete path integral in curvilinear coordinates

$$W(v) - W(v_0) = \sum_{e \in P} [\text{grad}\,W](e)\,dl(e) \tag{7.5}$$

We will define the discrete path integral for an arbitrary discrete edge field F by this same expression, that is, $\sum F(e)\,dl(e)$.

Now let us go on to the divergence and curl. Note that we were able to generalize the gradient because we had a coordinate-independent description of it ("the change in W per unit length") that made sense in any coordinate system. As a result, the important integral theorem involving the gradient (Eq. 7.5, that the path integral of grad W gives the change in W), is satisfied **exactly** in any curvilinear coordinate system. Can we do this for the divergence? We defined it as a sum over

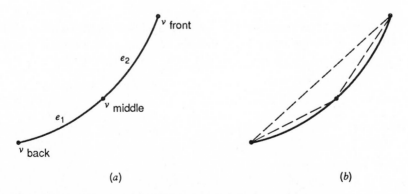

(a) (b)

Figure 7.10 (a) An edge divided into two. (b) Straight-line length.

the faces of a cube

$$[\operatorname{div} j](c) \equiv d\tau(c)^{-1} \sum_{f \text{ of } c} j(f)\, da(f) \qquad (7.6)$$

where the sum is over the six outward-oriented faces of the cube c. In cartesian coordinates, $da/d\tau = dr^2/dr^3 = dr^{-1}$ for every cell; we write it here as $da/d\tau$ so that it is describable in a way that makes no explicit reference to the coordinate system—the last sum is a discrete surface integral, so the divergence is the "surface integral per unit volume." Thus, we can use Eq. 7.6 as the definition of the divergence in **any** coordinate system. Thinking of the surface integral of j as the outward flux of particles from the cell, the divergence is the outward flux per unit volume.

▶   **Problem 7.1**

The most important integral theorem involving the divergence is the divergence theorem, stating that the volume integral of div j is the surface integral of j (Eq. 6.27). Verify that this theorem is exactly true in any curvilinear coordinate system if the divergence is defined by Eq. 7.6. *Hint*: The proof we gave for cartesian coordinates works in any coordinate system, except for the pictures. You can correct this by xeroxing the proof onto a plastic transparency and baking it at 325° for 10 minutes. Alternatively, you can redraw the pictures.   ◀

We will now look for a coordinate-independent interpretation of the curl. In cartesian coordinates it was (Eq. 6.12)

$$[\operatorname{curl} F](f) = dr^{-2} \sum_e F(e)\, dr \qquad (7.7)$$

where the sum is over the four edges e of the face f (Figure 7.11). We included in the sum a factor dr [which is the length $dl(e)$ of the edge] so it could be construed as a path integral around a path consisting of these four edges. Since dr^2 is the area $da(f)$ of the face, we can write

$$[\operatorname{curl} F](f) = da(f)^{-1} \sum_e F(e)\, dl(e) \qquad (7.8)$$

The path integral of an edge field F around a closed curve is often called the "circulation" of F. So the curl can be thought of as the "circulation per unit area" around a small closed curve; this description is not dependent on any particular

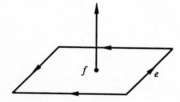

Figure 7.11   The four edges of a face, used in defining the curl.

coordinate system. It is natural, therefore, to **define** the curl in curvilinear coordinates by Eq. 7.8, as the circulation per unit area.

▶   **Problem 7.2**

Verify that the integral theorem involving the curl (Stokes's theorem) is true in any coordinate system (for a hint, see Problem 7.1).   ◀

SECTION 7.3   GRADIENT, CURL, AND DIVERGENCE OF A FUNCTION

Just as in cartesian coordinates, we can define continuum cell, face, edge, and vertex fields in curvilinear coordinates, some of which are related to scalar or vector functions. This is done formally exactly as before: a continuum vertex field is a consistent rule giving a discrete vertex field for each lattice, and so on. One difference is that, whereas in cartesian coordinates the lattice was determined by one number dr (and a choice of origin), in curvilinear coordinates it is determined by three numbers (dr, $d\phi$, and dz in cylindrical coordinates). The continuum limit is the limit in which all of these increments approach zero: $dr \to 0$, $d\phi \to 0$, and $dz \to 0$. We will sometimes denote this limit in a shorthand fashion by "$d \to 0$."

A scalar function in cylindrical coordinates has the form $W(r, \phi, z)$. We can associate it with a continuum vertex field (a rule giving a discrete field W_d for each choice of lattice spacings) by setting $W_d(c) = W(c_r, c_\phi, c_z)$ where c_r, c_ϕ, and c_z are the cylindrical coordinates of the center of the cell.

A vector function $\mathbf{F}$ in cylindrical coordinates is specified by giving the components of $\mathbf{F}$ along the unit vectors $\hat{\mathbf{r}}$, $\hat{\boldsymbol{\phi}}$, and $\hat{\mathbf{z}}$, which are defined to point in the directions in which r, ϕ, and z increase most rapidly (Figure 7.12). So we can write

$$\mathbf{F}(\mathbf{r}) = \mathbf{F}(r, \phi, z) = F_r(r, \phi, z)\hat{\mathbf{r}} + F_\phi(r, \phi, z)\hat{\boldsymbol{\phi}} + F_z(r, \phi, z)\hat{\mathbf{z}} \qquad (7.9)$$

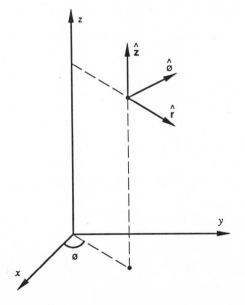

Figure 7.12   The unit vectors in cylindrical coordinates.

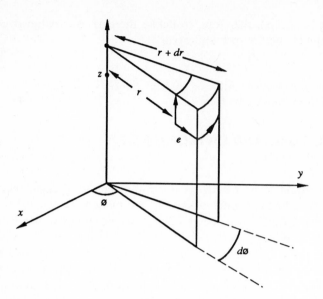

Figure 7.13 A radial edge in cylindrical coordinates.

To associate **F** with a continuum edge field (a rule giving a discrete edge field for any lattice) we must determine $F(e)$ for each edge e. Fortunately, the edges of our lattice are all along the directions $\hat{\mathbf{r}}$, $\hat{\boldsymbol{\phi}}$, and $\hat{\mathbf{z}}$ so we can identify the r-component F_r, for example, with the discrete field $F(e)$ along a radial edge e (Figure 7.13). To construct a consistent discrete edge field $F(e)$ we would have to average F_r over the edge, in analogy with Eq. 6.36. As in cartesian coordinates, however, we can approximate $F(e)$ by the value of F_r at the edge center

$$F(e) \cong F_r\left(r + \tfrac{1}{2}dr, \phi, z\right) \tag{7.10}$$

with an error that vanishes in the continuum limit $d \to 0$. Similarly, for the leftmost z-edge of the cell in Figure 7.13

$$F(e) \cong F_z\left(r, \phi, z + \tfrac{1}{2}dz\right) \tag{7.11}$$

and for the lower right ϕ-edge,

$$F(e) \cong F_z\left(r + dr, \phi + \tfrac{1}{2}d\phi, z\right) \tag{7.12}$$

We can get the vector function **F** back from the discrete fields by taking the $d \to 0$ limit:

$$F_r(r, \phi, z) = \lim_{d \to 0} F(e) \tag{7.13}$$

where the e's are r-edges whose positions approach the point (r, ϕ, z). The limit for ϕ-edges or z-edges gives F_ϕ or F_z.

Note that there is a set of points at which this correspondence with vector functions breaks down: the points with $r = 0$ (i.e., those on the z-axis). There we cannot define a unique $\hat{\mathbf{r}}$ or $\hat{\boldsymbol{\phi}}$ as in Figure 7.12. Even the notion of a **scalar** field breaks down here: a perfectly well-behaved differentiable function of r and ϕ may be discontinuous at $r = 0$ (for example, $\sin \phi$). When we need to work near such a singularity of the coordinate system, we will usually find it best to avoid trying to describe continuum fields by functions.

We can now define the gradient, grad W, of a scalar function W exactly as we did in cartesian coordinates, by taking the gradients of the discrete fields determined by its associated continuum vertex field and recovering a vector function by taking a limit. To determine the r-component of the gradient in cylindrical coordinates, for example, the simplest sequence of r-edges to use are all centered at (r, ϕ, z), with different lengths dr. Then v_{front} (Figure 7.14) is $(r + \frac{1}{2}dr, \phi, z)$ and v_{back} is $(r - \frac{1}{2}dr, \phi, z)$, so we get

$$[\text{grad } W]_r(r, \phi, z) = \lim_{dr \to 0} [\text{grad } W_d](e)$$

$$= \lim_{dr \to 0} \frac{W\left(r + \frac{1}{2}dr, \phi, z\right) - W\left(r - \frac{1}{2}dr, \phi, z\right)}{dr} \qquad (7.14)$$

$$= \frac{dW}{dr}$$

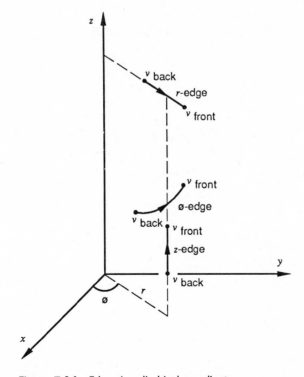

Figure 7.14   Edges in cylindrical coordinates.

by the definition of the derivative. The ϕ component of the gradient is computed by considering a sequence of ϕ-edges such as that shown in Figure 7.14. In this case, the length $dl(e)$ is $r\,d\phi$ (this is basically the definition of the radian measure of $d\phi$; all of the edge lengths are tabulated in Table 7.1). Again denoting the center of the edge by (r, ϕ, z), we have

$$[\text{grad } W]_\phi(r, \phi, z) = \lim_{d\phi \to 0} \frac{W\left(r, \phi + \frac{1}{2}d\phi, z\right) - W\left(r, \phi - \frac{1}{2}d\phi, z\right)}{r\,d\phi} \tag{7.15}$$

$$= r^{-1} \frac{dW}{d\phi}$$

The corresponding limit for z-edges gives

$$[\text{grad } W]_z(r, \phi, z) = dW/dz \tag{7.16}$$

These three formulas for the components of the gradient are listed for quick reference inside the back cover of this book, along with others we will derive below.

The procedure in spherical coordinates is exactly the same as in cylindrical coordinates, but the edge lengths $dl(e)$ are a little more complicated. As shown in Figure 7.15, the radial edge again has length dr, and the edge in the θ direction has length $r\,d\theta$ because it is traced out by the end of a line of length r swinging through an angle $d\theta$. The edge in the ϕ direction has length $r \sin \theta\, d\phi$ because it is traced out by the end of a line from the z-axis of length $r \sin \theta$, swinging through the angle $d\phi$.

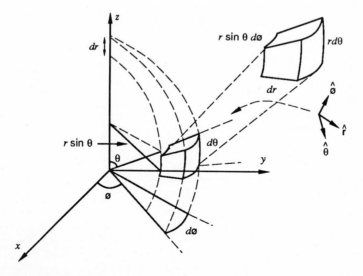

Figure 7.15 Edge lengths in spherical coordinates.

▶   **Problem 7.3**

Derive formulas for the r, θ, and ϕ components (along the unit vectors shown in Figure 7.15) of the gradient of a scalar function $W(r, \theta, \phi)$ in spherical coordinates. Check your answers with the formulas inside the back cover of this book.   ◀

Next we will define the divergence of a vector function $\mathbf{j}(r, \phi, z)$ in cylindrical coordinates. The discrete divergence operates on face fields, so we must first identify the function $\mathbf{j}$ with a continuum face field (i.e., a rule giving a discrete field $j(f)$ for any lattice). As in cartesian coordinates, we get $j(f)$ from the component of $\mathbf{j}$ normal to the face f; each face normal is along one of the unit vectors $\hat{\mathbf{r}}$, $\hat{\boldsymbol{\phi}}$, and $\hat{\mathbf{z}}$ in Figure 7.12. (Note that the unit vectors depend on position: we evaluate them at the face center here.) We can identify j_r with the discrete field $j(f)$ at the face f shown in Figure 7.16, which is normal to the r-direction. One might refer to this face as being "in the ϕz plane," or more precisely "in the ϕz cylinder." In analogy with Eq. 6.39, this discrete field is the integral of j_r over the face. However, with an error that vanishes in the continuum limit, we can evaluate the function $j_r(r, \phi, z)$ at the center of this face [denoted by (r_f, ϕ, z)], giving

$$j(f) = j_r(r_f, \phi, z) \tag{7.17}$$

To compute the divergence of this discrete field, we need to know the area of each face. To compute these, note that both of the curvilinear coordinate systems we are considering are **orthogonal** coordinate systems (i.e., the edges all intersect at right angles), hence the cells are rectangular solids in the $d \to 0$ limit. So, we can calculate the face areas da from the lengths dl of their edges simply by multiplying two together (length × width), and we can find $d\tau$ by multiplying three edge lengths

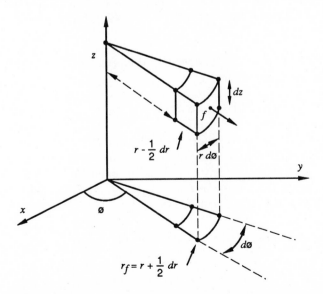

Figure 7.16 A cell for computing the divergence in cylindrical coordinates.

dl. This is correct in the $d \to 0$ limit; more precisely, if you computed the area exactly you might find extra terms with higher powers of $dl(e)$, which are negligible in this limit. The results for all three coordinate systems are shown in Table 7.1.

Table 7.1 *Edge lengths, face areas, and cell volumes in three coordinate systems (to **lowest** order in dx, dr, dϕ, etc.)*

	Coordinate System									
	Cartesian			Cylindrical			Spherical			
Edges										
Axis along e	x	y	z	r	ϕ	z	r		θ	ϕ
Length $dl(e)$	dr	dr	dr	dr	$r\,d\phi$	dz	dr		$r\,d\theta$	$r\sin\theta\,d\phi$
Faces										
Axis normal to f	x	y	z	r	ϕ	z	r		θ	ϕ
(Axes not normal)	(yz)	(xz)	(xy)	(ϕz)	(rz)	$(r\phi)$	$(\theta\phi)$		$(r\phi)$	$(r\theta)$
Area $da(f)$	dr^2	dr^2	dr^2	$rd\phi\,dz$	$dr\,dz$	$r\,drd\phi$	$r^2\sin\theta\,d\theta\,d\phi$		$r\sin\theta\,dr\,d\phi$	$rd\theta\,dr$
Cells										
Volume $d\tau(c)$		dr^3			$r\,dr\,d\phi\,dz$			$r^2\sin\phi\,dr\,d\phi\,d\theta$		

▶ **Problem 7.4**

a. Show that there are no such extra terms in cylindrical coordinates, nor for ϕ-faces or θ-faces in spherical coordinates (i.e., that these formulas in Table 7.1 are exact).

b. Calculate exact formulas in spherical coordinates for da at an r-face [in terms of the face center coordinates (r, ϕ, z)], and for $d\tau$ in terms of the cell center coordinates. Show that they reduce to the formulas of Table 7.1 in the $d \to 0$ limit, and to the proper values in the opposite limit $\theta = \pi/2$, $d\theta = \pi$, $d\phi = 2\pi$, $r = dr$ (for $d\tau$) and $dr/2$ (for da). ◀

Returning to the computation of the divergence, the area $da(f)$ of the face in Figure 7.16 is $r_f\,d\phi\,dz$, so it contributes

$$j(f)\,da(f) = r_f\,d\phi\,dz\,j_r\big(r_f, \phi, z\big) \tag{7.18}$$

to the surface integral in the divergence. For the opposite face [centered at $(r - \frac{1}{2}dr, \phi, z)$], the outward normal from the cell points **oppositely** to the unit vector $\hat{\mathbf{r}}$, which is the conventional orientation used in defining j_r. Denoting this outward-oriented face by f', this means $j(f') = -j_r(r - \frac{1}{2}dr, \phi, z)$. Thus the contribution of f' to the divergence has a minus sign, and we have

$$[\mathrm{div}\,j](r, \phi, z) = \lim_{d\to 0} d\tau(c)^{-1}\Big[\big(r + \tfrac{1}{2}dr\big)\,dz\,d\phi\,j_r\big(r + \tfrac{1}{2}dr, \phi, z\big)$$

$$-\big(r - \tfrac{1}{2}dr\big)\,dz\,d\phi\,j_r\big(r - \tfrac{1}{2}dr, \phi, z\big)\Big] \tag{7.19}$$

$+$ terms from other faces

Using $d\tau = r\,dr\,d\phi\,dz$ from Table 7.1, the limit above becomes

$$\lim r^{-1}\left[\left(r + \tfrac{1}{2}dr\right) j_r\left(r + \tfrac{1}{2}dr, \phi, z\right) - \left(r - \tfrac{1}{2}dr\right) j_r\left(r - \tfrac{1}{2}dr, \phi, z\right)\right]/dr \tag{7.20}$$

$$= r^{-1}d\left[rj_r\right]/dr$$

involving the partial derivative with respect to r of the function $rj_r(r, \phi, z)$. We will not write down the calculations for the other faces in detail; they are a little easier in that $da(f)$ is the same for both faces normal to the ϕ-direction, and for both faces normal to the z-direction. One just has to plug $d\tau(c)$ and $da(f)$ from Table 7.1 into the divergence (Eq. 7.6). The result is the divergence of a vector function j in cylindrical coordinates:

$$[\operatorname{div}j](r, \phi, z) = r^{-1}\frac{d}{dr}(rj_r) + r^{-1}\frac{dj_\phi}{d\phi} + \frac{dj_z}{dz} \tag{7.21}$$

▶   **Problem 7.5**

Check the j_ϕ and j_z terms in Eq. 7.21 by evaluating the limits explicitly.   ◀

An exactly similar computation of the divergence in spherical coordinates, using Figure 7.15, gives

$$[\operatorname{div}j](r, \theta, \phi) = \lim_{d\to 0} d\tau(c)^{-1}\sum_f j(f)\,da(f)$$

$$= \lim_{d\to 0}\left[r^2 \sin\theta\,dr\,d\theta\,d\phi\right]^{-1}$$

$$\times\Big[\left\{ j_r(r', \theta, \phi)r'^2 \sin\theta\,d\theta\,d\phi\right\}_{r+dr/2}$$

$$- \left\{ j_r(r', \theta, \phi)r'^2 \sin\theta\,d\theta\,d\phi\right\}_{r-dr/2} \tag{7.22}$$

$$+ \left\{ j_\theta(r, \theta', \phi)r \sin\theta'\,dr\,d\phi\right\}_{\theta+d\theta/2}$$

$$- \left\{ j_\theta(r, \theta', \phi)r \sin\theta'\,dr\,d\phi\right\}_{\theta-d\theta/2}$$

$$+ \left\{ j_\phi(r, \theta, \phi')r\,d\theta\,dr\right\}_{\phi+d\phi/2}$$

$$- \left\{ j_\phi(r, \theta, \phi')r\,d\theta\,dr\right\}_{\phi-d\phi/2}\Big]$$

where the subscript $r \pm \tfrac{1}{2}dr$ after the curly bracket indicates that the expression is to be evaluated at $r' = r \pm \frac{1}{2} dr$; and similarly for θ and ϕ. Thus

$$
[\operatorname{div}\mathbf{j}](r, \theta, \phi)
$$

$$
= \lim r^{-2} \Big[\{ r'^2 j_r(r', \theta, \phi) \}_{r+dr/2} - \{ r'^2 j_r(r', \theta, \phi) \}_{r-dr/2} \Big] / dr
$$

$$
+ \lim [r \sin \theta]^{-1} \Big[\{ \sin \theta' j_\theta(r, \theta', \phi) \}_{\theta + d\theta/2}
$$

$$
- \{ \sin \theta' j_\theta(r, \theta', \phi) \}_{\theta - d\theta/2} \Big] / d\theta \qquad (7.23)
$$

$$
+ \lim [r \sin \theta]^{-1} \Big[\{ j_\phi(r, \theta, \phi') \}_{\phi + d\phi/2} - \{ j_\phi(r, \theta, \phi') \}_{\phi - d\phi/2} \Big] / d\phi
$$

$$
= r^{-2} \frac{d}{dr} (r^2 j_r) + (r \sin \theta)^{-1} \frac{d}{d\theta} (\sin \theta j_\theta) + (r \sin \theta)^{-1} \frac{d}{d\phi} j_\phi
$$

which is the correct formula in spherical coordinates.

Finally, we will define the curl of a vector function $\mathbf{F(r)}$, which is associated with a continuum edge field. Its curl is associated with a continuum face field; each component must be evaluated as a limit of the discrete curl (Eq. 7.8) at a sequence of faces. For the calculation of the r-component of the curl in cylindrical coordinates, such a face f is shown in Figure 7.17. It is centered at (r, ϕ, z) and has area

$da(f) = r\, d\phi\, dz$, so

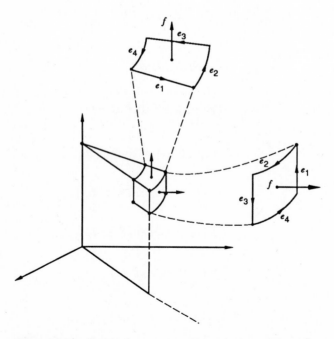

Figure 7.17 Detail of an r^- face (right) and a z-face (top) in cylindrical coordinates, for computation of the curl.

$[\operatorname{curl}\mathbf{F}]_r(r,\phi,z)$

$$= \lim_{d\to 0}[r\,d\phi\,dz]^{-1}\Big[\{F_z\,dz\}_{e_1} - \{F_z\,dz\}_{e_3} + \{F_\phi r\,d\phi\}_{e_4} - \{F_\phi r\,d\phi\}_{e_2}\Big]$$

$$= \lim r^{-1}\big[F_z(r,\phi+\tfrac{1}{2}d\phi,z) - F_z(r,\phi-\tfrac{1}{2}d\phi,z)\big]/d\phi \qquad (7.24)$$

$$-\lim\big[F_\phi(r,\phi,z+\tfrac{1}{2}dz) - F_\phi(r,\phi,z-\tfrac{1}{2}dz)\big]/dz$$

$$= r^{-1}\frac{dF_z}{d\phi} - \frac{dF_\phi}{dz}$$

For a face normal to $\hat{\mathbf{z}}$ (Figure 7.17), we get

$[\operatorname{curl}\mathbf{F}]_z(r,\phi,z)$

$$= \lim_{d\to 0}[r\,dr\,d\phi]^{-1}\Big[\{F_r\,dr\}_{e_1} - \{F_r\,dr\}_{e_3} + \{F_\phi r\,d\phi\}_{e_2} - \{F_\phi r\,d\phi\}_{e_4}\Big] \qquad (7.25)$$

$$= r^{-1}\frac{d}{dr}(rF_\phi) - r^{-1}\frac{d}{d\phi}F_r$$

For a face normal to $\hat{\boldsymbol{\phi}}$, we get

$$[\operatorname{curl}\mathbf{F}]_\phi = \frac{d}{dz}F_r - \frac{d}{dr}F_z \qquad (7.26)$$

▶   **Problem 7.6**
Verify Eq. 7.26 in the manner of Eq. 7.24.   ◀

▶   **Problem 7.7**
Compute the corresponding limits to get a formula for the curl of a function in spherical coordinates. Check your answer with the expression inside the back cover of this book. ◀

CHAPTER EIGHT

SYMMETRY

SECTION 8.1   SYMMETRY IN DYNAMICS

In this chapter we will discuss the role of **symmetry** in dynamical systems. The essential fact we want to establish is that if a particular symmetry property is possessed by the evolution equations of a dynamical system and by its initial conditions, then the future evolution of the system has that symmetry as well. We will find this result very useful in calculating solutions to electrodynamic and electrostatic problems. However, we must first define what we mean when we say a system or its evolution equations have a symmetry property.

The simplest example of a symmetry is one we encountered (but whose symmetry escaped our notice) when we discussed 1D diffusive systems. For example, the diffusing population field in Table 4.4 has a left–right mirror symmetry about its center, resulting from the fact that the initial population had this symmetry. We can prove that this will happen in general by using the mathematical induction method. To do this, we need only show that the evolution equations "preserve the symmetry": that is, **if** the system has the mirror symmetry at a time t and earlier times, **then** it has it at $t + \frac{1}{2}dt$. Then we are done. The symmetry of the initial condition implies that the system has symmetry at all later times: symmetry at $t = 0$ implies symmetry at $t = \frac{1}{2}dt$, which implies symmetry at $t = dt$, and so on.

In general, when we speak of a **symmetry** in physics, we have in mind a particular operation that we can do on a system (or more precisely, on the fields that describe a system), a symmetry operation, which doesn't change the fields. The particular operation we are dealing with here (reflection in a mirror perpendicular to the x-axis and located at $x = 0$) is one that takes the coordinate x into $-x$. As an operation on a density field, it gives a new field whose value at x is the value of the old field at $-x$. This leaves a density field ρ unchanged if

$$\rho(-x) = \rho(x) \tag{8.1}$$

for all x. This equation is what we mean when we say "ρ has reflection symmetry." [In the example of Table 4.4, the mirror was at $x = 2\,dr$, but we could easily redefine the origin so it was at zero.] To prove that the evolution equation giving the current j preserves the symmetry (i.e., that j has it at $t + \frac{1}{2}dt$ if ρ has it at t) we have to define what it means for j to have this reflection symmetry. Clearly it does **not** mean $j(-x, t) = j(x, t)$, as a glance at Table 4.4 will convince you. In fact, the

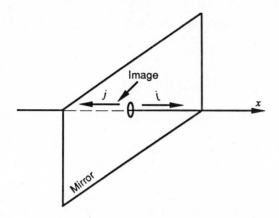

Figure 8.1   A mirror at $x = 0$, reflecting the current direction.

currents there have opposite signs on the left and right: they satisfy

$$j(-x, t) = -j(x, t) \tag{8.2}$$

for all x. You can understand this intuitively by noting that a rightward current, seen in our mirror, looks like a leftward current (Figure 8.1). So let us take Eq. 8.2 as our definition of symmetry for j, and ask whether it is satisfied at $t + \frac{1}{2}dt$ if ρ has symmetry at t. The current at $-x$ is determined by Eq. 4.33 as

$$j\left(-x, t + \tfrac{1}{2}dt\right) = -D[\text{grad } \rho](-x, t)$$

$$= -D\left[\rho\left(-x + \tfrac{1}{2}dr, t\right) - \rho\left(-x - \tfrac{1}{2}dr, t\right)\right] \tag{8.3}$$

Using the assumption (Eq. 8.1) of symmetry at time t, this becomes

$$j\left(-x, t + \tfrac{1}{2}dt\right) = -D\left[\rho\left(+x - \tfrac{1}{2}dr, t\right) - \rho\left(+x + \tfrac{1}{2}dr, t\right)\right] = -j\left(x, t + \tfrac{1}{2}dt\right) \tag{8.4}$$

using Eq. 4.33 for j at x. We have thus proved that the equation for j preserves the symmetry. We need also prove that the equation for updating ρ does. The new ρ is given by $\rho(x, t + \frac{1}{2}dt) = \rho(x, t - \frac{1}{2}dt) + dt(d\rho/dt)(x, t)$, and the first term was assumed to have symmetry, so we need only show that $d\rho/dt$ has the correct symmetry. However, Eq. 4.32 yields

$$[d\rho/dt](-x, t) = -[\text{div } j](-x, t)$$

$$= -\left[j\left(-x + \tfrac{1}{2}dr, t\right) - j\left(-x - \tfrac{1}{2}dr, t\right)\right]/dr \tag{8.5}$$

which, because of the assumed symmetry of j (Eq. 8.2), is

$$= -\left[-j\left(+x - \tfrac{1}{2}dr, t \right) - -j\left(+x + \tfrac{1}{2}dr, t \right) \right]/dr$$

$$= -[\operatorname{div} j](+x, t) = [d\rho/dt](+x, t) \tag{8.6}$$

This completes the proof that the evolution equations preserve the reflection symmetry.

Although it is useful to prove these symmetry relations once in detail as we have done, there is a shortcut we can use to formally determine the proper sign in Eq. 8.2. Note that ρ is even under the reflection operation $x \to -x$ (i.e., its sign doesn't change.) Then $j = -D \operatorname{grad} \rho = -D \, d\rho/dx$ is the ratio of something even (ρ) to something odd (x), and j is therefore odd (meaning that it satisfies Eq. 8.2). This conclusion is consistent with $d\rho/dt = -\operatorname{div} j = -dj/dx$, because dj/dx is the ratio of two odd quantities j and x and is therefore even, as we know ρ should be. This evenness or oddness property is called **parity** in physics; ρ is said to have even parity under the reflection operation; and j is said to have odd parity.

Let us now look at a slightly more complicated example involving a 2D diffusive system. Suppose the initial density is as shown in Figure 8.2a. It is unchanged under reflection in a mirror along the y-axis, that is, under the same

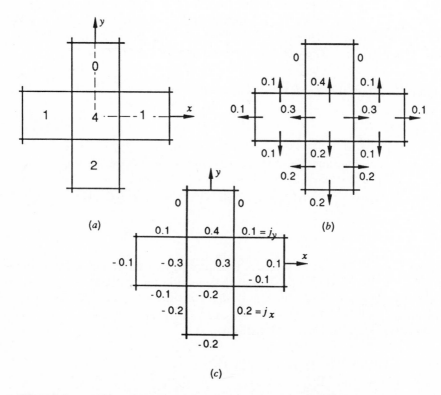

Figure 8.2 (a) The initial density in a 2D system. (b) The resulting current at $t = dt/2$. (c) The current components j_x, j_y at $t = dt/2$.

operation $x \rightarrow -x$ we discussed in 1D. Thus

$$\rho(-x, y, t) = \rho(x, y, t) \tag{8.7}$$

for $t = 0$ and arbitrary x, y. [Note that ρ does **not** have symmetry under $y \rightarrow -y$.] To see the symmetry of the current, let us use $j = -D \operatorname{grad} \rho$ (Eq. 6.29 in 2D) to calculate it at $t = \frac{1}{2}dt$. Using $dt = 1$, $D = 0.1$, and $dr = 1$, we get the results in Figure 8.2b—the particles flow out of the high-density cell. To see what the components j_x and j_y are, it is better to draw a figure in which all the faces are oriented in the conventional way (right and up, as in Figure 8.2c). From this figure it is clear that the symmetry condition for the x-component is the same as in one dimension:

$$j_x(-x, y, t) = -j_x(x, y, t) \tag{8.8}$$

(We are using j_x here to denote the value of the discrete face field j at an x-face.) But the symmetry of j_y is different:

$$j_y(-x, y, t) = +j_y(x, y, t) \tag{8.9}$$

So the complete condition for ρ and j to have x-reflection symmetry is given by Eqs. 8.7 through 8.9. Note that we could have guessed these from parity considerations as we did in one dimension: $j_y = -D \, d\rho/dy$, the ratio of two quantities that are even under x-reflection, so j_y is even.

▶ **Problem 8.1**

    a. Given the evolution equation (from Eq. 6.29)

$$j_x(x, y, t + \tfrac{1}{2}dt) = -D\big[\rho(x + \tfrac{1}{2}dr, y, t) - \rho(x - \tfrac{1}{2}dr, y, t)\big]/dr$$

    and the analogous equation for j_y, verify the numbers in Figure 8.4.

    b. Prove that if ρ has the x-reflection symmetry (Eq. 8.7) at time t, then j_x and j_y have this symmetry (Eqs. 8.8 and 8.9) at time $t + \frac{1}{2}dt$. ◀

▶ **Problem 8.2**

    Write the evolution equation for ρ (Eq. 6.23) for this system, and prove that if j has the x-reflection symmetry at time t and ρ has it at $t - \frac{1}{2}dt$, then ρ has it at $t + \frac{1}{2}dt$. ◀

▶ **Problem 8.3**

    Calculate ρ at $t = 1$ and j at $t = 1.5$ (provide sketches such as Figures 8.2a, b, c). Verify that they have the correct symmetry. ◀

▶ **Problem 8.4**

    Figure 8.3 shows a 2D diffusive system with **two** reflection symmetries, in mirrors at $x = 0$ and $y = 0$ respectively—the initial density field is unchanged by the operation $x \rightarrow -x$, and also by $y \rightarrow -y$.

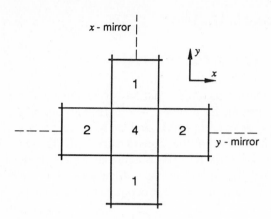

Figure 8.3 An initial density with two mirror symmetries.

a. Write the quantitative condition for ρ to have the y-reflection symmetry (i.e., an equation for $\rho(x, -y)$ analogous to Eq. 8.7). Write equations analogous to Eqs. 8.8 and 8.9 that state that j has y-reflection symmetry.

b. Compute the evolution of this system up to $t = 1.5 \, dt$ (use $dr = 1$, $dt = 1$, and $D = 0.1$). Check that both reflection symmetries are preserved numerically. ◄

Another useful type of symmetry is **translational** symmetry. Consider a 3D diffusive system in which the initial density ρ is given by an infinite stack of copies of Figure 8.2. That is, the density is independent of the z coordinate. If the whole system is moved along the z-axis (by a multiple of dr, of course, so the lattice is preserved) the density is unchanged. The density field is said to be **invariant** under the operation of translation along the z-axis. We can write the condition for this symmetry as

$$\rho(x, y, z + n \, dr) = \rho(x, y, z) \tag{8.10}$$

(for any integer n). When we compute the current at time $\frac{1}{2} dt$, clearly we will get the values shown in the two-dimensional Figure 8.2c for the x and y faces, independently of z; the component of grad ρ that determines them involves only cells with the same z (i.e., in the same layer of the stack) and looks just like the 2D formula. Thus

$$j_x(x, y, z + n \, dr) = j_x(x, y, z) \tag{8.11}$$

$$j_y(x, y, z + n \, dr) = j_y(x, y, z) \tag{8.12}$$

At a z-face centered at (x, y, z) the cells in front and in back of the faces are at $(x, y, z \pm \frac{1}{2} dr)$. The difference (and hence the gradient) vanishes for this face because ρ is independent of z. Thus, the current $j_z(x, y, z)$ is exactly zero. Let us guess that this will remain true, and define the conditions for translational symmetry of ρ and j to be Eqs. 8.10 through 8.12 and

$$j_z(x, y, z) = 0 \tag{8.13}$$

As in the previous cases, we can prove that a system whose initial density has translational symmetry will always have it, by mathematical induction: we need only show that each evolution equation preserves it. The above paragraph proves it for $j = -D$ grad ρ. We need also show that if j has this symmetry, $d\rho/dt = -$ div j has it as well. But of the six faces summed over in the divergence, the two z-faces have $j = 0$, and j_x and j_y at the other four faces are independent of z (Eqs. 8.11 and 8.12). Thus $d\rho/dt$ is independent of z, which we needed to show. Therefore, the 3D density field will remain translationally symmetric forever. [A trivial extension of the arguments used in 1D and 2D shows that it will retain the x-reflection symmetry as well.]

Notice that this translationally symmetric 3D system obeys essentially 2D equations; none of the fields depends on z, so we can essentially forget about the z-coordinate. We will refer to such a translationally symmetric system as a **pseudo-two-dimensional** system. We will find such systems very useful in electrodynamics; one advantage is that we can depict such systems easily on 2D paper (as in Figure 8.2) or 2D computer monitors.

Consider now a diffusive system with a **rotational** symmetry, for example a 3D system of diffusing objects, with the objects initially spread along the z-axis (Figure 8.4). This initial density is unchanged by rotation through any angle about the z-axis. We could compute its evolution using cartesian coordinates as in the previous case—the initial discrete density field is then similar to Figure 8.3, except that the density is nonzero only for the central cell of each layer. However, the result clearly wouldn't be invariant under rotations (except rotations by $\pi/2$), because the cartesian coordinate system itself is not invariant. To treat this symmetry properly, we need a coordinate system with the same symmetry as the physical system. This is, of course, a cylindrical coordinate system. Our problem is a pseudo-2D one (the initial condition, and therefore all later fields, are independent of z) so we need only draw the xy-plane of the coordinate system (i.e., a 2D polar coordinate system), as shown in Figure 8.5. This is a plane through cell centers of the 3D cylindrical lattice, so the radial lines in the figure are cross sections of ϕ-faces, and the circular segments are cross sections of r-faces.

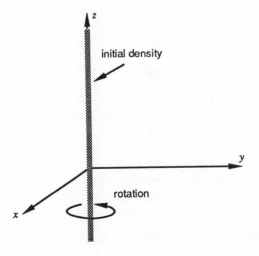

Figure 8.4   A density with rotational symmetry.

(a) ρ at *t* = 0

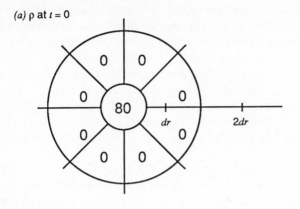

(b) j at t = *dt* / 2   *(c)* ρ at *t* = *dt*

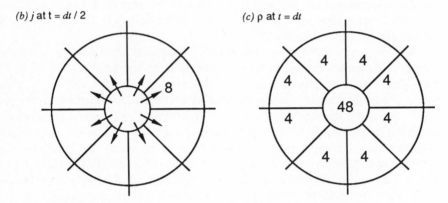

Figure 8.5   The evolution of a rotationally symmetric initial condition.

Note that this cell lattice is different from the ones we drew in Section 7.1, in that it has vertices at $0.5\,dr, 1.5\,dr, \ldots$, instead of at $0, dr, 2\,dr, \ldots$. We do this so that the origin will be a cell center instead of a vertex, so we can start with all the particles there. All of the equations we derived in Chapter 7 work just as well for this lattice, which is in fact exactly the dual of the previous lattice (i.e., vertices and cells have exchanged roles.)

In this coordinate system, we can denote the density field (which is independent of z) by $\rho(r, \phi, t)$. For now this is a discrete field, but we can use the same notation in the continuum limit for a scalar function. We can denote the components of the current field by $j_r(r, \phi, t)$ and $j_\phi(r, \phi, t)$ (for now, discrete fields defined at r-faces and ϕ-faces, respectively).

Let us begin by using the discrete evolution equations to evolve this system. We will use the initial condition shown in Figure 8.5*a*; there are 80 particles in the central cell and none in the others. We have used $d\phi = \frac{1}{4}\pi$ in the figure; we will take $dr = 1$, $dt = 1$, $D = 0.1$, and $dz = 1$ for simplicity. Then we can compute the current density at each of the r-faces at $r = \frac{1}{2}dr$, at time $\frac{1}{2}dt$:

$$j_r\left(\tfrac{1}{2}dr, \tfrac{1}{2}dt\right) = -D\,\text{grad}\,\rho\left(\tfrac{1}{2}dr, 0\right) = -D\left[\rho(dr, 0) - \rho(0, 0)\right]/dr = +8 \quad (8.14)$$

as shown in Figure 8.5*b*. The current density vanishes at all other faces, because $\rho = 0$ on both sides. At this point we can make an observation about the symmetry of j_r: it is independent of ϕ.

$$j_r(r, \phi) = j_r(r) \qquad (8.15)$$

Now we will calculate the density at time dt. The cell at $r = dr$ has volume $d\tau = r\, d\phi\, dr\, dz = \frac{1}{4}\pi$ (from Table 7.1), and only one of the faces contributing to the divergence has nonzero current. This is at $r = dr/2$, and has $j(f) = -j_r(\frac{1}{2}dr) = -8$ (the minus sign is because $\hat{\mathbf{r}}$ points **inward** to this cell). Its area is $r\, d\phi\, dz = (\frac{1}{2}dr)(\frac{1}{4}\pi) = \frac{1}{8}\pi$; so

$$\frac{d\rho\left(r, \frac{1}{2}dt\right)}{dt} = -\operatorname{div} j = -d\tau^{-1}\Sigma j(f)\, da = -\left(\frac{\pi}{4}\right)^{-1}(-8)\left(\frac{\pi}{8}\right) = +4$$

Adding $dt = 1$ times this to the old ρ (which was zero) gives a density $+4$, as shown in Figure 8.5*c*.

The central cell ($r = 0$) is a special case, in that it has more than four faces. But the same formula works, as long as we sum over all eight faces. The sum is eight times the $j(f)\, da$ we found above, and has a $+$ sign because $\hat{\mathbf{r}}$ is **outward** here:

$$\left(\frac{d\rho}{dt}\right)\left(r = 0,\ t = \frac{1}{2}dt\right) = -d\tau^{-1}\Sigma j(f)\, da = -\left(\frac{1}{4}\pi\right)^{-1} \times 8 \times 8 \times \left(\frac{1}{8}\pi\right) = -32$$

Here $d\tau = \pi(\frac{1}{2}dr)^2 = \frac{1}{4}\pi$; by coincidence all the cells in the figure have the same volume. Thus the new ρ here is $80 - 32 = 48$, as shown.

It is clear that the density at $t = dt$ has the rotational symmetry we started with, in that ρ is independent of ϕ:

$$\rho(r, \phi) = \rho(r) \qquad (8.16)$$

We will not explicitly calculate the current at $t = 1.5\, dt$. However, we will note that the current across each ϕ-face at $r = dr$ (between two cells with $\rho = 4$ in Figure 8.5*c*) vanishes because grad ρ does; ρ is the same in front and in back. Thus

$$j_\phi = 0 \qquad (8.17)$$

everywhere. Note also that the symmetry we found before (Eq. 8.15) is still satisfied; the radial current is independent of ϕ because the density from which it is computed is independent of ϕ. It turns out that Eqs. 8.15 through 8.17 provide a complete definition of what it means for ρ and j to have rotational symmetry, in terms of which we can prove that the symmetry is preserved forever.

▶   **Problem 8.5**

Show that the evolution equations (Eqs. 6.23 and 6.29) preserve rotational symmetry, using the cylindrical-coordinates form of the divergence and gradient as we have above. That is, consider arbitrary faces centered at (r, ϕ) and show that the $j(f, \frac{1}{2}dt)$ given by the evolution equations vanishes at ϕ-faces and is independent of ϕ at r-faces, if ρ has rotational symmetry. Also show that $d\rho/dt$ at the cell centered at (r, ϕ) is independent of ϕ, if j has rotational symmetry. (Note that you need not worry about the special case of the cell at the origin, because symmetry is automatic here: it just requires the density to be equal to itself.)   ◄

▶ **Problem 8.6**

Compute the evolution of the system shown in Figure 8.5 for $t \leq 2\,dt$. Verify that the rotational symmetry is preserved.   ◄

We can think of the system in Figure 8.5 as a truly 2D system (instead of as a cross-section of a 3D one) if we like, in which the particles are confined to the xy plane and all start at the origin. All of the calculations we have done for the 3D case can be converted to 2D by just leaving off the factors of dz (or equivalently, setting $dz = 1$). In 2D polar coordinates, the "volume" of a 2D cell is $d\tau(c) = r\,dr\,d\phi$ and the "area" of an r-face is $da(f) = r\,d\phi$. We retain the notation $d\tau$, da (even though these now have units of area and length respectively) so the definition of the divergence will look the same in any dimensionality. The continuum limit of the initial density in Figure 8.5a is then related to a 2D Dirac delta function δ. Like any continuum field, δ is a rule giving a discrete field δ_d for any lattice. The delta function describes a single object at the origin, so we require that the population of the cell containing the origin be one, and that of any other cell be zero. We wrote δ_d for cartesian coordinates in Eq. 6.42; the obvious generalization to an arbitrary coordinate system is

$$\delta_d(c) = 1/d\tau(c) \qquad \text{if } c \text{ contains the origin}$$
$$= 0 \qquad\qquad \text{if not} \tag{8.18}$$

This formula can actually be used in any dimensionality, as long as we interpret $d\tau$ as the "volume" of a cell in that dimensionality (the length in 1D, area in 2D, and so on). In the 2D polar coordinate system of Figure 8.4, evidently δ_d is independent of ϕ and

$$\delta_d = 4/\pi \qquad \text{if } r = 0$$
$$= 0 \qquad\quad \text{if } r > 0 \tag{8.19}$$

▶ **Problem 8.7**

a. Show that the 2D continuum field $\rho = A\delta$ (a rule giving a discrete field ρ_d for each choice of lattice) gives exactly the initial-density field in Figure 8.5a for the polar lattice shown. What value must A have?

b. The exact continuum solution for this initial condition can be described by the function

$$\rho(r, t) = Bt^{-1}\exp(-r^2/4Dt) \tag{8.20}$$

for $t > 0$. (For $t = 0$, it cannot, of course, be described by a function.) Show that this satisfies the evolution equations (Eqs. 6.23 and 6.29, using the equations inside the back cover for the divergence and gradient of a function—the equations for cylindrical coordinates work in 2D polar coordinates if you simply leave out the d/dz terms.) What is $j_r(r, t)$?

    c. By requiring that this agree with Figure 8.5*a* as $t \to 0$, find the constant B. (To get a discrete field from this function, you need to integrate over a cell.)

    d. Evaluate the function $\rho(r, dt)$ at each cell center in Figure 8.5*c*. How close are these values to those shown in the figure for the discrete field? [In the limit $dr \to 0$, they would agree exactly.]　　　　　　　　◂

SECTION 8.2　SYMMETRY IN STATICS

An important use of symmetry is to facilitate the calculation of static limits. Consider the 2D diffusive system of the previous section as an example. If we start with a fixed number of particles, as in Figure 8.5*a*, they will spread out over a larger and larger region (if the system is infinite) and the population of each cell will approach zero. To get a nontrivial static limit, we must have a steady source, as we did in 1D in Chapter 5. If we put a steady delta-function source at the origin, with a 2D time-independent source density

$$s(r, \phi, t) = A\delta(r) \tag{8.21}$$

the system still has rotational symmetry, and we should use polar coordinates as in the previous section. The equations governing the system are then Eq. 6.29 for j:

$$j_r(r, \phi, t + \tfrac{1}{2}dt) = -D[\text{grad } \rho]_r(r, \phi, t) \tag{8.22}$$

$$j_\phi(r, \phi, t + \tfrac{1}{2}dt) = -D[\text{grad } \rho]_\phi(r, \phi, t) \tag{8.23}$$

and Eq. 6.23 for $d\rho/dt$, which is zero in the static limit:

$$\frac{d\rho(r, \phi, t)}{dt} = -[\text{div } j](r, \phi, t) + s(r, \phi, t) = 0 \tag{8.24}$$

We may drop the time argument because we are looking for a time-independent static limit. These are then three coupled partial differential equations in two variables r and ϕ (in the continuum limit) which may be, in general, very difficult to solve. However, in this case we are saved by our knowledge of the symmetry of the system. We know that the updating equations preserve the rotational symmetry, so we know that the fields have this symmetry at all times (as long as the initial conditions and the source term have it, which they do in this case) and therefore the long-time (static) limit must have it. So we know from Eqs. 8.15 and 8.16 that the fields are independent of ϕ, and from Eq. 8.17 that j_ϕ vanishes. This leaves us with two equations for the fields in the static limit:

$$j_r(r) = -D[\text{grad } \rho]_r(r) \tag{8.25}$$

$$0 = -[\text{div } j](r) + s(r) \tag{8.26}$$

Since ρ does not appear in the second equation, the problem of finding j has been reduced to one differential equation in one variable. This is easy to solve, either on a lattice or in the continuum limit. In terms of a function $j(r)$, Eq. 8.26 becomes

(using the formula for the divergence given inside the back cover)

$$(d/dr)[rj_r(r)] = 0 \qquad (8.27)$$

for $r > 0$. We have replaced the source $s(r)$ by zero because the equation only makes sense away from the origin, and the delta function is zero there. This equation says that rj_r is independent of r—we will call it B. So

$$j_r(r) = B/r \qquad (8.28)$$

is an explicit formula for j_r.

▶ **Problem 8.8**

Show that Eq. 8.28 is an exact expression for the discrete face field j as well. Start by writing Eq. 8.26 for a cell centered at an arbitrary r, using the discrete divergence. ◀

To find B in terms of the source strength A, we must use Eq. 8.26 to relate j to the source s. We cannot do this in terms of functions, because s is not a function. We must go back to the definition of the delta "function" as a continuum cell field, a rule for determining a discrete cell field δ_{dr} for each lattice. A continuum face field j (which determines a discrete j_{dr} for each lattice) is a solution of the static Eq. 8.26 if

$$s_{dr}(r) = [\operatorname{div} j_{dr}](r) \qquad (8.29)$$

for each lattice. We have already used the equation at $r > 0$ to conclude that $j_{dr}(r) = B/r$; the only information we haven't used is the equation at $r = 0$. This is

$$s_{dr}(0) = A\delta_{dr}(0) = A \, d\tau^{-1} = d\tau^{-1} \sum_f j(f) \, da(f) \qquad (8.30)$$

where $d\tau = \pi(\frac{1}{2}dr)^2$ is the 2D "volume" of the cell at the origin, and the sum is over faces at $r = \frac{1}{2}dr$. These all have the same current density $j = B/r = 2B/dr$, so we can factor $j(f)$ out of the sum. The remaining sum is just the total 2D "area" (the circumference of a circle) $2\pi(\frac{1}{2}dr)$, so

$$A \, d\tau^{-1} = d\tau^{-1}(2B/dr)2\pi\left(\tfrac{1}{2}dr\right)$$

Thus $A = 2\pi B$, and

$$j_{dr}(r) = A/2\pi r \qquad (8.31)$$

is our exact solution for the static current density. Fortunately, this is the same for any lattice (i.e., it is independent of dr), so it is easy to take the $dr \to 0$ limit to get a function; the function j_r is given by Eq. 8.31. As a vector function, we can write the current density as

$$\mathbf{j}(\mathbf{r}) = (A/2\pi r)\hat{\mathbf{r}} \qquad (8.32)$$

We can extract a useful mathematical fact from the above analysis: Eqs. 8.21, 8.24, and 8.32 imply that in 2D,

$$\mathrm{div}\,(\hat{\mathbf{r}}/r) = 2\pi\delta(r) \tag{8.33}$$

This is an exact statement about continuum fields; in a region not containing the origin, both sides can be represented as functions and it reduces to $\mathrm{div}\,(\hat{\mathbf{r}}/r) = 0$.

▶ **Problem 8.9**

Use the formula for the divergence in 2D polar coordinates (which you can get from the cylindrical coordinates formula inside the back cover by ignoring the d/dz term) to verify that $\mathrm{div}\,(\hat{\mathbf{r}}/r) = 0$ for all $r > 0$. ◀

▶ **Problem 8.10**

The 2D divergence theorem relates the "volume" integral of the divergence of any face field j over a 2D region R to the "area" integral of j itself over the boundary S of R:

$$\sum_{c \in R} [\mathrm{div}\,j](c) = \sum_{f \in S} j(f)\,da(f) \tag{8.34}$$

Show that Eq. 8.31 can be obtained from the divergence theorem, without solving Eq. 8.26 explicitly. [Hint: Use a disk of radius r as the region R.] ◀

▶ **Problem 8.11**

  a. Solve Eq. 8.25 for a continuum function $\rho(r)$, in terms of a constant of integration. Use Eq. 8.32 for j. (This gives a static solution to the problem, but a peculiar one because ρ is negative in some places—in fact the total population is $-\infty$.)
  b. Show that $\rho(r, t)$ in a system with a delta-function source and zero initial density can never approach this static limit. [This is true even though the current density j **does** approach the static limit $(A/2\pi r)\hat{\mathbf{r}}$ in this case. In fact, the density in this system continues to increase forever, but more and more slowly, as in the 1D case (Problem 5.6). In 3D, the density for a delta-function source has a true static limit.] ◀

Another system in which we can calculate static limits with the help of symmetry is the "trampoline," the continuum limit of the network of masses and springs in Section 6.9. The equations governing this system (Eqs. 6.70 and 6.71) can be written

$$dv/dt = K\,\mathrm{div}\,s + \mu^{-1}f_{\mathrm{ext}} \tag{8.35}$$

$$ds/dt = \mathrm{grad}\,v \tag{8.36}$$

where s is a 2D edge field (the strain in the springs—be careful not to confuse it with the source s in a diffusion problem), v is the velocity field (a vertex field), and μ is the mass per unit area of the trampoline. All displacements are transverse, that is, normal to the plane of the lattice. The earlier equations were written in terms of the external force F_{ext} on each discrete mass; we have written them here in terms of the external force per unit area f_{ext}, defined below, because this makes more sense in the continuum limit.

These equations become, in the static limit in which we drop time-derivative terms,

$$0 = K \operatorname{div} s + \mu^{-1} f_{\text{ext}} \tag{8.37}$$

$$0 = \operatorname{grad} v \tag{8.38}$$

Eq. 8.38 says that v does not change from vertex to vertex, that is, it is a constant throughout the lattice. If the trampoline is finite, the boundary condition will force $v = 0$ everywhere. In an infinite trampoline, nonzero v would just mean that the whole system is moving at a constant speed; we can make $v = 0$ by looking at it from a reference frame that is moving along with the trampoline. Thus, the only important equation is Eq. 8.37 for the stress field. However, this equation has exactly the same form as the one we solved in the diffusive system: if you replace f_{ext} by minus the source density s and the stress (unfortunately also denoted by s) by the current density j, you will obtain what is essentially Eq. 8.26. The only difference is that to turn strain into current density we must turn faces into edges, and to turn the external force into a density, we must turn vertices into cells. But this replacement is by now familiar, and gives us the **dual** lattice (Figure 8.6). (Recall that we needed to introduce the dual lattice in Section 6.9 just to define what we meant by the divergence of an edge field.) Here the dual lattice also allows us to attach a precise meaning to the force per unit area f_{ext}: it is defined at each vertex v by $f_{\text{ext}} \, da(v) = F_{\text{ext}}$, where $da(v)$ is the area of the cell of the dual lattice centered at v. The mass m at that vertex is related to the mass per unit area by $m(v) = \mu \, da(v)$.

Thus, to find a static field for a trampoline, we need only borrow the corresponding field for the diffusive system. In particular, we can solve for the static

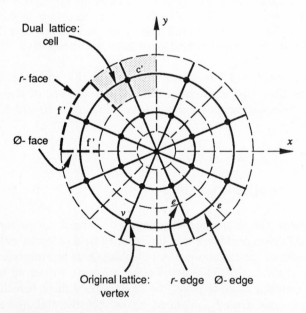

Figure 8.6 The dual lattice (dashed lines) and the original lattice (solid lines) in polar coordinates.

stress field in the presence of a delta-function "source" (i.e., external force) at the origin. If the source is

$$f_{\text{ext}}(r) = W\delta(r) \tag{8.39}$$

then the result is

$$s_r(r) = -W/2\pi\mu Kr \tag{8.40}$$

This can be thought of as the stress field produced by a pointlike person of weight W standing at the center of the trampoline.

▶   **Problem 8.12**

There is no static limit for the displacement field of the infinite trampoline, because it satisfies the same equation as the density in the diffusion case (see Problem 8.11). However, we can find a static limit for a **finite** trampoline, in the continuum limit. Assume that the displacement vanishes on the boundary, a circle of radius R, and find the function $w(r)$ describing the displacement for all $0 < r \le R$, in terms of A.   ◀

The static fields we have found have a **power law** form: the current or strain is proportional to the inverse first power of r, r^{-1}. We can think of the 1D result (current = constant $\propto r^0$) as a power law in which the exponent is zero. It will turn out that the 3D version of this calculation is identical to the calculation of the static electric field of a point charge, and yields a power law r^{-2}.

PART B

Electric and Magnetic Fields

CHAPTER NINE

MAXWELL'S EQUATIONS

SECTION 9.1 THE LOGICAL STRUCTURE OF ELECTRODYNAMICS

Man's attempts to understand and learn to predict the behavior of the natural world, which comprise the scientific endeavor, involve two types of activity. The first is the accumulation of experimental facts. For many predictive purposes this is sufficient; one can predict what will happen in a given situation by remembering what happened in similar situations previously. However, the accumulated information will very quickly exceed the capability of a human being to remember it, making the further accumulation of information less and less useful. There is thus a need for the other form of scientific activity, namely **synthesis**: the discovery of general principles from which large bodies of observed facts can be deduced. This clears the way for future progress by making it unnecessary for future generations to memorize a large mass of facts, thereby freeing their time and memory capacity for further investigation. Such a synthesis has the effect of compressing the predictive power of many pages of data into a few lines. One familiar example of such a synthesis is Newton's theory of particle motion: instead of memorizing hundreds of examples of particle motion, the modern student can learn three simple laws from which they can all be deduced. This does not free him or her from the need to work out any special cases, of course; the techniques of deduction must be learned. But it does make it unnecessary to **memorize** the special cases.

The mass of data with which this book is concerned has to do with electric and magnetic fields and their interaction with matter on a macroscopic scale. These particular phenomena are conveniently packaged together into a single course in the physics curriculum because they were tied together by a particularly elegant and useful synthesis, now known as "Maxwell's equations," by the Scottish physicist James Clerk Maxwell in 1865. Maxwell's equations are a set of four (or two, depending on how you count) equations that completely describe the electric and magnetic fields (and therefore the electric and magnetic forces on all moving and stationary objects) created by an arbitrary time-varying distribution of macroscopic ("classical") charges and currents. Bits and pieces of the equations, and the other ideas that led to Maxwell's great synthesis, had been developed gradually over 100 years by many people. But it was not until Maxwell that anyone realized that all the phenomena involving electric and magnetic fields could be deduced from the four equations now known as the Maxwell equations. The logical structure of the field is something like this:

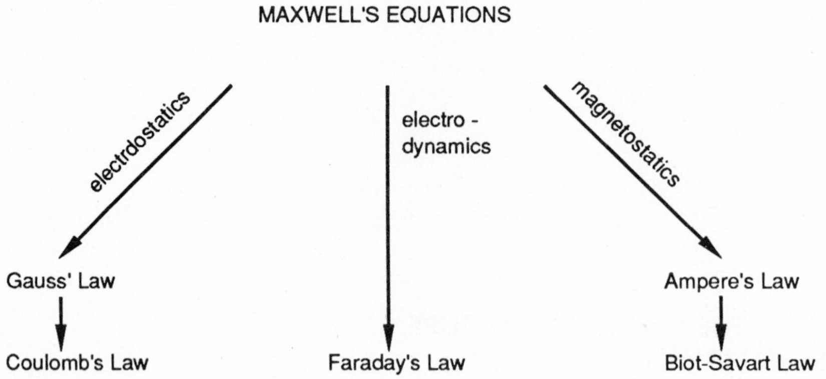

MAXWELL'S EQUATIONS

electrostatics   electro - dynamics   *magnetostatics*

Gauss' Law   Ampere's Law

Coulomb's Law   Faraday's Law   Biot-Savart Law

Figure 9.1   The logical structure of electrodynamics.

In the other field that has a simple logical structure based on a great synthesis, classical particle mechanics, all of the textbooks follow what is generally thought to be a logical approach of beginning by stating Newton's laws and deducing various consequences. One might think, therefore, that all textbooks of electricity and magnetism would start by stating Maxwell's equations and then discuss their various logical consequences in turn. However, most textbooks follow a "historical" approach, describing the bits and pieces of the theory in the order in which they were discovered. Thus, Coulomb's law (which yields the electric field of a point charge) is presented as an independent experimental observation, even though it can be derived from Maxwell's equations. The historical structure of electrodynamics looks something like its logical structure turned upside down:

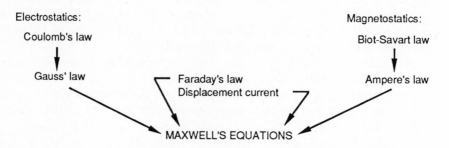

Electrostatics: Magnetostatics:

Coulomb's law   Biot-Savart law

Gauss' law   Faraday's law   Ampere's law
                    Displacement current

MAXWELL'S EQUATIONS

Figure 9.2   The historical structure of E & M.

The disadvantage of learning electrodynamics from the historical point of view is that each of the bits and pieces must be memorized as an isolated fact; one doesn't find out how they fit together until the very end. It is as though one learned mechanics by memorizing separate formulas for 1D uniformly accelerated motion, 2D parabolic trajectories in a gravitational field, circular motion with a centripetal force, etc., and then learned at the end that these could all be easily deduced from Newton's laws. When one approaches electrodynamics from the "logical" point of view (starting from Maxwell's equations), each of the laws is a direct consequence of something that's already known. Most people find things easier to understand and easier to remember when they seem to fit together logically.

Why, then, is the historical approach so often used? One can think of several possible reasons for this. First, there is something to be said for teaching students something about the history of science, and this can most easily be done by introducing concepts in their historical order, in which Coulomb preceded Maxwell. Presenting physics only in a clean, deductive way conveys a misleading impression of how science is done: it ignores the dead ends and the importance of partial syntheses in building understanding. However, most attempts to teach physics in a historical context end up distorting the history somewhat anyway. For example, most texts suggest that Coulomb set out to measure the force between two charges, and experimentally arrived at the inverse-square law that bears his name; Gauss' law is then derived from it. In fact, the inverse square law had already been suggested, because it follows from Gauss's law, and Coulomb just checked it experimentally. In mechanics, historical accuracy is rarely even attempted; it would require exploring all the dead ends of pre-Newtonian mechanical ideas. Demanding such accuracy would defeat the purpose of Newton's great synthesis. In general, the history of science is probably best studied after the principles and phenomena are understood.

The second conventional reason for beginning with Coulomb's law is probably the more important one. It is that an inverse square law can be written using only algebra with which most students feel fairly comfortable. Beginning a course with Maxwell's partial differential equations and their unfamiliar differential operators would be rather intimidating. One ends up tolerating some muddling of the logical structure of the field to avoid beginning with the most difficult mathematics. However, the use of a discrete formulation alters this tradeoff considerably: the discrete versions of Maxwell's equations are easier to understand and not as scary. The balance then favors the "logical" over the "historical" approach.

Our decision not to present the laws of electricity and magnetism in the context of the historical experiments in which they were supposedly derived does not mean we should present them abstractly, with no experimental context at all. It is very important to maintain contact with the real world by explaining how the laws could be experimentally verified, or even deduced. However, it turns out that the human race stumbled upon these laws in a very circuitous and difficult way, through no fault of its own. This was a result of a very unfortunate mismatch between the time, distance, and hence velocity scales we normally deal with (seconds, meters, meters/sec., and so on) and the velocity scale on which the basic electromagnetic phenomena take place (the speed of light, 3×10^8 m/s). If the pioneers in electricity and magnetism had been able to make measurements on a time scale of 10^{-8} seconds, it is quite clear that Maxwell's equations would have been discovered centuries earlier, and with much less mental anguish. There doesn't seem to be any particular virtue in forcing modern students to re-live this anguish. Therefore the laws of electricity and magnetism are presented here in the context of a series of experiments through which they might have been deduced by a species much more nimble than ours. This is not as dishonest as it seems, because all the experiments that are described could easily be carried out using modern high-speed measuring equipment; it is only because of an accident of history that this was not available to Coulomb and his contemporaries.

To make it clear what a difference the time scale makes, I will begin by describing an analogous situation involving a phenomenon that can be grasped intuitively more easily than electromagnetic phenomena, namely the vibrations of masses on springs, in the next section.

SECTION 9.2   THE PARABLE OF THE TORTOISES AND THE HARES

Once upon a time there was a colony of tortoises and one of hares. The members of both colonies were highly intelligent, but because they lived some distance apart, they did not communicate with each other. Each species, however, communicated freely with its own kind and engaged in frequent scientific discussions. Each colony was near a discarded bedspring and a trampoline, so one of the most frequently-discussed topics was the behavior of systems of masses on springs in 2D, and their continuum limit. Initially neither species knew anything about the physics of such systems, but they were curious and could crawl under the bedspring or the trampoline to do various measurements. It was dark underneath, so it was hard for them to measure the displacements of the masses at the vertices of the bedspring. But the springs creaked when they stretched, so by listening carefully to each spring the hares and the tortoises could measure the strain field $s(e, t)$ at each edge (spring position) e. Also, they were very sensitive to wind (which is why they spent so much time under bedsprings) and could measure the velocity $v(r, t)$ of each mass by detecting the resulting downward air motions (Figure 9.3). Being believers in discrete methods, the hares and the tortoises measured the strain field at multiples of a convenient time interval dt (0.01 seconds for the hares, 1 minute for the tortoises). When they discovered the loose wire hanging down from the leftmost mass in the figure, which allowed them to impose an external force F_{ext}, they decided to do some experiments to determine the equations governing the motion of the masses and springs.

We will first describe the experiments of the hares. They stationed one hare under each spring at an edge e, to listen to the creaks and record $s(e, t)$ at times 0, dt, $2\,dt$, etc., and one hare under each mass at a vertex r, to detect the breeze and record the velocity $v(r, t)$ at $t = 0.5\,dt, 1.5\,dt, \ldots$. The hare under the mass at r_0 which had a loose wire attached had the job of pulling the wire, in addition to that of recording the breeze. This imposed an external force $F_{ext}(r_0, t = 0)$ during the

Figure 9.3   Hares under bedspring (left to right): exerting external force, detecting v, detecting s.

time interval centered at $t = 0$ (i.e., extending from $-\frac{1}{2}dt$ to $+\frac{1}{2}dt$). The first consequence of this was noted at time $\frac{1}{2}dt$: there was a strong breeze at vertex r_0, and nowhere else. The hare at r_0 wrote down the value of $v(r_0, \frac{1}{2}dt)$ just before succumbing to hypothermia.

The coroner's jury returned a verdict of accidental self-inflicted death, finding little doubt that the sudden increase in $v(r_0, \frac{1}{2}dt)$ was a consequence of the external force $F_{ext}(r_0, 0)$. They even proposed an equation for the increase in v:

$$dv(r, t)/dt = m^{-1}F_{ext}(r, t) \tag{9.1}$$

where m^{-1} is a constant, whose value they calculated from the recorded values of $F_{ext}(r_0, 0)$ and $v(r_0, \frac{1}{2}dt)$, and the known value of dt. One juror suggested that the left hand side might be v instead of dv/dt, by analogy with the current in Fick's law. But he was converted to the majority view by the observation that v continued to have a large value for several dt's after F_{ext} became zero due to the collapse of the hare pulling the wire, whereas his suggestion would imply that v would return to zero as soon as F_{ext} did.

After a suitable period of mourning for their colleague who had given his life for science, the surviving hares returned to the analysis of their recorded data. At the next recording time $t = dt$, a stress $s(e, dt)$ was recorded in each outward edge e adjoining the vertex r_0 (Figure 9.4). They could have attributed this either to $F_{ext}(r_0, 0)$ or to $v(r_0, \frac{1}{2}dt)$. However, it seemed reasonable to attribute it to the cause closest to it in time [i.e., $v(r_0, \frac{1}{2}dt)$]; they also noted that the stress at these edges continued to increase for $t > dt$, which it would not have done if it had been due to the transient $F_{ext}(r_0, 0)$. Thus their equation for $ds(e, \frac{1}{2}dt)/dt$ had a term proportional to $v(r_0, \frac{1}{2}dt)$. Because of the obvious symmetry of the bedspring under reflection in the center of edge e (which interchanges r_0 with the vertex r_1 shown in Figure 9.4), r_1 should clearly appear in the equation also [although it had no effect in this instance, because $v(r_1, \frac{1}{2}dt)$ was zero.] Because e is an **inward** edge at r_1 (and reversing the edge orientation changes the sign of an edge field), the r_1 term must have the opposite sign. Thus ds/dt is proportional to the difference, that is, the gradient of v:

$$ds(e, \tfrac{1}{2}dt)/dt = C[\text{grad } v](e, \tfrac{1}{2}dt) \tag{9.2}$$

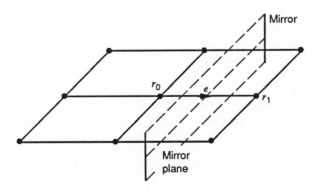

Figure 9.4   Vertex r_0 and four outward edges, showing mirror interchanging r_0 and r_1.

where C is a constant. Not wishing to have to remember another constant, the hares adjusted the units in which they measured s so that C was exactly 1. Assuming also that the system has time-translation invariance (so the equation is the same at any time t as at $\frac{1}{2}dt$), this gave

$$ds(e, t)/dt = [\text{grad } v](e, t) \qquad (9.3)$$

The hares now had updating equations for v (Eq. 9.1) and s (Eq. 9.3), and examined the data at the next time, $t = 1.5\,dt$, to confirm them. They were disappointed, because the hare at vertex r_1 had recorded a nonzero $v(r_1, 1.5\,dt)$, which was fortunately smaller than $v(r_0, 0.5\,dt)$, so he survived, whereas Eq. 9.1 predicted v would still be zero there, because $F_{\text{ext}}(r_1, dt) = 0$. Clearly there must be another term in the equation for dv/dt; because $s(e, dt)$ was the closest nonzero variable in both space and time, it was the leading candidate for this term. The fourfold symmetry of the bedspring (i.e., its invariance under rotations by $\pi/2$ about a vertical axis through a vertex) required all four inward edges at r_1 to be treated equally, so the term must be the sum of the strains at these edges. This is proportional to the 2D divergence (Eq. 6.69), so the hares' revised equation (at an arbitrary time t and vertex r) was

$$m\,dv(r, t)/dt = F_{\text{ext}}(r, t) + K[\text{div } s](r, t) \qquad (9.4)$$

Now the hares tested Eqs. 9.3 and 9.4 again, with data at $t = 2\,dt$, $2.5\,dt$, and so on. This time they found no large discrepancies, because these equations are the same ones we found from Newton's laws (Eqs. 6.70 and 6.71). They found small discrepancies, of course, because they were using a nonzero dt. When they decreased dt, however, the equations agreed better and better with their measurements. Thus, they adopted these equations as the correct ones for describing bedspring motions.

The hares then began a series of experiments on the trampoline, marking off a lattice underneath it with chalk and stationing hares at vertices and edges. These experiments went very similarly to those under the bedspring, except that the hare at vertex r_0, who pulled down on the trampoline, dressed warmly enough to avoid the fate of his predecessor. The results agreed better and better with Equations 9.3 and 9.4 as dt and dr were decreased, leading the hares to conclude that the continuum limits of these equations were the correct partial differential equations for the motion of the trampoline.

The hares' understanding of the physics of trampolines enabled them to enrich their lives with a number of useful inventions. For example, Figure 9.5 shows Heinrich Hares sending a message to a friend using pseudo-1D pulses on a trampoline.

Now that the hares understood the dynamics of stress and velocity fields on a trampoline, they were curious about what would happen if they applied a small delta-function external force for a very long time. That is, they wanted to know the **static** stress due to a static point source $f_{\text{ext}} = W\delta(r)$. We have already computed this in polar coordinates (Eq. 8.40). It is the power law

$$s_r(r) = -W/2\pi\mu Kr \qquad (9.5)$$

Figure 9.5   Pseudo-1D pulses on a trampoline (or real 1D pulses on an elastic string).

for this case. This was a very simple calculation for the hares, once they understood the equations governing the dynamics, and they easily verified it experimentally.

The hares now possess a complete understanding of the physics of their world, and with it the ability to accurately predict all trampoline phenomena. They have proceeded to discussion of theology, in which we leave them as we check on the progress of the tortoises. There's no need to hurry; they haven't even finished breakfast yet. Eventually, however, they get around to wondering whether there are simple laws governing the motions of bedsprings and trampolines. Their first experiment is the same as that of the hares: they station tortoises at various vertices and edges under the bedspring to record the velocities and strains. Fortunately, the tortoise at r_0, who pulls down on the mass there, pulls so weakly that the mass achieves only a very small velocity and he is spared the martyrdom of his hare counterpart. In fact, the tortoises don't notice the velocity (or the breeze it creates) at all. Their dt is much larger than that of the hares, so they can achieve the same displacement (and hence the same strain) with much smaller velocities. But for the same reason, by the time $t = 0.5\,dt$ rolls around, many hare-dt's have passed, and the stress has essentially reached its static value at the nearby faces, given by Eq. 9.5 in polar coordinates (it can also be computed in cartesian coordinates.) Thus, to the tortoises, the strain produced by the external force appears everywhere instantaneously. Similarly, when the tortoise at r_0 **stops** pulling, the strains disappear in a short time compared to $dt/2$, instantaneously from their point of view. Though they can fairly easily guess and check the inverse-first-power law (Eq. 9.5) for the stress, it gives them no inkling of the dynamics behind it, or indeed that there **are** any dynamics, in the sense of the fields spreading out in space after the "source" f_{ext} is turned on. The technological opportunities that were open to the hares because of their ability to observe these dynamics remain unavailable to the tortoises. It appears to them that this instantaneous "action-at-a-distance" inverse power law may be the most fundamental description there is.

To sum up the tortoises' situation: Even though they can predict static strain fields (which are all they can measure), they have no idea of the simple dynamics behind them. For the tortoises, the study of physics consists in the tedious memorization of power laws. It may be many centuries before a reptilian J. Clerk Maxwell comes along and discovers that the power laws are consequences of simple, "local" (meaning that one variable affects only those closest to it in space and time) dynamical laws. And what is worse, by that time the preeminence of the static limit may be so ingrained into their thought patterns and traditions that they continue to memorize power laws even **after** they understand the dynamics.

SECTION 9.3   THE ADVENTURES OF TACHMAN

Heeding the moral lesson of the Tortoises–Hares Fable, we look for a fast-moving species in our own environment of electric and magnetic phenomena, from whose experience we may learn more easily than from our own. We find such a species in *homo tachiens* (from the Greek *takhus*, "swift"; common name *tachman*). The average intelligence among tachmen is somewhat lower than among humans, and they tend toward impulsive behavior. However, they have the distinct advantage that the time scale on which they think and move, instead of being of the order of 1 second as it is for humans, is about 10^{-9} seconds (one nanosecond). Their distance scale is comparable to ours, say $dr = 1$ meter. The time scale of electromagnetic phenomena on this distance scale is dr/c, where c is the speed of light; this is comparable to the tachmen's time scale. They are thus admirably suited to investigate these phenomena, and it may be instructive to watch them do so.

Let us therefore hearken back to the earliest days of the tachmen, when they first learned to make electric charges by rubbing amber and fur together. It was evident that the amber and fur had somehow changed: there were unexplained forces on them. If one moved such a "charged" object, after a short delay other charged objects in its vicinity would move. Furthermore, certain types of uncharged rocks (lodestones) would twist around after charges were moved (though this requires moving the charges quite rapidly, as only a tachman can). They soon realized that charges came with both positive and negative signs, which produced exactly opposite effects on other charges and lodestones. The effect of the charge on a fur depends on how big the fur is and how hard one rubs it with amber. To attach a numerical value to an amount of charge the tachmen must establish a standard unit of charge. We will assume they use the same unit humans do, the **coulomb** (we will use the SI, or Système International, system of units). But the exact definition of the coulomb depends on concepts we have not come to yet, so we will assume for the moment that they have made some arbitrary choice of unit of charge, which will just happen to turn out to be identical with our coulomb. So the tachmen's coulomb is defined as the amount of charge that can be put onto the skin of a standard animal by rubbing the skin with amber until it won't take any more charge. This standard animal should be one whose skin can quickly be given a large charge. This property is possessed, of course, by the tachyderm (*dermatitis maximus*, Figure 9.6). Obtaining a tachyderm hide was fairly easy, due to the naive and trusting nature of these animals, so the tachmen chose this as their standard animal.

The tachmen could then describe the distribution of charges in space by counting how many coulombs were in various cells of a three-dimensional cubic lattice, with cube side dr. Like the hares and the tortoises, the tachmen were believers in discrete methods; they constructed for their experiments a cubic lattice

Figure 9.6   Tachyderm.

of logs as shown in Figure 9.7. They defined a cell scalar field $\rho(c, t)$, the **electric charge density**, obtained from the net charge in the cell (positive minus negative) by dividing by the volume $d\tau(c) = dr^3$. By counting how much charge crossed the faces f of these cubes and dividing by the face area $da(f) = dr^2$, they defined a face vector field j, the electric **current density**. If the tachmen follow our convention of defining ρ at integer multiples of dt, j will be given for intervals centered at half-multiples: $j(f, 0.5\, dt)$, $j(f, 1.5\, dt)$, $j(f, 2.5\, dt)$, and so on. Note that ρ and j here have different units [c $(=$ coulombs$)/\text{m}^3$ and c$/\text{m}^2$ s] from those in the diffusive systems of earlier chapters $(\text{m}^{-3}, \text{m}^{-2}\ \text{s}^{-1})$; the latter were **number** densities, not **charge** densities. We can write the unit of electric current density as ampere$/\text{m}^2$ or a$/\text{m}^2$, where the **ampere** is defined for the moment as 1 c/s.

The tachmen observed that charges were created only as cancelling **pairs** of positive and negative charges (the amber gets a negative charge equal in magnitude to the positive charge on the fur.) Electric charge is **conserved**, in the sense that the net charge in a cell changes only if a charge crosses one of the faces of the cell. This observation can be summarized in an equation for the change in the

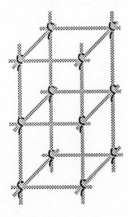

Figure 9.7   A cubic lattice.

discrete density:

$$d\rho(c, t)/dt = -[\operatorname{div} j](c, t) \qquad (9.6)$$

where the left hand side is the discrete time derivative, involving the change in ρ between $t - \frac{1}{2}dt$ and $t + \frac{1}{2}dt$. This is called the **equation of continuity** for electric charge density, and is exactly analogous to the equation of continuity for the number density of ants or hares we encountered before. It states simply that the change in the amount of charge in cell c is exactly equal to the amount entering through the faces of the cell. There is no source term in this equation because electric charge is absolutely conserved: it cannot be created or destroyed. Eq. 9.6 is a statement of the **law of conservation of electric charge**.

The tachmen are now ready to investigate the mechanism whereby the charges exert forces on other charges. Having some experience with dynamical systems such as masses on springs and diffusing ants, they decide to try to regard the charge density as part of a dynamical system. Clearly there are some other variables involved: for one thing, when they put a charge somewhere there are forces **F** on other nearby charges ("test charges")—forces they can measure. They might consider using these forces **F** as variables, but they would quickly discover that **F** depends on the size q of the test charge. In fact they are proportional: **F**/q is independent of q. So the sensible thing is to use **F**/q as the dynamical variable. In the continuum limit, this is evidently a vector field (i.e., a vector quantity defined at each point of space, as in Chapter 6), which we will call the **electric field** and denote by **E**. As we saw in Chapter 6, there are two ways to discretize a vector field, as a face field or as an edge field. We used the former for the current density in a diffusive system, and the latter for force fields. The fact that we have defined the electric field in terms of a force field suggests discretizing it as an edge field. We will in fact do that later, in discussing electrostatic potential. However, we will soon see that the dynamics of E are closely related to the electric current density j, and for that reason it is more convenient here to regard it as a face field.

Using a **face** vector field at least lends itself to being measured conveniently: the tachmen can stretch animal skins on the faces of their log lattice (Figure 9.8). A small charge on each skin can serve as a test charge. It turns out that this scheme even measures the correct component of the vector field: The charge can only move in the direction normal to the face (if the skin is tight) and its displacement will be proportional to the normal component of the force. The tachmen therefore take the normal component of **F**/q at the face f as their discrete electric field $E(f)$. Its units are newtons/coulomb.

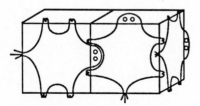

Figure 9.8   Animal skins used as test charges.

A reader who is bothered or confused by the introduction of a new abstract concept, the "electric field," will be gratified to know that James Clerk Maxwell himself did not think in these terms. To Maxwell and many of his contemporaries, what we call here the "electric field" was a perfectly concrete mechanical displacement of a hypothetical material medium, the "aether". Modern terminology retains a vestige of this viewpoint: E is sometimes referred to as "electric flux density", a phrase originally meaning a flow of aether ("flux" is Latin for "flow"). Later physicists abandoned the aether concept in their embarassment at not being able to detect it experimentally. In a way this is unfortunate, because the displacement fields of the aether theory are more concrete and easier to visualize than abstract electric fields, but after using electric fields for a century we are probably stuck with them.

Now the tachmen have enough discrete variables to describe the forces on the stationary charges. However, it soon becomes apparent that the behavior of the lodestones is not determined by the electric field; for example, a stationary lode-stone creates no electric field, but exerts a torque on another "test" lodestone. This torque tends to twist the lodestone to point in a certain direction. The magnitude of the torque depends on the strength of the test lodestone, but the direction does not. Any long, narrow test lodestone, such as a compass needle, placed at a given point will point in the same direction. So we have a magnitude (the torque on some standard-strength compass needle) and a direction. This almost defines a vector, but the orientation (left versus right) is still ambiguous. The tachmen resolve this by labeling one end of their standard compass needle N (for North, because it points north in the magnetic field of the earth) and the other end S. Then all the other compass needles can be labeled by comparing their directions to that of the standard one. Now we can define our vector to point from the S to the N pole of a compass; the resulting vector is called the magnetic-field vector, **B**. Like the electric field, it is most conveniently visualized as a vector arrow, but to study its dynamics the tachmen must define a **discrete** magnetic field. Should they make it a face vector field, or an edge vector field? The faces are getting cluttered, so they put it on the edges. For each oriented edge e, they define the discrete magnetic field $B(e)$ to describe the tendency for the N pole of a compass needle to point in the positive direction along e. Quantitatively, it is proportional to the torque on a compass needle mounted on an edge as in Figure 9.9. The needle is mounted on an axis perpendicular to the edge and forced to point in the third mutually perpendicular direction. You may note that we have not defined the numerical magnitude of B, but only said what it is **proportional** to. We will return later to establish a numerical value for B, that is, to define a unit of magnetic field.

Note that the electric and magnetic fields exchange roles when we switch from the lattice of Figures 9.8 and 9.9 to the dual lattice (Figure 6.28). The centers of the faces of a lattice are the centers of the edges of its dual lattice. The tachmen could have used this dual lattice and measured B on faces and E on edges. We will see later that this "duality" between E and B even extends to the equations of motion, with minor deviations. The theory of electromagnetism is said to be (almost) **self-dual** in this sense.

Now the tachmen have defined their electric and magnetic fields, and found ways to measure them. They can control the apparent **sources** of these fields, namely the current j and the density ρ. This situation is analogous to that of the hares in

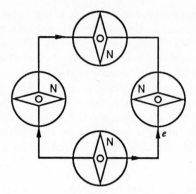

Figure 9.9 *B*-meters on edges.

the previous section, for whom the source was the external force f_{ext} on a trampoline and the resulting fields were the stress and velocity. Like the hares, the tachmen wish to design experiments to determine the equations that govern the dynamics of the fields. They adopt the same strategy the hares did: try to set up situations in which as few variables as possible are nonzero, so the effects of those variables can be easily sorted out. The sources they have at their disposal are pairs of charges (amber and fur); the simplest experiment that creates nonzero source fields ρ and j (electric charge density and electric current) is to create a pair of charges by rubbing. The charges cancel each other to leave $\rho = 0$ as long as they are in the same cell, so the tachmen must move one to an adjacent cell, through a small hole in the skin covering the face. (Note that there is no way to create ρ without creating j, because of the continuity equation relating them.) If the charge is moved during the time interval from 0 to dt, the earliest nonzero variable is j for this interval, that is, $j(f, \frac{1}{2}dt)$. It is nonzero for only **one** face f (Figure 9.10a). Of course, the density $\rho(c, dt)$ will be nonzero at $t = dt$ in these two cells; but that is a later time. We evaluate densities at integer multiples of dt and currents between integer multiples.

What effect on the field variables $E(f)$ and $B(e)$ do they then measure? Well, the skin on the face f we have just crossed is billowing like a sail in a breeze (Figure 9.10b). The tachman assigned to measure $E(f)$ at that face does so at the next

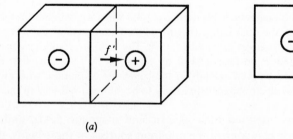

(a)

(b)

Figure 9.10 (*a*) Positive charge moved past face *f*. (*b*) Test charge indicating leftward electric field.

possible time, $t = dt$; we will continue to measure E at multiples of dt. So $E(f, dt)$ has a large, negative value. There's not much doubt what previous variable to attribute it to: the only nonzero one was $j(f, \frac{1}{2}dt)$ at the **same** face. Assuming the dependence is linear, they can write

$$dE\left(f, \tfrac{1}{2}dt\right)/dt = -\epsilon_0^{-1}j\left(f, \tfrac{1}{2}dt\right) \tag{9.7}$$

where ϵ_0^{-1} is just a proportionality constant, which we write in this peculiar way in order to be consistent with human notation. The proportionality constant ϵ_0 can be determined by the tachmen from their measured E and j; the value they obtain is

$$\epsilon_0 = 8.85 \times 10^{-12}\ \mathrm{c^2/N\,m^2} \tag{9.8}$$

Note that we wrote an equation for the change in E (the discrete derivative is $[E(f, dt) - E(f, 0)]/dt$) rather than for E itself. We could make the tachmen deduce this from experiment, as we made the hares deduce that the equation for the velocity field should have this form, but it is easier just to let the tachmen assume it. At these early times the question is academic, because $E(f, 0) = 0$ anyway. But later it will matter.

At this point, you may feel an urge to blurt out, "The E-field at the center of Figure 9.10b isn't caused by the **current**, it's caused by the **charges** that are now to its right and left." This would be inadvisable, as it would alert the tachmen to the presence of a human in the vicinity, one afflicted with the prejudices which that race has acquired by staring too long at static limits. Your argument would imply that the fields on the **other** faces of the cell with the positive charge should be comparably large, which is not true until later, when we approach the static limit. Also, for the tachmen's equations to be useful in updating E, $E(f, dt)$ must be expressed in terms of **earlier** variables; this rules out the charge density $\rho(c, dt)$.

The results the tachmen have obtained so far from their experiment are summarized in Table 9.1, which indicates what variables are nonzero at each time:

Table 9.1

Time	Sources	E, B Fields
$t = 0$	(none)	(none)
$t = 0.5\ dt$	$j(f, dt/2)$ (one face)	(none)
$t = dt$	$\rho(c, dt)$ (two cells)	$E(f, dt)$ (one face)
$t = 1.5\ dt$	(none)	$B(e, 1.5\ dt)$ (4 edges)

We have included in the table the next time, $t = 1.5\ dt$. At this time the tachmen notice a torque on some of the compass needles, namely those on the four edges e of the face f (Figure 9.11). So they measure the magnetic field at this time, $B(e, 1.5\ dt)$, and at subsequent odd half-multiples of dt. Due to the symmetry of the experiment (which looks the same when rotated 90° about the left–right axis through the centers of the cells), the fields $B(e)$ measured by these four tachmen are the same. Let us focus on the tachman farthest from us, whose lodestone points **upward**. He has no idea why; scanning the table above for something at $t = dt$ to blame

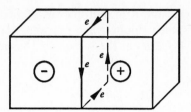

Figure 9.11   Edges at which magnetic field is observed.

$B(e, 1.5\,dt)$ on, he sees $\rho(c, dt)$ for the two cells on either side of f, and $E(f, dt)$ for f itself. The face center f is closer to e (distance $0.5\,dr$) than the cell center (distance $0.707\,dr = dr/\sqrt{2}$), so it seems more reasonable to blame it on $E(f, dt)$. Accordingly, he writes for the change in B:

$$dB(e, dt)/dt = -k_E E(f, dt) \tag{9.9}$$

where we have inserted a minus sign because the left side is positive but $E < 0$; k_E is a positive proportionality constant whose numerical value depends on what units the tachmen use for B. Of course, only the face f appears because that's the only face with a nonzero electric field. There are four faces with the same geometrical relationship to e (Figure 9.12a). Call the others f_1, f_2, f_3. They are oriented faces pointing counterclockwise around e as indicated. Evidently each of $E(f_1)$, $E(f_2)$, and $E(f_3)$ should appear with the same coefficient $-k_E$, because of the assumed rotational symmetry of the dynamic equations. This combination is defined as the curl of the face field E:

$$\text{curl } E(e, dt) = \frac{E(f, dt) + E(f_1, dt) + E(f_2, dt) + E(f_3, dt)}{dr} \tag{9.10}$$

In Chapter 6 we defined the curl for an **edge** vector field; it may not be obvious that Eq. 9.10 is the same thing until you look at the **dual** lattice (Figure 9.12b), in which the centers of the faces f are now centers of **edges**, and the center of e is the center

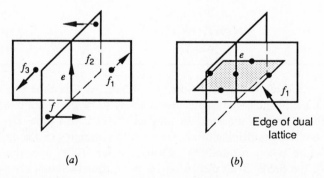

(a)   (b)

Figure 9.12   (a) Faces involved in the curl of a face field. (b) Face of dual lattice (shaded) corresponding to edge e of original lattice.

of the shaded face. In the dual lattice, Eq. 9.10 is exactly the sum over the four edges of the shaded face, which is a curl. In the original lattice, the sum is over a sort of "paddle wheel" configuration of faces (Figure 9.12a).

So our expression for the change in the magnetic field B, shifted to an arbitrary time t, is

$$dB(e, t)/dt = -k_E \, dr \, \text{curl} \, E(e, t) \tag{9.11}$$

The tachmen can now exploit their freedom to choose their unit of magnetic field to simplify this equation, by insisting that the value of $k_E \, dr$ be 1.0. They can even make it dimensionless, in which case you can see from Eq. 9.11 that the units of B are $(\text{s}/\text{m}) \cdot (\text{N}/\text{c})$ or $\text{N}/\text{a} \cdot \text{m}$. This is referred to as a **tesla**: $1 \, \text{T} = 1 \, \text{N}/\text{a} \cdot \text{m}$. The equation is then

$$dB(e, t)/dt = -\text{curl} \, E(e, t) \tag{9.12}$$

This is the first of Maxwell's equations.

> If the tachmen had been very careful they might have detected effects of E on further B's than these, analogously to the coefficients K' in the diffusive system of Section 4.5, which described the influence of non-neighboring cells on a current. However, as in that case, equations involving such further terms are equivalent to the simpler ones in the continuum limit, so there is little incentive to worry about them; we will discuss them in Section 9.4.

We now come to the time $t = 2 \, dt$. We observe that several additional tachmen now detect changes in the discrete electric fields they are monitoring: Those at the front, back, top, and bottom faces of the two cells we have drawn, and also the faces extending outward from the edges of f (all shaded in Figure 9.13). Interestingly, these are exactly the faces touching the edges with nonzero $B(e, 1.5 \, dt)$. These new

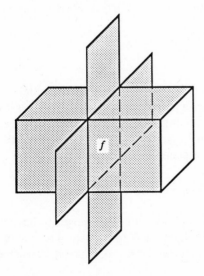

Figure 9.13   Faces with nonzero E at $t = 2 \, dt$.

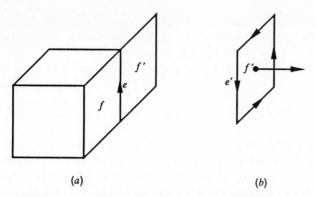

(a)   (b)

Figure 9.14 (a) One of the faces f' with nonzero E. (b) Edges of face f', oriented for computation of the curl.

E's are all nearly equal. Let us focus on one which is the farthest from the source disturbance (so as few as possible of the previously-disturbed variables will be near it.) The face f' farthest from us (Figure 9.14a) has only one disturbed variable near it, namely, $B(e, 1.5\, dt)$, at a distance $\frac{1}{2} dr$. [The closest other disturbed variable is $E(f, dt)$, which is at distance dr and at an earlier time than B.] So the tachmen assume $E(f', 2\, dt)$ (which they measure to be negative, that is, leftward) is influenced mostly by $B(e, 1.5\, dt)$:

$$dE(f', 1.5\, dt)/dt = -k_B B(e, 1.5\, dt) \qquad (9.13)$$

where k_B is yet another positive proportionality constant. Since B is measured in tesla ($= $ N s/c m) and E in N/c, k_B has units m/s^2 and its numerical value can be calculated by the tachmen from their measured values of $E(f', 2\, dt)$ and $B(e, 1.5\, dt)$.

If the equation for dE/dt is to have rotational symmetry under rotations about an axis normal to f' and through its center, it must include B-terms for all four edges of f' (Figure 9.14b). The sum of these was defined in Section 6.3 as a curl:

$$\text{curl } B(f') = \sum_{e'} B(e')/dr \qquad (9.14)$$

where e' is one of the four edges (related to f' by the right-hand rule) in Figure 9.14b. The edge e in Figure 9.14a is oriented oppositely to these, so when we express Eq. 9.13 in terms of e', the sign changes: $-k_B B(e) = +k_B B(-e)$ where $-e$ is one of the edges e' in Figure 9.14b. Symmetry requires these to all appear with a plus sign, giving the curl. Thus at a general time t,

$$dE(f', t)/dt = +k_B\, dr\, \text{curl } B(f', t) \qquad (9.15)$$

Of course, we already know (Eq. 9.7) that this equation is supposed to have a source term in it; the current at $t = 1.5\, dt$ just happened to vanish. The tachmen can therefore rewrite this equation with both terms in it; to make it consistent with human notation let us denote the proportionality constant $k_B dr$ by c^2.

$$dE(f, t)/dt = c^2 \text{ curl } B(f, t) - \epsilon_0^{-1} j(f, t) \qquad (9.16)$$

The tachmen measure the constant c^2 to be 9.0×10^{16} m^2/s^2. This is not such an astronomical value to them, because they measure time in nanoseconds (1 ns $= 10^{-9}$ s): $c^2 = 0.09$ m^2/ns^2. These are all the terms we need—Eq. 9.16 is the second Maxwell equation. With the other Maxwell equation (Eq. 9.12) it allows us to calculate E and B at all times.

It is interesting that all of the terms found by the tachmen relate variables at the closest possible distances, $\frac{1}{2}dr$ and 0, and the smallest possible time difference $\frac{1}{2}dt$ ($\frac{1}{2}dr$ is the distance between a cell center and its bounding face, or between a face and its bounding edge.) Furthermore, all possible interactions at that distance and time difference have been included. The equations are in a certain sense uniquely determined by these conditions; this makes the discrete equations easy to remember. If you were told to assume an E defined at faces and integer times and B defined at edges at staggered times, and asked to write the simplest dynamical equations coupling them, you would have to let $E(f)$ depend on j and B at the preceding time and the closest position. It is then obvious from the geometry that there are four such closest B's, at the edges of the face f, and you would be led to Maxwell's equation (Eq. 9.16).

It is remarkable that the equations governing electromagnetic phenomena are so simple that they can be deduced from as simple-minded an experiment as that performed by the tachmen. It is fortunate for the tachmen, because they might have had a much harder time deducing any more complicated equations. It is perhaps even more fortunate for us humans, because we had a much harder time finding Maxwell's equations than the tachmen did. If they had been more complicated, we would probably never have found them at all.

As a first example of the use of the Maxwell equations (Eqs. 9.12 and 9.16), let us calculate the fields encountered in the tachmen's experiment explicitly for a particular choice of the charge q and dr and dt. The current $j(f, \frac{1}{2}dt)$ (Figure 9.10a) is the charge passing f per unit time per unit area:

$$j\left(f, \tfrac{1}{2}dt\right) = q/dr^2 dt \qquad (9.17)$$

Taking $q = 1$ microcoulomb (1 μc), for example, and $dr = 1$ m, $dt = 1$ ns, we get

$$j\left(f, \tfrac{1}{2}dt\right) = 1 \,\mu\text{c/m}^2 \text{ ns} = 1 \,\text{kA/m}^2 \qquad (9.18)$$

(we have defined 1 kA = 1 kiloamp $\equiv 10^3$ amp $= 1\,\mu$c/ns), as shown in Figure 9.15a. Then the electric field is given by the Maxwell equation (Eq. 9.16), which can be written

$$E(f, dt) - E(f, 0) = -dt \,\epsilon_0^{-1} \, j\left(f, \tfrac{1}{2}dt\right) \qquad (9.19)$$

(because $B = 0$ at time $dt/2$) giving

$$E(f, dt) = -1 \text{ ns}(0.11 \text{ N m}^2/\mu\text{c}^2)1\,\mu\text{c/m}^2 \text{ ns}$$

$$= -0.11 \text{ N}/\mu\text{c} \qquad (9.20)$$

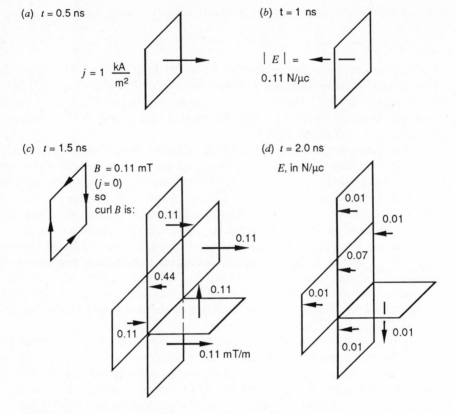

Figure 9.15 The evolution of the tachmen's E and B fields.

as in Figure 9.15b. To compute the magnetic fields on the edges e of this face (Figure 9.11), at the next time $t = 1.5\,dt$, from the other Maxwell equation (Eq. 9.12), we require curl E:

$$\text{curl } E(e, dt) = dr^{-1}E(f, dt) = -0.11\,\text{N m}^{-1}\,\mu\text{c}^{-1} \tag{9.21}$$

at each of the four edges, because the sum over the faces in a "paddlewheel" like that in Figure 9.12 has only one nonzero term. This gives

$$B(e, 1.5\,dt) = B(e, 0.5\,dt) - dt\,\text{curl } E(e, dt)$$

$$= 0 - (1\,\text{ns})(-0.11\,\text{N m}^{-1}\,\mu\text{c}^{-1}) = +0.11\,\text{mT} \tag{9.22}$$

as in Figure 9.15c, where 1 mT = 1 millitesla $\equiv 10^{-3}$ T = 1 N kA^{-1} m^{-1}. The electric field at time $2\,dt$ at the further-out face f' (Figure 9.14) is given by Maxwell's equation (Eq. 9.16) as

$$E(f', 2\,dt) = E(f', dt) + dt\,c^2 B(-e, 1.5\,dt)/dr$$

$$= 0 + (1\,\text{ns})(0.09\,\text{m}^2\,\text{ns}^{-2})(-0.11\,\text{mT})/1\,\text{m}$$

$$= -0.01\,\text{mT m/ns} = -0.01\,\text{N}/\mu\text{c} \tag{9.23}$$

where we have used Figure 9.14a to compute curl B. The electric field has this value at 12 faces, 5 of which are shown in Figure 9.15d. The current j dropped out because it is zero at time 1.5 dt, and $E(f', dt)$ was also zero. When we compute the field $E(f, 2\,dt)$ at the central face, however, the previous value $E(f, dt)$ is not zero (Eq. 9.20), and all four terms in curl B are nonzero. We get

$$E(f, 2\,dt) = -0.11 \text{ N}/\mu\text{c} + 4 \times 0.01 \text{ N}/\mu\text{c} = -0.07 \text{ N}/\mu\text{c} \qquad (9.24)$$

So the magnitude of the central E is now decreasing; we can think of E as "spreading out" to neighboring faces. We will examine this process further in Section 10.1.

Note that the tachmen's units m, ns, N, μc, kA/m^2, N/μc, and mT form a consistent set of units for distance, time, force, charge, current density, electric field, and magnetic field, in the sense that the products of them that arise in using Maxwell's equations are again among this set. The same is true of the SI units m, s, N, c, A/m^2, N/c, and T, of course, but the tachmen's units have the advantage of not requiring astronomical numbers in calculations.

▶ **Problem 9.1**

Consider discrete fields $E(f, t)$ and $B(e, t)$, satisfying the discrete Maxwell equation with no current ($j = 0$). The initial conditions are $E(f, 0) = 0$ and $B(e, \frac{1}{2}dt) = 0$ except at a single edge e', where $B(e', \frac{1}{2}dt) = B_0 = 1$ mT. (We'll see later that this initial condition is not physically possible, but it makes a good "textbook problem.") Use $dr = 1$ m, $dt = 1$ ns. Compute $E(f, dt)$ and $B(e, 1.5\,dt)$ at each face and edge at which they are nonzero, and indicate their directions and magnitudes in a sketch.   ◀

SECTION 9.4   SLOWLY VARYING ELECTRIC AND MAGNETIC FIELDS

The discrete Maxwell equations (Eqs. 9.12 and 9.16) that we have written down are not the only ones that correctly describe slowly-varying electric and magnetic fields. The situation is just as it was for the diffusive system in Chapter 4: we can add terms involving other edges and faces, but they will give the same results for smooth fields (slowly varying in time and space). For example, there could be effects due to non-nearest-neighbor faces or edges. Perhaps the current $j(f, dt/2)$ causes a small field $E(f', dt)$ at the face f' (Figure 9.14) as well as the large $E(f, dt)$ we discussed above. And perhaps $E(f, dt)$ causes a small $B(e', 1.5\,dt)$ at the furthest-away edge in Figure 9.14 (at a distance 1.5 dr from the center of f); if so, this effect would be very hard to sort out from the effect of the smaller (but closer to e') $E(f', dt)$. Fortunately, the tachmen don't need to determine all these coefficients, if they are interested in calculating the evolution of fields that are slowly varying in time and space. This is true for the same reason that it was in the 1D diffusive system of Section 4.6. There, the current created by a smoothly varying density depended only on the gradient of the density, that is, the coefficient of r in a linear expression $\rho(r) = \rho_0 + gr$ that approximates the density near a particular point. The only property of a discrete equation for the current j that mattered was the ratio j/g (the diffusivity D). Any two discrete equations with the same D would give the same j for smooth densities.

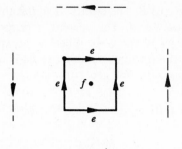

Figure 9.16 Edges whose B-fields might influence E(f).

The present situation is more complicated, but the idea is the same. The only information the tachmen really **needed** to get from the experiments we described was the symmetry; any equation with that symmetry will give the same results for slowly-varying fields as the simple equation, as long as the constants c^2 and ϵ_0 are the same. To see what we mean by "the same symmetry," let's look at an example. In our simple Maxwell equations, $dE(f)/dt$ depends on B at the four edges (labeled e in Figure 9.16) of the face f. These appear with the same coefficient k_B (in the notation of Eq. 9.15). If we wanted to allow $dE(f, t)/dt$ to depend also on B at the edge labeled e', we would have to include terms involving the dotted edges, which are equivalent to e', to maintain the 90° rotational symmetry.

Suppose we did include these extra terms, with coefficients $k_{B'}$. What effect would this have on the result for dE/dt? Suppose that $B(e)$ is represented by a vector function $\mathbf{B}_f(x, y, z)$, which is given by a linear formula

$$B_{fy}(x, y, z) = gx \tag{9.25}$$

and $B_{fz} = B_{fx} = 0$. The corresponding discrete edge field is shown in Figure 9.17. Then the simplest discrete Maxwell equation for updating E,

$$dE/dt = k_B \, dr \operatorname{curl} B - \epsilon_0^{-1} j \tag{9.26}$$

gives

$$dE_z/dt = (k_B \, dr) g \tag{9.27}$$

This can be obtained from the continuum formula for the curl (Chapter 6) but it is

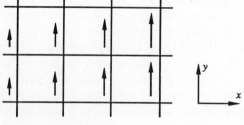

Figure 9.17 A linear B field.

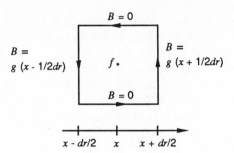

Figure 9.18 Computing the curl of a linear B field.

actually exactly true for finite dr, not just in the $dr \to 0$ limit. This can be seen from Figure. 9.18, where

$$\text{curl } B(f) = dr^{-1} \Sigma B(e)$$

$$= dr^{-1}\left[g\left(x + \tfrac{1}{2} dr \right) - g\left(x - \tfrac{1}{2} dr \right) \right]$$

$$= dr^{-1}\left[2g \tfrac{1}{2} dr \right] = g \qquad (9.28)$$

So dE/dt depends only on the "B-field gradient" g. Our new equation including $k'_B B(e')$ for the edges e' in Figure 9.16 would give an extra k'_B $[2g$ $1.5 \, dr] = 3gk'_B \, dr$. This is still proportional to g; the result is the same as though the simple equation had been used with a larger value of k_B. Given any such equation, it is only the ratio

$$(dE/dt)/g = (k_B + 3k'_B) \, dr \qquad (9.29)$$

that matters in determining the evolution of smooth fields; it is this ratio that we denote in general by c^2, as in Eq. 9.16.

Once the tachmen have established the symmetry of the Maxwell equations, they can in principle determine a precise value of c^2 by setting up a uniform B-gradient as in Eq. 9.25 and measuring dE/dt. (In practice, there is an easier way, related to the fact that c is the speed of light, as we will see later.)

The corresponding procedure for measuring ϵ_0^{-1} would be to set up a system with a uniform j (and zero B). This is because the contribution of the ϵ_0 term to dE/dt for smooth fields depends only on the value of j locally (and not on its gradient); all equations will give the same results for smooth fields if they have the same ratio $(dE/dt)/j$, which we define as ϵ_0^{-1}. Because we chose the coefficient in the equation for dB/dt to be 1, the ratio $(dB/dt)/g$ in a system with an E-field gradient g (analogous to Eq. 9.25) tells us the correct unit for B; it calibrates the tachmen's B-meters.

SECTION 9.5 MAXWELL'S EQUATIONS FOR HUMANS

The rest of this text will be based on the Maxwell equations the tachmen have just deduced. It is therefore worth summarizing (from a human point of view) what they say. They describe the time-evolution of two fields, the electric field E, and the

magnetic field *B*. In a cartesian coordinate system divided into cubical cells of
length *dr*, it is convenient to specify the electric field at the faces *f* of the cubes, at
times *t* that are integer multiples of *dt*: $E(f, dt)$, $E(f, 2\,dt), \ldots,$. The magnetic
field is intimately related to the **change** in the electric field between two multiples of
dt, so it is most convenient to specify it at a time between these, that is, at
half-integer multiples of *dt*. The magnetic field *B* is related to the **curl** of *E*, which is
most naturally defined on edges if *E* is defined on faces, so we specify *B* at edges *e*:
$B(e, 0.5\,dt)$, $B(e, 1.5\,dt)$, $B(e, 2.5\,dt), \ldots,$. (It should not be inferred from this that
E "is zero" at times like 1.5 *dt* between multiples of *dt*; it is simply not defined at
such times.)

We will always assume that all fields are zero up until some specific time ("the
beginning of an experiment") at which they are created by a source. The source is
the electric current density *j*, which is defined at each face *f*, at half-integer
multiples of *dt*. It is related to the charge density ρ, which does not appear in the
dynamical Maxwell equations, by the continuity (charge conservation) equation:

$$d\rho(c, t)/dt = -\operatorname{div} j(c, t) \tag{9.30}$$

(where *t* is an odd half-multiple of *dt*).

Once we know the source current density $j(f, t)$ at all places (*f* 's) and all
times (*t* 's), we can calculate the electric and magnetic fields at all places and times,
from Maxwell's equations (Eqs. 9.12 and 9.16). We will rewrite them here, giving
them their conventional names (we will discuss the origin of the names in later
chapters)

$$dE(f, t)/dt = c^2 \operatorname{curl} B(f, t) - \epsilon_0^{-1} j(f, t) \qquad \text{Ampere's law} \tag{9.31}$$

$$dB(e, t)/dt = -\operatorname{curl} E(e, t) \qquad \text{Faraday's law} \tag{9.32}$$

where *t*/*dt* is an integer in the first equation and a half-integer in the second.
Starting at the first time (say *dt*/2) when *j* is nonzero, we use the first Maxwell
equation (with *t* = *dt*/2) to calculate $E(f, dt)$, the second (with *t* = *dt*) to calculate
$B(f, 1.5\,dt)$, the first again (with *t* = 1.5 *dt*) to calculate $E(f, 2\,dt)$, and so on. Note
that *E* and *B* are **never** calculated at the same time in this formulation.

It is an experimental fact, confirmed by more than a century of observations by
thousands of people, that these equations correctly predict the evolution of the fields
in the continuum limit. (At least in "classical" physics—that is, on a macroscopic
scale, where quantum-mechanical corrections are not important.)

We will use Eqs. 9.31 and 9.32 to describe continuous fields as well as discrete
ones. The relationship between these, and the usefulness of continuum fields, was
discussed for diffusive systems in Section 4.7. Continuum fields described by
continuous functions are useful in electromagnetism when the actual physical
current is approximately given by a vector function $\mathbf{j}(\mathbf{r}, t)$ and the initial conditions
for the fields are approximately given by $\mathbf{E}(\mathbf{r})$ and $\mathbf{B}(\mathbf{r})$. If we can find time-dependent vector functions $\mathbf{E}(\mathbf{r}, t)$ and $\mathbf{B}(\mathbf{r}, t)$ that satisfy the continuum versions of the
Maxwell equations then these will approximately describe the system at later times.

We should point out that Faraday's and Ampere's laws are usually written
slightly differently, with the curl terms on the left-hand side. (There are historical

reasons for this, which we will explain later.) However, we will follow the usual convention for writing differential equations for the time-evolution of a quantity, for example, $df/dt = 2f(t)$, in which the time derivative is written on the left. Physically, this convention corresponds to putting the effect to be calculated on the left, and its "cause" on the right. There is, of course, another well-known violation of this convention: Newton's second law is still usually written $F = ma$, despite the quixotic efforts of several textbook authors who have tried to popularize $a = F/m$. The $F = ma$ notation doesn't really do any harm, in the sense that most mechanics students realize that the force causes the acceleration, and not the other way around. In the case of Faraday's law for dB/dt, however, many students who write curl $E = -dB/dt$ think of dB/dt as somehow "causing" curl E. This creates conceptual problems, because dB/dt involves a time **later** than that at which curl E is evaluated (perhaps only slightly later, as $dt \rightarrow 0$, but nonetheless later). The student is left trying to make sense of an "effect" that precedes its "cause." This interpretation violates the **principle of causality**, which states that causes should precede their effects. Violations of this principle clearly grate against common sense as well as our physical intuition. Theories that violate our intuition aren't necessarily wrong or useless (look at the theory of relativity, which violates the intuitive notion of simultaneity) but no one has ever found a theory that violates the principle of causality to be helpful in understanding any physical phenomenon. Most physicists, therefore, regard the principle of causality to be a fundamental law of nature. So we will insist that curl E "causes" dB/dt, and write dB/dt on the left to remind ourselves of this.

CHAPTER TEN

PSEUDO-TWO-DIMENSIONAL SYSTEMS

SECTION 10.1   OUT-OF-PLANE CURRENTS

In dealing with bedsprings and diffusive systems, we began with one- and two-dimensional versions, because these have all the essential features of a three-dimensional system (at least a 2D one does), and they are much easier to visualize and illustrate. Unfortunately, we couldn't do this with Maxwell's equations because they are intrinsically three-dimensional, in the sense that the 2D analog does not have all the important features. Consider for example an electric field at a face f and a magnetic field at an edge e, which are separated by $\frac{1}{2}dr$ so each affects the other through Maxwell's equations, as shown in Figure 10.1. Three directions are involved: the normal to the face (i.e. the direction of the E vector), the direction of the edge, and the direction of the vector of length $\frac{1}{2}dr$ separating the centers of the face and the edge. These three directions are all mutually perpendicular. There is no way to reduce the problem entirely to a two-dimensional one without losing something important. This can be thought of as being due to the essential three-dimensionality of the curl: note that in each term of the continuum formula for the x-component of the curl, both y and z components appear. Every term involves all three axes.

By contrast, this is **not** true of diffusion. The normal to a face and the direction to the center of the cell whose density interacts with the current at that face are in the **same** direction. Only one direction is involved, so we could (and did, in Chapter 4) discuss the phenomenon of diffusion in one dimension without losing any of its essential properties.

We can, however, set up 3D electrodynamic systems that can be easily **depicted** in 2D by insisting that the fields depend on only two coordinates, say x and y. Then every cross-section of the system parallel to the xy-plane is the same, so we can completely describe the system by drawing only one. We will refer to such a system as "pseudo-2D." We can think of this requirement on the fields as a **translational symmetry** (Chapter 8) of the system: the fields are unchanged if we translate them all by the same distance (a multiple of dr, of course, so the lattice is preserved) in the z-direction. All of the fields are determined by the source field, which is the current, so we begin by imposing this symmetry on the current. We will show that this forces the resulting electric and magnetic fields to have the same symmetry.

For simplicity, let us consider first what would happen if the current j had only a z-component, that is, $j(f, t)$ was nonzero only at faces normal to the

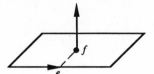

Figure 10.1   The three dimensions of the curl.

z-direction, as shown in Figure 10.2. That is, j can be described by $j_z(x, y)$. Then the j/ϵ_0 term in dE/dt (Eq. 9.16) would cause E at later times to have a z-component, that is, it would give $E_z(x, y)$. The curl of E would then be nonzero at the edges of these faces, which would give nonzero B at these edges, along the x and y axes in the figure, because $dB/dt = -\text{curl } E$. Because E is independent of z, curl E is, so the new B is: we can write it as $B_x(x, y)$, $B_y(x, y)$. These B's would influence E subsequently through the curl B term in dE/dt; it looks from Fig. 10.2 as if curl B may be nonzero at any face. But consider the face labeled f, which has vertical B's at two of its edges. These edges have the same x and y coordinates, and therefore the same $B_y(x, y)$. They contribute oppositely to the counterclockwise path integral around f, so $[\text{curl } B]_x = 0$. The same argument shows that curl B has no y-component. Thus this term in dE/dt leaves E with only a z-component, as it was before. You can go through these same arguments at each subsequent updating step: at each step the nonzero components are $j_z(x, y)$, $E_z(x, y)$, $B_x(x, y)$, and $B_y(x, y)$ (i.e., j and E are out of the xy-plane and B is in the plane), and each updating equation preserves this condition.

▶   **Problem 10.1**

Assuming that only these components are nonzero, show that the space-continuous form of the Maxwell equation (involving partial derivatives d/dx, etc.) gives dE/dt and dB/dt with these same nonzero components. (The advantage of the discrete formulation is that it allows us to use mathematical induction to infer that these relations will remain true at all times.)   ◀

This system, therefore, remains pseudo-two-dimensional, because everything depends only on x and y. We can draw pictures of the evolution of this system without drawing the z-axis. The cross-sections parallel to the xy-plane through vertices contain all the faces normal to the z-axis (where E and j are nonzero) and

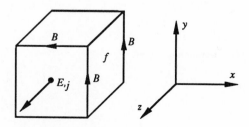

Figure 10.2   An out-of-plane current and the resulting fields.

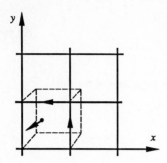

Figure 10.3   2D cross-section of a pseudo-2D system.

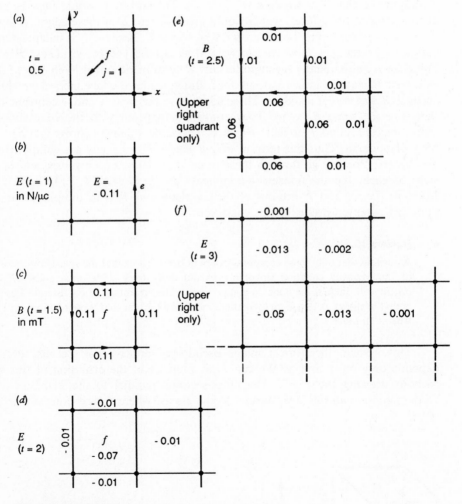

Figure 10.4   Evolution of a system with an out-of-plane current.

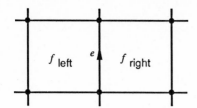

Figure 10.5 Faces involved in the curl at e.

all the x and y edges (where B is nonzero). And the cross-sections are all the same, because of the translational symmetry; we need only draw one, as in Figure 10.3.

As an example of the evolution of a system with out-of-plane currents, suppose that there is a transient current density across one face during the interval centered at $\frac{1}{2}dt$, $j(f,\frac{1}{2}dt) = 1\ \mu\text{c/m}^2\text{ns}$, as shown in Figure 10.4a. The currents at all the other faces in the figure are zero. But remember that because this is a pseudo-2D system, all faces displaced out of or into the paper from this one have $j = 1$ also; this field could describe the current in an infinitely long wire perpendicular to the paper. To avoid having to draw orientation arrows pointing perpendicularly to the paper, in pseudo-2D systems we will always orient these faces out of the paper; fields into the paper are negative. If we use $dr = 1$ m, $dt = 1$ ns, then Ampere's law (Eq. 9.16) gives the electric field $E(f, 1.0\ \text{ns}) = -0.11\ \text{N}/\mu\text{c}$, as shown in Figure 10.4. We calculate the curl of E at the edges of this square by summing $E(f)$ over the four faces of a paddlewheel, but only the two in the plane of the paper can have $E \neq 0$. In general, the curl at an arbitrary edge e (Figure 10.5) is

$$[\text{curl } E](e) = \big[E(f_{\text{left}}) - E(f_{\text{right}})\big]/dr \qquad (10.1)$$

in such a pseudo-2D system, as you can verify from the right-hand rule. At the edge e in Figure 10.4b, the curl is $[-0.11 - 0]/1$, so that Faraday's law [Eq. 9.12] gives $B(e, 1.5\ dt) = -dt\ \text{curl } E = +0.11$ mT. So far, the evolution of this system is the same as that of the 3D system the tachmen studied (Section 9.3). However, at $t = 2$ they found a nonzero E field at faces touching e but perpendicular to the paper (Figure 9.13); in the pseudo-2D system these all vanish. By $t = 2.5$ even the fields in the plane are different from those of the 3D system. Figure 10.4 shows the evolution out to $t = 3$ ns; clearly it is easier to draw and to visualize than the evolution of a 3D system.

Qualitatively, you can see that the fields are gradually moving away from the source current. The wire carrying the current can be regarded as a simple antenna; what we are seeing is just the outward propagation of an electromagnetic wave.

▶ **Problem 10.2**

Use Maxwell's equations (Eqs. 9.31 and 9.32) to compute the evolution of the fields in the pseudo-2D system whose initial fields $E(f,0)$ and $B(e, dt/2)$ are zero except at a single edge e' where $B(e', dt/2) = 1$ mT, until $t = 2$ ns. Use $dt = 1$ ns, $dr = 1$ m, $j = 0$ again. (As in Problem 9.1, this is not a physically possible initial condition.) At what time do the results first differ from those of Problem 9.1 (the 3D case)? ◀

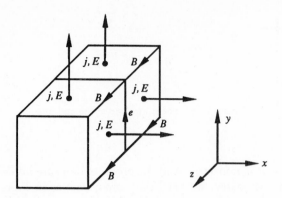

Figure 10.6 Pseudo-2D system with in-plane currents.

SECTION 10.2   IN-PLANE CURRENTS

Let us now consider a pseudo-2D system having nonzero $j_x(x, y)$ and $j_y(x, y)$, the in-plane current components we assumed to vanish in the previous section. This time we will assume there is no out-of plane current: $j_z = 0$. (We can handle a general case by separating the current into an in-plane part and an out-of-plane part, evolving them separately and combining them at the end.) So, the discrete $j(f, t)$ is nonzero only at the faces perpendicular to the paper in Figure 10.6. Then the j/ϵ_0 term in dE/dt produces nonzero $E_x(x, y)$ and $E_y(x, y)$ as shown. The curl of E is nonzero, producing a B, only at the edges perpendicular to the paper as shown. The curl at a vertical edge e involves the sum over a paddlewheel consisting of two faces parallel to the paper (where $E = 0$) and the two x-faces shown in Figure 10.6. The two E's at these x-faces are equal because of the translational symmetry, and cancel out of the counterclockwise sum. The curl at an x-edge vanishes for a similar reason. The only nonzero component of B is thus $B_z(x, y)$. The curl of such a B can be nonzero on x and y-pointing faces, but at a z-face (in the plane of the paper) the edges summed over in the curl all have $B = 0$, and the curl vanishes. Thus the c^2 curl B term in dE/dt gives nonzero E only where we already had it. The nonzero fields remain j_x, j_y, E_x, E_y, and B_z, and this property is preserved in all subsequent updates by Maxwell's equations.

To depict such fields in a 2D drawing, we should use a cross-section in a plane through the centers of faces and edges with nonzero fields (Figure 10.7). In interpreting a 2D picture like Figure 10.8 (below), we must remember that the squares are not faces, but cross-sections through cells. The lines bounding the squares are not edges, but cross-sections through faces. The corners are not vertices, but cross-sections through edges.

As an example of a pseudo-2D system with an in-plane current, consider the current shown in Figure 10.8. This is similar to the current we obtained in Chapter 9 by moving a 1 μc charge from the left-hand to the right-hand cell, but here the translational symmetry requires such a charge in every cell directly out of the paper from these cells. You can think of this as a **line** charge, strung out along a line perpendicular to the paper, with a linear charge density $\lambda = 1\ \mu c/m$ so that each

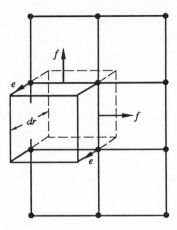

Figure 10.7 2D cross-section of a pseudo-2D system with in-plane current.

1-meter cell has exactly 1 μc of charge. Thus, Figure 10.8 shows the current at $t = 0.5$ ns resulting from the movement of a line charge from left to right. (If you like, you can think of the out-of-plane current of Figure 10.4 as the result of moving such a line charge perpendicularly to the paper.) The picture at $t = 1$ ns shows the charge density: the positive line charge is now in the right-hand cell, leaving a negative line charge on the left. The E-field at $t = 1$ is the same as in Chapter 9. The B-field at $t = 1.5$ ns differs in that it is nonzero only on edges perpendicular to the paper; the paddlewheel of faces used to compute the curl of E looks like Figure 10.9 in the cross section we are using. To calculate E at $t = 2$ ns, we need curl B, which is

$$[\text{curl } B](f) = \big[B(e_{\text{left}}) - B(e_{\text{right}}) \big] / dr \qquad (10.2)$$

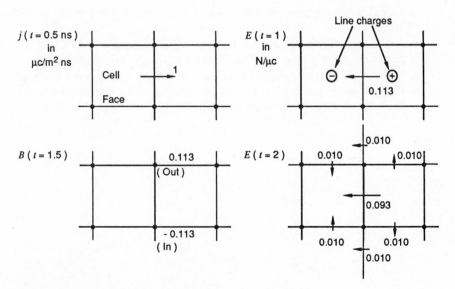

Figure 10.8 Evolution of a system with an in-plane current.

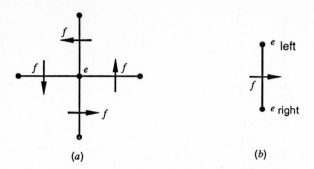

Figure 10.9 Cross-sections of (a) paddlewheel used to compute curl E, and (b) a face and its edges.

in terms of the labeling in Figure 10.9b. Figure 10.9b depicts a cross section of a face f perpendicular to the paper and the cross sections of its edges; it has edges parallel to the paper as well, but $B = 0$ there. The only nonzero terms in the curl involve the edges shown, both assumed to be outwardly oriented. (Note the similarity to Eq. 10.1 for the curl of an out-of-plane E. In a sense, these 2D projections of the curl operation resemble the gradient.) Ampere's law for dE/dt then gives the $E(t = 2)$ shown in Figure 10.8; the electric field "leaks out" from the central face. The value at the central face is larger than that for the 3D (or out-of-plane-current 2D) calculation because there are only two edges for it to "leak past" here, whereas before there were four.

A program called MAXWELL, which is used to compute the evolution of pseudo-2D fields, is described in the program guide.

▶   **Problem 10.3**

   a. Calculate the evolution of the system in Fig. 10.8, to $t = 3$ ns, to an accuracy of ± 0.001.

   b. Check Fig. 10.8 and your answer to part (a) using program MAXWELL. You may need to multiply j (and therefore E and B) by a constant to make them visible.

   c. Use MAXWELL to compute E at the central face at time $t = 5$ ns. Indicate on a diagram where you put the central face relative to the system boundaries. Do you think the positions of the boundaries affected the answer significantly?   ◀

CHAPTER ELEVEN

PSEUDO-ONE-DIMENSIONAL SYSTEMS

SECTION 11.1 TRANSVERSE CURRENTS: PLANE WAVES

A pseudo-1D system is one in which the fields depend only on **one** coordinate, which we will take to be x. As in the case of pseudo-2D systems, we can consider fields generated by currents along the x-axis (usually called **longitudinal** currents) separately from the those generated by currents perpendicular to that axis (called **transverse** currents). We consider the transverse case first. Figure 11.1 shows a pseudo-1D system whose current has a nonzero y-component $j_y(x)$, at the horizontal faces. As usual, the j/ϵ_0 term in dE/dt makes E nonzero at these faces at the next time. The system has translational symmetry in both the y and z directions: all the horizontal faces whose centers lie in each plane perpendicular to the x-axis have the same E-field. This causes curl E to vanish on x-edges (each paddlewheel has two vanishing E's and two canceling E's). The paddlewheels at vertical (y) edges involve only vertical faces, at which E vanishes. Thus curl E is nonzero only at z-edges (out of the paper), such as that at coordinate x shown in Figure 11.2a. The paddlewheel used for computing the curl here has two vertical faces (one of which is shown), at which $E = 0$. The sum over the remaining two faces is:

$$[\text{curl } E]_z(x) = \left[E_y\left(x + \tfrac{1}{2}dr \right) - E_y\left(x - \tfrac{1}{2}dr \right) \right] \big/ dr \tag{11.1}$$

We will denote the discrete spatial derivative on the right side by $dE_y(x)/dx$ (it is essentially what we called a discrete gradient in the truly 1D systems of Chapter 3.) Thus, Faraday's law becomes

$$dB_z(x, t)/dt = -dE_y(x, t)/dx \tag{11.2}$$

in this pseudo-1D system and B_x and B_y vanish.

To update E, we need curl B. At a horizontal face with coordinate x (Figure 11.2b), the curl is a sum over four bounding edges. At the two edges parallel to the paper, $B(e) = B_x = 0$; the other two give

$$[\text{curl } B]_y(x) = \left[B_z\left(x - \tfrac{1}{2}dr \right) - B_z\left(x + \tfrac{1}{2}dr \right) \right] \big/ dr$$
$$= -dB_z(x)/dx \tag{11.3}$$

191

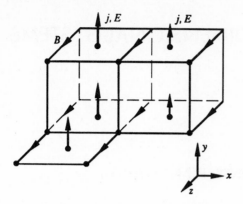

Figure 11.1 A pseudo-1D system with j along y.

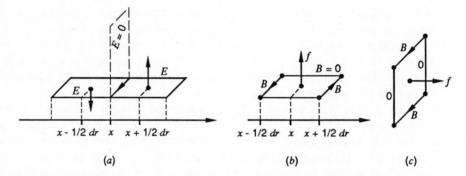

Figure 11.2 Computation of the curl of (a) a face field, (b) a edge field at a y-face, and (c) an edge field at an x-face.

The z-component of the curl involves four vanishing B's and the x-component (Figure 11.2c) has two vanishing B's and two that are equal because of the translational symmetry, and cancel from the counterclockwise sum. Thus, Ampere's law reads

$$dE_y(x, t)/dt = -c^2\, dB_z(x, t)/dx - j_y(x, t)/\epsilon_0 \qquad (11.4)$$

and E_x and E_z remain zero.

Our final result is that if only $j_y(x, t)$ is nonzero in a pseudo-1D system, only $B_z(x, t)$ (for x an edge position, usually a multiple of dr) and $E_y(x, t)$ (for x a face position) will be generated by Maxwell's equations, which take the form of Eqs. 11.2 and 11.4. In displaying these fields, we need only draw a one-dimensional cross-section through the system, such as the dashed line in Figure 11.3, which passes through the centers of the faces where $E \neq 0$ and the edges where $B \neq 0$. Each face or edge on this line represents a whole plane full of faces or edges, extending in the y and z directions, each of which has the same field value. Rather than draw the

Figure 11.3   1D cross-section of a pseudo-1D system.

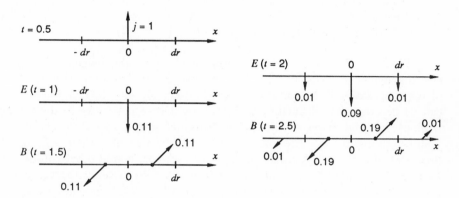

Figure 11.4   Evolution of the pseudo-1D fields generated by a transverse current.

faces explicitly, we will often just indicate the values of E and B by arrows of the appropriate direction and length, as in Figure 11.4.

As an example, let us compute the fields generated by a current at a single face, as we did for 2D and 3D systems. We will take the face to be at $x = 0$ as shown in Figure 11.4 so the resulting fields will be symmetrical about the origin. (Note that this differs from our usual convention of putting the origin at a vertex; any labeling system is acceptable as long as we're consistent.) We have used $dr = 1$ m, $dt = 1$ ns. We can compute the fields from Eqs. 11.2 and 11.4; for example,

$$E_y(dr, 2) = E_y(dr, 1) + dt \left[dE_y(dr, 1.5)/dt \right]$$

$$= 0 - dt\, c^2 [0 - (-0.11)]/dr = -0.01$$

However, you may find it easier to get the signs to come out correctly by using the 3D Maxwell equations directly and using the right-hand rule (Section 6.3) to compute the curls.

▶   **Problem 11.1**

Compute the evolution of the fields and draw a figure like Figure 11.4 for the case where $dt = 2$ ns instead of 1 ns.   ◀

If you compare Eqs. 11.2 and 11.4 to those governing the evolution of the strain s and velocity v in a chain of masses laid out along the x-axis (Section 2.5) by replacing E and B by $-v$ and s respectively, you will find they are exactly equivalent. The source term involving j becomes an external force F_{ext} (as in Eq. 6.70). (Note, however, that a **2D** network of masses is not equivalent to a pseudo-2D electromagnetic system. The analogy works in 1D only because the 1D projections of the gradient, divergence, and curl are all the same.)

▶ **Problem 11.2**

Determine the relation between k/m in the chain of masses and c^2 in the pseudo-1D Maxwell equations. Use program MAX1 to check your answer to Problem 11.1, starting with $B(t = 0.5) = 0$ and $E(t = 1.0) > 0$. You may want to rescale it to make it visible; note that you must shift the time origin by 0.5. Give $E(5.0\ \text{ns})$. ◀

We therefore already know a great deal about the behavior of pseudo-1D electromagnetic systems. In particular, we know that they support the propagation of **traveling waves**. For a certain choice of dt, any strain field s and velocity field v can be decomposed into delta-pulses (Section 2.6), which move to the left or right without changing shape. A delta-pulse, translated into E and B instead of v and s, is shown in Figure 11.5. The initial conditions shown are $E(\frac{1}{2}dr, 0) = E_0$ [i.e., E has this value at a plane of faces having $x = \frac{1}{2}dr$; $E(x, 0) = 0$ at other values of x] and $B(dr, \frac{1}{2}dt) = B_0$ [and $B(x, \frac{1}{2}dt) = 0$ at other x]. We assume the current is zero everywhere. We can easily deduce from Eqs. 11.2 and 11.4 the requirements for this pulse to propagate as shown. For Eq. 11.4 to "turn on" E_0 at $r = 1.5\ dr$ during the interval from $t = 0$ to dt, we need

$$E_0/dt = dE_y\left(1.5\ dr, \tfrac{1}{2}dt\right)/dt$$

$$= -c^2\left[B_z\left(2\ dr, \tfrac{1}{2}dt\right) - B_z\left(dr, \tfrac{1}{2}dt\right)\right]/dr$$

$$= +c^2 B_0/dr$$

so this requirement is

$$B_0 = \left(dr/c^2\ dt\right)E_0 \qquad\qquad (11.5)$$

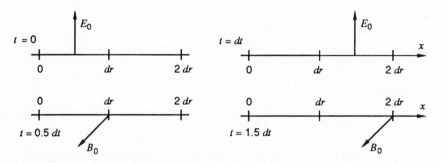

Figure 11.5 An electromagnetic delta-pulse.

Then the new E at $x = \frac{1}{2}dr$ is

$$E_y\left(\tfrac{1}{2}dr, dt\right) = E_y\left(\tfrac{1}{2}dr, 0\right) + dt\left(dE/dt\right)$$

$$= E_0 + dt\left(-c^2 B_0/dr\right) = E_0 - E_0 = 0$$

so it is "turned off" as the pulse passes, as shown in the figure. Another require-ment is obtained from Eq. 11.2 for updating B: it "turns on" B at $r = 2\,dr$ if

$$B_0/dt = dB_z(2\,dr, dt)/dt = -[0 - E_0]/dr - [0 - E_0]/dr$$

$$= \left(dr/c^2\,dt^2\right)E_0$$

or

$$dr = c\,dt \tag{11.6}$$

You can easily show that it then "turns off" B behind the pulse as well. Using Eq. 11.6, we can rewrite the other condition (Eq. 11.5) for delta-pulse propagation as

$$E_0 = cB_0 \tag{11.7}$$

It is clear that c is the wave speed because this pulse moves a distance $c\,dt$ during the time dt. From the value measured by the tachmen (Chapter 9) for c^2, we get $c = 3 \times 10^8$ m/s, which happens to be just the speed one measures for light waves. It is this coincidence between the speed of electromagnetic waves and that of light that led Maxwell in the 1860s to conclude that light is an electromagnetic wave.

You should keep in mind in looking at the E-field depicted in Figure 11.5 (at $t = 0$ for example) that the E-vector represents a field that is constant over a plane perpendicular to the direction of pulse propagation, as shown in Figure 11.6. This planar slab of nonzero E moves to the right at the speed c; for this reason waves constructed by superposing such pulses are called **plane waves**.

In Section 2.6, we showed that an arbitrary 1D wave can be decomposed into leftward and rightward traveling waves, the form of which does not change with

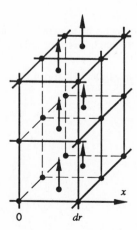

0   dr **Figure 11.6** A delta-pulse depicted in 3D.

time. The equations governing their motion are the same, so this is true of arbitrary pseudo-one-dimensional electromagnetic waves (plane waves) as well.

We have so far dealt only with fields generated by the y-component of the current. But we can do exactly the same thing with a transverse current in the z-direction; in place of the updating equations (Eqs. 11.2 and 11.4) we would get

$$dB_y(x, t)/dt = +dE_z(x, t)/dx \tag{11.8}$$

$$dE_z(x, t)/dt = +c^2 dB_y(x, t)/dx - j_z(x, t)/\epsilon_0 \tag{11.9}$$

All the pictures we have drawn would be rotated by $\frac{1}{2}\pi$ about the x-axis. The fields generated by the two current components j_y and j_z evolve independently of each other.

▶ **Problem 11.3**

Draw a figure such as Figure 11.5 to describe a **leftward** pulse. Show that Eqs. 11.2 and 11.4 correctly turn the field on and off as the pulse passes. Hint: To get the correct initial conditions at $t = 0$ and $dt/2$, rotate those in Figure 11.5 by π about a vertical axis. ◀

▶ **Problem 11.4**

(Planar antenna) Suppose the electric and magnetic fields in a pseudo-1D system are initially zero, but there is a transient current $j(x, \frac{1}{2}dt) = 1 \ \mu c/m^2$ ns for $x = \pm \frac{1}{2}dr$ as shown in Figure 11.7. The current vanishes at later times.

a. If $dr = 1$ m, compute the dt for which simple delta pulses are possible.

b. For this dt, compute the evolution of the system via Eqs. 11.2 and 11.4 up to $t = 3dt$. Sketch its fields at each time, as in Figure 11.4.

c. You should now be able to decompose the fields into delta pulses. What are E_0 and B_0? Check Eq. 11.7.

d. Check your answer to (b) using MAXWELL or MAX1. You will need to use the value you computed in (a) for dt. If you use Maxwell, note that a pseudo-1D system is a special case of a pseudo-2D system. Choose a value for j, and put it in the positions shown in Figure 11.7 and on each face above and below them, until you reach the boundary. ◀

▶ **Problem 11.5**

a. Using the general formulas for the cartesian components of the curl of a function, compute curl E and curl B for the pseudo-1D case in which there are only $E_y(x)$ and $B_z(x)$.

b. Put these expressions into the 3D Maxwell equations (Eqs. 9.12 and 9.16) and show that they reduce to Eqs. 11.2 and 11.4. ◀

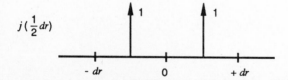

Figure 11.7 Transient current in a planar antenna.

► **Problem 11.6**

(Sinusoidal plane wave) Suppose $E_y(x, t) = E_0\cos(kx - \omega t)$ for some constants E_0, k, and ω. Assuming $j = 0$, determine a condition on k and ω so that $E_y(x, t)$ satisfies the continuum limits of Eqs. 11.2 and 11.4, for some $B_z(x, t)$. Find the function $B_z(x, t)$. ◄

SECTION 11.2 LONGITUDINAL CURRENTS

We have so far considered pseudo-1D systems with transverse currents, either j_y or j_z or both. We will now consider a system with a longitudinal (along the axis of variation, x) current, for which $j_x(x)$ is the only nonzero component. This system has nonzero j (and therefore E) on the faces indicated in Figure 11.8. When we take the curl of such a field, we find ourselves adding up fields on faces of paddlewheels such as those in Figure 11.9; in all cases the fields are either zero or they cancel because of the translational symmetry in the y and z directions. Thus, curl $E = 0$, and therefore $B = 0$ by Faraday's law. The equation for updating the field is very simple:

$$dE_x(x, t)/dt = -j_x(x, t)/\epsilon_0 \qquad (11.10)$$

The fields do not change at all except while the current is flowing. The behavior of this longitudinal-current system is much less interesting than in the transverse-current case: for example, wave propagation is not possible. However, the fields in one very important physical device, the parallel-plate capacitor, can be described by this equation. Suppose we create a positive and a negative charge, $+\sigma\,dr^2$ and $-\sigma\,dr^2$, in each of a plane of cells at $x = 0$ (Figure 11.10). Thus, σ is the charge per unit area on the plane. We then move each of the positive charges to the right a distance

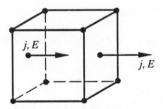

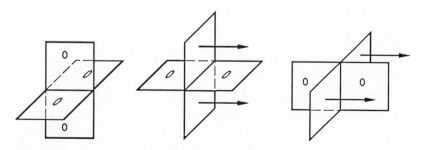

Figure 11.8 Pseudo-1D system with longitudinal current.

Figure 11.9 The curl of a longitudinal pseudo-1D field vanishes.

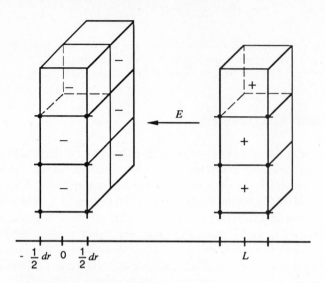

Figure 11.10   The parallel-plate capacitor and its electrostatic field.

L, as shown, during the time interval from 0 to dt. Although these charged planes are infinite, they make a good model for a real capacitor composed of parallel conducting plates whose transverse dimensions are large compared to L. The current is

$$j_x\left(x, \tfrac{1}{2}dt\right) = \text{charge}/dr^2\,dt = \sigma/dt \qquad (11.11)$$

for each x-face with $0 < x < L$. The electric field at such a face is given by Ampere's law (Eq. 11.10):

$$E_x(x, dt) = dt\left(dE_x/dt\right) = -dt\,j/\epsilon_0 = -\sigma/\epsilon_0 \qquad (11.12)$$

At later times $j = 0$, so E never changes again: this is an example of an electrostatic field, of which we will meet many more in Part C.

▶   **Problem 11.7**

    a. If a real capacitor plate has area $A = 0.1$ m^2 and the plates are separated by $L = 0.1$ mm, compute the electric field inside when their charges are ± 2 μc.

    b. If the time dt during which the planes moved apart was 100 ns, find the current density j.   ◀

▶   **Problem 11.8**

Suppose that instead of remaining at $x = L$ (so $j = 0$ at $t = 1.5\,dt$), the positive charges moved almost all the way back, to $x = dr$, between $t = dt$ and $t = 2\,dt$.

    a. Compute the current $j_x(x, 1.5\,dt)$ for all face positions x.

    b. Compute the field $E_x(x, 2\,dt)$ for all x.

CHAPTER TWELVE

MAGNETIC FORCES AND CURRENTS

SECTION 12.1   THE LORENTZ FORCE

You may have noticed that we were rather vague in Chapter 9 about precisely what the effects of a magnetic field **B** were. We defined it there in terms of the torque with which it twists compass needles to point along the magnetic field vector **B**. However, a detailed treatment of torques on magnets is rather complicated, and we will not get to it until Chapter 23. We will start by looking at the force exerted by a magnetic field on a single charged particle. We have already determined the force on such a particle due to the electric field **E**; it is $\mathbf{F} = q\mathbf{E}$, where q is the charge on the particle. If we are going to follow the trajectory of this particle, we need to know the field at each possible position. That is, we must consider the field in the continuum limit.

> As in Chapter 4, the (space) continuum limit means $dr \ll b$, where b is the distance over which the fields change significantly: then we can attach an unambiguous value to E at any position **r** because there are faces f (at which E is defined) very close to **r** and they have nearly the same value of $E(f)$. Although we will regard the particle as a "point" particle of zero size here, everything we do will be valid as long as it is much smaller than b.

It turns out that the magnetic field has no effect on the force on a stationary particle. However, if the particle moves with a velocity **v**, we can measure an additional force (the **Lorentz force**) $q\mathbf{v} \times \mathbf{B}$. Thus the total force on a moving charged particle is

$$\mathbf{F} = q\mathbf{E} + q\mathbf{v} \times \mathbf{B} \tag{12.1}$$

Recall that the cross-product $\mathbf{v} \times \mathbf{B}$ is a vector perpendicular to the plane of **v** and **B**, with magnitude $vB \sin \theta$, where θ is the angle between **v** and **B**. The direction of $\mathbf{v} \times \mathbf{B}$ is given by a right-hand rule: if you stick out your right thumb and orient your hand so you can swing your fingers palmward from pointing along **v** to pointing along **B**, your thumb points along $\mathbf{v} \times \mathbf{B}$. (The cross-product can also be expressed in terms of the components of **v** and **B**; see Section 6.7.) Thus the Lorentz force is perpendicular to both **B** and **v**, as shown in Figure 12.1 for the case in which **v** and **B** are in the plane of the paper. The particle is never accelerated along its

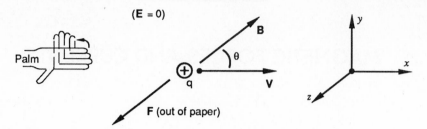

Figure 12.1 The right-hand rule for the Lorentz force.

direction of motion by the Lorentz force, but only perpendicularly to it. If it happens to be moving along the magnetic field, it is not accelerated at all (because $\theta = 0$).

▶ **Problem 12.1**

a. Compute the Lorentz force (magnitude and direction) on a particle of charge 1 μc moving at 0.1 m/ns to the east, in a magnetic field 0.05 mT pointing north. (This is approximately the magnitude of the Earth's magnetic field.)

b. Compare this to the force in an electric field $E = 0.05$ N/μc. What velocity would be necessary to make the magnetic and electric forces equal in magnitude? Is this velocity possible? ◀

SECTION 12.2 PARTICLE TRAJECTORIES IN MAGNETIC FIELDS: CYCLOTRON MOTION

The simplest magnetic field is a uniform one, in which **B** is independent of position. Consider a uniform B-field in the z-direction (out of the paper), in which a particle of charge q and mass m is moving at a speed v in the y-direction (Figure 12.2). The Lorentz force on the particle (Eq. 12.1) points to the right, and has magnitude qvB. The acceleration is qvB/m, so the trajectory curves to the right as shown in the

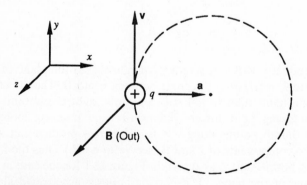

Figure 12.2 A circular cyclotron orbit.

Figure 12.2. If the acceleration vector were constant, such as that of a particle in free fall, it would have a parabolic trajectory. However, in this case the acceleration vector turns when the velocity vector does: they must remain perpendicular. You can guess the form of the trajectory if you remember that in uniform circular motion the acceleration always points toward the center of the circle and has magnitude v^2/r. If we choose r so this matches the known Lorentz acceleration

$$v^2/r = qvB/m$$

or

$$r = mv/qB \qquad (12.2)$$

then the circular trajectory has the correct acceleration, at least at the point depicted in Figure 12.2. However, to describe the acceleration at any other point of the circle, we need only rotate the figure. Thus the acceleration matches that of the circular trajectory at **all** points, so this must be the true trajectory. It is called a **cyclotron orbit**, after an early type of accelerator in which electrons or protons follow such orbits in a uniform magnetic field. The radius given by Eq. 12.2 is called the **cyclotron radius**.

We will also give an alternate derivation of the cyclotron orbit based on solving coupled differential equations, which is more precise but more mathematical. Suppose the trajectory is described by a position vector $\mathbf{r}(t)$, which is a function of time, with components $x(t)$, $y(t)$, and $z(t)$. Then the velocity components are

$$v_x(t) = \frac{dx(t)}{dt}$$

$$v_y(t) = \frac{dy(t)}{dt} \qquad (12.3)$$

$$v_z(t) = \frac{dz(t)}{dt}$$

and Newton's second law ($F = ma$) says

$$\frac{d\mathbf{v}}{dt} = \left(\frac{q}{m}\right)\mathbf{v} \times \mathbf{B}$$

or in terms of components (using $B_z = B$, $B_x = B_y = 0$)

$$\frac{dv_x}{dt} = \left(\frac{qB}{m}\right)v_y$$

$$\frac{dv_y}{dt} = -\left(\frac{qB}{m}\right)v_x \qquad (12.4)$$

$$\frac{dv_z}{dt} = 0$$

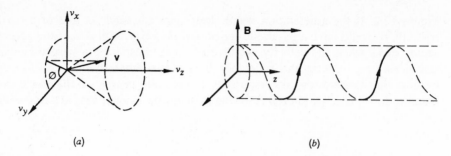

(a)   (b)

Figure 12.3   (a) Rotation of velocity vector. (b) A helical cyclotron orbit.

The solution for v_z is very simple: the velocity along the B-field is just a constant. The equations for v_x and v_y are a coupled pair of ordinary differential equations (in the $dt \to 0$ limit) equivalent to those of the harmonic oscillator (Eqs. 1.1 and 1.2; just replace v_y by v, v_x by qBx/m, and qB/m by $(k/m)^{1/2}$). The solution is therefore the same:

$$v_x(t) = A \sin(\omega t + \phi) \tag{12.5}$$

where $\omega = qB/m$ (this is called the **cyclotron frequency**), and A and ϕ are constants. We can get v_y from Eqs. 12.4,

$$v_y(t) = A \cos(\omega t + \phi) \tag{12.6}$$

and you can verify that the velocity given by Eqs. 12.5 and 12.6 satisfies the differential Eqs. 12.4. These are the equations of a circle: the velocity vector has constant magnitude ($v^2 = A^2 + v_z^2$) and rotates with angular speed ω about the z-axis (Figure 12.3a). We can get the position coordinates by integrating with respect to time:

$$x(t) = -r\cos(\omega t + \phi) + \text{constant}$$

$$y(t) = r\sin(\omega t + \phi) + \text{constant} \tag{12.7}$$

$$z(t) = v_z t + \text{constant}$$

where $r = A/\omega$ is the cyclotron radius. This trajectory is more general than that shown in Figure 12.2, where we assumed $v_z = 0$. The projection of the trajectory onto the xy-plane is still a circle of radius r, but the actual trajectory is a helix (Figure 12.3b).

▶   **Problem 12.2**

A particle with velocity $v = 10^6$ m/s and mass $m = 10^{-15}$ kg follows a circular path of radius $r = 5$ m in a uniform magnetic field $B = 1$ T. Compute its charge q.   ◀

▶   **Problem 12.3**

A particle of charge q moves in uniform crossed electric and magnetic fields (v, **E**, and **B** are all mutually perpendicular.) For what speed does the particle move in a straight line? Compute this v if $E = 10$ N/μc, $B = 1$ T.   ◀

SECTION 12.3   FORCE ON A SMOOTH CURRENT DENSITY

When charges move, there is a nonzero current density. In the case of a single point charge, this current density is rapidly varying in space. In fact, for any given time interval it is nonzero at one, or at most a few, faces; it is not very useful to try to express the Lorentz force on the charges in terms of the current density. On the other hand, one frequently deals with smooth current densities (slowly varying in space) involving large numbers of charged particles. This is the case, for example, in a metal wire through which a current is flowing. We would like to deduce the force on such a current from Lorentz's formula for the force on a single particle.

We must first relate the current to the density of particles (which is proportional to the charge density.) Suppose we have a slowly varying density of charged particles, so each of the cells has nearly the same density ρ. The amount of charge in each cell is then $\rho \, d^3 r$. Suppose further that the particles all have the same charge q, and that they are all moving with the same velocity $\mathbf{v}$. Then the particles that cross the face f during dt are precisely those in the parallelopiped shown in Figure 12.4. Assuming the density is slowly varying even on a much smaller scale than dr (this requires that the number of particles per cell $N = \rho \, d^3 r / q$ be very large), the amount of charge Q in the parallelopiped is ρ times its volume. The volume is the product of the area $da = dr^2$ of the base (i.e., the face f) by the height normal to f, $v_x \, dt$, so

$$Q = \rho v_x \, dt \, da$$

By definition of the current density $j(f)$ through this face,

$$Q = j(f) \, da \, dt$$

from which it follows that

$$j(f) = \rho v_x$$

When the current is represented by a vector function $\mathbf{j}$, this $j(f)$ is the x-component j_x. Making this same argument for the other two components as well gives the vector equation

$$\mathbf{j}(\mathbf{r}, t) = \rho(\mathbf{r}, t)\mathbf{v}(\mathbf{r}, t) \tag{12.8}$$

where ρ and $\mathbf{v}$ may vary (slowly) with position and time.

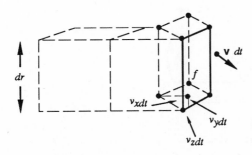

Figure 12.4   Parallelopiped of particles about to cross the face f of a cubical cell.

We can now calculate the Lorentz force on this current. Each particle has a force $q\mathbf{v} \times \mathbf{B}$ on it. Thus the total force on a cell of volume $d\tau = dr^3$ (containing $\rho\, d\tau/q$ particles) is

$$d\mathbf{F} = (\rho\, d\tau/q)q\mathbf{v} \times \mathbf{B} = \rho\mathbf{v} \times \mathbf{B}\, d\tau$$

or

$$d\mathbf{F} = \mathbf{j} \times \mathbf{B}\, d\tau \tag{12.9}$$

Of course, this assumes that the current arises from a particular (and rather unlikely) distribution of particle velocities in which every particle has exactly the same velocity. But it turns out that the Lorentz force on a current density doesn't depend on the details of the particle motions, but only on the total current $\mathbf{j}$. It is possible to calculate the force on a distribution of particles with many different velocities. If we group those with the same velocity together, we can write an Eq. 12.9 for each group, giving the force per unit volume on that group. Adding all these equations gives the total force per unit volume, and (because Eq. 12.9 is linear in the current density) the right-hand side is just (total current) $\times \mathbf{B}\, d\tau$. Thus Eq. 12.9 gives the total force in terms of the total current for this case as well.

As a simple example of the application of Eq. 12.9, let us compute the force on a straight wire whose length and orientation are given by $\mathbf{l}$ in Figure 12.5. Assume a uniform current density $\mathbf{j}$ and a cross-sectional area A. Equation 12.9 gives for the force

$$\mathbf{F} = j\hat{\mathbf{x}} \times \mathbf{B}lA = (jA)\mathbf{l} \times \mathbf{B} \tag{12.10}$$

The product jA (current per unit area times area) is, of course, the (total) **current**, denoted I. This is the total amount of charge passing any cross-section of the wire per unit time, and its unit is the ampere (one coulomb per second). In terms of I,

$$\mathbf{F} = I\mathbf{l} \times \mathbf{B} \tag{12.11}$$

In Figure 12.5, we chose $\mathbf{B}$ to point out of the paper and $\mathbf{l}$ to point to the right; the force then points down, with magnitude IlB.

Reversing the direction of the current reverses this force: if the current pointed to the left, the force would point up. If I is sufficiently large, this magnetic force will

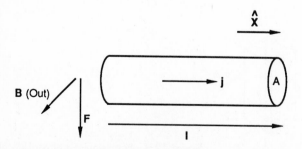

Figure 12.5 Force on a wire with uniform current.

support the wire: if the mass of the wire is M, its weight Mg is exactly balanced by the Lorentz force if

$$IlB = Mg$$
$$I = Mg/lB$$

$$(12.12)$$

If $B = 5 \times 10^{-5}$ T (the earth's field) and $M/l = 10^{-2}$ kg/m, this gives $I = 2 \times 10^3$ amperes.

▶   **Problem 12.4**

What magnetic field would be necessary for a wire of mass 10^{-2} kg/m to be self-supporting with a current $I = 1$ ampere?   ◀

▶   **Problem 12.5**

The earth's magnetic field is horizontal only at the equator. At a latitude at which it is at an angle 30° downward from the horizontal (still with magnitude 5×10^{-5} T; its horizontal component is still north, of course), what are the magnitude and direction of the Lorentz force on a one-meter section of a horizontal wire carrying $I = 20$ a exactly northward?   ◀

SECTION 12.4   NONUNIFORM CURRENT DENSITIES

When the current density (in the wire in Figure 12.7, for example) is not spatially uniform, we can still calculate the total current carried across a given surface S. Consider first a discrete current density j on a lattice like that shown in Figure 12.6; S is an oriented surface composed of oriented faces f. The amount of charge passing a face f during an interval of length dt centered at time t is $j(f, t)\, da(f)\, dt$ (this is the definition of j) where $da(f)$ is the area of face f. Define the **total current** through S for the interval t, denoted $I_S(t)$, to be the charge crossing S per unit time, so the total charge crossing is $I_S(t)\, dt$. Clearly, this is the sum of the amounts crossing the faces; dividing by dt gives

$$I_S(t) = \sum_{f \in S} j(f, t)\, da(f)$$

$$(12.13)$$

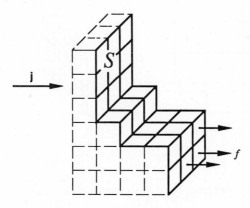

Figure 12.6   Discrete surface S for computing total current.

which is just a discrete surface integral (Section 6.3) of the current density. In the continuum limit, we can define j by a vector function $\mathbf{j}(\mathbf{r}, t)$. The limit of Eq. 12.13 is usually written

$$I_S(t) = \int \mathbf{j} \cdot d\mathbf{a} \tag{12.14}$$

This notation arises from thinking of the area of a face as a **vector**, denoted $d\mathbf{a}$, normal to the face, with magnitude $da(f)$. Then the dot product $\mathbf{j} \cdot d\mathbf{a}$ picks out the right component of the vector $\mathbf{j}$, that is, for a face normal to the x-direction it gives $\mathbf{j} \cdot (\hat{\mathbf{x}}\, da) = j_x\, da$, which is the correct term in the discrete surface integral. The dot product notation has the advantage that $\mathbf{j} \cdot d\mathbf{a}$ is the flux across a small area even if the area is not normal to one of the coordinate axes.

■ **EXAMPLE 12.1**   Consider a wire of square cross section and area b^2, shown in Figure 12.7a. The current function is $j_x(x, y, z) = Ay$ for $0 < y < b$ and $0 < z < b$ (i.e., inside the wire) and $j_x = 0$ outside the wire; $j_y = j_z = 0$ everywhere. The current is independent of time. The surface S is the plane $x = 0$. To evaluate Eq. 12.14 we must usually go back to Eq. 12.13 and set up a discrete lattice, as shown in Figure 12.7b. At each face f centered at $(0, y, z)$,

$$j(f)\, da = j_x(0, y, z)\, dz\, dy$$

and the sum in Eq. (12.13) is over y and z such that $0 < y < b$, $0 < z < b$. This can be expressed as a repeated sum, first over y and then over z, whose limit is the repeated integral

$$\int_0^b dz \int_0^b dy\, j_x(0, y, z) = \int_0^b dz \left(\int_0^b dy\, Ay \right) = \int_0^b dz \left(\tfrac{1}{2}Ab^2 \right) \tag{12.15}$$

so the total current is

$$I_S = \tfrac{1}{2}Ab^3 \tag{12.16}$$

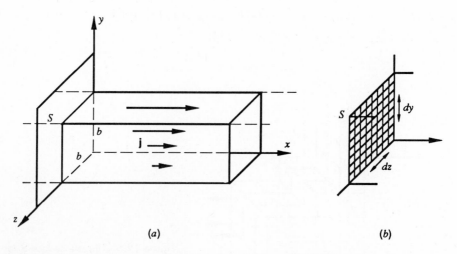

(a)   (b)

Figure 12.7   (a) A square wire cut by a surface S. (b) a discretization of S.

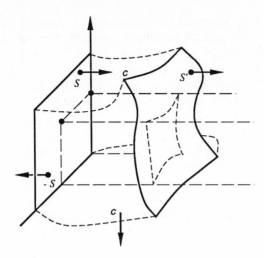

Figure 12.8 An arbitrary surface S' cutting a wire.

We would expect the total current in the above example to be the same for any surface S that cuts through the wire. This is easy to see for other surfaces $x = $ constant (parallel to the yz-plane), because $j_x(x, y, z)$ is independent of x so the integral is the same. It is not obvious for an arbitrary surface S' such as that in Figure 12.8, but we can prove it by the following general argument. We can connect S and S' together to form a **closed** surface, as suggested by the dashed lines in Figure 12.8. For this surface to be **orientable** (as defined in Section 6.3), we must reverse the orientation of one of the surfaces, say S; we refer to the oppositely oriented surface as $-S$. We can then apply the divergence theorem (Section 6.3) to this closed surface consisting of S', $-S$, and the connecting surface, say C, and the region V it encloses:

$$\int_{S'} \mathbf{j} \cdot d\mathbf{a} + \int_{-S} j \cdot d\mathbf{a} + \int_C \mathbf{j} \cdot d\mathbf{a} = \int_V \operatorname{div} \mathbf{j}\, d\tau \qquad (12.17)$$

We have put the connecting surface C **outside** the wire, so $j = 0$ there and the third integral vanishes. We can reverse the orientation of the surface in the second integral, changing its sign (in this context we can think of this as being because the area element vector $d\mathbf{a}$ reverses), and obtain

$$\int_{S'} \mathbf{j} \cdot d\mathbf{a} - \int_S \mathbf{j} \cdot d\mathbf{a} = \int_V \operatorname{div} \mathbf{j}\, d\tau \qquad (12.18)$$

In general, the right-hand side is $-\int (d\rho/dt)\, d\tau$ because of the continuity equation; this is just the rate of decrease of the charge in the volume V between the surfaces, and Eq. 12.18 is a statement of the conservation of charge. In a steady-state system, the charge density should not change: $\operatorname{div} \mathbf{j}$ should vanish. (In the present case, we can easily check this: $\operatorname{div} \mathbf{j} = dj_x/dx = 0$ because $j_x = Ay$ is independent of x.) Then Eq. 12.18 becomes

$$\int_{S'} \mathbf{j} \cdot d\mathbf{a} = \int_S \mathbf{j} \cdot d\mathbf{a} \qquad (12.19)$$

which is what we wanted to prove. The total currents I_S and $I_{S'}$ are the same; the subscript S is redundant. We can use just I to denote "the current in the wire," as long as it is understood that the integral is to be taken over a surface that cuts all the way across the wire and that the system is in a steady state; it is independent of the surface. We do have to be careful to specify an orientation along the wire (which becomes the orientation of the surface), or there is a sign ambiguity.

▶   **Problem 12.6**

Compute the total current in the wire of Figure 12.7 if $j_x(x, y, z) = Cz(b - z)y(b - y)$ and $j_y = j_z = 0$ inside the wire, and $\mathbf{j} = 0$ outside. ◀

▶   **Problem 12.7**

Compute the total current in a **cylindrical** wire, in which (in cylindrical coordinates) $j_z(r, \phi, z) = A(b^2 - r^2)$ for $r < b$, $j_z = 0$ outside the wire (for $r > b$), and $j_r = j_\phi = 0$. ◀

SECTION 12.5   ELECTROMAGNETIC PULSE (EMP)

In this section we will work out an extended example involving the use of both Maxwell's equations and the properties of cyclotron orbits: the electromagnetic pulse (EMP) produced by a high-altitude nuclear explosion. We will assume a 1 megaton detonation at 400 km altitude, such as might occur in a "Star Wars" battle in earth orbit. [Don't try this experiment unless Mom or Dad is there to help if something goes wrong.] Such a detonation produces both neutrons and gamma rays, but it is the gamma rays that are responsible for the EMP. There are about 10^{25} of these, emitted over a period of roughly 30 ns. (The data used in this example are from an article on EMP by Conrad L. Longmire, IEEE Transactions on Antennas and Propagation vol. AP-26, p. 3, 1978, in which you can also find more detailed calculations.) The gamma rays travel outward at the speed of light, in a spherical shell of thickness $c \times 30$ ns $= 9$ m. When this shell strikes the upper atmosphere, the gamma rays can interact with atoms, knocking electrons out of the atoms; this is called the **Compton effect**. The Compton electrons, initially moving almost at the speed of light in the same direction as the gammas (i.e., downward), produce a current density j that is the source of the EMP. This current is greatest at an altitude of about 30 km, at which the shell of gammas has a radius of 370 km and can be well approximated by a horizontal planar slab (see Figure 12.9). Thus, assuming a high-altitude detonation allows us to treat this as a pseudo-1D problem. The gamma-ray area density on the shell (gamma rays per square meter) can be calculated by dividing the total number of gammas by the area $4\pi r^2$ of the sphere, giving about 6×10^{12} m^{-2}. (In terms of rads, the standard unit of radiation dosage, this is about 0.6 rads; for comparison, 100 rads is usually considered the threshold for radiation sickness, and 300 for 50% mortality in the absence of medical treatment.) We will use a cubic lattice with $dr = 100$ m; the air density at 30 km is such that the fraction of the gammas that are absorbed in traversing one 100 m slab of air is about 0.00(The   air density at sea level is larger by about the same factor, so that almost all of the prompt gammas from a surface detonation are absorbed in a few hundred meters.)

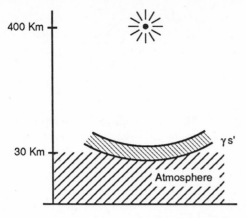

Figure 12.9 Shell of γ-rays (not to scale) hitting atmosphere.

The current density produced by the Compton electrons depends on how far they go before stopping (their mean range). Fortunately, this happens to also be about 100 m at 30 km altitude, so the Compton electrons that cross a horizontal plane are those that were created in the 100 m slab above it (Figure 12.10). The number of Compton electrons crossing a horizontal plane is therefore $0.007 \times 6 \times 10^{12}$ per square meter of lateral area. The charge per square meter crossing this plane is the electron charge (to be found on the inside front cover of this book) times this, or $(1.6 \times 10^{-19}$ c$)(4 \times 10^{10}$ m$^{-2}) = 6 \times 10^{-9}$ c/m^2. The current flows across this plane only while the gamma-ray slab is crossing it, which is for 30 ns, so the instantaneous current density is about 0.2 a/m^2. However, to make the calculations simpler we will use $dt = dr/c = 333$ ns, and to make the current smoother we will split the crossing charge between two intervals dt (the latter is really to avoid exciting the corrugation wave of Section 2.7, which is always a danger when $dr = c\,dt$.) Thus, our discrete current density at 30 km is $(6 \times 10^{-9}$ c/m$^2)/(666 \times 10^{-9}$ s$) = 0.01$ a/m^2. (Admittedly we are making a lot of approximations, and you have only my word for it that they don't change the answer qualitatively. But these sorts of numerical approximations have the enormous advantage over analytical approximations, such as leaving terms out of formulas, that you can easily improve them: just decrease dr and dt. See, for example, Problem 12.9.) During each interval dt, the current density is nonzero at two adjacent planes (Figure 12.11). The

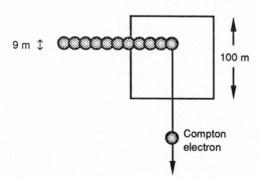

Figure 12.10 Downward current of Compton electrons.

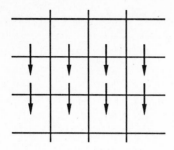

Figure 12.11 Nonzero discrete current field for a time interval when the γ-ray slab is at 30 km altitude.

gamma-ray slab moves down at the speed of light, one cell per dt, so at the next interval the current density has moved down by dr.

The intensity of this Compton current source changes smoothly from zero above the atmosphere to the maximum value we calculated above and then back to zero at an altitude below 25 km when all of the gammas have been absorbed and all the Compton electrons stopped (see Figure 12.12). But (again to simplify the calculation) we will approximate the current density by a step function, 0.01 a/m² while the source is in a 10-km thick slab extending from 25 to 35 km, and zero elsewhere, as shown in the figure. (Again, if you wanted to improve on this approximation, you could do so quite easily.)

Having settled on a model of the current density, we are now ready to calculate the fields. Here our quest for EMP runs into a snag: our pseudo-1D system has a longitudinal current, as defined in Section 11.2, so we can never get a magnetic field or electromagnetic radiation. The gamma ray slab leaves an electric field behind it, which we can easily calculate from the Maxwell equation for dE/dt: each of the two 0.01 a/m² current densities at a given face increases E by $j\,dt/\epsilon_0$, leaving $E = 800$ N/c. There will be no electric field below the current cutoff (at 25 km altitude); no electromagnetic effects at all will be observed at the earth's surface.

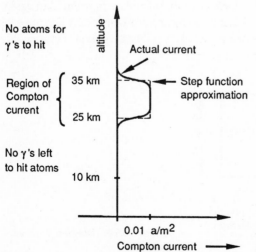

Figure 12.12 Maximum Compton current as a function of altitude.

A calculation similar to the above was done in 1962 by physicists preparing for the high-altitude nuclear test code-named Starfish, except that they did not approximate the spherical shell of gammas by a slab, so the system was not exactly pseudo-1D and they did not get exactly zero for the EMP amplitude. But they got a relatively small number, which they used in designing their EMP detectors. During the actual test, some of the detectors went off-scale and gave no usable reading. Those who had done the calculation concluded that they must have left something out, as we have done also. It will not surprise you (given the title of this chapter) that what has been left out is a magnetic force, the Lorentz force on the Compton electrons that is due to the Earth's magnetic field. It might appear to you that the effects of this tiny (5×10^{-5} T) field, which we walk around in every day without apparent ill-effects, must be negligible compared to those unleashed by a nuclear explosion. Evidently it also appeared this way to the people who thought about it in 1962. But it turns out that this is not true; if you thought it, you are only the latest in a long and distinguished list of people, beginning with the Viking King Canute, who have overestimated the power of human contrivances in comparison with the forces of nature. In fact, the Earth's magnetic field twists the trajectories of the Compton electrons from the straight lines we assumed above into circles (see Figure 12.13; we assume we are over the equator, so **B** is horizontal). Their radius (the cyclotron radius) is about 50 m for the electron velocities involved here. This changes the current in a very important way: it now has a nonzero transverse (horizontal in the figure) component. According to Section 11.1, this allows the generation of electromagnetic waves.

It is possible to do a detailed calculation of the horizontal current density. However, for our present order-of-magnitude calculation, it is enough to observe from Figure 12.13 that the Compton electrons travel nearly one cell width horizontally before being stopped, so the horizontal current is about the same as the vertical current we calculated before: at each face it is 0.01 a/m^2 for each of two consecutive time intervals; because it moves down dr during each dt, for a given time interval it is nonzero at two faces, as shown in Figure 12.14.

The electric and magnetic fields are zero until the Compton current starts, which is when the gamma ray slab hits the top of our step-function "atmosphere" at 35 km altitude. We set the time origin so that $t = \frac{1}{2}dt$ when the current is nonzero at the top two faces of the atmosphere in Figure 12.14. (To be entirely consistent, there is also current at the top face at $t = -\frac{1}{2}dt$, but this is a small effect and we will ignore it here.) The evolution of the system for several time intervals is shown in Figure 12.14.

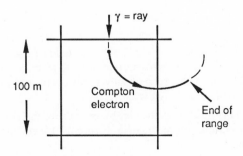

Figure 12.13   View of a Compton electron trajectory from the north (so *B* points out of the paper).

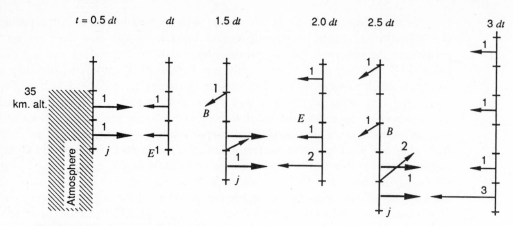

Figure 12.14 Evolution of the EMP. Heavy arrows (→) are current density j (in units $0.01\, a/m^2$), light ones are E (units 380 N/c) and B (units 1.3×10^{-6}T). View is from south.

It is easy to guess from Figure 12.14 what the fields look like after t/dt intervals: there are t/dt delta-pulses of amplitude 1 moving backward (upward) and one, of amplitude t/dt, moving forward. We can understand this easily in terms of Problem 11.4, in which a transient current at a single time launches one delta-pulse forward and one backward. In the present case, we have such pulses emitted at **every** time, from a **moving** source. The forward pulses are all created on top of each other, so they merge into one very large forward pulse. The backward pulses move away from the source, so they are all separate. It is the constructive interference of the forward pulses that makes the EMP such an important effect. It is exactly analogous to the bow shock that builds up on the nose of an airplane breaking the sound barrier. The amplitude of the forward EMP can become quite large; when the γ-rays have all been absorbed, at 25 km altitude, they have gone through 10 km/100 m = 100 cells, so the forward EMP has $E = 100 \times (800 \,\text{N/c}) = 8 \times 10^4$ N/c. Actually this is now a slight **overestimate** of the amplitude, because we have still left out an important effect. The collisions with atoms, that slow down the Compton electrons, also knock other electrons ("secondary electrons") out of the atoms. When these electrons feel the large electric field we calculated above, they move, creating their own current, which we have ignored. In effect, the air becomes a **conductor**; we will discuss these effects later, in Chapter 21.

▶   **Problem 12.8**

   a. Use the Maxwell equations to compute for $t \leq 2.5\, dt$ the fields generated by a smoother current, having at a given time nonzero current at **three** faces, of magnitude $\frac{1}{2}$, 1, and $\frac{1}{2}$, in units of 0.01 a/m². (The current moves down one cell per dt, as before.) Give your results in a sketch such as Figure 12.14.

   b. Check your result using the superposition principle, by expressing the current as a superposition of a constant multiple of the one in Figure 12.14 and a spatially translated copy of this multiple.   ◀

▶   **Problem 12.9**

Improve the numerical calculation done in this section by decreasing dt, as follows.

a. The actual duration of the horizontal current at a given altitude is about the lifetime of the Compton electron, 300 ns, so $j \approx 0.02$ a/m^2. Write or borrow a program for solving the pseudo-1D Maxwell equations (MAX1 would work if modified to allow a source term j), and compute the fields using $dt = 150$ ns, $dr = 100$ m (so that j has the above value for two time intervals at each face, as it did in Figure 12.14.) What is the electric field amplitude in the EMP after the source has moved 10 km (i.e., after 200 dt)?

b. Try $dt = 75$ ns, $dr = 50$ m, so the current is spread over four intervals; what is the amplitude after 10 km now?   ◀

▶   **Problem 12.10**

a. Show that if f is any function of one variable, $j(x, t) = 2f(x - ct)$, and we define $g(x, t) \equiv \int_0^t f(x + ct - 2ct') \, dt'$, then

$$E_y(x, t) = -tf(x - ct)/\epsilon_0 - g(x, t)/\epsilon_0$$

$$B_z(x, t) = -tf(x - ct)/c\epsilon_0 + g(x, t)/c\epsilon_0$$

is an exact solution of the continuum Maxwell equations (Eqs. 11.2 and 11.4).

b. Sketch j, E, and B for the case that f is a Dirac delta function, so g is a step function, for some time $t > 0$. Compare to Figure 12.14 and to Problem 12.8, if you did it.   ◀

CHAPTER THIRTEEN

ENERGY IN ELECTROMAGNETISM

SECTION 13.1   WORK DONE BY AN ELECTRIC FIELD

The law of conservation of energy is a very important one in mechanics. Clearly, mechanical energy is not conserved when electromagnetic fields are present: when a radio wave hits an antenna, for example, the electric field exerts a force on the electrons in the antenna, accelerating them and giving them kinetic energy, though there was no mechanical energy before. The only way to resurrect the law of conservation of energy is to assume that the **fields** themselves have energy and interpret the above process as a transfer of energy from the fields to the electron. In this chapter, we will look for a suitable definition of the energy of the fields, such that the law of conservation of energy still holds.

We will think of the electric and magnetic field together as constituting an electromagnetic **system**, whose state is specified by giving these fields and whose energy is some function of the fields. This system can do work on external objects (say charged particles) by exerting electric and magnetic forces, such as $q\mathbf{E}$ in Figure 13.1.

This work results in an increase in the energy of these external objects (kinetic energy of the charged particle in this case). This is consistent with conservation of energy only if there is a corresponding **decrease** in the energy of the electromagnetic system (the EM energy), which we will denote by U:

$$\Delta U = -W \tag{13.1}$$

where W is the work done by the EM system.

As a step toward finding a formula for the EM energy U, let us compute the work done by the EM system. Consider a discrete EM system with smooth fields (so it can describe a real system.) We will start with an electric field and a current pointing in the x-direction, E_x and j_x, as in Figure 13.2, which do not vary significantly over the figure. We get this current by having a charge density ρ that is moving to the right with velocity v_x: $j_x = \rho v_x$ (Eq. 12.8). The force on the charge $\rho\, d\tau$ in each cell is $\rho\, d\tau\, E_x$. The charge moves a distance $v_x\, dt$ in the direction of the force during an interval dt, so the work done on it is

$$dW = \text{force} \times \text{distance} = \rho\, d\tau\, E_x v_x\, dt = j_x E_x\, d\tau\, dt \tag{13.2}$$

and the rate at which work is done (i.e., the power) is

$$\frac{dW}{dt} = j_x E_x\, d\tau \tag{13.3}$$

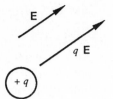

Figure 13.1   A particle subject to an electric force.

Now, note that if **E** had had y or z components, they would have had no effect: forces perpendicular to the direction of motion do no work. However, if **j** had also had a y-component, the motion in the y-direction would do work against E_y, giving a term $j_y E_y \, d\tau$ in the power. Adding a z-term for the same reason gives

$$\frac{dW}{dt} = \mathbf{j} \cdot \mathbf{E} \, d\tau \tag{13.4}$$

for the rate at which the EM system does work on the charges in the volume element $d\tau$. The corresponding rate for the whole system is the sum over such elements, whose continuum limit is the volume integral

$$\frac{dW_{\text{total}}}{dt} = \int \mathbf{j} \cdot \mathbf{E} \, d\tau \tag{13.5}$$

We should also consider the work done by the **magnetic** field. However, it is a general rule that magnetic fields never do work on charged particles or currents. This is because the magnetic force (the Lorentz force) $q\mathbf{v} \times \mathbf{B}$ is always perpendicular to **v**, and hence to the particle displacement. The work (which is the dot product of the force and the displacement, as in Eq. 13.2) therefore vanishes.

Although Eq. 13.5 for the work, like Maxwell's equations, applies to the real world only in the smooth-field (continuum) limit, we can define concepts of "power" and "work" in a discrete system that reduce correctly to Eq. 13.5 in the continuum limit and are useful in visualizing the flow of energy in electromagnetic systems. These are quite analogous to the concepts of energy and work we discussed

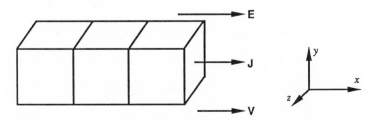

Figure 13.2   System with a field [E] doing work on a charge density.

for a 1D system of masses and springs in Section 2.8; it may be worth reviewing that section if what follows is hard to understand.

In a discrete system, the j and E in the work are defined at faces; the j_x and E_x in Eq. 13.2 for the work correspond to $j(f)$ and $E(f)$ for a face normal to the x-direction. It is therefore reasonable to think of this work $j(f)E(f)\,dr^3\,dt$ as being done **by** the field $E(f)$ at this particular face. The difficulty in doing this is that j and E are defined at different times, differing by $\frac{1}{2}dt$. We encountered a similar problem in Section 2.8, which we solved by defining a discrete notion of "work" for time intervals of length $\frac{1}{2}dt$ instead of dt. If j is defined at t, we define the work done between $t - \frac{1}{2}dt$ and t by

$$dW\left(f, t - \tfrac{1}{4}dt\right) = j(f, t)E\left(f, t - \tfrac{1}{2}dt\right) dr^3\, dt/2$$

$$\equiv P\left(f, t - \tfrac{1}{4}dt\right) dr^3\, dt/2 \tag{13.6}$$

(we label the interval by its midpoint $t - \frac{1}{4}dt$). The P defined by Eq. 13.6 has units of work per unit volume per unit time; it is a **power density**. The work done between t and $t + \frac{1}{2}dt$ is

$$dW\left(f, t + \tfrac{1}{4}dt\right) = j(f, t)E\left(f, t + \tfrac{1}{2}dt\right) dr^3\, dt/2$$

$$\equiv P\left(f, t + \tfrac{1}{4}dt\right) \quad dr^3\, dt/2 \tag{13.7}$$

This defines a discrete power density field $P(f, t)$, for each face f and each odd multiple t of $\frac{1}{4}dt$. A general formula for P is

$$P(f, t) = j(f, t_j)E(f, t_E) \tag{13.8}$$

where t_j is the time at which j is defined that is closest to t (strictly speaking, we should write this as a function of t: $t_j(t)$), and t_E is the time at which E is defined that is closest to t (Figure 13.3). The appearance of multiples of $\frac{1}{4}dt$ should look familiar from the discussion of energy in chains of masses and springs (Section 2.8). The power we have just defined is the exact analog of the power transmitted to an external force on a chain of masses.

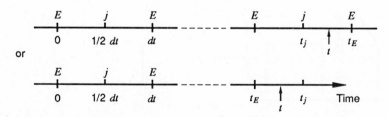

Figure 13.3   Times used in defining the power density.

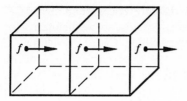

Figure 13.4 Faces normal to x-axis, summed over in Eq. 13.10.

The total work done by the field in all space during such an interval of length $\frac{1}{2}dt$ is the sum over all the faces

$$dW_{\text{total}}(t) = \sum_f dW(f, t)$$

so the work per unit time, substituting Eqs. 13.6 and 13.7 for $dW(f, t)$, is

$$dW_{\text{total}}(t)/(dt/2) = \sum_f j(f, t_j) E(f, t_E) \, dr^3 \qquad (13.9)$$

To be sure our definition of work in a discrete system is consistent, we should check that the total work (Eq. 13.9) reduces in the continuum limit to the correct result (Eq. 13.5). When j and E are defined by vector functions $\mathbf{j}(\mathbf{r}, t)$ and $\mathbf{E}(\mathbf{r}, t)$, Eq. 13.9 can be broken down into three sums, the first involving the f's normal to the x-direction (Figure 13.4), the next involving f's normal to $\hat{\mathbf{y}}$, and similarly for z. The first sum is

$$\sum j_x(\mathbf{r}, t_j) E_x(\mathbf{r}, t_E) \, dr^3 \qquad (13.10)$$

The continuum limit of this discrete sum is a volume integral. It is not **exactly** like the ones we defined in Section 6.5, because the sum is over x-faces instead of over cells. However, we can define a lattice displaced to the right by $\frac{1}{2}dr$ from that of Figure 13.4, whose cell centers lie where j and E are defined. Then the limit is the volume integral

$$\int j_x(\mathbf{r}, t) E_x(\mathbf{r}, t) \, d\tau$$

(Note that t_j and t_E approach t as $dt \to 0$.) Adding the contributions from the other f's gives the integral of $j_x E_x + j_y E_y + j_z E_z = \mathbf{j} \cdot \mathbf{E}$, so the limit is exactly Eq. 13.5.

SECTION 13.2 ELECTRIC FIELD ENERGY

We will begin our search for the energy of an electromagnetic system by considering a system having only an **electric** field. (One such system is the parallel-plate capacitor of Section 11.2.) Then the electromagnetic energy is just that of the electric field. Having $B = 0$ also simplifies Ampere's law for dE/dt:

$$dE/dt = -j/\epsilon_0 \qquad (13.11)$$

We will denote the total energy of the discrete EM system at time t by $U(t)$. We suppose this to be a function of the electric fields, so we define it (for now, at least) only at times (usually multiples of dt) at which $E(f, t)$ is defined. Conservation of energy demands that the change in the total energy $U(t)$ during a time interval from $t - \frac{1}{2}dt$ to $t + \frac{1}{2}dt$ be equal to the work done on the system. The latter can be expressed by adding up (over all faces) the work done at a face f, defined by Eqs. 13.6 and 13.7.

$$U\left(t + \tfrac{1}{2}dt\right) - U\left(t - \tfrac{1}{2}dt\right) = \text{Total work}$$

$$= -\sum_f \left[dW\left(f, t + \tfrac{1}{4}dt\right) + dW\left(f, t - \tfrac{1}{4}dt\right)\right]$$

$$= -\sum_f j(f, t)\left[E\left(f, t + \tfrac{1}{2}dt\right)\right.$$

$$\left. + E\left(f, t - \tfrac{1}{2}dt\right)\right] dr^3 \, dt/2$$

[The $-$ sign is because Eqs. 13.6 and 13.7 gave the work done **by** the fields, not **on** them.] Using Ampere's law ($j\,dt = -\epsilon_0\,dE$),

$$U\left(t + \tfrac{1}{2}dt\right) - U\left(t - \tfrac{1}{2}dt\right) = \sum_f \tfrac{1}{2}\epsilon_0\left[E\left(f, t + \tfrac{1}{2}dt\right) - E\left(f, t - \tfrac{1}{2}dt\right)\right]$$

$$\cdot\left[E\left(f, t + \tfrac{1}{2}dt\right) + E\left(f, t - \tfrac{1}{2}dt\right)\right] dr^3$$

$$= \sum_f \tfrac{1}{2}\epsilon_0 E\left(f, t + \tfrac{1}{2}dt\right)^2 dr^3$$

$$- \sum_f \tfrac{1}{2}\epsilon_0 E\left(f, t - \tfrac{1}{2}dt\right)^2 dr^3 \tag{13.12}$$

So the change in U is exactly equal to the change in $\sum \frac{1}{2}\epsilon_0 E^2 \, dr^3$. Except for an additive constant, they are therefore equal:

$$U(t) = \sum_f \tfrac{1}{2}\epsilon_0 E(f, t)^2 \, dr^3 \tag{13.13}$$

(The constant must be zero if we started at $t = 0$ with zero fields and zero energy.)

Conservation of energy does not tell us how this energy is distributed among the faces, or indeed that it makes sense to think of it as being localized at a face. However, the form of Eq. 13.13 suggests that we should think of

$$U_E(f, t) \equiv \tfrac{1}{2}\epsilon_0 E(f, t)^2 \, dr^3 \tag{13.14}$$

as the energy of the field $E(f, t)$ at a particular face f. If we do this, conservation of energy is true **locally** as well as globally. That is, the change in the energy of a face is exactly the work done at that face:

$$U_E\left(f, t + \tfrac{1}{2}dt\right) - U_E\left(f, t - \tfrac{1}{2}dt\right)$$

$$= -j(f, t)E\left(f, t + \tfrac{1}{2}dt\right) dr^3\, dt/2$$

$$-j(f, t)E\left(f, t - \tfrac{1}{2}dt\right) dr^3\, dt/2 \tag{13.15}$$

(The derivation of this formula is exactly like that leading to Eq. 13.12, but without summing over faces.) This implies that there can be no transport of energy: energy put into the system at a particular place just stays there. We will see in the next section that transport of energy **is** possible if we allow a B field.

We can compute the continuum limit of the total electric field energy (Eq. 13.13) just as we did of the work done by a field, Eq. 13.9. When the electric field is specified by a vector function $\mathbf{E}(\mathbf{r}, t)$, Eq. 13.13 can be broken up into three sums, over faces normal to x, y, and z. The sum over x-faces is $\int \tfrac{1}{2}\epsilon_0 E_x(\mathbf{r}, t)^2\, d\tau$ in the limit, an integral over all of space. Adding the y and z sums gives

$$U(t) = \int \tfrac{1}{2}\epsilon_0 E(\mathbf{r}, t)^2\, d\tau \tag{13.16}$$

where $E^2 = E_x^2 + E_y^2 + E_z^2$. If we define the continuum electric-field **energy density**

$$u_E(\mathbf{r}, t) \equiv \tfrac{1}{2}\epsilon_0 E(\mathbf{r}, t)^2 \tag{13.17}$$

then the total energy is

$$U(t) = \int u_E(\mathbf{r}, t)\, d\tau \tag{13.18}$$

■ **EXAMPLE 13.1 THE PARALLEL PLATE CAPACITOR**   Compute the electric field energy in a parallel-plate capacitor of area A (Figure 13.5), with charge $\pm Q$ on the plates.
Solution: We will assume that L is much smaller than the lateral dimensions of the plates. Then except near the edges, the field is almost the same as that of infinite plates (Section

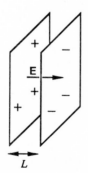

Figure 13.5   Parallel-plate capacitor.

11.2), namely $E = \sigma/\epsilon_0$ where $\sigma = Q/A$ is the charge per unit area. Then Eq. 13.18 gives

$$U = \int d\tau \left(Q^2/2\epsilon_0 A^2 \right) = LQ^2/2\epsilon_0 A \qquad (13.19)$$

because $\int d\tau$ is the total volume LA.

■ **EXAMPLE 13.2**   One common use of the concept of electric-field energy is in the computation of forces on charged objects. If we were to pull the plates in Figure 13.5 apart a distance dL with a force F that just balanced their attraction, we would do mechanical work $F\,dL$. Due to conservation of energy, this must be equal to the increase in electric field energy $dU = (dU/dL)\,dL$. Thus, the force must be dU/dL. For the case of Eq. 13.19, this is

$$F = Q^2/2\epsilon_0 A$$

Note that this result could not easily be calculated from the usual formula for the electric force, $Q\mathbf{E}$, because the electric field $\mathbf{E}$ is **discontinuous** at the plate, going from $Q/\epsilon_0 A$ to zero. (Evidently we would get the right answer if we split the difference and used $Q/2\epsilon_0 A$ for E, but this is by no means obvious *a priori*.)

▶   **Problem 13.1**

   Calculate the electric field energy in a parallel-plate capacitor whose plates have area $A = 0.1$ m^2 and charge $\pm Q$ where $Q = 1$ μc, and are separated by $L = 0.1$ mm.   ◀

▶   **Problem 13.2**

   Show that the derivation of Eq. 13.12 works in curvilinear coordinates if we replace $jE\,dr^3$ by $j(f)E(f)\,da(f)\,dl(f)$ in the definitions of work and power (Eqs. 13.6 to 13.8). Here $dl(f)$ is the length of the edge of the dual lattice which shares its center with the face f.   ◀

SECTION 13.3   ENERGY TRANSFER BETWEEN *E* AND *B*

In the previous section, we applied the principle of conservation of energy in the absence of a B-field to deduce a formula for the energy of an electric field. We now want to apply the same principle in the presence of a B-field to deduce a formula for the energy transferred from the B-field.

Let us first look at the effect of a B-field on the relation we obtained (Eq. 13.15) between the change in electric-field energy at a face f and the work done on the EM system there. The change in this energy during an interval of length dt centered at t_B (a time at which B is defined) is

$$dU_E(f, t_B) \equiv \tfrac{1}{2}\epsilon_0 E\left(f, t_B + \tfrac{1}{2}dt\right)^2 dr^3 - \tfrac{1}{2}\epsilon_0 E\left(f, t_B - \tfrac{1}{2}dt\right)^2 dr^3$$

$$= \tfrac{1}{2}\epsilon_0 \left[E\left(f, t_B + \tfrac{1}{2}dt\right) - E\left(f, t_B - \tfrac{1}{2}dt\right)\right]$$

$$\times \left[E\left(f, t_B + \tfrac{1}{2}dt\right) + E\left(f, t_B - \tfrac{1}{2}dt\right)\right] dr^3 \qquad (13.20)$$

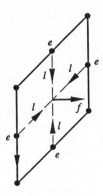

Figure 13.6 The edges e of a face f, each connected to f by a "linkage" l.

In the previous section we used Ampere's law to replace the first quantity in brackets, dE, by $-j\,dt/\epsilon_0$. We can still use Ampere's law, but we must now include the curl B term: $dE = c^2\,\mathrm{curl}\,B\,dt - j\,dt/\epsilon_0$ gives

$$dU_E(f, t_B) = \epsilon_0 c^2\,dr^{-1}\sum B(e, t_B)$$

$$\times\left[E\left(f, t_B + \tfrac{1}{2}dt\right) + E\left(f, t_B - \tfrac{1}{2}dt\right)\right]\,dr^3\,dt/2$$

$$-j(f, t_B)E\left(f, t_B + \tfrac{1}{2}dt\right)\,dr^3\,dt/2$$

$$-j(f, t_B)E\left(f, t_B - \tfrac{1}{2}dt\right)\,dr^3\,dt/2 \qquad (13.21)$$

where the sum $\sum$ is over the four edges of the face f (Figure 13.6). The two $-jE$ terms are the ones we had before; they represent the work done on the EM system at this face during the two intervals from t_B to $t_B + \tfrac{1}{2}dt$ and from $t_B - \tfrac{1}{2}dt$ to t_B. We now have additional terms in the energy change involving B at neighboring edges. The easiest way to retain conservation of energy is to interpret these as energy **transfers** from these edges, in analogy with the transfers of energy across the linkages between masses and springs in Section 2.8. We therefore define an energy-transfer field

$$S(l, t) \equiv \epsilon_0 c^2 E(f, t_E)B(e, t_B) \qquad (13.22)$$

at each edge-to-face linkage l in Figure 13.6 and each time interval of length $dt/2$ centered at an odd multiple t of $dt/4$. As in Eq. 13.8, t_E is the closest time to t at which E is defined, and t_B, which was called t_j previously, is the closest time at which B is defined; they are depicted in Figure 13.3. In terms of the energy-transfer field S, the change in energy at f (Eq. 13.21) is

$$dU_E(f, t_B) = \sum_l S\left(l, t_B - \tfrac{1}{4}dt\right)\,dr^2\,dt/2 + \sum_l S\left(l, t_B + \tfrac{1}{4}dt\right)\,dr^2\,dt/2$$

$$-P\left(f, t_B - \tfrac{1}{4}dt\right)\,dr^3\,dt/2 - P\left(f, t_B + \tfrac{1}{4}dt\right)\,dr^3\,dt/2 \quad (13.23)$$

where we have replaced jE by the power density P (Eq. 13.8). The two sums over l can be interpreted as energy transfers over the two intervals of length $\frac{1}{2}dt$ centered at $t_B \pm \frac{1}{4}dt$; S is the energy transferred per unit area per unit time. Thus, Eq. 13.23 is the statement of energy conservation at the face f: the change in energy is exactly equal to the energy transferred from the neighboring edges plus the external work done on the face. Note that the energy change (Eq. 13.23) breaks up naturally into two parts representing the energy changes during the first half of the interval (centered at $t_B - \frac{1}{4}dt$) and the second half ($t_B + \frac{1}{4}dt$). It is therefore easy to define the electric field energy U_E at a time t_B at which B but not E is defined, by

$$U_E(f, t_B) \equiv U_E\left(f, t_B - \tfrac{1}{2}dt\right) + \sum_l S\left(l, t_B - \tfrac{1}{4}dt\right) dr^2\, dt/2$$

$$- P\left(f, t_B - \tfrac{1}{4}dt\right) dr^3\, dt/2 \tag{13.24}$$

Rewritten in terms of the center $t = t_B - \frac{1}{4}dt$ of this time interval, this equation becomes

$$U_E\left(f, t + \tfrac{1}{4}dt\right) \equiv U_E\left(f, t - \tfrac{1}{4}dt\right)$$

$$+ \sum_l S(l, t)\, dr^2\, dt/2 - P(f, t)\, dr^3\, dt/2 \tag{13.25}$$

By subtracting Eq. 13.24 from Eq. 13.23, you can show that Eq. 13.25 holds for the other times ($t = t_B + \frac{1}{4}dt$, at which E is defined) as well. We can therefore forget about Eqs. 13.23 and 13.24 and use the simpler Eq. 13.25 as the statement of local energy conservation.

▶ **Problem 13.3**

Consider an electromagnetic delta-pulse (Section 11.1) traveling in the x-direction, which has $dr = c\,dt$, nonzero $E_y(x, t = 0) = E_0$ only at $x = \frac{1}{2}dr$, and nonzero $B(x, t = \frac{1}{2}dt) = E_0/c$ only at $x = dr$.

a. Compute the energy transferred to each face from its edges, between $t = 0$ and $t = \frac{1}{2}dt$. Compare to the total energy of the face (when nonzero).

b. Compute $E(x, t = dt)$ and determine the energy transferred between $t = \frac{1}{2}dt$ and $t = dt$.   ◀

▶ **Problem 13.4**

The generalization to curvilinear coordinates described in Problem 13.2 can be extended to the energy transfers by replacing $EB\,dr^2$ by $E(f)B(e)\,dl(f)\,dl(e)$. Show that the derivation of Eq. 13.23 then still works.   ◀

▶ **Problem 13.5**

(Analog of Problem 2.17) Show that Eq. 13.24 is equivalent to defining $U_E = \frac{1}{2}\epsilon_0 E^2$, where E is the geometric mean of $E(r, t \pm \frac{1}{2}dt)$.   ◀

SECTION 13.4 MAGNETIC FIELD ENERGY

We now know the energy transferred away from an edge e, but not the actual energy of the magnetic field $B(e, t)$ there: we will denote this energy by $U_B(e, t)$ in analogy with the electric field energy $U_E(f, t)$ at a face f. However, conservation of

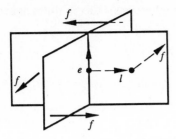

Figure 13.7 The four faces f touching an edge e, and the linkage l to one of them.

energy tells us the **change** in energy at e: it must be the total energy transferred to e from the neighboring faces f (Figure. 13.7). Let us compute the change during an interval between two times at which B is defined, say $t \pm \frac{1}{2}dt$:

$$U_B\left(e, t + \tfrac{1}{2}dt\right) - U_B\left(e, t - \tfrac{1}{2}dt\right) = -\sum_l S\left(l, t - \tfrac{1}{4}dt\right) dr^2 \, dt/2$$

$$- \sum_l S\left(l, t + \tfrac{1}{4}dt\right) dr^2 \, dt/2 \quad (13.26)$$

Here the transfer from f to e is $-S$ because S was defined by Eq. 13.22 as the energy transfer from e to f. The sum is over the four linkages l to the four faces in Figure 13.7; this is essentially the same as summing over the four faces. To determine the energy as a function of B, we must express this energy change in terms of B. Using the definition of S, we get

$$U_B\left(e, t + \tfrac{1}{2}dt\right) - U_B\left(e, t - \tfrac{1}{2}dt\right)$$

$$= -\sum_f \epsilon_0 c^2 E(f, t) B\left(e, t - \tfrac{1}{2}dt\right) dr^2 \, dt/2$$

$$- \sum_f \epsilon_0 c^2 E(f, t) B\left(e, t + \tfrac{1}{2}dt\right) dr^2 \, dt/2$$

$$= -\epsilon_0 c^2 \left[B\left(e, t + \tfrac{1}{2}dt\right) + B\left(e, t - \tfrac{1}{2}dt\right)\right] dr^{-1} \sum_f E(f, t) \, dr^3 \, dt/2$$

$$= -\epsilon_0 c^2 \left[B\left(e, t + \tfrac{1}{2}dt\right) + B\left(e, t - \tfrac{1}{2}dt\right)\right] \operatorname{curl} E(e, t) \, dr^3 \, dt/2 \quad (13.27)$$

If we use Faraday's law to replace $-\operatorname{curl} E \, dt$ by dB, the energy change becomes

$$\tfrac{1}{2}\epsilon_0 c^2 \left[B\left(e, t + \tfrac{1}{2}dt\right) + B\left(e, t - \tfrac{1}{2}dt\right)\right]$$

$$\times \left[B\left(e, t + \tfrac{1}{2}dt\right) - B\left(e, t - \tfrac{1}{2}dt\right)\right] dr^3$$

$$= \tfrac{1}{2}\epsilon_0 c^2 B\left(e, t + \tfrac{1}{2}dt\right)^2 dr^3 - \tfrac{1}{2}\epsilon_0 c^2 B\left(e, t - \tfrac{1}{2}dt\right)^2 dr^3 \quad (13.28)$$

The energy change is the same as the change in $\frac{1}{2}\epsilon_0 c^2 B^2 \, dr^3$, so except for a possible additive constant (which we may as well take to be zero), this **is** the magnetic field energy:

$$U_B(e, t) \equiv \tfrac{1}{2}\epsilon_0 c^2 B(e, t)^2 \, dr^3 \qquad (13.29)$$

For historical reasons, the coefficient in this formula is denoted by $\tfrac{1}{2}\mu_0^{-1}$, where

$$\mu_0 \equiv 1/\epsilon_0 c^2 \qquad (13.30)$$

is called the **magnetic permeability of space**. It has the numerical value

$$\mu_0 = 12.57 \times 10^{-7} \mathrm{T}^2 \, \mathrm{m}^3/j \qquad (13.31)$$

(we will see later it is exactly $4\pi \times 10^{-7}$). The magnetic field energy is

$$U_B(e, t) = \left[B(e, t)^2/2\mu_0 \right] dr^3 \qquad (13.32)$$

Just as we did for the electric field energy $U_E(f, t)$, we can define the magnetic field energy $U_B(e, t)$ for times when B is not defined (when we can't use Eq. 13.29.) Noting that the change (Eq. 13.26) in $U(e, t)$ between $t - \tfrac{1}{2}dt$ and $t + \tfrac{1}{2}dt$ breaks up neatly into terms that can be interpreted as the changes during the two halves of the interval, we can define

$$U_B(e, t) \equiv U_B\left(e, t - \tfrac{1}{2}dt\right) - \sum_l S\left(l, t - \tfrac{1}{4}dt\right) dr^2 \, dt/2 \qquad (13.33)$$

at a time t for which B is not defined. Then Eq. 13.33 is true for both even and odd multiples t of $\tfrac{1}{2}dt$, and supercedes Eq. 13.26 as the statement of local energy conservation at an edge.

We can now write the total energy of our electromagnetic system,

$$U(t) = \sum_f U_E(f, t) + \sum_e U_B(e, t) \qquad (13.34)$$

▶ **Problem 13.6**

Consider the delta-pulse of Problem 13.3. Compute the magnetic energy of each edge at which B is nonzero, at $t = \tfrac{1}{2}dt$. Show that it is exactly the energy transferred to that edge between $t = 0$ and $t = \tfrac{1}{2}dt$ (which you calculated in Problem 13.3.)   ◀

▶ **Problem 13.7**

Show that the discrete Maxwell equations conserve energy exactly. Start by computing the change dU in total energy by adding Eqs. 13.24 and 13.33 for the change in energy at a face or an edge, over all faces and edges. When you write a Σ, be very explicit about what is being summed over: all edges touching a given face, all linkages attached to a given face, and so on. Show that dU is exactly the total external work done on the system.   ◀

▶ **Problem 13.8**

Calculate the total electromagnetic field energy in the tachmen's experiment (Section 9.3), at times $t = 1$ ns and $t = 2$ ns. The current density was $j(\tfrac{1}{2}dt) = 1$ $\mu c/m^2$ ns at one face only, and we used $dr = 1$ m, $dt = 1$ ns.   ◀

▶   **Problem 13.9**

We can extend the results of this section to curvilinear coordinates by replacing $B\,dr^3$ by $B(e)\,dl(e)\,da(e)$, where $da(e)$ is the area of the face of the dual lattice sharing its center with the edge e, in addition to the replacements $EB\,dr^2 \to E(f)B(e)\,dl(f)\,dl(e)$, $E^2\,dr^3 \to E(f)^2\,da(f)\,dl(f)$, and $jE\,dr^3 \to j(f)E(f)\,da(f)\,dl(f)$ given in Problems 13.2 and 13.4. Show that the derivation of Eq. 13.28 then works in curvilinear coordinates.   ◀

The form of the magnetic field energy in the continuum limit is exactly analogous to that of the electric field energy (Eqs. 13.16 to 13.18). When B is determined by a vector function $\mathbf{B}(\mathbf{r}, t)$, the discrete field $B(e, t)$ for an x-pointing edge centered at $\mathbf{r}$ is approximately $B_x(\mathbf{r}, t)$. The sum of the energies (Eq. 13.32) over all x-pointing edges is

$$\sum_{x\text{-edges}} \left[B_x(\mathbf{r}, t)^2 / 2\mu_0 \right] dr^3$$

whose limit is an integral over all space

$$\int \left[B_x(\mathbf{r}, t)^2 / 2\mu_0 \right] d\tau$$

and the sum including y and z edges as well is the total magnetic-field energy

$$U_B(t) = \int \left[B(\mathbf{r}, t)^2 / 2\mu_0 \right] d\tau \tag{13.35}$$

The integrand is called the **magnetic energy density** and is denoted u_B:

$$u_B(\mathbf{r}, t) = B(\mathbf{r}, t)^2 / 2\mu_0 \tag{13.36}$$

The total EM energy density is

$$u(\mathbf{r}, t) = u_B(\mathbf{r}, t) + u_E(\mathbf{r}, t) \tag{13.37}$$

and its integral is the total EM field energy

$$U(t) = U_E(t) + U_B(t) = \int u(\mathbf{r}, t)\, d\tau \tag{13.38}$$

■ **EXAMPLE 13.3 BOXCAR PULSE**   Consider a pseudo-1D system in which E and B depend only on x and are transverse ($E_x = B_x = 0$). We know that if $dr = c\,dt$ there is an exact delta-pulse solution (Section 11.1) of the discrete Maxwell equations in which E_y is nonzero at only one x (i.e., $E(f) \neq 0$ on all horizontal faces centered at this x) at any given time t and B_z is nonzero at $x + \frac{1}{2}dr$ at $t + \frac{1}{2}dt$. If the nonzero E is E_0, the nonzero B is

$$B_0 = E_0/c \tag{13.39}$$

Instead of a delta pulse, we will use a superposition of many such pulses, which is easier to

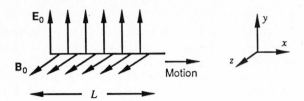

Figure 13.8   A boxcar pulse.

relate to the continuum limit. We show in Figure 13.8 a superposition of delta pulses, each displaced by dr from the previous one. This "boxcar pulse" (the graph of E versus x resembles a railroad boxcar) has uniform E and B fields in a slab of width L. The extent of the fields in the y and z directions is infinite in a true pseudo-1D system, but to have a finite total energy, we will assume it is merely very large (of area A in the yz plane) and we will stay far away from the edges so that we can use the pseudo-1D Maxwell equations. The fields then occupy a volume $V = AL$. Because of the relation (Eq. 13.39) between E_0 and B_0, the electric and magnetic field energy densities (Eqs. 13.17 and 13.36) are exactly the same: $u_B = B_0^2/2\mu_0 = \frac{1}{2}c^2\epsilon_0 B_0^2 = \frac{1}{2}\epsilon_0 E_0^2$. The total energy density is $u = \epsilon_0 E_0^2$, and the energy of the entire boxcar pulse is

$$U = \int u\,d\tau = Vu = V\epsilon_0 E_0^2 \qquad (13.40)$$

It is interesting to examine what happens when two such boxcar pulses collide, as in Figure 13.9. The "before" picture shows the pulse of Figure 13.8 (on the left) and the same pulse rotated 180° about a vertical axis (on the right). Each is moving toward the other, without changing shape. So when they collide (i.e., occupy the same space), the total field is just the superposition of the fields of the pulses. The electric fields reinforce each other, producing $E = 2E_0$. But the magnetic fields cancel; the magnetic field energy is zero during the collision. We can check that the **total** energy is still conserved:

$$U = V\tfrac{1}{2}\epsilon_0(2E_0)^2 = 2V\epsilon_0 E_0^2 \qquad (13.41)$$

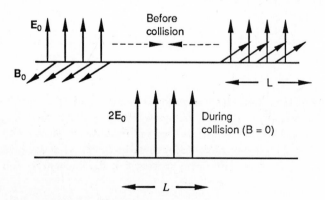

Figure 13.9   Collision of boxcar pulses.

which is exactly the total energy before the collision, that is, twice the energy of each pulse (Eq. 13.40).

▶ **Problem 13.10**

Construct two approaching boxcar pulses so that the **electric** field vanishes during the collision (make a sketch such as Fig. 13.9). Compute the total energy when the pulses are superimposed and verify that energy is conserved.   ◀

SECTION 13.5   ENERGY CONSERVATION IN A FINITE REGION: POYNTING'S THEOREM

In the previous section we proved that the electromagnetic energy of a system extending over all space is conserved. The formal proof was left as Problem 13.7, but the result essentially follows from the local conservation of energy at each edge and face: the energy changes only due to mechanical work jE done at faces and energy transfers $S = EB/\mu_0$ between edges and faces. Extending the principle of energy conservation to a finite region is easy, once we specify what we mean by a region. For the purposes of Poynting's theorem, a **region** R will be any set of edges and faces; part of such a set is shown in Figure 13.10. The obvious definition of the total electromagnetic energy in the region R is then

$$U_R(t) \equiv \sum_{f \in R} U_E(f, t) + \sum_{e \in R} U_B(e, t) \tag{13.42}$$

We can compute the rate of change of U_R by adding up the rates of change of the face and edge energies (Eqs. 13.25 and 13.33). This gives a term describing the work done in the region, and terms involving transfers $S(l, t) = E(f, t_E)B(e, t_B)/\mu_0$ across linkages l between edges e and faces f. The same type of cancellation occurs as that which we saw in proving the divergence theorem (Section 6.5). If e and f are both in the region R, then the linkage l connecting them appears exactly twice in the sum, once in the change in the energy at f, as $+S(l, t)$, and once in the change in the energy at e, as $-S(l, t)$. The terms cancel unless one end of the linkage is in R and the other is not. We will call the set of such non-cancelling linkages the **boundary** of R, denoted ∂R (see Fig. 13.10).

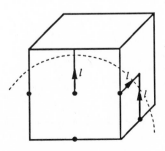

Figure 13.10   A region (a set of faces and edges, marked at their centers by large dots) R and its boundary ∂R (a set of linkages labeled l). Only the four edges and one face on one cell of the lattice are shown. The dotted line is a continuum rendition of the boundary.

To be perfectly precise, we need to define the concept of an **oriented** linkage, and insist that the linkages in ∂R be **outwardly** oriented. The linkages l we have pictured thus far have all been oriented with the edge in back and the face in front, but by analogy with our treatment of faces in Chapter 6 we can denote by $-l$ the oppositely-oriented linkage, and define $S(-l, t) \equiv -S(l, t)$.

The noncancelling terms in the rate of change of U_R are then

$$dU_R(t)/dt \equiv \left[U_R\left(t + \tfrac{1}{4}dt \right) - U_R\left(t - \tfrac{1}{4}dt \right) \right] \Big/ \tfrac{1}{2} dt$$

$$= - \sum_{f \in R} P(f, t)\, dr^3 - \sum_{l \in \partial R} S(l, t)\, dr^2 \qquad (13.43)$$

where $P = jE$ (Eq. 13.8) and $S = EB/\mu_0$ (Eq. 13.22). This is the discrete form of Poynting's theorem: it states that the rate of change of the energy in a region is exactly the mechanical work done inside minus the outward energy flux through the boundary.

The energy flux can be defined for any discrete surface A (any set of linkages, not necessarily the boundary of a region); we will denote it by F_A:

$$F_A(t) = \sum_{l \in A} S(l, t)\, dr^2 \qquad (13.44)$$

The continuum limit of this flux must be a surface integral over a continuum surface, say A_c. Such a surface can be approximated arbitrarily closely by a surface composed of flat pieces normal to the coordinate axes, such as the piece labeled A' in Figure 13.11. We have chosen a lattice of spacing dr; for each dr the discrete "surface" A to use in Eq. 13.44 is just the set of linkages l crossing the continuum surface A_c. The energy flux $F_{A'}$ across the piece A' shown has one term $S(l, t)\, dr^2$

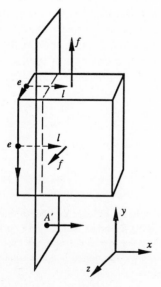

Figure 13.11   Two of the linkages carrying energy across a surface A.

for each of the linkages crossing A'. If the vector fields **E** and **B** are smooth, clearly many of the $S(l, t)$'s are nearly the same: in particular, each one is the same as all those obtainable from it by translating along the y or z axis by dr. However, you cannot get **all** of the l's this way; the two shown in Figure 13.11 are not related by such a translation. Half of the transfers, those across linkages on horizontal faces such as the upper one in Figure 13.11, involve $E(f)B(e) = E_y B_z$, and the other half (on faces in the plane of the paper) involve $E(f)B(e) = E_z(-B_y)$. The $-$ sign is because this e is opposite to the y-direction. There is one linkage of each type for each cell intersecting the surface A'; those for one cell are shown in Figure 13.11. Thus,

$$F_{A'}(t) \approx \sum dr^2 \left[E_y(\mathbf{r}, t) B_z(\mathbf{r}, t) - E_z(\mathbf{r}, t) B_y(\mathbf{r}, t) \right] / \mu_0 \qquad (13.45)$$

where the sum is over these cells. In the continuum limit, each of these two terms becomes a surface integral. It is not exactly the same as the surface integrals of face fields we considered in Section 6.2, because the integrand is located at a linkage l rather than a face center. However, it is at the center of a face of a slightly displaced lattice, so this doesn't matter in the continuum limit. The surface integral is

$$F_{A'}(t) = \int_{A'} da \left[E_y(\mathbf{r}, t) B_z(\mathbf{r}, t) - E_z(\mathbf{r}, t) B_y(\mathbf{r}, t) \right] / \mu_0 \qquad (13.46)$$

We have proved this under rather restrictive assumptions, but it is straightforward to verify that it works more generally. For example, if the surface A' cut the right part of the cell instead of the left (imagine Figure 13.11 rotated by π about the y-axis), the rightward transfer would involve $S(-l) = -S(l)$ instead of $S(l)$. But the sign is canceled by the fact that e now points along $-z$, so $B(e)$ changes sign and we still get Eq. 13.45.

The expression in the brackets is the x-component of a cross product; using the conventional notation $d\mathbf{a}$ for $\hat{x} da$, the vector of magnitude da in the direction normal to the surface A', we can write

$$F_{A'}(t) = \int_{A'} d\mathbf{a} \cdot (\mathbf{E} \times \mathbf{B}) / \mu_0 \qquad (13.47)$$

We can prove this for surfaces A' normal to the y or z axes as well, either by repeating the above argument with permuted axes, or (because the coordinates don't appear explicitly) by rotating our coordinate system so that any given surface A' is normal to the x axis. In fact, the coordinate-free formula (Eq. 13.47) frees us from the requirement that the area element $d\mathbf{a}$ be normal to an axis at all. By adding it up over all the pieces A' of a continuum surface A_c, we see that Eq. 13.47, with A' replaced by A_c, gives the power transferred across an arbitrary continuum surface.

The cross product in Eq. 13.47 is the energy flux vector, referred to as the Poynting vector and conventionally denoted by **S**

$$\mathbf{S}(\mathbf{r}, t) = \mathbf{E}(\mathbf{r}, t) \times \mathbf{B}(\mathbf{r}, t) / \mu_0 \qquad (13.48)$$

The Poynting vector points in the direction of energy flow; its magnitude is the power per unit area transmitted across a surface normal to that direction. It is the continuum limit of the discrete energy transfer field $S(l, t)$.

■ **EXAMPLE 13.4**   Evaluate the Poynting vector for the "boxcar pulse" of Example 13.3. In the region of length L in which the fields are nonzero, $\mathbf{E} = E_0\hat{\mathbf{y}}$ and $\mathbf{B} = E_0\hat{\mathbf{z}}/c$. Thus, $\mathbf{S} = \mathbf{E} \times \mathbf{B}/\mu_0 = E_0^2\hat{\mathbf{x}}/c\mu_0$: the energy is moving in the x-direction (as is the pulse, of course) and the power passing an area A is $E_0^2 A/c\mu_0$. It passes for a total time L/c, so the total energy that passes is $E_0^2 AL/c^2\mu_0 = \epsilon_0 E_0^2 V$ (V is the volume of the pulse). But this should be the total energy of the pulse (Eq. 13.40), and indeed it is.

Finally, we can write the continuum version of Poynting's theorem, from Eqs. 13.43, 13.47, and 13.48:

$$dU_R(t)/dt = - \int_R \mathbf{j} \cdot \mathbf{E} \, d\tau - \int_{\partial R} \mathbf{S} \cdot d\mathbf{a} \tag{13.49}$$

PART C

Electrostatics

CHAPTER FOURTEEN

GAUSS' AND COULOMB'S LAWS

SECTION 14.1 EVOLUTION OF STATIC FIELDS

Having learned in Chapter 9 what electromagnetic phenomena look like to tachmen, let us ask what these phenomena usually look like to humans. Without high-speed measuring equipment, we humans can usually only detect the **static** behavior of electric and magnetic fields. We will look first at the problem of **electrostatics**, which is the study of the time-independent electric fields that exist a long time after the currents stop and the charges no longer move (i.e., are static). Our first guess might be that the E and B fields after a long time are zero. We might expect this because we are assuming the source term (j) in the equations for E and B is "transient," that is, $j = 0$ everywhere after a certain time. And our experience with transient sources in infinite systems is that the fields produced are also transient. In diffusion, for example, the particles produced by a transient source eventually spread out infinitely thinly and the density and current both approach zero. It requires a steady source to produce a nonzero static field. It might seem that the same should happen with Maxwell's equations if the source j is transient: after a long time the E and B fields should die away. However, we already know a counterexample: in Section 11.2 we charged up an infinite parallel-plate capacitor and found a nonzero electric field that lasts forever. It will turn out that B always goes to zero eventually, but there will sometimes be a static electric field even when the current is transient. We will see in Section 14.4 that this static field is associated with the presence of a nonzero **charge** density after the current stops. It also turns out that the static field depends only on the static charge distribution (and not on the details of the currents that created it); we will also see the reason for this later. For now, however, let us assume the static electric field is uniquely determined by the charge distribution, and try to calculate it in some more cases.

In principle, we can calculate the electrostatic field of a charge distribution by evolving Maxwell's equations for a transient current distribution that creates that charge distribution. This is made difficult in practice by the fact that the fields spread eventually throughout all of space; in contrast with calculating the fields after a few time steps, when they are nonzero at a fairly small number of faces and edges, we now have to calculate an infinite number of values for the fields. We will calculate electrostatic fields for two types of problems in which this difficulty can be avoided: problems in which the fields are confined by physical boundaries (usually conducting surfaces) to a finite volume, and continuum problems in which the fields are specified throughout an infinite volume by rules that can be expressed on a finite amount of paper. We will begin in the next section with the continuum case.

SECTION 14.2   THE ELECTROSTATIC FIELD
OF A POINT CHARGE: COULOMB'S LAW

You probably have an intuitive notion of what a point charge is: a charged particle with zero size, which occupies a mathematical point. There is no way of determining experimentally whether such things exist; all experiments can do is to probe a particle on a scale $s > 0$, and decide whether the particle is smaller than s. Scattering experiments on the proton show clearly that it is not a point charge; its size is about 10^{-13} cm. Similar experiments on the electron show it to be smaller than about 10^{-14} cm (i.e., no spreading-out is detected on a scale of 10^{-14} cm); zero size is not ruled out, but it can't be proven. A "point charge" is therefore an idealization, but a very useful one because most electromagnetic phenomena we study involve scales much larger than these.

From our point of view, a point charge is a special continuum charge distribution. In general, a continuum charge distribution is a rule giving a discrete charge density field ρ for any cell lattice. If the electrostatic fields of these discrete charge densities approach a limit as $dr \to 0$, we will refer to this limit as "the electrostatic field of the continuum charge distribution." The particular rule that defines a point charge of magnitude q at a position $\mathbf{r}'$ is that $\rho(c) = q/d\tau(c)$ if $\mathbf{r}'$ is in the cell c, and 0 if it is not; this is q times the Dirac delta function.

Therefore, to determine the electrostatic field of a point charge, we must consider a discrete charge distribution such as that in Figure 14.1a having a charge q in one cell, which we may as well take to be centered at the origin of coordinates. To find the fields associated with it, we must know the current density that created it from the assumed initial condition $E = B = 0$. One way to create it is to create a positive and negative charge $\pm q$ in the cell at the origin, and move $-q$ very far to the right ("to infinity") during the interval (labeled $\frac{1}{2}dt$) from 0 to dt, so it won't affect the rest of our experiment. Then we get a current

$$j\left(f, \tfrac{1}{2}dt\right) = -q/dr^2\, dt \qquad (14.1)$$

across each of the faces to the right, as shown in Figure 14.1. (Note that this procedure is equivalent to creating $\pm q$ at infinity and moving $+q$ to the origin: both the current density and the resulting charge distribution are the same.) We can

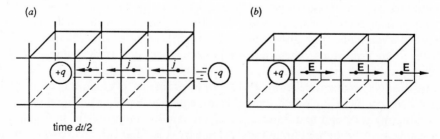

Figure 14.1   (a) A discrete EM system with a point charge, produced by removing its negative counterpart to infinity. (b) The electric field at time dt produced by the current in a.

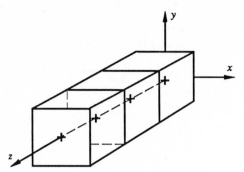

Figure 14.2   A line charge perpendicular to the paper.

now calculate the E and B fields, using this current (which vanishes for $t > dt$) as the source. Maxwell's equation for E (Eq. 9.15) gives a nonzero electric field

$$E(f, dt) = q/\epsilon_0 \, dr^2 \tag{14.2}$$

at these same faces, as shown in Figure 14.1b.

As mentioned above, we expect the static electric field to depend only on the charge distribution and not on the details of the current (in the case of Figure 14.1, this means the path $-q$ took to infinity). In particular, this means it should look the same in any direction from the point charge: Figure 14.1b clearly doesn't possess this symmetry. To see if it does at later times, we should let the field evolve for a while according to Maxwell's equations.

Instead of doing this for a point charge, which would require drawing three-dimensional pictures, we will do it first for the simpler (but qualitatively similar) case of a **line** charge (Figure 14.2). We can also think of the line charge problem as a continuum problem; the discrete density is given by the rule $\rho(c) \, d\tau(c) = \lambda \, dl(c)$, where $dl(c)$ is the length of the segment of the line inside cell c; λ is then the charge per unit length. In our case of a straight line in the z-direction, $dl(c) = dr$, so the amount of charge per cell along the line is $q = \rho \, d\tau = \lambda \, dr$. Just as we did for the point charge, we create a positive and negative line charge through the origin and move the negative one to infinity in the x-direction.

The line charge problem is pseudo-2D (ρ depends only on x and y); the current at $t = \frac{1}{2}dt$ and the electric field at $t = dt$ are shown in Figure 14.3. We are using $q = 1 \ \mu c$, $dr = 1$ m, $dt = 1$ ns, so the E in Eq. 14.2 is 0.11 N/μc. The edges of each face with $E \neq 0$ in Figure 14.3 will acquire a B-field at time $t = 1.5 \, dt$. The

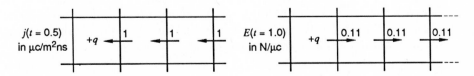

Figure 14.3   The pseudo-2D current associated with moving a negative line charge to infinity, and the resulting E field at $t = dt$.

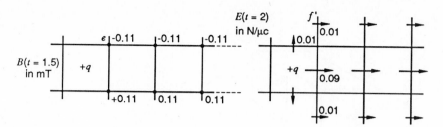

Figure 14.4   The evolution of the system of Figure 14.3.

fields are initially almost the same as those for a current at a single face (Section 10.2), except that now more faces are involved. Maxwell's equation for dB/dt (Eq. 9.32) gives

$$B(e, 1.5\, dt) = -E(f, dt)\, dt/dr = -0.11\text{ mT} \qquad (14.3)$$

at the edge e in Figure 14.4 (the negative sign means **B** points into the paper). At $t = 2\, dt$, the electric field has spread out slightly; at the face labeled f' in the figure, Eq. 9.15 gives

$$E(f', 2\, dt) = E(f', dt) + c^2\, dt\ \text{curl}\ B(f', dt)$$

$$= 0 - c^2\, dt(-0.11\text{ mT})$$

$$= +0.01\text{ N}/\mu\text{c} \qquad (14.4)$$

The field at f (along the charge's trajectory) decreases to 0.09 N/μc.

▶   **Problem 14.1**

Calculate the field $E(f, 2\, dt)$ at all the faces shown in Figure 14.4 for an actual 3D system. That is, the charge q is **really** a point charge (as in Figure 14.1b), and not a line charge. *Hint:* $E(f, dt)$ and $B(e, 1.5\, dt)$ are just as before (Eqs. 14.2 and 14.3), except that now there are also B's on edges parallel to the paper, which will affect $E(f, 2\, dt)$. ◀

Clearly going much further by hand would be very tedious. However, the MAXWELL program will do it for us. We will find that the E field spreads out evenly around the charge, although it takes a long time for this to happen far from the charge because E has a long way to go. The magnetic field B, although necessary to redistribute the E field, eventually gets smaller and disappears. The resulting static E field is perfectly symmetrical, as it must be if it is to be independent of our choice of direction in which to move $-q$ away to infinity. The field would get symmetrical faster if it started **out** more symmetrical, that is, if the current distribution were more symmetrical.

▶   **Problem 14.2**

Construct a more symmetrical pseudo-2D current by dividing the charge in each cell into four $-0.25\ \mu$c pieces and moving each of them away along a different axis, $90°$

apart. They move all the way to infinity during the first time interval; use $dt = 1$ ns, $dr = 1$ m.

a. Compute $j(f, \frac{1}{2}dt)$, $E(f, dt)$, $B(f, 1.5\, dt)$, and $E(f, 2\, dt)$ for the four faces and edges closest to the charge. Will E ever change from this value?

b. Check your answer using MAXWELL.   ◄

However, we can't achieve complete symmetry, allowing a rotation through any angle, because the cartesian coordinate system itself isn't symmetrical. To do that, we need a coordinate system that's as symmetrical as the charge distribution. For the case of the point charge, this is a spherical coordinate system (Figure 7.7) centered at the point charge, which we can divide into cells by choosing dr, $d\theta$, and $d\phi$, as in Chapter 7. A current distribution with spherical symmetry is one in which j depends only on r and not on the angles θ and ϕ. Also, the charge should move outward from the point charge, so no charge should pass the faces whose planes pass through the origin (faces in planes of constant θ and ϕ respectively). So the current is only nonzero at radial faces (i.e., ones whose normal points radially); it has only an r-component $j_r(r, t)$. The charge passing each face f during the interval t is $j(f, t)\, da(f)\, dt$ (by the definition of j), so the total passing a sphere of radius r is $\sum j(f, t)\, da(f)\, dt$, where the sum is over the radial faces f into which the sphere is divided. If we assume all the charge passes this spherical surface in the interval $t = \frac{1}{2}dt$ (i.e., between 0 and dt), this total charge is just $-q$:

$$\sum_f j\left(f, \tfrac{1}{2}dt\right) da(f)\, dt = -q \tag{14.5}$$

However, the current density at each of these faces is the same, because it depends only on r: call it $j_r(r, \frac{1}{2}dt)$. We can bring it and dt out of the sum; the remaining sum $\sum da(f)$ is the sum of the areas of the faces, which is just the area $4\pi r^2$ of the sphere. So

$$j_r\left(r, \tfrac{1}{2}dt\right) dt \times 4\pi r^2 = -q$$

$$j_r\left(r, \tfrac{1}{2}dt\right) = -q/4\pi r^2\, dt \tag{14.6}$$

Knowing the spherical symmetry of the current source, we can determine what spherical symmetry means for the resulting E and B fields, and show that this symmetry is preserved by Maxwell's equations, much as we did for diffusive systems in Chapter 8. We will begin by computing the electric field at $t = dt$, from Maxwell's equation

$$dE_r\left(r, \tfrac{1}{2}dt\right)/dt = c^2\left[\operatorname{curl} B\left(r, \tfrac{1}{2}dt\right)\right]_r - j_r\left(r, \tfrac{1}{2}dt\right)/\epsilon_0$$

$$\left[E_r(r, dt) - 0\right]/dt = 0 - \left(-q/4\pi\epsilon_0 r^2\, dt\right) \tag{14.7}$$

$$E_r(r, dt) = +q/4\pi\epsilon_0 r^2$$

At $t = 1.5\, dt$ we should calculate B from the curl of E, by adding up E's over paddlewheels. But for B's at radial edges (along $\hat{r}$) these E's (E_ϕ and E_θ) are zero,

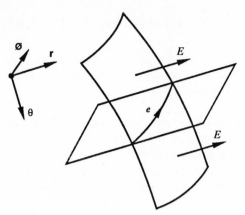

Figure 14.5 Faces contributing to curl E at a ϕ-edge e.

and for the edges along $\hat{\theta}$ and $\hat{\phi}$ the E's cancel (this is shown for $\hat{\phi}$ explicitly in Figure 14.5). If we think of the curl in terms of a "torque on the paddlewheel" analogy, the symmetrical current puts the same force on each side so there is no net torque. Actually, to be quite precise, the expression for curl E (Chapter 7) also involves the lengths of edges in the dual lattice. But you can check that these are the same. So there is no $B(e, 1.5\,dt)$; this is true of any spherically symmetric system with a radial current. But then (because $j(f, t) = 0$ for $t > dt$) Maxwell's equation (Eq. 9.31) gives $dE/dt = 0$: E never changes again, and we have found the exact static field! At any time $t \geq dt$,

$$E_r(r) = q/4\pi\epsilon_0 r^2 \qquad \text{(Coulomb's law)} \qquad (14.8)$$

This is Coulomb's law, for the static field of a point charge. We have derived it for a discrete lattice, but the continuum limit is trivial because the formula is independent of dr; its limit as $dr \to 0$ is again Eq. 14.8.

So our nonhistorical approach, which started with Maxwell's equations (1865), has finally reached Coulomb's law, which dates from the 1700s. You will note that it has a power-law form, just as the static stress field in the Tortoises–Hares parable did.

From Coulomb's law we can calculate the force $F = QE$ on a test charge Q at a distance r from our point charge q. It points radially away from q and has magnitude

$$F = Qq/4\pi\epsilon_0 r^2 \qquad (14.9)$$

▶ **Problem 14.3**

It turns out that to stay in good enough shape to do electrostatics experiments with whole tachyderm hides (Section 9.3), a tachman has to spend a lot of time pumping iron. Verify this by computing the force between two standard charges (1 coulomb each) separated by 1 m. What is this in pounds?   ◀

▶ **Problem 14.4**

Use the procedure we used to get Coulomb's law to calculate the static field of a line charge with charge per unit length λ (Figure 14.2). Here the coordinate system with the same symmetry as the charge distribution is a cylindrical one (Figure 7.4) with the line charge along the z-axis.

a. Compute the current density $j_r(r, \frac{1}{2}dt)$, assuming the charge moves radially away from the z-axis so the charge $-\lambda\, dz$ on a length dz of the axis all passes through each of many concentric circular bands of radial faces of length dz during this interval.

b. Compute $E_r(r, dt)$ and show that $B = 0$ at time $1.5\, dt$. ◄

▶ **Problem 14.5**

Note that the $E(t = dt)$ in Eq. 14.7 had the same spherical symmetry that j did: it was radial and depended only on r. Show that this is not accidental, nor is the fact that $B = 0$ at all times, but both are because the discrete Maxwell equations preserve this spherical symmetry. That is, show that if the source $j(f, t)$ has this symmetry and the initial $E(f, 0)$ and $B(f, \frac{1}{2}dt)$ have it (the symmetry condition on B is just $B = 0$), then E and B have it at all times. ◄

SECTION 14.3 APPLYING COULOMB'S LAW
TO AN ARBITRARY CHARGE DISTRIBUTION

Although we obtained Coulomb's law for a point charge located at the origin of our coordinate system, we can, of course, generalize it for a point charge at an arbitrary position $\mathbf{r}'$ (Figure 14.6). We denote the "field point" (the point at which we wish to determine the electric field) by $\mathbf{r}$. The vector displacement between the point charge and the field point is $\mathbf{s} = \mathbf{r} - \mathbf{r}'$. We can use our old result (Eq. 14.8) for E in a coordinate system centered at $\mathbf{r}'$, in which $|s|$ is the radial coordinate. The static electric field points along $\mathbf{s}$; let us define a unit vector in this direction, $\hat{\mathbf{s}} \equiv \mathbf{s}/|s|$. Then the continuum electric field at $\mathbf{r}$ in the presence of the charge at $\mathbf{r}'$ is

$$\mathbf{E}(\mathbf{r}) = q\hat{\mathbf{s}}/4\pi\epsilon_0|s|^2 \tag{14.10}$$

This can be further generalized to a formula for the field of several point charges, using the principle of superposition. This principle holds for solutions of Maxwell's equations because Maxwell's equations are linear, just as it did for the solutions of the linear equations in Section 2.3. It says the following: If you know the fields E_i and B_i produced by several current distributions j_i ($i = 1, 2, \ldots, n$),

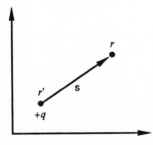

Figure 14.6 A point charge at vector position r', and a field point r.

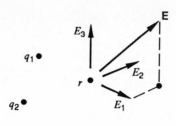

Figure 14.7 The electrostatic field of three point charges.

simply add the dynamic Maxwell equations for the E_i's to get

$$\sum_{i=1}^{n} \frac{dE_i}{dt} = \sum_i c^2 \,\mathrm{curl}\, B_i - \sum_i \frac{j_i}{\epsilon_0}$$

or

$$\frac{d}{dt}\left(\sum_i E_i\right) = c^2 \,\mathrm{curl}\left(\sum_i B_i\right) - \frac{(\sum_i j_i)}{\epsilon_0} \tag{14.11}$$

With the corresponding equation for $d(\sum B_i)/dt$, this states that the evolution equation for the fields in the presence of the total current $\sum j_i$ is satisfied by the sums of the fields, $\sum E_i$ and $\sum B_i$. That is, $\sum E_i$ is the electric field that would be produced by this total current.

Now suppose that each current j_i produced a point charge q_i (Figure 14.7), so that after a long time E_i is the electrostatic field of q_i. The total current $\sum j_i$ therefore produces a charge distribution in which all n charges are present simultaneously, and the electrostatic field E of this distribution is just $\sum E_i$. The final result is that by using the principle of superposition, we can write the field of a collection of charges $q_1, \ldots, q_n$ at positions $\mathbf{r}'_1, \mathbf{r}'_2, \ldots, \mathbf{r}'_n$:

$$\mathbf{E(r)} = \sum_{i=1}^{n} \frac{q_i}{4\pi\epsilon_0 |s_i|^2} \hat{\mathbf{s}}_i \tag{14.12}$$

where $\mathbf{s}_i \equiv \mathbf{r} - \mathbf{r}'_i$ and the unit vector $\hat{\mathbf{s}}_i = \mathbf{s}_i/|s_i|$.

▶ **Problem 14.6**

Consider two charges of equal magnitude q, separated by a distance d (Figure 14.8).

a. Compute the electric field vector (give both magnitude and direction) at the point $\mathbf{r}$ a distance z above the midpoint.

b. Compute the field at $\mathbf{r}$ for the case that the right-hand charge is $-q$ and the left-hand one is still q.   ◀

We now know how to calculate the electrostatic fields of a particular class of continuum charge distributions, namely superpositions of point charges. In fact, we can extend Coulomb's law to give the electrostatic field of an arbitrary charge distribution by taking the limit of the electrostatic fields of the corresponding

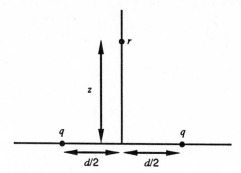

Figure 14.8   Two point charges.

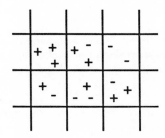

Figure 14.9   A system with a continuum charge distribution, with a lattice of spacing dr.

discrete charge distributions as $dr \to 0$. Consider a continuum charge distribution (by definition a rule giving a discrete distribution for each dr) whose discrete distribution $\rho(c)$ for a particular dr is shown in Figure 14.9. The charge in each cell is $\rho(c)\,d\tau(c)$. Here $d\tau(c)$ is the volume of cell c; it is dr^3 in a rectangular coordinate system. We can approximate the electrostatic field of that cell's charge alone by assuming it is a point charge at the cell center, as in Figure 14.10; this involves moving it a distance less than dr, and is a negligible change in the $dr \to 0$ limit. Thus, Eq. 14.12 becomes

$$\mathbf{E}(\mathbf{r}) = \sum_{\text{cells } c} \rho(c)\,d\tau(c)\,\hat{\mathbf{s}}\Big/4\pi\epsilon_0|s|^2 \tag{14.13}$$

where $\mathbf{s} \equiv \mathbf{r} - \mathbf{c}$, and c is the vector position of the center of the cell c.

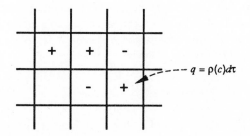

Figure 14.10   Point-charge approximation to the continuum charge distribution shown in the previous figure.

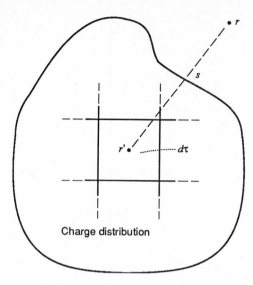

Figure 14.11 Continuum charge distribution whose electrostatic field is given by Eq. 14.14.

If the continuum charge distribution is given by a continuous function $\rho_f(\mathbf{r}')$, the discrete charge density $\rho(c)$ can be approximated by the value of $\rho_f(\mathbf{c})$ and the electrostatic field approaches a limit

$$\mathbf{E}(\mathbf{r}) = \int \rho_f(\mathbf{r}')\, d\tau\, \hat{\mathbf{s}} / 4\pi\epsilon_0 |s|^2 \qquad (14.14)$$

(As indicated in Figure 14.11, $d\tau$ is a volume element at $\mathbf{r}'$, the integration variable; also $\mathbf{s} = \mathbf{r} - \mathbf{r}'$ again). This formula gives the static field of a charge distribution specified by a continuous function.

This electrostatic field $\mathbf{E}(\mathbf{r})$ that coexists, after a long time, with the static charge distribution $\rho(r)$ is often referred to as "the electrostatic field of $\rho(r)$." This terminology is fine, as long as you remember that it isn't produced instantaneously when ρ is moved into position: this is easy for a human to forget, since it is produced very rapidly on our time scale.

■ **EXAMPLE 14.1** Consider a line charge with charge per unit length λ on a straight line of length $2L$, as in Figure 14.12. We can calculate the electric field anywhere (at any point $\mathbf{r}$). However, it is rather difficult except at points with special symmetry. One such point is at a height h along the bisector of the line charge, the dashed line in the figure. We can't use Eq. 14.14 for this example, because this charge distribution can't be represented by a continuous function ρ_f. We must return to the discrete formulation; a cubic lattice divides the line charge into segments of length dx, each containing a charge $\lambda\, dx$. We can then use Eq. 14.13 (with $\rho\, d\tau$ replaced by $\lambda\, dx$) to get in the $dx \to 0$ limit

$$\mathbf{E}(r) = \lambda \int dx\, \hat{\mathbf{s}} / 4\pi\epsilon_0 |s|^2 \qquad (14.15)$$

We can replace $\hat{\mathbf{s}}$ by $\mathbf{s}/|s|$, and $s_x = r_x - x = -x$ and $s_y = r_y = h$. Thus, $|s|^2 = x^2 + h^2$,

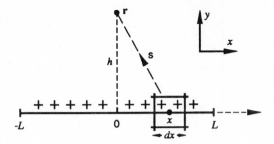

Figure 14.12   Calculation of the field of a finite line charge.

and

$$E_x(r) = \lambda \int_{-L}^{L} dx(-x)/4\pi\epsilon_0(x^2 + h^2)^{3/2} \tag{14.16}$$

$$E_y(r) = \lambda \int_{-L}^{L} dx\, h/4\pi\epsilon_0(x^2 + h^2)^{3/2} \tag{14.17}$$

If you work out the x-component (Eq. 14.16) you will get zero; this is because the x-components of the contributions to the field from symmetrically placed cells at $+x$ and $-x$ exactly cancel. The integral for E_y is

$$E_y(r) = (\lambda h/4\pi\epsilon_0) \int_{-L}^{L} dx(x^2 + h^2)^{-3/2}$$

$$= (\lambda h/4\pi\epsilon_0)xh^{-2}(x^2 + h^2)^{-1/2}\Big|_{-L}^{L} \tag{14.18}$$

$$= (2\lambda L/4\pi\epsilon_0 h)(L^2 + h^2)^{-1/2}$$

As a check, we can look at the limit in which the line is very short ($L \ll h$) and resembles a point charge $q = 2\lambda L$. We get $E \approx 2\lambda L/4\pi\epsilon_0 h^2$, in agreement with Eq. 14.8.

▶   **Problem 14.7**

Compute the electric field of the charged line segment of Example 14.1, but at a height h over the **end** (instead of the midpoint). As in the example, check your result in the limit $L \ll h$.   ◀

▶   **Problem 14.8**

Calculate the electric field at a height h above the center of a square loop, whose four sides of length a each have linear charge density λ (Figure 14.13). You may use the results of Example 14.1.   ◀

▶   **Problem 14.9**

Find the electric field at a height h over the center of a circle of radius R, whose circumference has linear charge density λ. Again, check the limit $R \ll h$.   ◀

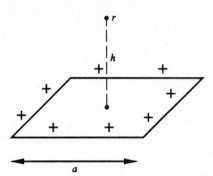

Figure 14.13   A charged square loop.

▶   **Problem 14.10**

Find **E** at a height h over the center of a disk of radius R, having uniform **surface** charge σ.   ◄

▶   **Problem 14.11**

a. Find the electric field at a distance r from the center of a charged spherical surface of radius R, having charge per unit area σ. Consider $r < R$ and $r > R$. *Hint*: Use spherical coordinates and write s in terms of the coordinates R and θ of the source point using the law of cosines.

b. Consider a sphere of radius R whose interior is filled with a uniform **volume** charge density ρ. Calculate **E** at any distance r from the center. You may use the result of **a**, or you may use Eq. 14.14 (note that this is the only one of our problems involving a volume integration for which Eq. 14.14 can be used directly).   ◄

SECTION 14.4   GAUSS' LAW

We have found for some particularly simple charge distributions that after a long time there is only an electric field (the electrostatic field) and it seems to depend only on the final charge distribution, independently of the current distribution that flowed to create the charge distribution. We would like to understand the relation between ρ and E better, and find a general way to deduce the electrostatic field E from ρ. Can we deduce from Maxwell's equations a relation between the charge density ρ and the static field E? Well, a Maxwell equation relates E to j, and the continuity equation relates div j (not j itself) to ρ. We can combine these equations and eliminate j by taking the divergence of the Maxwell equation (Eq. 9.31)

$$[\operatorname{div} dE/dt](c, t) = c^2 \operatorname{div} \operatorname{curl} B(c, t) - \operatorname{div} j(c, t)/\epsilon_0 \qquad (14.19)$$

The term involving B, which vanished accidentally before (Fig. 14.5), now vanishes in general, for **any** choice of B! This can be seen by evaluating the divergence of the curl for a cell c: it is the sum of curl B evaluated at the three oriented faces shown in Figure 14.14 (and three others that are hidden). When each of these is expressed as the sum of $B(e)$ at four edges, it is clear that there are two terms involving the edge labeled e, one from the top face and one from the right face. In addition, because the directions of the oriented edges involved are opposite, the terms cancel.

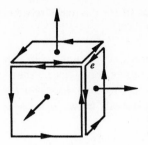

Figure 14.14 A cell c at which div curl B is evaluated.

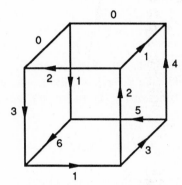

Figure 14.15 Discrete B field on the edges of one cell.

This is true for every edge, so the final result for div curl B is zero. (This is a special case of a much more general mathematical theorem that states that any two differentiations of certain types give zero. Another special case states that the curl of a gradient is zero, as we saw in Section 6.2.)

▶ **Problem 14.12**

Consider the discrete B field shown in Figure 14.15. Compute its curl for each of the faces of the cell shown, and show it on a sketch. Compute div curl B for this cell. ◀

Now we have a relation between div E and ρ:

$$\frac{d}{dt}\,\text{div}\,E(c, t) = \frac{d}{dt}\,\rho(c, t)/\epsilon_0 \tag{14.20}$$

[Note that we have interchanged the time derivative and the divergence. You can show that the order doesn't matter in the discrete case, even though this is not always true in the continuum case.]

▶ **Problem 14.13**

Consider any discrete vector field $E(f, t)$. Show that

$$\frac{d}{dt}[\text{div}\,E(c, t)] = \text{div}\left[\frac{d}{dt}\,E(c, t)\right]$$

Hint: Just write out both sides using the definitions of the discrete time derivative and divergence.   ◄

In the static limit, a statement about time derivatives isn't going to be very useful (both sides of Eq. 14.20 approach zero). We'd rather have $\mathrm{div}\, E$ than its time derivative. We can integrate this equation with respect to time, however (discretely, this just involves adding up the equations for all intervals between times 0 and t):

$$\mathrm{div}\, E(c, t) - \mathrm{div}\, E(c, 0) = \rho(c, t)/\epsilon_0 - \rho(c, 0)/\epsilon_0 \tag{14.21}$$

But we assumed there were initially no charges or fields, so

$$\mathrm{div}\, E(c, t) = \rho(c, t)/\epsilon_0 \qquad \text{(Gauss' law)} \tag{14.22}$$

As it stands, this is a statement about discrete fields. Of course, we can apply it to fields determined by continuous functions and take the $dr \to 0$ limit, whereupon it becomes a statement about the continuous functions $\rho_f(\mathbf{r})$ and $\mathbf{E}_f(\mathbf{r})$ representing the density and the electric field:

$$\nabla \cdot \mathbf{E}_f(r, t) = \rho_f(r, t)/\epsilon_0 \qquad \text{(Gauss' law)} \tag{14.23}$$

While we are at it, we may as well do the same thing to the Maxwell equation for B ($dB/dt = -\mathrm{curl}\, E$, Eq. 9.32) that we did for E. The only difference is that there are no source terms. The result is

$$\mathrm{div}\, B(v, t) = 0 \qquad \text{(TANSTAAMM law)} \tag{14.24}$$

for every vertex v.

Note that deriving Eq. 14.24 involves the divergence of curl E. The curl is most easily defined for an **edge** vector field, so it is easiest to think of this in the dual lattice, in which E is an edge vector field and we can use Figure 14.14 to see that $\mathrm{div}\,\mathrm{curl}\, E = 0$.

► **Problem 14.14**

Show that $\mathrm{div}\,\mathrm{curl}\, E = 0$ **without** using the dual lattice.
Hint: In the original ("direct") lattice, $[\mathrm{div}(\mathrm{curl}\, e)](v)$ is a sum of $(\mathrm{curl}\, E)(e)$ over six edges e oriented outward from a vertex v. Draw a sketch to show that the E at each face f enters twice, with opposite signs.   ◄

These two new laws, Gauss' law and the TANSTAAMM ("there ain't no such thing as a magnetic monopole") law, are exact consequences of our original two Maxwell equations (with charge conservation and the initial absence of fields and charge in a system). However, they turn out to be particularly useful in electrostatics and magnetostatics, because they don't involve time derivatives. Because they are so useful, and because they were discovered before the fundamental Maxwell equations for the dynamics, they are generally given "equal billing," and all four are referred to as Maxwell's equations. To keep things straight, we will refer to the more fundamental equations describing the time evolution (i.e., "dynamics") of E and B

as the "dynamic Maxwell equations," and the less fundamental, derived equations (Gauss' and TANSTAAMM laws) as "Maxwell's divergence equations."

▶   **Problem 14.15**

Why are the initial conditions of Problem 9.1 (having B nonzero at only one face) not physically possible? Is the corresponding pseudo-2D-field (Problem 10.2) consistent with the divergence equations?   ◀

We will explore the application of Gauss' law to an arbitrary problem in the next section; in general, we need to combine it with other information from the Maxwell equations to compute the electrostatic field. However, for charge distributions of very high symmetry, we can deduce the electrostatic field from Gauss' law alone. To do this, we must introduce the so-called "integral form" of Gauss' law, obtained by applying the divergence theorem to Eq. 14.22. For discrete fields, this amounts to summing Gauss' law over all the cells in a region R (Figure 14.16)

$$\epsilon_0 \sum_c [\operatorname{div} E](c)\, d\tau(c) = \sum_c \rho(c)\, d\tau(c) \tag{14.25}$$

or by the definition of the divergence (Chapter 7)

$$\epsilon_0 \sum_c \sum_{f \text{ of } c} E(f)\, da(f) = q \tag{14.26}$$

where the right-hand side q is just the total charge in the region R. The inner sum is over the faces f of the cell c. Each internal face f is a face of **two** cells, and the outward normals are in opposite directions, so each such $E(f)$ is counted twice with opposite signs. Thus, the only terms in Eq. 14.26 that don't cancel are those referring to the external faces, that is, those that form the boundary S of the region R (Figure 14.16).

Gauss' law becomes

$$\epsilon_0 \sum_{f \in S} E(f)\, da(f) = q \qquad \text{(discrete integral form of Gauss' law)} \tag{14.27}$$

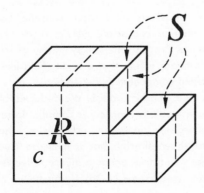

Figure 14.16 A region R composed of cells c (dotted lines) and its boundary S.

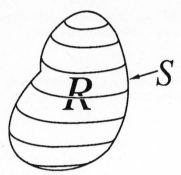

Figure 14.17 A continuum region R and its boundary S.

where the sum is over the f's that make up the surface S (i.e., the sum is a discrete surface integral). In the continuum limit we can we define $E(f)$ from a continuous vector function $\mathbf{E}_f(\mathbf{r})$. If R is an arbitrary continuum region of space bounded by a closed surface S (Figure 14.17), the continuum divergence theorem (Section 6.4) gives

$$\epsilon_0 \int_S \mathbf{E} \cdot d\mathbf{a} = q \qquad \text{(continuum integral form of Gauss' law)} \qquad (14.28)$$

The sum in Eq. 14.27 or the integral in Eq. 14.28 is called the **electric flux** through the surface S.

The set of charge distributions whose electrostatic fields can be computed by Gauss' law alone is very limited: they are the distributions that are so symmetrical that a surface S can be found on which they are exactly uniform. These distributions are the point charge (whose symmetry is such that any point on a sphere centered at the charge is equivalent to any other), the line charge (for which any two points on a cylinder are equivalent), and the plane charge (for which all points of a parallel plane are equivalent). The surface S used in Gauss' law is sometimes called a Gaussian surface.

■ **EXAMPLE 14.2 OBTAINING COULOMB'S LAW FROM GAUSS' LAW.** We calculated the electrostatic field of a point charge q in Section 14.2 using the dynamical Maxwell equation for dE/dt. It can also be obtained by purely electrostatic considerations, from Gauss' law. It is convenient to again use a spherical coordinate system. The method relies on the assumption that the electrostatic field has the full spherical symmetry of the charge distribution. We saw in Section 14.2 what it means to say that a vector field like E (or j) has spherical symmetry; it means that the radial component E_r depends only on r and not on θ and ϕ, and that the nonradial components E_θ and E_ϕ vanish. We deduced this before from physical considerations (the charge moves radially, hence j is radial). However, one can make a more abstract argument involving the operations under which the system (including the fields) is to be invariant, to show directly that it must be true of the electrostatic field. Spherical symmetry implies invariance under a mirror reflection in any plane containing the origin and the field point $\mathbf{r}$ (see Figure 14.18). Under such a reflection

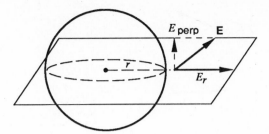

Figure 14.18 Mirror plane, used to show that E is radial for a point charge.

the component E_{perp} of **E** normal to this plane changes sign; invariance requires $E_{perp} = -E_{perp}$, so $E_{perp} = 0$. We can choose the plane so $E_{perp} = E_\theta$, or $E_{perp} = E_\phi$; thus both of these vanish. Invariance under rotations about the point charge then requires that the remaining (radial) component be equal at points that can be rotated into each other, that is, points at the same distance r from the point charge.

Using Gauss' law, in the form of Eq. 14.27 with a sphere of radius r as the Gaussian surface, the sum is over the faces comprising the surface of the sphere, and the field $E_r(r)$ is constant and can be factored out. The sum of the areas is $4\pi r^2$ as in Section 14.2, and Gauss' law becomes

$$\epsilon_0 E_r(r)4\pi r^2 = q$$

or

$$E_r(r) = q/4\pi\epsilon_0 r^2 \qquad \text{(Coulomb's law)} \qquad (14.29)$$

which of course agrees with the previous result. This result is correct for a discrete field; because it is independent of dr, taking the $dr \to 0$ limit is trivial, and Eq. 14.29 is also true of a continuous field $\mathbf{E}_f$.

Equation 14.29 applies to a somewhat more general charge distribution than just the point charge; note that we could make the same assumptions for any charge distribution with spherical symmetry, such as a spherical shell of charge or a number of concentric shells. The field would still be given by Eq. 14.29, as long as we understood q to mean the total charge inside the sphere of radius r.

■ **EXAMPLE 14.3** Compute the field of a thin, uniformly charged shell of radius R and total charge q. For $r < R$, the charge inside r is zero, so Eq. 14.29 gives $E_r(r) = 0$. For $r > R$, the charge inside r is q, so in general

$$E_r(r) = q/4\pi\epsilon_0 r^2 \qquad \text{for } r > R$$

$$0 \qquad \text{for } r < R$$

This field is shown in Figure 14.19. [If you did Problem 14.11, you now appreciate the usefulness of Gauss' law.]

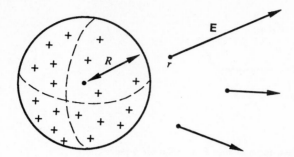

Figure 14.19   Uniformly charged spherical shell and its electrostatic field.

▶   **Problem 14.16**

Compute the electrostatic field of a line with charge per unit length λ using Gauss' law, with a cylindrical coordinate system.

a. Show that symmetry requires two of the three components of E in cylindrical coordinates to vanish. [*Hint*: find mirror planes as we did in the spherical case.]

b. Compute the third component, at arbitrary r, ϕ, and z. If you did Problem 14.4, compare the answers.   ◄

▶   **Problem 14.17**

Compute the electrostatic field of an infinitely long cylinder of radius R, whose surface has a uniform charge σ per unit area. Give the field as a function of r, the distance of a radial face from the axis of the cylinder.   ◄

One other type of charge distribution has enough symmetry that its electrostatic field can be computed from Gauss' law: the infinite plane charge. This case is a bit tricky, though, because it is one in which the electrostatic field is **not** independent of the currents that set up the charge distribution. We can see this by using the dynamic Maxwell equations to compute the fields for two different choices of current. The first choice involves creating a positive and a negative plane of charge (with charge per unit area $\pm\sigma$) at $x = 0$ and moving the negative one an infinite distance to the right (Figure 14.20a) during the interval around $\frac{1}{2}dt$. The current $j(f, \frac{1}{2}dt)$ points to the left at all faces with $x > 0$; this is a special case of the longitudinal-current quasi-1D systems we discussed before. The electric field at $t = dt$ points to the right at all these faces and never changes thereafter (because $B = 0$). The electrostatic field is as in Figure 14.20b: it has a uniform value $E_x = \sigma/\epsilon_0$ in the region where $x > 0$ and is zero for $x < 0$.

Obviously we could construct another current that gives rise to the same charge distribution by moving the negative charge to the left instead of the right. The corresponding pictures would look just like Figure 14.20, reflected in a mirror in the $x = 0$ plane: $E_x = 0$ for $x > 0$, and $E_x = -\sigma/\epsilon_0$ for $x < 0$. Which is the correct field? Unfortunately, both are; the field just isn't uniquely determined by the charge distribution. However, this isn't really the question we want to ask. An infinite plane charge is not physically obtainable; presumably we are interested in it as a simple

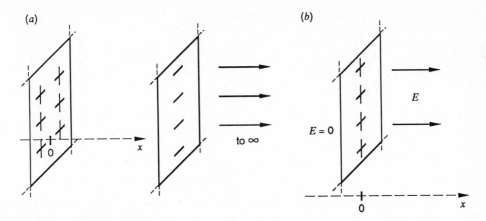

Figure 14.20 (*a*) Creating a positively charged plane by moving its negative counterpart to infinity. (*b*) Resulting electrostatic field.

model of some real charge distribution. This could be a large (but finite) positively charged plate, such as that actually drawn in Figure 14.20*b*. What we really want to know is the electrostatic field of this finite charge. If it is the left-hand plate of a parallel-plate capacitor, then Figure 14.20*b* shows the correct field. Suppose, on the other hand, it is far from any negative charge (i.e., we have moved the negative charge away a distance large compared to the lateral size L of the plate). The electric field will initially exist only between the plates, as in Figure 14.20*b*. However, because the plates are finite, E will not be uniform near the edges, so dB/dt will not be zero. In fact, if we look at the system on a large enough scale, the plates look like point charges, and the system resembles Figure 14.4, in which a point charge was created. In that case the B field caused the E field to spread around the point charge so it eventually had the same symmetry (spherical) as the charge distribution. In the present case, the charged plate doesn't have spherical symmetry, but it does have a left–right mirror symmetry (reflection in the plane $x = 0$). It seems reasonable (and we could verify by evolving the fields numerically) that the E field will spread around to the left until it has the same mirror symmetry as the charge distribution. Thus the true electrostatic field points away from the plate on both sides, as in Figure 14.21*a*.

It turns out that this symmetry allows us to compute the field using Gauss' law. This requires finding a surface S over which the field is uniform. We can use a pillbox-shaped surface as shown in Figure 14.21*b* with two flat sides of area A in the planes $x = d$ and $x = -d$. The fields at these two sides are the same because of the left–right symmetry. We also require the field to be uniform over each surface; this requires assuming that the charged plane looks like an infinite plane from the distance d. So we assume the pillbox is far from the boundaries of the charge distribution, and $d \ll L$, where L is the linear dimension of the plane. Then the electric field has the symmetry of the infinite plane: it depends only on x, and has no y or z components ($E_y = 0$ follows from the mirror symmetry in a plane $y = $ constant, via $E_y = -E_y$). When we break up the surface of the pillbox into faces (Figure 14.21*c*), only those on the left and right sides of the pillbox will have

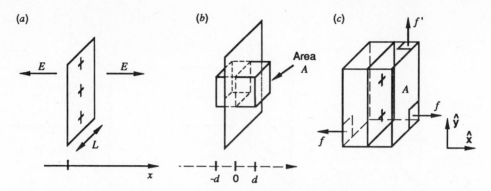

Figure 14.21 (*a*) Electrostatic field of a finite charged plate. (*b*) Gaussian pillbox for computing the value of the electrostatic field of a planar charge. (*c*) Dividing the pillbox into faces to use Gauss' law.

nonzero $E(f)$. The E field points along the outward orientation indicated for f, so $E(f)$ has the same positive value E at all these faces. The faces on the narrow sides perpendicular to the charged plane will have $E(f') = E_y$ (or E_z) = 0. So the integral form of Gauss' law (Eq. 14.27) gives

$$\epsilon_0 \sum_f E(f)\, da(f) = q = \sigma A$$

$$\epsilon_0 E(2A) = \sigma A$$

$$E = \sigma/2\epsilon_0 \qquad \text{(field of a charged plane)} \qquad (14.30)$$

where $\sum da(f) = 2A$ is the total area of the left and right surfaces. Note that we could also have obtained this result from the dynamic Maxwell equations if we had chosen a current distribution with the left–right symmetry of the charge distribution, namely one in which half of the negative charge $(-\sigma/2)$ moves away to infinity to the right and the other half moves to the left.

> This example is a little unsetting in that it forces us to wonder whether the other fields we have calculated by assuming a particular current also depend on the current, or only on the final charge distribution. It turns out that this problem only arises when we move an infinite amount of charge to infinity. In a certain sense the field is still determined uniquely by the charge distribution, if we regard an infinite negative charge at "right infinity" as being different from one at "left infinity." The important point is that the electrostatic field attributable to a **finite** charge approaches zero as the charge moves to infinity (as you can see by letting $r \to \infty$ in Coulomb's law), but that of an **infinite** charge does not always. (Although it does if it is a "small" infinite charge: we moved an infinite **line** charge to infinity in Section 14.2 without running into this problem.) At any rate the problem is not worth worrying about, because it is entirely academic. Real physical systems never have infinite charges.

The electrostatic field of **two** infinite oppositely charged planes with charge densities $\pm\sigma$ (a model of a parallel-plate capacitor) has already been calculated

(Section 11.2) by evolving the dynamic Maxwell equations. We found that it was zero except between the plates, and we calculated its value between the plates. Gauss' law won't tell you that it's zero outside the plates, but if you can remember that much you can get the numerical value inside from Gauss' law. Simply set up the Gaussian pillbox of Figure 14.21c around the left-hand plate, which has charge density $+\sigma$. The field on the left-hand surface of the pillbox is then zero, we only need to integrate over the area A of the right-hand surface, and Eq. 14.30 becomes $E = \sigma/\epsilon_0$, the same value we found before.

SECTION 14.5   FIELD LINES

Gauss' law makes possible a very useful way of depicting electric fields, using the "flux" picture of an electric field that exploits the analogy between an electric field and a flowing current of particles. In the case of a steady current, one can show that the continuity equation ($\mathrm{div}\, j = 0$) implies that we can draw continuous flow lines such that the number of lines crossing an oriented face f is proportional to $j(f)\, da(f)$. In a region with no charge, Gauss' law does the same for us: we can draw **field lines** such that the number crossing f in the positive direction is proportional to the electric flux $E(f)\, da(f)$. We can choose any proportionality constant we want, that is, given any constant E_0 we can draw field lines so that each line represents an amount E_0 of field, and the number of lines is $E(f)/E_0$. (There is then one line for each $E_0\, dr^2$ of flux; we are using cartesian coordinates so $da(f) = dr^2$.) To see this, note that Gauss' law can be multiplied by $dr/\epsilon_0 E_0$ to give

$$\rho(c)\, dr/\epsilon_0 E_0 = \mathrm{div}\, E(c)\, dr/E_0 = \sum_f E(f)/E_0 \qquad (14.31)$$

which is exactly the net number of lines leaving the cell c (if $E(f)/E_0$ is positive it is the number of lines leaving through the face f, and if it is negative its magnitude is the number entering.) So if there is no charge in the cell ($\rho(c) = 0$, as in the top cell of Figure 14.22) the number leaving is equal to the number entering, and they can be connected together so there are no loose beginnings or ends inside. If $\rho(c) \neq 0$, as in the lower cell of Figure 14.22, there will be loose beginnings or ends.

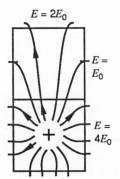

$E = 2E_0$

$E = E_0$

$E = 4E_0$

Figure 14.22   Field lines in a discrete pseudo-2D system.

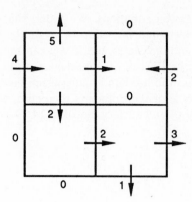

Figure 14.23   A discrete electric field.

Equation 14.31 gives the number of loose beginnings (lines that start in the cell and pass outward through a face) if it is positive; if it is negative its magnitude is the number of loose ends. Thus, the number of loose beginnings is exactly $\rho(c)\,dr/\epsilon_0 E_0$; this is consistent with assuming that lines begin on positive charges. (You can see by setting the right-hand side of Eq. 14.31 equal to unity that a single line begins on a charge $q = \rho(c)\,dr^3 = \epsilon_0 E_0\,dr^2$, or ϵ_0 times the flux associated with the line.) Similarly, lines **end** on **negative** charges, meaning that the arrows point toward the charges.

The figure actually shows a pseudo-2D field; all lines crossing a face perpendicular to the paper are shown projected onto the plane of the paper. Given a discrete pseudo-2D field $E(f)$ and a charge distribution $\rho(c)$, satisfying Gauss' law, it is straightforward to draw these lines: just draw the correct number of lines crossing each face (with arrows indicating the direction of "flow"); you will automatically be able to connect them up smoothly just as in Figure 14.22.

▶   **Problem 14.18**

Figure 14.23 shows a discrete electric field (pseudo-2D), the numbers are $E(f)/E_0$.

a. Compute the charge density field $\rho(c)$ from Gauss' law, in terms of E_0 and dr. Indicate $\rho(c)$ on a diagram.

b. Draw the field lines in all four cells.

c. Count the loose ends in each cell and verify that the result is properly related to ρ.   ◀

In the continuum limit, in which the field is presented by a continuous vector function $\mathbf{E}_f(r)$, the field lines will be parallel to the vector $\mathbf{E}_f$. We will show this for a pseudo-2D field. Choose a distance x_0 along the x-direction (Figure 14.24a) so that $\mathbf{E}_f$ doesn't change much over the distance x_0, and use $dr \ll x_0$. The electric field lines that cross it in the 2D view are those that in the real three-dimensional system (Figure 14.24b) cross a rectangle perpendicular to the paper with one side x_0 and extending dr out of the paper. This rectangle contains many faces, so many lines will cross it; the actual number is $E_y x_0\,dr/E_0\,dr^2$. (Here E_y is the y-component of $\mathbf{E}_f$, and $E_y x_0\,dr$ is the flux through the rectangle.) The density of field lines along the x-direction is the same at any height y, so they must be parallel as shown in Figure 14.24a. Similarly, the number crossing the vertical rectangle of length y_0

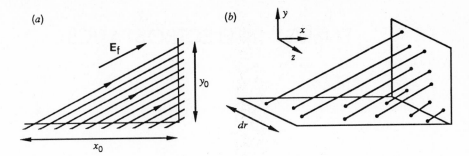

Figure 14.24   (*a*) Electric field lines for a smooth pseudo-2D field. (*b*) 3D view.

is $E_x y_0 / E_0\, dr$. But these are the same field lines that cross the horizontal line, so

$$E_y x_0 / E_0\, dr = E_x y_0 / E_0\, dr$$

$$y_0 / x_0 = E_y / E_x$$

(14.32)

Thus the slope of the lines is equal to the slope of the vector $\mathbf{E}_f$: the field line is **tangent** to the electric field vector at each point.

Many computer programs exist for drawing the field lines of an arbitrary system of point charges; see for example the book by Gould and Tobochnik referred to in the bibliography.

> In 3D discrete problems, the procedure used in Figure 14.22 does not uniquely determine how to connect up the field lines; they can be connected in many different ways without crossing. We must define the field lines in three dimensions by the continuum requirement that they be parallel to the electric field vector at every point. Once we know a single point on a field line (say where it crosses a given surface) the rest of the line is then determined. The choice of the first point is inherently arbitrary.

CHAPTER FIFTEEN

ENERGY IN ELECTROSTATICS

SECTION 15.1 THE ELECTROSTATIC POTENTIAL

In Chapter 14 we looked at how charges can give rise to electrostatic fields without discussing the effects the fields might have on the charges. We now reverse our point of view: assume the existence of an electrostatic field, and look at its effects on a charge q that happens to be in its vicinity. To avoid getting confused and thinking that this charge q is the one that **created** the field, we will assume that it is very small (infinitesimal) and refer to it as a **test charge**. The situation we have in mind is shown in Figure 15.1—we can move the charge around, and at each point $\mathbf{r}$ it is subject to an electrostatic force $\mathbf{F}(\mathbf{r}) = q\mathbf{E}(\mathbf{r})$. In Chapter 6 we studied an important special type of force, the **conservative** force. We established a number of equivalent criteria for a force field to be conservative, one of which was that its curl vanish everywhere. Another, which simplifies many calculations in the case of a conservative force, is that the work done by the force when the charge moves from a to b (perhaps along the path shown in Figure 15.1) is independent of the path. We have not used this concept thus far in the case of the electric force, because the electric force is **not** in general conservative. For the electric force $q\mathbf{E}$ on a charge q to be conservative, we would need curl $\mathbf{E} = 0$; according to Faraday's law (Eq. 9.32) this is not zero, but $-d\mathbf{B}/dt$. In electrostatics, however, it is zero, since the fields do not change with time. Thus

The electro**static** force **is** conservative.

One of the properties we demonstrated for conservative forces in Chapter 6 was that they allow the definition of a **potential energy** function, whose negative gradient is the force. In the present context this is called the **electrostatic potential energy** of the test charge. Just as the electric field (force per unit charge) turned out to be more convenient for calculations than the electric force because it is independent of the magnitude of the test charge, it is convenient here to deal with the electrostatic potential energy per unit charge, usually called the **electrostatic potential** (sometimes even just "potential"). We will denote the electrostatic potential by V. Dividing the negative gradient $q\mathbf{E}$ of the electrostatic potential energy by q then gives

$$\mathbf{E} = -\operatorname{grad} V \tag{15.1}$$

as the equation defining the electrostatic potential. The results of Chapter 6 imply

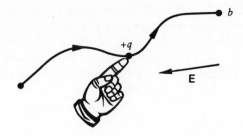

Figure 15.1 Moving a test charge in an electrostatic field.

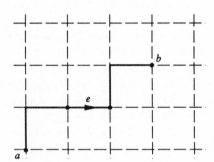

Figure 15.2 A discrete path from a to b.

that V can always be defined, because the curl of an electrostatic field is zero. It is unique in the sense that any two choices can differ only by a constant.

These results were proved in Chapter 6 using a force field discretized as an edge field; the path from a to b in Figure 15.1 then becomes a set of edges e (Figure 15.2). This creates a slight complication when we apply these results to the electrostatic force, because we have usually been thinking of the electrostatic field as a **face** field. We can easily solve this problem, however, by using the **dual** lattice (Section 6.5) in discussing electrostatic energy. The dual lattice is shown by dotted lines in Figure 15.3. The electric field assigns a number $E(f)$ to each face in the original lattice (sometimes called the "direct" lattice). The faces of the direct lattice are in one-to-one correspondence with the edges of the dual lattice (the center of

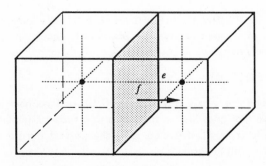

Figure 15.3 The dual lattice (dotted lines), showing the edge e corresponding to the face f of the direct lattice.

each direct face is also the center of a dual edge) so we can regard E as an edge field that assigns this same number to the edge corresponding to f.

In the present context, the central discrete results of Chapter 6 were that the vanishing of curl E implies the existence of a vertex field V on the dual lattice (i.e., a cell field on the direct lattice) satisfying $E = -\operatorname{grad} V$, and that the value of V at any vertex v could be computed from E by the formula

$$V(v) = V(v_0) - \sum_e E(e)\, dl(e) \tag{15.2}$$

where $V(v_0)$ is an arbitrarily chosen value for the potential at a reference vertex v_0, and the sum (a discrete path integral) is over edges e forming a path from v_0 to v.

To write the continuum limit of Eq. 15.2 we must consider a continuum path from a reference point $\mathbf{r}_0$ to an arbitrary point $\mathbf{r}$ such as that in Figure 15.1. It can be approximated by discrete paths like that in Figure 15.2, whose edges e are parallel to the coordinate axes. At an edge e parallel to the x-axis, the $E(e)$ in the discrete path integral becomes the component E_{fx} of the vector function $\mathbf{E}_f$; if we define the vector $d\mathbf{l}$ to have magnitude dl and to point along the x-axis, we can replace $E(e)\, dl(e)$ by the dot product $\mathbf{E}_f \cdot d\mathbf{l}$. This allows us to write the sum in Eq. 15.2 without explicit reference to the axes. Its limit as $dr \to 0$ is a continuum path integral. It is sometimes denoted by an integral sign with the letter "C" superimposed, but we will rely on the context (and the $\cdot d\mathbf{l}$) to distinguish it from other types of integral. Thus the continuum potential function is given by

$$V(\mathbf{r}) = V(\mathbf{r}_0) - \int_{\mathbf{r}_0}^{\mathbf{r}} \mathbf{E}(\mathbf{r}) \cdot d\mathbf{l} \tag{15.3}$$

(we omit the f subscripts here; all quantities are assumed to be functions).

> This trick of defining a vector $d\mathbf{l}$ actually does more than simplify the notation in the continuum limit; it also frees us from the requirement that our discrete approximations to the continuum path involve only pieces $d\mathbf{l}$ along coordinate axes. It is not hard to show, by approximating a short straight segment $d\mathbf{l}$ (not along an axis) by smaller pieces along the axes, that $\mathbf{E} \cdot d\mathbf{l}$ correctly gives the contribution to the path integral from $d\mathbf{l}$.

We can calculate the continuum electrostatic potential exactly in certain cases, of which one example follows.

■ **EXAMPLE 15.1 THE ELECTROSTATIC POTENTIAL OF A POINT CHARGE Q**   We know that the components of the electric field of a point charge are particularly simple in spherical coordinates. Only one component is nonzero:

$$E_r(r,\theta,\phi) = Q/4\pi\epsilon_0 r^2$$

If we pick a reference point along the z-axis, say $\mathbf{r}_0 = (r_0, 0, 0)$, and connect it to an arbitrary point (r,θ,ϕ) by a curve that goes first up the z-axis to a radius r (Figure 15.4) and then out in the θ direction to (r,θ,ϕ), it is clear only the part of the path going radially outward contributes to the path integral.

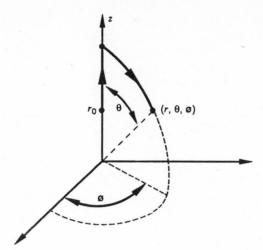

Figure 15.4   Integration path from r_0 to r.

If we choose a discretization such that there is a line of vertices (of the dual lattice) going up the z-axis, spaced dr apart, the sum in Eq. 15.2 whose limit is the path integral in Eq. 15.3 is

$$\sum_e E(e)\, dl(e) = \sum_e E_r(r)\, dr$$

But the limit of this sum is

$$\int_{r_0}^{r} E_r(r)\, dr = \int_{r_0}^{r} (Q/4\pi\epsilon_0 r^2)\, dr = (-Q/4\pi\epsilon_0 r) - (-Q/4\pi\epsilon_0 r_0)$$

Thus Eq. 15.3 becomes

$$V(r, \theta, \phi) = V(r_0) + Q/4\pi\epsilon_0 r - Q \lesssim 4\pi\epsilon_0 r_0$$

Clearly the simplest choice for $V(r_0)$ is $Q \lesssim 4\pi\epsilon_0 r_0$, so that

$$V(r, \theta, \phi) = Q/4\pi\epsilon_0 r \qquad \text{(Coulomb's law for } V\text{)} \qquad (15.4)$$

which is evidently only a function of r, and not of θ and ϕ.

   If we think of this choice of additive constant in terms of choosing a reference point to have zero potential [$V(r_0) = 0$], evidently our reference point must be $r_0 = \infty$. We will usually use this reference point in problems involving an infinite volume, because the other logical choice ($r_0 = 0$) has infinite potential.

■ **EXAMPLE 15.2 THE POTENTIAL OF A CHARGED SPHERICAL SURFACE**   Consider a spherical surface of radius R carrying a uniform surface charge density σ, so that its total charge is $Q = 4\pi R^2 \sigma$. We computed the electrostatic field of this charge distribution in Example 14.3; it is zero for $r < R$ (inside the sphere) and $Q/4\pi\epsilon_0 r^2$ for $r > R$. This time we will choose our reference point to be $r_0 = \infty$ from the beginning. The path along the z-axis

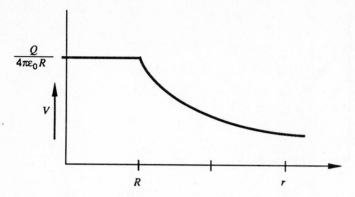

Figure 15.5   The potential of a charged spherical surface.

from r_0 to r then consists of inward-pointing radial edges, at which $E(e) = -E_r(r)$. The limit of the path integral from r_0 to r is then

$$\int_r^\infty [-E_r(r)]\, dr = (-Q/4\pi\epsilon_0 r_0) - 0$$

as long as $r > R$. If $r < R$, we must stop integrating at R (inside which E vanishes) instead of at r. Equation 15.3 then gives

$$V(r) = Q/4\pi\epsilon_0 R \qquad (r < R)$$

$$Q/4\pi\epsilon_0 r \qquad (r > R) \tag{15.5}$$

This potential is shown in Figure 15.5.

▶   **Problem 15.1**

Compute the electrostatic potential inside and outside a sphere of radius R with a uniform volume charge density ρ.   ◀

▶   **Problem 15.2**

Consider an infinitely long line charge with linear charge density λ. Show that the potential due to λ is $-(\lambda/2\pi\epsilon_0)\ln r +$ constant, at a distance r from the line. See Problem 14.16 for $E(r)$.   ◀

We can use the potential (Eq. 15.4) of a point charge to deduce the electrostatic potential of a set of point charges by the principle of superposition. We derived the principle of superposition for electrostatic fields in Chapter 14: it says that the field of a collection of charges, say $q_1, q_2, \ldots, q_n$, is exactly the sum

$$\mathbf{E(r)} = \sum_{i=1}^n \mathbf{E}_i$$

where $\mathbf{E}_i$ is the field due to q_i alone. This can easily be extended to electrostatic

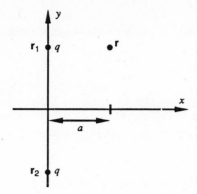

Figure 15.6 Two point charges.

potentials: if $V_i(\mathbf{r})$ is the potential for q_i (so $\mathbf{E}_i = -\nabla V_i$) then the potential

$$V(\mathbf{r}) = \sum_{i=1}^{n} V_i(\mathbf{r})$$

satisfies

$$-\nabla V(\mathbf{r}) = -\sum_i \nabla V_i = \sum_i \mathbf{E}_i = \mathbf{E}$$

and is therefore a possible potential for the collection of charges. If we require the potentials to vanish as $|\mathbf{r}| \to \infty$, it is unique.

Thus the potential of a collection of point charges $q_1, q_2, \ldots, q_n$ at $\mathbf{r}_1, \mathbf{r}_2, \ldots, \mathbf{r}_n$ is

$$V(\mathbf{r}) = \sum_{i=1}^{n} q_i / 4\pi\epsilon_0 |\mathbf{r} - \mathbf{r}_i| \tag{15.6}$$

■ **EXAMPLE 15.3** Calculate the potential at the point $\mathbf{r} = (a, a, 0)$ due to two equal point charges q at $\mathbf{r}_1 = (0, a, 0)$ and $\mathbf{r}_2 = (0, -a, 0)$ (Figure 15.6). Setting $q_1 = q_2 = q$ in Eq. 15.6,

$$V(\mathbf{r}) = q / 4\pi\epsilon_0 |\mathbf{r} - \mathbf{r}_1| + q / 4\pi\epsilon_0 |\mathbf{r} - \mathbf{r}_2|$$

$$= q / 4\pi\epsilon_0 a + q / 4\pi\epsilon_0 5^{1/2} a$$

$$= q(1 + 5^{-1/2}) / 4\pi\epsilon_0 a \approx 1.447 q / 4\pi\epsilon_0 a$$

Note that this is easier than the problem of determining **E** for this charge configuration (Eq. 14.12) because V is not a vector.

▶ **Problem 15.3**

Compute the potential at the point **r** in Figure 15.6 if the charge at $\mathbf{r}_2$ were $-q$ instead of $+q$. Give your answer as a decimal number times $q / 4\pi\epsilon_0 a$. ◀

▶ **Problem 15.4**

Compute the potential at the center of the **side** of a square (side length $= a$) having four charges $+q$ at the corners.   ◀

▶ **Problem 15.5**

a. Compute the potential at the point **r** in Figure 14.8. Compute the z-component of its gradient and compare with your answer to Problem 14.6a.

b. Similarly check the answer to Problem 14.6b by differentiating a potential.   ◀

▶ **Problem 15.6**

A cage is constructed of four infinite line charges at the corners of a square of side d, each with charge density $+\lambda$, as shown in Figure 15.7.

a. Compute the electric field at the center (labeled a in the figure).

b. Compute the field and the potential at point b, halfway between two adjacent lines. You may use the result of Problem 15.2. Use point a as the reference point (i.e., set $V = 0$ there).

c. If a positive test charge q is placed at point a, what will happen to it? What will happen if it is displaced slightly toward b?   ◀

If we have a charge distribution described by the discrete density $\rho(c)$ (for example, that in Figure 14.11), we can approximate it by a set of point charges at the cell centers c; then Eq. 15.6 gives an approximation to the potential,

$$V(\mathbf{r}) \approx \sum_c \rho(c)\, d\tau(c)/4\pi\epsilon_0|\mathbf{s}| \tag{15.7}$$

where $\mathbf{s} \equiv \mathbf{r} - \mathbf{c}$. This approximation becomes more and more accurate as dr decreases, because the distance we have to move the charge in the cell c to reach the center decreases with dr. In the continuum limit in which the density is described by a continuous function $\rho(\mathbf{r}')$, Eq. 15.7 becomes the exact equation

$$V(\mathbf{r}) = \int d\tau \rho(\mathbf{r}')/4\pi\epsilon_0|\mathbf{r} - \mathbf{r}'| \tag{15.8}$$

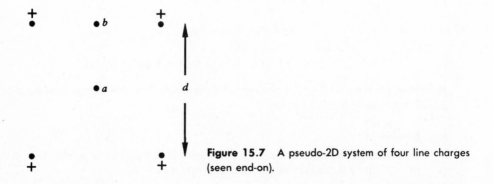

Figure 15.7 A pseudo-2D system of four line charges (seen end-on).

where ρ and V are scalar functions. Just as in the calculation of $\mathbf{E}$ (Section 14.3), we can generalize this to a surface charge density or a linear charge density by replacing $\rho\,d\tau(c)$ by $\sigma\,da(c)$ or $\lambda\,dl(c)$ respectively. This solution (Eq. 15.8) of the Poisson equation, in the form of an integral over the inhomogeneous function ρ on the right-hand side of the Poisson equation, is called a **Green's function** solution. The function $G(\mathbf{r}, \mathbf{r}') = 1/4\pi\epsilon_0|\mathbf{r} - \mathbf{r}'|$ is called the **Green's function** for the Poisson equation.

■ **EXAMPLE 15.4 THE POTENTIAL OF A CHARGED SPHERICAL SHELL (AGAIN)** Consider a sphere of radius R whose surface has a uniform charge density σ. We computed its potential in Example 15.2 the easy way (from the electrostatic field given by Gauss' law); we will now do it the hard way, using Eq. 15.8 with $\sigma\,da$ replacing $\rho\,d\tau$. The cell lattice determined by any dr, $d\phi$, and $d\theta$ (as in Chapter 7) divides the sphere into small pieces. The center of each cell containing a piece has the same radial coordinate (close to R) but has ϕ ranging from 0 to 2π and θ ranging from 0 to π. The area of the piece in a cell centered at (r, ϕ, θ) is given in Table 7.1 as $da = R^2 \sin\theta\,d\theta\,d\phi$ (this was actually computed as the area of a face of a cell, but it also gives the area of the piece of the sphere in a cell). The integral in Eq. 15.8, the limit of a sum over these pieces (a double sum over ϕ and θ), becomes the double integral

$$V(\mathbf{r}) = \int_0^{2\pi}\int_0^{\pi} \sigma R^2 \sin\theta\,d\theta\,d\phi / 4\pi\epsilon_0 s$$

$$= (\sigma R^2/4\pi\epsilon_0)\int_0^{\pi} d\theta\,\sin\theta\,s^{-1}\int_0^{2\pi} d\phi$$

because s (Figure 15.8) is independent of ϕ. The integral over ϕ is just 2π, and we can express s in terms of R and θ using the law of cosines ($s^2 = R^2 + r^2 - 2Rr\cos\theta$), giving

$$V(\mathbf{r}) = (\sigma R^2/2\epsilon_0)\int_0^{\pi} d\theta\,\sin\theta\,(R^2 + r^2 - 2Rr\cos\theta)^{-1/2}$$

$$= (\sigma R^2/2\epsilon_0)\left[R^{-1}r^{-1}(R^2 + r^2 - 2Rr\cos\theta)^{1/2}\right]_0^{\pi}$$

$$= (\sigma R/2r\epsilon_0)\left[(R^2 + r^2 + 2Rr)^{1/2} - (R^2 + r^2 - 2Rr)^{1/2}\right]$$

$$= (\sigma R/2r\epsilon_0)\left[(R + r) - (R - r)\right] = \sigma R/\epsilon_0 \tag{15.9}$$

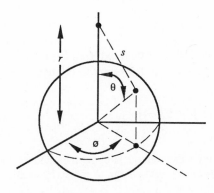

Figure 15.8 The separation s between the field point $(r, 0, 0)$ and the source point (R, ω, ϕ).

This is the value we got before, **inside** the sphere. What do we have to do to get the answer outside? When we took the square root of $(R - r)^2$ to get $(R - r)$, we implicitly assumed $(R - r) \geq 0$, that is, that the field point r was inside. If $(R - r) < 0$, the positive square root is $-(R - r)$, turning Eq. 15.9 into

$$V(\mathbf{r}) = \sigma R^2 / \epsilon_0 r \qquad (15.10)$$

which is the correct answer outside.

We can use this result for V to calculate the electrostatic field from $\mathbf{E} = -\text{grad} V$. Inside the sphere V is constant so $\mathbf{E} = 0$, and outside

$$E_r(r) = -(\text{grad} V)_r = -dV(r)/dr = \sigma R^2 / \epsilon_0 r^2 \qquad (15.11)$$

where we have used the formula from the inside back cover of this book for the gradient. It is usually easier to compute E from V in this way than to evaluate it directly (Eq. 14.14). (In this particular case there is a still easier way, of course, using Gauss' law.)

There are relatively few cases for which Coulomb's law (in the form of Eq. 15.8) is a tractable way to compute the potential; usually the general methods we will describe in Chapter 16 are necessary. Some cases that can be solved by Coulomb's law are the following:

▶   **Problem 15.7**

Calculate the electrostatic potential inside a sphere with a uniform volume charge density ρ, from Eq. 15.8. Check your answer with that of Problem 15.1.   ◀

▶   **Problem 15.8**

A solid cylinder of length L and radius R has uniform charge density ρ inside.

a. Compute the potential at a point $(0, 0, z)$ on the axis of the cylinder, but outside it (the origin is at the center of the cylinder, so this means $z > \frac{1}{2}L$).

b. Use this potential to calculate the electric field at $(0, 0, z)$.   ◀

SECTION 15.2   CALCULATING THE WORK DONE ON A TEST CHARGE

In the previous section we defined the electrostatic potential V of an electrostatic field $\mathbf{E}$ from the electrostatic potential energy of a test charge q, by

$$qV = \text{electrostatic PE of } q \text{ in } \mathbf{E} \qquad (15.12)$$

This equation allows us to calculate the work done by an external force (such as that in Figure 15.1) that moves the test charge from a to b as the change in potential energy

$$W = q[V(b) - V(a)] \qquad (15.13)$$

It is very important here to state precisely which charge is being regarded as the test charge in any given problem, and to allow only that charge to move. If we allow the

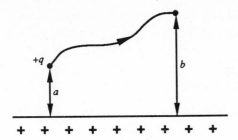

Figure 15.9 Moving a test charge in the field of a line charge.

charges that **created** the field **E** to move, **E** is no longer a static field and this entire approach fails. In such a case it is still possible to calculate the work done by calculating an energy change, but we must use the more general formulas of Chapter 13, which are not restricted to static fields. In static problems, evaluating V is a lot easier than integrating E^2 as in Chapter 13.

■ **EXAMPLE 15.5**   Let V be the (continuum) field of a point charge Q at the origin, and let a test charge q be a distance r away. The electrostatic potential energy of q in the field of Q is given by Eq. 15.4,

$$qV(r) = qQ/4\pi\epsilon_0 r$$

The work necessary to move q from $r = a$ to $r = b$ is the increase in potential energy

$$(qQ/4\pi\epsilon_0)(b^{-1} - a^{-1})$$

▶   **Problem 15.9**

Compute the work necessary to move a test charge q that is initially a distance a from an infinite line with charge per unit length λ_0, to a distance b, as in Figure 15.9. See Problem 15.2 for $V(r)$.   ◀

▶   **Problem 15.10**

Do the same for a charge q moving from a distance a to a distance b from a charged **plane** (charge per unit area σ) when there is another plane with charge density $-\sigma$, further away than b. (Use Figure 15.9 again, but regard the line at the bottom as a cross section of a plane, and put a negatively charged plane at the top.)   ◀

▶   **Problem 15.11**

(Equivalence of Eq. 13.16 and Eq. 15.13 for computing work) Consider a large parallel-plate capacitor formed by two planes at $x = L$ and $x = -L$, with charge densities σ and $-\sigma$, as in Figure 15.10. Between them, at $x = x'$, there is another plane with a small uniform positive test charge $\sigma' \ll \sigma$. Let the area of the three charged planes be A, but consider it infinite in calculating the fields (i.e., use the results in Section 14.4).

a. Compute the potential $V(x)$ between the planes σ and $-\sigma$ (in the absence of σ').

b. Compute the electrostatic potential energy of the plane of charge σ' in the field of $\pm\sigma$, as a function of x'.

c. Compute the work done in moving the plane σ' from 0 to x' from the result of **b**.

d. Compute the electric field $E(x)$ both above and below σ'.

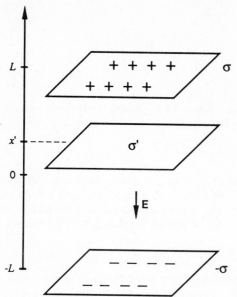

Figure 15.10 Moving a test-charge plane between two capacitor plates.

e. Compute the total energy U_E of the electric field to first order σ' (Eq. 13.16: you must integrate over the entire volume of the capacitor), as a function of x'.

f. Compute the work required to move σ' as the change in the total energy, $U_E(\sigma'$ at $x') - U_E(\sigma'$ at 0). Compare to the result of **c**.   ◀

The expression (Eq. 15.12) for the electrostatic potential energy of an infinitesimal charge in a potential can be generalized to an infinitesimal discrete test charge distribution $\rho(c)$. Approximating the charge in a cell c by a point charge at the center, the electrostatic PE of this charge $\rho(c)\,d\tau(c)$ is $V(c)\rho(c)\,d\tau(c)$, so we can define the total

$$\text{PE of } \rho \text{ in } V = \sum_c V(c)\rho(c)\,d\tau(c) \tag{15.14}$$

This can be used to approximately calculate the work required for any change in ρ, as long as V is unchanged. It becomes exact in the continuum limit, where the density and potential are represented by functions ρ_f and V_f:

$$\text{PE of } \rho_f \text{ in } V_f = \int V_f(r)\rho_f(r)\,d\tau \tag{15.15}$$

■ **EXAMPLE 15.6**   A small charge q is spread evenly throughout a spherical shell of inner radius a and outer radius $3a$, which has a point charge Q at its center, as shown in Figure 15.11. Compute the work necessary to compress the charge q to be uniform in a shell of inner and outer radii a and $2a$. The initial volume is $(4\pi/3)[(3a)^3 - a^3]$ so the density is $\rho = 3q/(26 \times 4\pi a^3)$. The potential V is $Q/4\pi\epsilon_0 r$. Because V and ρ depend only on r, $V(c)$ and $\rho(c)$ are the same for all the cells in a thin spherical shell of thickness dr,

Figure 15.11   A cloud of test charge surrounding a point charge Q.

extending from $r - \frac{1}{2}dr$ to $r + \frac{1}{2}dr$. The total volume of these cells is $\sum d\tau(c) = 4\pi r^2 dr$ (for small dr), so Eq. 15.14 becomes a sum over shells of different radii r

$$\sum_r V(r)\rho(r)4\pi r^2 dr \tag{15.16}$$

In the continuum limit, this is the integral

$$\int_a^{3a} V(r)\rho(r)4\pi r^2 dr$$

$$= (3qQ/26 \times 4\pi a^3\epsilon_0)\int_a^{3a} r\, dr$$

$$= (3qQ/26 \times 4\pi a^3\epsilon_0)[(3a)^2 - a^2]/2$$

$$= (3qQ/26 \times 4\pi a^3\epsilon_0)8a^2/2 \tag{15.17}$$

This is the initial PE; the final one has $2a$ replacing $3a$, so $3^3 - 1 = 26$ becomes $2^3 - 1 = 7$, $3^2 - 1 = 8$ becomes $2^2 - 1 = 3$, and the PE is

$$(3qQ/7 \times 4\pi a^3\epsilon_0)3a^2/2 \tag{15.18}$$

The work is the change in PE,

$$(qQ/4\pi\epsilon_0 a)(3/2)(3/7 - 8/26)$$

$$= (qQ/4\pi\epsilon_0 a)(3/2)(0.429 - 0.308)$$

$$= 0.181(qQ/4\pi\epsilon_0 a)$$

Note that we had to assume q small because we are ignoring forces exerted by parts of the shell on other parts (these give rise to energies of order q^2). Compressing q would require work even if $Q = 0$. ◄

▶   **Problem 15.12**

Consider a sphere of radius a having a charge Q uniformly distributed throughout its volume. Compute the work required to move a test charge q that is initially at the center of the sphere to a distance $2a$ from the center. You may use the result of Problem 15.1 for the potential. ◄

SECTION 15.3   ELECTROSTATIC POTENTIAL ENERGY OF A SYSTEM OF POINT CHARGES

Consider a point charge Q at the origin, which produces a potential $V(r) = Q/4\pi\epsilon_0 r$. The electrostatic energy (Eqs. 15.4 and 15.12) of another charge q, a distance r away, in the potential V is

$$qV(r) = qQ/4\pi\epsilon_0 r \tag{15.19}$$

(Eq. 15.12 usually requires that q be infinitesimal, lest it exert forces on and move the charges that created V. This is not necessary in the present case, because Q is assumed immobile.) Equation 15.19 can be thought of as the work necessary to pull the charges together from a reference configuration in which they are far apart. We obtained it by assuming that Q was stationary and q moved, but it is clear from the symmetry of the result (it's unchanged if we interchange q and Q) that it doesn't matter which we move. In fact, it turns out we would get the same result if we moved them both, as long as the final separation was r and we moved them **slowly** (so the field could be regarded at all times as being static; if we moved them fast, we would lose energy to radiation).

It is useful to talk about the work needed to put together an **arbitrary** charge distribution in this way (starting with all charges at infinity). This work can be calculated analytically when the distribution consists of point charges; we have already done it for the case of **two** point charges. If there are three, say q_1, q_2, and q_3, at positions $\mathbf{r}_1$, $\mathbf{r}_2$, and $\mathbf{r}_3$, then we can put together the distribution one charge at a time. Putting q_1 at $\mathbf{r}_1$ (from infinity) costs no energy because there is no electrostatic force on q_1. (Note that the field of the charge q itself must be omitted from the V in Eq. 15.12—it exerts no force on itself.) Bringing q_2 in requires an amount of work that we calculated before: $q_1q_2/4\pi\epsilon_0 r_{12}$ where $r_{12} = |\mathbf{r}_1 - \mathbf{r}_2|$ is the distance between the first two charges. This work is also the change in potential energy of q_2 in the field of q_1. To bring in q_3 requires further work, which we compute by computing the change in potential energy of q_3 in the field of q_1 and q_2. The potential of q_1 and q_2 is (at an arbitrary point $\mathbf{r}$)

$$V(r) = q_1/4\pi\epsilon_0|\mathbf{r} - \mathbf{r}_1| + q_2/4\pi\epsilon_0|\mathbf{r} - \mathbf{r}_2| \tag{15.20}$$

from the principle of superposition. The electrostatic energy of the charge q_3 at $\mathbf{r}_3$ in this potential V is

$$q_3V(r_3) = q_3q_1/4\pi\epsilon_0|\mathbf{r}_3 - \mathbf{r}_1| + q_3q_2/4\pi\epsilon_0|\mathbf{r}_3 - \mathbf{r}_2| \tag{15.21}$$

The corresponding energy when q_3 was at $\mathbf{r}_3 = \infty$ was zero, so this is also the **change** in energy, that is, the work done by mechanical forces to pull q_3 in. The **total** work to assemble the charge distribution consisting of q_1, q_2, and q_3 is the sum of the values we have calculated:

$$\text{Work} = q_1q_2/4\pi\epsilon_0 r_{12} + q_1q_3/4\pi\epsilon_0 r_{13} + q_3q_2/4\pi\epsilon_0 r_{23} \tag{15.22}$$

Clearly, we can repeat this argument for **any** number of charges, say $q_1, q_2, \ldots, q_n$.

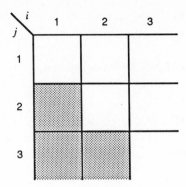

Figure 15.12 Table for visualizing the terms of Eq. 15.24.

The work necessary to move q_j from infinity in the field of $q_1, q_2, \ldots, q_{j-1}$ is

$$\sum_{i=1}^{j-1} q_i q_j / 4\pi\epsilon_0 |\mathbf{r}_i - \mathbf{r}_j| \tag{15.23}$$

and the sum of these expressions for all j is

$$\text{Total work} = \sum_{j=1}^{n} \sum_{i=1}^{j-1} q_i q_j / 4\pi\epsilon_0 r_{ij} \tag{15.24}$$

This is the sum over all pairs (i, j) such that $i < j$. To keep track of the terms in Eq. 15.24, it is convenient to make a table like Figure 15.12. The shaded boxes (below the diagonal, so $i < j$) are the ones included in the sum. Evidently the terms above the diagonal are the same [(i, j) gives the same term as (j, i)]. So we would get twice the work if we included them; we can rewrite the total work as

$$\tfrac{1}{2} \sum_{i, j(i \neq j)} q_i q_j / 4\pi\epsilon_0 r_{ij} \tag{15.25}$$

where the sum over (i, j) pairs now excludes only $i = j$. [It had better: the $i = j$ terms are infinite.]

We will refer to this work as "the potential energy of the charge distribution consisting of $q_1, \ldots, q_n$." It must be kept in mind that this is a completely different concept from "the potential energy of a test charge q **in a potential** V" discussed in the previous section. The work to assemble $q_1, \ldots, q_n$ is **not** equal to the sum of the potential energies of the charges in any potential. And certainly not in the potential

$$V(\mathbf{r}) = \sum_j q_j / 4\pi\epsilon_0 |\mathbf{r} - \mathbf{r}_i| \tag{15.26}$$

of the charges themselves, because this gives each of the charges an infinite energy ($V(\mathbf{r}_j)$ has $|\mathbf{r}_j - \mathbf{r}_j|$ in the denominator). The relation between the work to assemble a charge distribution and its own electrostatic field is best discussed from a discrete point of view, in which infinities can be avoided; see Eq. 15.28.

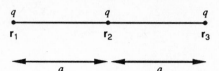

Figure 15.13 Three point charges.

Any potential energy has an arbitrary additive constant which is fixed by selecting a reference configuration to have zero potential energy. When we regard Eq. 15.25 as a potential energy, this reference configuration is that in which all charges are infinitely far from each other (i.e., they've "gone to infinity" in different directions).

■ **EXAMPLE 15.7** Compute the work necessary to assemble (from infinity) three point charges along a straight line, separated by a distance a, as shown in Figure 15.13. Plugging $q_1 = q_2 = q_3 = q$, $|r_1 - r_2| = a$, $|r_3 - r_2| = a$, $|r_3 - r_1| = 2a$ into Eq. 15.22, we get

$$\text{Work} = q^2/4\pi\epsilon_0 a + q^2/4\pi\epsilon_0 a + q^2/4\pi\epsilon_0 2a = \tfrac{5}{2}q^2/4\pi\epsilon_0 a$$

▶   **Problem 15.13**

a. Compute the work necessary to assemble from infinity four equal point charges q at the corners of a square of side a.

b. Compute the work which would be necessary if the charges alternated in sign: $+q, -q, +q, -q$ (the two $+q$'s are at opposite ends of a diagonal).   ◀

SECTION 15.4 THE POTENTIAL ENERGY OF A CHARGE DISTRIBUTION

In this section we will consider a generalization of the problem just considered (computing the work required to assemble a group of point charges). We will consider a discrete system with an arbitrary charge distribution $\rho(c)$, as in Figure 15.14. We will also suppose that we have calculated the correct electrostatic field $E(f)$ (either by evolving Maxwell's equations for a long time, or by the methods of Chapter 16) and the correct potential $V(c)$ (e.g., from $V = -\int E\,dl$, with the reference potential $V = 0$ at infinity).

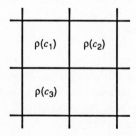

Figure 15.14 A discrete charge distribution.

▶   **Problem 15.14**

By $V = 0$ at infinity, of course, we mean that $V(\mathbf{r}) \to 0$ as $|\mathbf{r}| \to \infty$. If $V(\mathbf{r})$ has a nonzero limit V_∞, then we can always achieve this by subtracting V_∞ from $V(\mathbf{r})$. However, $V(\mathbf{r})$ does not necessarily have a limit at all, as you can see by considering an infinite parallel-plate capacitor, in which the limit of V depends on which direction we go to infinity in. Show from Coulomb's law (Eq. 15.8) that as long as the charge is confined to a finite volume (i.e., $\rho(c)$ is nonzero for a finite number of cells c), V approaches the same limit as $|\mathbf{r}| \to \infty$ in any direction.   ◀

We will compute the work required to assemble the discrete charge density ρ from Eq. 15.13 for the work $(q\Delta V)$ needed to move a charge q in a given potential V. Each cell c will have a charge $q = \rho(c)\, d\tau(c)$. We could construct the charge distribution by bringing these charges to their cells from infinity one at a time. The necessary work could be computed as the change in potential energy of q in the field of the other charges that had already been placed in their cells. This would involve a very large number of field calculations (one for each partially assembled charge distribution). Fortunately, there is a much easier way, depicted schematically in Figure 15.15. This involves filling all the cells concurrently, but doing it very

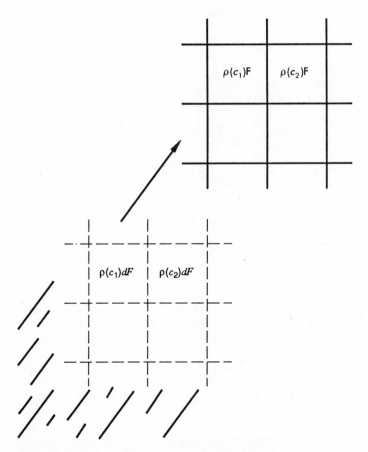

Figure 15.15   Assembling an arbitrary charge distribution.

slowly by bringing in a small fraction dF of each charge at each step (say 1% of it at a time, so $dF = 0.01$; we will eventually let $dF \to 0$). We will assemble the charge from an initial configuration in which all the charge is at infinity and infinitely spread out (perhaps spread over a sphere of radius a, as $a \to \infty$; see Problem 15.19 for details). Let F be the fraction that has been assembled at a particular step: that is, the charge distribution has increased to ρF. The amount added to the density in each cell in one step is $\rho(c)\, dF$. The advantage of this procedure is that the work required to move $\rho(c)\, dF$ can be computed from the potential energy of the charge $\rho(c)\, dF\, d\tau(c)$, in the potential of the already-assembled charge distribution ρF.

> You may object that we should also worry about the potential of the new charge $\rho(c')\, dF\, d\tau(c')$ in a cell c' if we brought it from infinity before the charge in cell c. However, the contribution of this to the work is proportional to $(dF)^2$ at each step, which when summed over the whole process (because $\Sigma\, dF = 1$) is proportional to dF. This goes to zero in the $dF \to 0$ limit.

In principle, this again involves the calculation of an infinite number of potentials, but these are all proportional: if we denote by V the potential of ρ itself, the potential of ρF is simply VF. This fact is related to the superposition principle; it is just a consequence of the linearity of the equations for determining V from ρ.

Then the work done to bring all the charges $\rho(c)\, dF\, d\tau(c)$ from infinity is

$$\sum_c \rho(c)\, dF\, d\tau(c)\, V(c) F \tag{15.27}$$

The total work of assembly is obtained by summing this expression over the various increments dF, which in the limit $dF \to 0$ becomes an integral:

$$
\begin{aligned}
\text{Total work} &= \int_0^1 \sum_c \rho(c) V(c)\, d\tau(c)\; F\, dF \\
&= \left[\sum_c \rho(c) V(c)\, d\tau(c) \right] \int_0^1 F\, dF \\
&= \tfrac{1}{2} \sum_c \rho(c) V(c)\, d\tau(c) \tag{15.28}
\end{aligned}
$$

because the sum in square brackets is independent of F and can be taken out of the integral, and the integral of F is $\tfrac{1}{2}F^2$.

> Note that this result is **not** independent of the reference potential; we must calculate $V(c)$ relative to infinity. Also, we have ignored the question of how one creates these charges at infinity. One can think of making them by separating positive and negative charges in amounts proportional to dF. The energy cost of this is proportional to $(dF)^2$, and we found above that $(dF)^2$ terms have no influence on the $dF \to 0$ limit. It is important that we are doing this charge separation in the discrete formulation; if the $d\tau \to 0$ (continuum) limit is taken before $dF \to 0$, it costs an infinite amount of energy to separate the charges.

The work (Eq. 15.28) necessary to assemble a charge distribution should be regarded as the "potential energy of the charge distribution ρ." You will note a formal similarity to Eq. 15.14 for the "PE of ρ in V"; they differ only by a factor of $\frac{1}{2}$. They are, however, completely different concepts. In Section 15.2, V had **nothing** to do with ρ; in the present case, V is exactly the potential of ρ. One can make sense of the factor $\frac{1}{2}$ by observing that if we erroneously used the earlier formula, we would count the work needed to bring $\rho(c_1)\,d\tau$ near $\rho(c_2)\,d\tau$ (via the potential of the latter) **and** that needed to bring $\rho(c_2)\,d\tau$ near $\rho(c_1)\,d\tau$ (which is really the **same** work), thus counting everything twice. The factor $\frac{1}{2}$ is needed to correct for this error.

In the continuum limit, when the charge density is described by a continuous function ρ_f we find that the potential energy of ρ_f (the $dr \to 0$ limit of Eq. 15.28) is

$$\frac{1}{2}\int \rho_f(\mathbf{r})V_f(\mathbf{r})\,d\tau \tag{15.29}$$

where $V_f(\mathbf{r})$ is the potential (relative to infinity) due to ρ_f itself.

■ **EXAMPLE 15.8**   Let us apply Eq. 15.29 to the computation of the potential energy of a sphere of radius R with a uniform charge density and a total charge Q. We have already computed the electrostatic potential of such a sphere (Problem 15.1), which is

$$V_f(r) = Q/4\pi\epsilon_0 r \qquad\qquad \text{for } r \geq R$$

$$= (Q/8\pi\epsilon_0 R^3)(3R^2 - r^2) \qquad \text{for } r \leq R \tag{15.30}$$

The charge density is $\rho = Q/(4\pi R^3/3)$ independently of r for $r < R$ and zero for $r > R$, so Eq. 15.29 gives for the energy

$$U = (3Q^2/16\pi R^6\epsilon_0)\int_0^R (3R^2 r^2 - r^4)dr = 3Q^2/20\pi\epsilon_0 R \tag{15.31}$$

■ **EXAMPLE 15.9**   Another continuum problem is that of an infinitely thin spherical **shell** of radius R, with a uniform surface charge density and a total charge Q. We can't use Eq. 15.29 because the charge density is not described by a continuous function ρ_f, so we must return to the discrete Eq. 15.28. If we use a lattice in spherical coordinates having cell centers at radius R as in Figure 15.16, the potential at these cells (the only ones with charge) is $V(c) = Q/4\pi\epsilon_0 R$, from Example 15.2. Equation 15.28 for the potential energy then gives

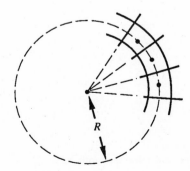

Figure 15.16   Cross section of lattice in spherical coordinates.

then gives

$$U = (Q/8\pi\epsilon_0 R)\sum \rho(c)\,d\tau(c) = Q^2/8\pi\epsilon_0 R \qquad (15.32)$$

because $\sum \rho(c)\,d\tau(c) = Q$.

It is interesting to calculate the potential energy of a point charge q at the origin; this is called the "self-energy" of the charge. The continuum charge density is a Dirac delta function, that is, the discrete density for a given cell size dr is given by $\rho(c_0)\,d\tau(c_0) = q$ for the cell c_0 containing the origin, and $\rho(c) = 0$ elsewhere. The potential energy of this discrete density is

$$U = \tfrac{1}{2}\sum \rho(c)\,d\tau(c)\,V(c) = \tfrac{1}{2}qV(c_0) \qquad (15.33)$$

We can calculate $V(c_0)$ explicitly on a lattice in spherical coordinates. The direct lattice is shown as a dotted line in Figure 15.17; the origin is at the center of a spherical cell. This is then a vertex in the dual lattice (solid line) in Figure 15.17. We can draw a path from infinity to an arbitrary dual vertex v along a radial line (horizontal in the figure.) All the dual edges e along the path are radial and are centered at $r = 0.5dr, 1.5dr, 2.5dr, \ldots$, so $E(e) = -E_r(0.5dr)$, and so on. (The minus sign is because the orientation of e is toward the origin, opposite to the radial direction.) The potential at any dual vertex (say at r) is given by Eq. 15.2 as a sum over edges between r and infinity:

$$V(r) = -\sum_e E(e)\,dl(e)$$

$$= +(q\,dr/4\pi\epsilon_0)\left[(r + 0.5dr)^{-2} + (r + 1.5dr)^{-2} + \cdots\right] \qquad (15.34)$$

In particular, at $r = 0$ we have

$$V(0) = (q/4\pi\epsilon_0\,dr)\left[\left(\tfrac{1}{2}\right)^{-2} + \left(\tfrac{3}{2}\right)^{-2} + \left(\tfrac{5}{2}\right)^{-2} + \cdots\right]$$

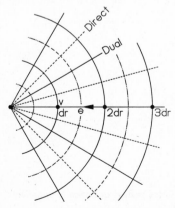

Figure 15.17   The direct and dual lattices.

The sum in square brackets is $\frac{1}{2}\pi^2$ (it is related to the Riemann zeta function $\xi(2)$ and can be found in *Handbook of Mathematical Functions*, AMS 55, ed. by Abramowitz and Stegun, National Bureau of Standards, 1965). So

$$V(0) = \pi q/8\epsilon_0 \, dr \qquad (15.35)$$

and the potential at a point charge diverges as $1/dr$ as the cell size $dr \to 0$. Therefore, the work needed to construct one,

$$U = \pi q^2/16\epsilon_0 \, dr \qquad (15.36)$$

(from Eq. 15.33) also diverges. The numerical coefficient depends on exactly how we discretize the problem (cartesian coordinates give a slightly different answer) but the potential energy diverges as dr^{-1} in any event: one cannot assign a finite value to the potential energy of a point charge. One can see this in another way from Eq. 15.31 or Eq. 15.32, the potential energy of a solid sphere or a spherical shell respectively. We can think of each of these as approaching a point charge as the radius $R \to 0$, and indeed they, too, diverge as $1/R$. This divergence does not have any physical consequences because no one has ever constructed a point charge. (The electron is thought to be a point charge, however; the theoretical difficulties associated with this are discussed in Section 18.3.) One can put charges on very small objects, though, and it is true that the work necessary to do so becomes very large as the object becomes smaller. This potential energy computation is a case in which the notion of idealizing a charge distribution that is localized in a small volume as a point charge, which usually simplifies the analysis of a problem, backfires. The problem instead becomes more complicated.

▶   **Problem 15.15**

Determine the "discretization error" in Eq. 15.34 for the potential of a point charge, by comparing it to the continuum limit ($V_f(r) = q/4\pi\epsilon_0 r$). Give the fractional error (error/V_f) within 10^{-3}, at $r = dr, 2\,dr$, and $3\,dr$.   ◀

"Discretization error" is a phrase commonly used to describe the difference between the discrete and continuum answers. The phrase is used reluctantly here because it suggests that the continuum answer is the "correct" one and the discrete one is just an approximation. In fact, there are many physical systems for which a discrete approximation is better than any solvable continuum approximation, so the difference could as properly be called the "continuumization error."

▶   **Problem 15.16**

a. Compute the potential energy of a pair of concentric charged thin spherical shells, with radii a and $3a$ and charges $-Q$ and $+Q$ respectively.

b. Do the same if each sphere has charge $+Q$.   ◀

▶   **Problem 15.17**

a. Compute the potential energy of a charge Q spread uniformly throughout a spherical shell of thickness $2a$, extending from radius a to radius $3a$. Compare it to the potential energy when q is spread throughout the entire sphere of radius $3a$.

b. Compute the work done in compressing it into a shell from a to $2a$. Why is the answer different from the one we got in Example 15.6?   ◄

▶   **Problem 15.18**

Use Eq. 15.28 to calculate the potential energy of a parallel-plate capacitor of area A and plate spacing d. Assume the charge is spread uniformly over the plates, with density $\sigma = \pm Q/A$, and the potentials are $\pm \sigma d/2\epsilon_0$ (so the potential difference is $\sigma d/\epsilon_0$). Compare your answer with the total energy of this capacitor computed in Chapter 13.   ◄

▶   **Problem 15.19   How to put charge at infinity**

a. Consider a spherical shell of radius a with a charge Q uniformly distributed over it. Compute the work required (in the continuum limit) to transfer the charge to another shell of (smaller) radius b, **without** using Eq. 15.28. That is, transfer $Q\,dF$ at a time and use Eq. 15.13 ($W = q\Delta V$) to compute the work done.

b. Show that the work necessary to assemble the b-sphere from the a-sphere has a limit as $a \to \infty$; we refer to this limit as the work to assemble the b-sphere "from infinity." The same method can be used to precisely define the work (Eq. 15.28) necessary to assemble an arbitrary charge distribution "from infinity": it is the limit of the work to assemble it from the a-sphere, as $a \to \infty$.   ◄

SECTION 15.5   ELECTROSTATIC ENERGY AS A SPECIAL CASE OF $\sum \frac{1}{2}\epsilon_0 E^2$

Before discussing electrostatics at all, we derived in Chapter 13 from Maxwell's equations a formula for the energy of a system of electric and magnetic fields:

$$U = \int \left[\tfrac{1}{2}\epsilon_0 E^2 + \tfrac{1}{2}B^2/\mu_0 \right] d\tau \qquad (15.37)$$

Applying this to an electrostatic situation in which there is a charge density ρ and an electric field E, but $B = 0$, gives

$$U = \tfrac{1}{2} \int \epsilon_0 E^2 \, d\tau \qquad (15.38)$$

This seems quite inconsistent with the formula obtained by considering the work done to assemble the charge (Eq. 15.29)

$$U = \tfrac{1}{2} \int \rho V \, d\tau \qquad (15.39)$$

In particular, the E^2 formula doesn't seem to depend on the charge at all. However, we know the field determines the charge via Gauss' law, so perhaps we can relate them. We can do this exactly in the discrete formulation in which the ρV formula is

$$U = \tfrac{1}{2} \sum_c \rho(c) V(c) \, d\tau(c) \qquad (15.40)$$

Gauss' law relates $\rho(c)$ to a sum of E's on the faces of the cell c, with certain signs.

A good way to avoid confusion here over the signs is to define a symbol

$$\text{sgn}(f, c) = \begin{cases} 1 & \text{if } f \text{ is an outward-oriented face of } c \\ -1 & \text{if } f \text{ is an inward-oriented face of } c \\ 0 & \text{if } f \text{ is not a face of } c \end{cases} \quad (15.41)$$

for the correct sign to use on $E(f)$ in the divergence, so that Gauss' law says

$$\rho(c) = \epsilon_0 \, d\tau(c)^{-1} \sum_f \text{sgn}(f, c) E(f) \, da(f) \quad (15.42)$$

Note that this sum can be taken over **all** faces f, not just the faces of c, because the factor $\text{sgn}(f, c)$ vanishes if f is not a face of c. It turns out that this symbol can also be used to keep track of signs in the gradient. The gradient, which was originally defined on the dual lattice, on which V is a vertex field (Chapter 6), involves $V(c_{\text{front}}) - V(c_{\text{back}})$ on the direct lattice, and can be written

$$\text{grad } V(f) = -dl(f)^{-1} \sum_c V(c) \, \text{sgn}(f, c) \quad (15.43)$$

where $dl(f)$ is the length of the edge dual to f (i.e., the edge of the dual lattice having the same center as f). This can be seen from Figure 15.18: the potential $V(c_{\text{back}})$ appears with sign $-\text{sgn}(f, c_{\text{back}}) = -1$ (because f points out of this cell), and $V(c_{\text{front}})$ appears with sign $-\text{sgn}(f, c_{\text{front}}) = +1$ (f points **into** c_{front}).

Using Gauss' law (Eq. 15.42) in the electrostatic energy (Eq. 15.40) gives

$$U = \tfrac{1}{2}\epsilon_0 \sum_c \sum_f \text{sgn}(f, c) E(f) \, da(f) V(c) \quad (15.44)$$

The advantage of having unrestricted sums over c and f is that we can interchange the order of summations to get

$$U = \tfrac{1}{2}\epsilon_0 \sum_f E(f) \, da(f) \sum_c V(c) \, \text{sgn}(f, c) \quad (15.45)$$

in which we recognize the gradient (Eq. 15.43) and write the energy as

$$U = \tfrac{1}{2}\epsilon_0 \sum_f E(f)^2 \, da(f) \, dl(f) \quad (15.46)$$

But this is exactly the discrete form of the volume integral in Eq. 15.38. Thus the

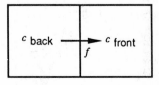

Figure 15.18 The cells involved in computing the gradient at the face f.

two apparently inconsistent methods for calculating the energy of an electrostatic system give exactly the same answer.

This remains true when we take the continuum limit; the integral formulas (Eqs. 15.38 and 15.39) give exactly the same answer for the electrostatic energy. This can also be shown directly by the continuum analog of the calculation we have just done, which is an integration by parts (see for example the book by Griffiths cited in the bibliography).

■ **EXAMPLE 15.10**   Let us recalculate the self-energy of a point charge in a spherical lattice system (Figure 15.17), which we have already calculated using the electrostatic potential method, from the more general formula (Eq. 15.46). Substituting the exact electric field $q/4\pi\epsilon_0 r$ gives

$$U = (q^2 dr/32\pi^2\epsilon_0) \sum da(f)/r^4 \tag{15.47}$$

where the sum is over all radial faces f (the others have $E = 0$). We can decompose this sum into an inner sum over the faces at a particular radius r, and an outer sum over $r = 0.5dr, 1.5dr, 2.5dr, \ldots$. The sum of the areas of the faces at a given radius r is of course $4\pi r^2$, so the sum in Eq. 15.47 becomes a sum over radii,

$$\sum_r 4\pi/r^2 = (4\pi/dr^2)[(\tfrac{1}{2})^{-2} + (\tfrac{3}{2})^{-2} + \cdots] \tag{15.48}$$

The sum in square brackets is the one we identified before as $\tfrac{1}{2}\pi^2$, so the total energy is

$$U = (q^2 dr/32\pi^2\epsilon_0)(4\pi/dr^2)\tfrac{1}{2}\pi^2 = \pi q^2/16\epsilon_0 dr \tag{15.49}$$

which agrees exactly, as it must, with the earlier result (Eq. 15.36).

CHAPTER SIXTEEN

SOLVING GENERAL ELECTROSTATICS PROBLEMS

SECTION 16.1 THE GENERAL EQUATIONS OF ELECTROSTATICS

In the previous chapter we computed the electrostatic fields of certain charge distributions using Coulomb's law (Eq. 14.5) and Gauss' law ($\rho = \epsilon_0 \operatorname{div} E$). We found that Coulomb's law is not usually a good way to calculate electrostatic fields, except for point charges; the integrals tend to be hard, for continuous distributions. In cases of very high symmetry, Gauss' law works better. In these cases we have, in effect, proved that ρ determines E: we know the real E field obeys Gauss' law, and we find a **unique** E that obeys Gauss' law and has the right symmetry. Can we do this for **any** charge distribution? The answer is no: in Section 14.2, for example, when we created a positive charge by moving its negative counterpart to infinity in one piece, the electric field satisfied Gauss' law at **all** times, but only after a very long time was it the correct static field. So E isn't uniquely determined by $\operatorname{div} E$. However, there is an additional equation satisfied by E—we haven't yet used all our information. This is the Maxwell equation $dB/dt = -\operatorname{curl} E$. In the static limit B isn't changing (actually it's usually zero), so $\operatorname{curl} E = 0$; this is not true at earlier times.

Is it possible that $\operatorname{div} E$ and $\operatorname{curl} E$ **together** determine E uniquely, in general? (In the high-symmetry cases of the previous sections, $\operatorname{curl} E$ was actually forced to vanish by the symmetry.) Clearly this can't be quite true: for example, we can add a constant x-component to E without changing $\operatorname{curl} E$ or $\operatorname{div} E$. However, this can be ruled out by placing **boundary conditions** on the fields; it turns out that if we impose the proper boundary conditions the field **is** uniquely determined by $\operatorname{curl} E$ and $\operatorname{div} E$. We will eventually do this in an infinite system, in which the boundary condition is that $E \to 0$ as $r \to \infty$. For now, however, we will look at problems that we can actually solve numerically in a finite amount of time. This requires that we have a system with finite volume. Physically, we can confine electrostatic fields to a finite volume by enclosing them in a conducting box. The conducting material then imposes requirements on the fields that play the role of boundary conditions; we discuss these in the next section.

279

SECTION 16.2 CONDUCTORS

The defining characteristic of an electrical conductor is that there are charged particles in it (electrons, in a metal) that are free to move around, so when there is an electric field (which exerts a force on each charge), the charges respond to the force by moving. We will study these motions and the current field that describes them in detail in Section 21.1. We can guess the eventual result here, however: the current j is along E, so Maxwell's equation $dE/dt = c^2$ curl $B - j/\epsilon_0$ (we ignore B for now) gives dE/dt opposite to E: E decreases in magnitude. If we wait a long time, this "negative feedback" effect will cause E to become zero. In the static limit

the electric field vanishes within a conductor.

   A simple argument by contradiction for the same conclusion goes as follows: If there were an electrostatic field in a conductor, there would be forces on the electrons, so they would move; but this contradicts the assumption that the situation was static. Therefore, the electrons must somehow rearrange themselves so the field vanishes. This argument is rather glib (it doesn't explain how the rearrangement is accomplished) but it is a useful mnemonic device.
   Metals and electrolytes (ions dissolved in water) are examples of conducting materials. Glass, most plastics, and oil, among others act as insulators because their electrons are not free to move.
   A conducting region is characterized by $E = 0$ so it is easiest to specify a discrete conducting region by saying which faces are in it. We will therefore define a discrete conducting region as a set of faces. A pseudo-2D system (an approximation to a line charge in a long duct extending perpendicularly to the paper) is shown in Figure 16.1a. The conducting region consists of the faces labeled by 0 (to indicate that $E = 0$ there) in Figure 16.1b. The discrete charge density (nonzero only in the cell on the left in Figure 16.1b) represents a line of charge pointing out of the paper, so we will represent it by its charge per unit length, denoted λ_0; this is related to the volume charge density ρ by $\lambda_0\,dr = \rho(c)\,dr^3$.

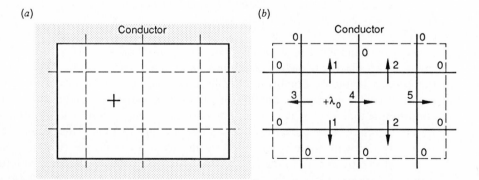

Figure 16.1   (a) A discretization of a line charge in a rectangular conducting duct. (b) Labeling of the inequivalent faces. The numbers are arbitrary integer labels, **not** values of $E(f)$: "1" is short for "f_1".

We would like to calculate the discrete electric field at each nonconducting face. The fields at the two faces labeled 1 in Figure 16.1b will be the same because of the horizontal mirror symmetry, as will those labeled 2. We can now write the two conditions div $E(c) = \rho(c)/\epsilon_0$ and curl $E(e) = 0$ explicitly. Note that we only know $\rho(c)$ for the two central cells; the cells in the conductor (by which we mean cells with at least one conducting face) may have charge (conducted from elsewhere in the conductor: it is said to be "induced" by the charge λ) but we don't know what it is. So we get two equations from Gauss' law, in which $E(f_1)$ is abbreviated to E_1:

$$\text{left } c: \quad dr^{-1}(E_1 + E_3 + E_1 + E_4) = \lambda_0/dr^2\epsilon_0$$

$$\text{right } c: \quad dr^{-1}(E_2 - E_4 + E_2 + E_5) = 0$$

and three from curl $E(e) = 0$

$$\text{upper left } e: \quad E_1 - E_3 = 0$$
$$\text{upper center } e: \quad E_2 - E_1 + E_4 = 0$$
$$\text{upper right } e: \quad E_5 - E_2 = 0$$

(the equations at the lower edges are the same by symmetry). These are five equations in five unknowns, which can be solved straightforwardly by the method of elimination. We can eliminate E_3 and E_5 by solving the third and fifth equations for them ($E_3 = E_1$, $E_5 = E_2$), leaving three equations in E_1, E_2, and E_4:

$$3E_1 + E_4 = \lambda_0/\epsilon_0 \, dr$$

$$3E_2 - E_4 = 0$$

$$E_2 - E_1 + E_4 = 0$$

We can similarly eliminate E_4 by solving the second equation for it ($E_4 = 3E_2$), leaving

$$3E_1 + 3E_2 = \lambda_0/\epsilon_0 \, dr$$

$$4E_2 - E_1 = 0$$

and finally eliminate $E_2 = E_1/4$, leaving

$$15E_1/4 = \lambda_0/\epsilon_0 \, dr$$

Solving for E_1 and successively plugging it into the expressions we wrote for E_2, E_4, and E_3, we get

$$E_1 = E_3 = (4/15)\lambda_0/\epsilon_0 \, dr$$

$$E_2 = E_5 = (1/15)\lambda_0/\epsilon_0 \, dr$$

$$E_4 = (3/15)\lambda_0/\epsilon_0 \, dr$$

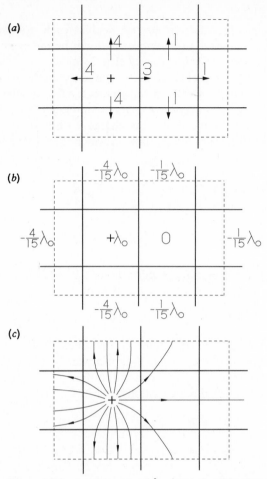

Figure 16.2   (a) E in units of $\lambda_0/15\epsilon_0 dr$. (Note that the numbers here are the values of E, unlike Figure 16.1b, where they were just integer labels.) (b) Charge density. (c) Electric field lines.

which is shown in a more intelligible form in Figure 16.2a. You can easily check with the figure that the divergence and the curl are correct.

Now that we know E, we can compute the charge density induced on the conducting surface, from Gauss' law. Consider the cell to the left of λ_0. The outward-oriented face at the right side of this cell is f_3 in the notation of Figure 16.1b, except for having the opposite orientation; its field is therefore $-E_3$. So

$$\rho(c) = \epsilon_0 \operatorname{div} E(c) = -\epsilon_0 \, dr^{-1} E_3$$

$$= -(4/15)\lambda_0/dr^2$$

To avoid this factor of dr^2, in pseudo-2D problems it makes sense to describe the charge distribution by the linear charge density $\lambda(c)$ (charge per unit length perpendicular to the paper, instead of per unit volume), defined by $\lambda(c)\, dr = \rho(c)\, dr^3$. Then $\lambda(c) = \lambda_0$ for the cell containing the line charge λ_0, and $\lambda(c) =$

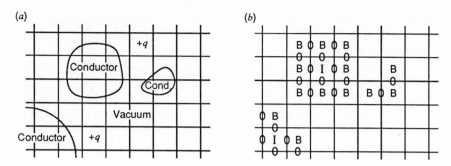

Figure 16.3 (*a*) A general discrete conducting-boundary problem. (*b*) Interior (I) and boundary (B) cells of the conductors. Conducting faces are labeled by zeroes.

$-(4/15)\lambda_0$ for the cell to its left. We can compute λ similarly at all the boundary cells, with the results shown in Figure 16.2*b*.

We can also draw the electric field lines for this case, using the rules given in Section 14.5. In Figure 16.2*c* we have drawn one line for each $\lambda_0\, dr/15\epsilon_0$ of electric flux. The total flux emanating from the charge $\lambda_0\, dr$ is $\lambda_0\, dr/\epsilon_0$, as required by Gauss' law.

We can generalize from this example to draw some general conclusions about fields and charges in conductors. Consider a problem in which there are conducting regions and vacuum regions, as shown in Figure 16.3*a*. There are also some charges fixed in various positions in the vacuum region (indicated by $+q$). We want to compute the fields in the vacuum region ($E = 0$ in the conductors, of course). For this to be a tractable problem, the vacuum region must be finite: suppose the conductor at the lower left extends all the way around the region shown (we haven't shown this, to avoid cluttering the figure).

In a discrete system a conducting region (which we indicated schematically in Figure 16.3*a* by enclosing it in curved boundaries) is defined as a set of conducting faces, which we indicate by 0's in Figure 16.3*b* to indicate that $E(f) = 0$ there. The other $E(f)$'s are unknown; we will calculate them from $[\text{curl } E](e) = 0$ (which is true at every edge e) and $\rho(c) = \epsilon_0[\text{div } E](c)$. The latter equation is true for every cell, but we know ρ only in the vacuum region: on the conductor there are unknown induced charges. Our criterion for a cell being "on the conductor" is that it has at least one conducting face; these cells are labeled B and I in Figure 16.3*b*. We use this criterion because it ensures (as we will show in Section 16.3) a unique solution to the problem on a discrete lattice. So in computing E we use Gauss' law only at the cells in the vacuum region (unlabeled in the figure). Once we have E everywhere, we can determine the induced charge in the conducting region. It is given by Gauss' law $\rho(c) = (\epsilon_0/dr)\Sigma E(f)$. For cells having all faces conducting (we will call these "interior" cells, labeled I in the figure) these E's all vanish, so $\rho = 0$:

**There can be no electrostatic charge
in the interior of a conductor**

We will refer to the other cells in the conducting region as "boundary" cells: these have at least one conducting face and at least one nonconducting face, and are labeled B in the figure. Only the boundary cells can have induced charge, which is determined by Gauss' law.

The method we have used in the above numerical example is perfectly general, and can be used to solve any electrostatics problem involving charges and conductors, even in three dimensions. However, it is not usually the most efficient method. We will describe in the Section 16.3 a method involving the concept of electrostatic potential, which will allow us to solve this problem with **two** equations in **two** unknowns, rather than five in five.

SECTION 16.3 POISSON'S EQUATION

Consider the general electrostatic problem of solving curl $E = 0$, div $E = \rho/\epsilon_0$ for a finite 3D system of N cells. Such a system has roughly $3N$ faces. (Each cell can be regarded as "owning" only three of its six faces to avoid double-counting. The other three belong to neighboring cells, except perhaps at the boundary of the system.) So we have a linear system of equations for $3N$ variables $E(f)$, which we could solve directly as in the previous section. However, there is an easy way to reduce the problem to N variables: the electric field is uniquely determined by the scalar electrostatic potential V, defined at each of the N cells. Since $E = -\operatorname{grad} V$, the conditions on V corresponding to the above two equations for E are curl grad $V = 0$ and

$$\operatorname{div} \operatorname{grad} V = -\rho/\epsilon_0 \qquad \text{(Poisson's equation)} \qquad (16.1)$$

The former is useless (it is an identity, true for every V; see Chapter 6). The latter is called **Poisson's equation**. In the special case that $\rho = 0$, it is called **Laplace's equation**, and the combination of operations on the left-hand side is called the **Laplacian**, denoted ∇^2

$$\nabla^2 V \equiv \nabla \cdot \nabla V \equiv \operatorname{div} \operatorname{grad} V \qquad (16.2)$$

We will first discuss the solution of Poisson's equation in a discrete system. As in the previous section, we will enclose our system in a conducting box. (In addition to making the problem finite, it turns out that this also ensures the uniqueness of the solution by determining the values of V on the boundary.) An electrostatics problem of this sort is specified by giving a lattice and by stating which faces are conducting; Figure 16.4 shows a pseudo-2D example. If a face f is conducting (so $E = 0$ there), then by definition of the gradient (Eq. 6.31), $V(c)$ is the same on both sides. We can therefore start at any cell in a conductor (having potential V_0, say) and move around to all the cells that can be reached by crossing conducting faces: $V(c) = V_0$ for each of them. We will call a collection of conducting cells that are derived this way from a single cell a **connected piece** of conductor. There is only one connected piece in Figure 16.4, and it is indicated by shading in Figure 16.5. The essential property of a conductor is therefore:

V is constant on a connected piece of conductor

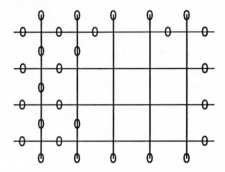

Figure 16.4 A pseudo-2D conducting box. The conducting faces are marked by zeroes.

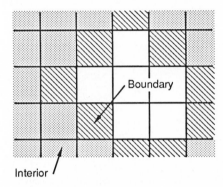

Figure 16.5 The connected piece of conductor in the system of Figure 16.4 (shaded). The lighter region is the boundary, and the darker cross-hatched region is the **interior** of the conductor (cells having all neighbors in the conducting region), where $\rho = 0$ (see Section 16.2).

On a cubic lattice, Poisson's equation takes the form

$$-\rho(c)/\epsilon_0 = [\text{div grad } V](c) = (dr)^{-1} \sum_f \text{grad } V(f)$$

$$= (dr)^{-1} \sum_f (dr)^{-1}[V(c_{\text{front}}) - V(c_{\text{back}})] \tag{16.3}$$

For each face f, c_{back} is just the cell c itself, and c_{front} is one of the six neighbors shown in Figure 16.6a. It is easier to label the cells in a 2D cross section (Figure 16.6b) displaying four of the six neighbors c', each of which is c_{front} in one term of Eq. 16.3. Equation 16.3 becomes

$$-\rho(c)/\epsilon_0 = (dr)^{-2} \sum_{c'} [V(c') - V(c)] \tag{16.4}$$

where the sum is now over the six neighbors c'. Another way to write Poisson's equation is to bring the $V(c)$'s out of the sum:

$$(dr)^{-2}\left[\sum_{c'} V(c') - 6V(c)\right] = -\rho(c)/\epsilon_0 \tag{16.5}$$

(a)   (b)

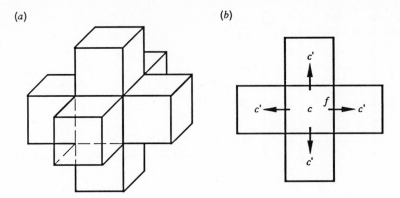

Figure 16.6 (a) The six neighbors of a (hidden) cell c. (b) The four neighbors of a cell c in a pseudo-2D system.

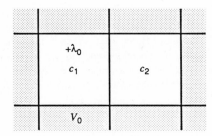

Figure 16.7 The two nonconducting cells in the pseudo-2D duct of Figure 16.1.

In a quasi-2D system (e.g., the duct in Figure 16.7), we need only include in Eq. 16.4 the four c' shown in Figure 16.6b because $V(c') - V(c) = 0$ for the cells c' into and out of the paper. Thus the 6 in Eq. 16.5 becomes a 4. Expressing ρ in terms of the linear charge density $\lambda(c) = \rho(c)\, dr^2$, we get

$$\sum_{c'} V(c') - 4V(c) = -\lambda(c)/\epsilon_0 \qquad \text{(2D Poisson equation)} \qquad (16.6)$$

where the sum is now over the four c' shown in Figure 16.6b.

■ **EXAMPLE 16.1**   Let us apply Poisson's equation to the pseudo-2D problem we did using E in Section 16.2, shown again in Figure 16.7. The shaded cells are part of the conductor, and have the same potential V_0. There are only two nonconducting cells, which we can call c_1 and c_2; their potentials are denoted V_1 and V_2. Then Eq. 16.6 becomes

$$\text{cell } c_1: \qquad 3V_0 + V_2 - 4V_1 = -\lambda_0/\epsilon_0$$

$$\text{cell } c_2: \qquad 3V_0 + V_1 - 4V_2 = 0$$

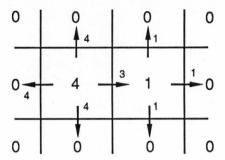

Figure 16.8 V in units of $\lambda_0/15\epsilon_0$, and E in units $\lambda_0/15\epsilon_0\,dr$.

We may choose the reference potential so $V_0 = 0$, and solve for V_1 and V_2:

$$V_1 = (4/15)(\lambda_0/\epsilon_0)$$

$$V_2 = (1/15)(\lambda_0/\epsilon_0)$$

The complete potential is shown in the cells of Figure 16.8.

The electric field, obtained by subtracting neighboring V's (Eq. 15.1), is shown on the faces. You can compare it to Figure 16.2a to verify that the potential method gives exactly the same result as solving div $E = \rho/\epsilon_0$ and curl $E = 0$ for E directly. It's just easier this way.

■ **EXAMPLE 16.2**   This example illustrates the use of symmetry to simplify the use of Poisson's equation. Like the previous example, it involves a line charge in a rectangular duct, but this time the line is in the center of the duct (Figure 16.9a). There are three nonconducting cells. The system is symmetric under reflections in a horizontal mirror plane, hence the highest nonconducting cell is equivalent to the lowest one. Therefore, they have the same potential, which we will denote by V_2. We will denote by V_1 the potential of the central cell (Figure 16.9b). The Poisson equations for the V_1 and V_2 cells are

$$2V_2 + 2V_0 - 4V_1 = -\lambda_0/\epsilon_0$$

$$V_1 + 3V_0 - 4V_2 = 0 \tag{16.7}$$

We can write this as a matrix equation if we like:

$$\begin{pmatrix} -4 & 2 \\ 1 & -4 \end{pmatrix}\begin{pmatrix} V_1 \\ V_2 \end{pmatrix} = \begin{pmatrix} -\lambda_0/\epsilon_0 - 2V_0 \\ -3V_0 \end{pmatrix} \tag{16.8}$$

As usual, we may set $V_0 = 0$; we can then easily solve the system of equations (Eqs. 16.7 or 16.8), obtaining

$$V_1 = 4\lambda_0/14\epsilon_0$$

$$V_2 = \lambda_0/14\epsilon_0$$

as shown in Figure 16.10a.

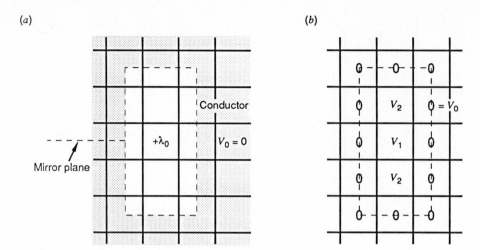

Figure 16.9 (a) A duct with a symmetrically placed line charge. (b) Labeling the potentials of the cells.

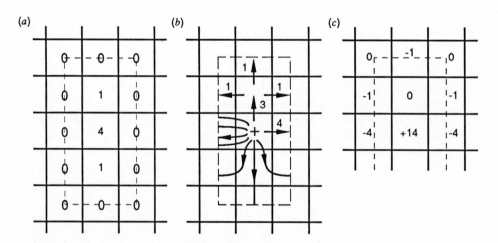

Figure 16.10   (a) V, in units $\lambda_0/14\epsilon_0$. (b) Upper right: electric field in units $\lambda_0/14\epsilon_0 dr$. Lower left: electric field lines. (c) Total charge density $\lambda(c)$ in units $\lambda_0/14$ (top half only; the bottom half is the same by symmetry.)

The electric field is shown in Figure 16.10b. The charge density can be calculated from E by Gauss' law

$$\lambda(c) = \epsilon_0 dr \sum_f E(f) \qquad \text{(pseudo-2D Gauss' law)} \qquad (16.9)$$

For cells c in the nonconducting region, this equation is redundant (it is equivalent to Poisson's equation, which we already know to be satisfied). In the interior of the conductor, it gives $\lambda(c) = 0$, as we observed in Section 16.2. However, on the boundary, there can

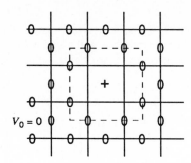

Figure 16.11   Pseudo-2D system used in Problem 16.2.

be induced charge, which we have shown in Figure 16.10c. Note that in this and the previous problem, the total charge (the fixed line charge λ_0 plus the induced charge on the conductor) is exactly zero. This is a consequence of the integral form of Gauss' law (Eq. 14.27) if the Gaussian surface S is taken to enclose the entire nonconducting region and lie in the interior of the outer conductor with potential V_0. Then $E = 0$ on S, and the total charge inside is $q = \epsilon_0 \int E \, da = 0$.

▶   **Problem 16.1**

Compute the charge density for Example 16.1. Draw a figure like Figure 16.8 showing $\lambda(c)$ for each cell.   ◀

▶   **Problem 16.2**

Compute the potential V, the field E at each nonconducting face, and the charge density λ at each cell for the discrete pseudo-2D system shown in Figure 16.11, having a single nonconducting cell with a charge λ_0 in it.   ◀

▶   **Problem 16.3**

Compute V, E, and λ for a pseudo-2D discrete system having four nonconducting cells, with a line charge λ_0 in the upper right cell (Figure 16.12). Are any of the cells equivalent by symmetry?   ◀

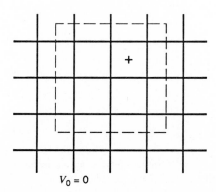

$V_0 = 0$

Figure 16.12   System for Problem 16.3.

▶ **Problem 16.4**

Solve the preceding problem using MAXWELL. Set up a transient current that carries charge from the conducting boundary to the upper right cell, and wait for the static field to evolve from Maxwell's equations. You will need to use absorbing boundaries and let B diffuse; see Appendix G.

▶ **Problem 16.5**

Consider a truly 3D analog of the pseudo-2D Example 16.1. There are only two nonconducting cells, one with charge $+q_0$ and one with charge zero. For an illustration of cross section through the center of the system see Figure 16.7. Compute V, E, and $\rho(c)$ everywhere, and sketch them in two inequivalent cross sections through q_0. ◀

SECTION 16.4   DISCRETE BOUNDARY VALUE PROBLEMS

The techniques we have described for solving Poisson's equation in a region bounded by a box do not depend on the fact that the potential is constant on the box. In this section we will solve some problems in which V takes on more than one value on the boundary of the region.

We will define a **discrete boundary value problem** to be specified by

1. giving a region R (a set of cells in a discrete lattice, such as that shown in Figure 16.13; R corresponds to the nonconducting region in Section 16.3) and
2. giving $\rho(c)$ for $c \in R$ and
3. giving the values of $V(c)$ on the complement $\overline{R}$ of R. (The **complement** of R is the set of cells not in R.)

It is called a boundary value problem because only the values of V on the boundary of $\overline{R}$ matter, that is, those cells of $\overline{R}$ that share faces with cells of R.

A **conducting-boundary** problem is a special type of boundary value problem in which one begins by specifying which faces are conducting (as in Figure 16.4, for example). This then determines the set of conducting cells $\overline{R}$ (each has at least one conducting face), which may consist of one connected piece (e.g., the shaded region in Figure 16.5) or more than one (e.g., the three in Figure 16.3b). The boundary value problem is then determined by specifying the charge density on the cells of R

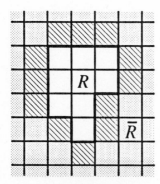

Figure 16.13   A set R of cells (white) and its complement $\overline{R}$ (shaded). The boundary of $\overline{R}$ is cross-hatched.

(the complement of $\overline{R}$) and the (constant) value of V on each connected piece of conductor.

■ **EXAMPLE 16.3**   Consider the conducting-boundary problem whose conducting faces are marked by zeroes in Figure 16.14a. The discrete conducting regions are indicated in Figure 16.14b; they have constant potentials V_0 and V_1. We will assume there is no charge in the vacuum region; the only charge is that on the conducting surfaces. (Of course there must be charge somewhere, or we won't get any field.) We will assume that some charge Q_1 was put on the center conductor a long time ago (so the fields have become static). The charge may have rearranged itself by moving around within the conductor (in particular, we have seen that it must all be at the boundary in the static limit) but the total charge Q_1 cannot change. Similarly, there may be a charge Q_0 on the boundary of the outer conductor. It is not constrained to remain constant like Q_1, because charge can move in from parts of the conductor not shown in the figure. However, it is constrained by the Gaussian-surface argument of Section 16.3: $Q_0 + Q_1 = 0$, so $Q_0 = -Q_1$. That is, just enough charge moves in from the rest of the outer conductor to cancel Q_1.

It might appear that the most natural way to specify the problem would be to specify the charge Q_1. We could then pick a reference potential (say $V_0 = 0$) and calculate the potential everywhere (and in particular V_1), in terms of Q_1. However, it turns out to be easier to solve problems in which the potential is known at all of the boundaries, that is, to start by assuming V_1 is known and to calculate Q_1. Having done this, it will be easy enough to solve for V_1 in terms of Q_1 if we happen to know Q_1 instead. Additionally, in many real experimental situations it makes more sense to assume the potentials are known: it is much easier to measure the potential of a conductor than to determine exactly how much charge is on it.

So we will assume we know $V_0(=0)$ and V_1. Label the nonconducting cells c_2, c_3, and c_4 as in Figure 16.14b. All the others are equivalent to these because of the 90° rotation symmetry: all the corner cells have the same potential V_4 as c_4, and so on. The two cells

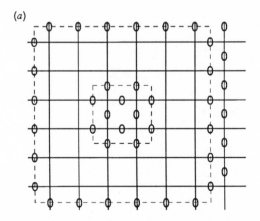

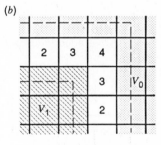

Figure 16.14   (a) The conducting region (conducting faces are marked with zeroes). The dashed lines are the boundaries of continuum conductors for which this might be a discretization. (b) Labels for nonconducting cells and conducting regions. Because of the 90° rotation symmetry, we only need to show the upper right quadrant.

labeled 3 are equivalent because of a mirror symmetry in a diagonal plane between them. Poisson's equation (Eq. 16.6) (really Laplace's because $\rho = 0$) is then:

cell 2: $V_1 + 2V_3 + 0 - 4V_2 = 0$

cell 3: $V_1 + V_2 + V_4 + 0 - 4V_3 = 0$

cell 4: $2V_3 + 0 - 4V_4 = 0$

This is equivalent to the matrix equation

$$\begin{pmatrix} -4 & 2 & 0 \\ 1 & -4 & 1 \\ 0 & -4 & 2 \end{pmatrix} \begin{pmatrix} V_2 \\ V_3 \\ V_4 \end{pmatrix} = \begin{pmatrix} -V_1 \\ -V_1 \\ 0 \end{pmatrix}$$

It is most easily solved by solving for V_4 in the last equation ($V_4 = \frac{1}{2}V_3$) and substituting this for V_4 in the other two equations. We can similarly use the second equation to eliminate V_3, and the first to obtain V_2:

$$V_2 = 11V_1/24$$

Evaluating the formulas for V_3 and V_4 then gives

$$V_3 = 10V_1/24$$

$$V_4 = 5V_1/24$$

This solution for the potential is shown in Figure 16.15, and the corresponding electric field in Figure 16.16.

In Figure 16.16 we have shown the induced charge λ, obtained from Gauss' law. With it we can calculate the total charge on each conductor (in our pseudo-2D system we really calculate a charge per unit length, which we will denote by Λ_0 or Λ_1). We get

$$\Lambda_0 = (-11 \times 4 - 10 \times 8 - 5 \times 8) \times \frac{\epsilon_0 V_1}{24} = -\frac{41}{6}\epsilon_0 V_1$$

$$\Lambda_1 = (13 \times 4 + 28 \times 4) \times \frac{\epsilon_0 V_1}{24} = +\frac{41}{6}\epsilon_0 V_1$$

which are equal and opposite, as expected.

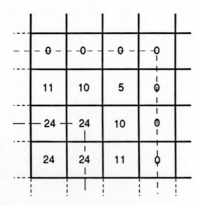

Figure 16.15   The potential V, in units $V_1/24$.

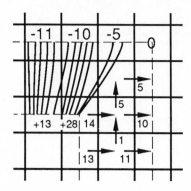

Figure 16.16 Lower right: E, in units $V_1/24\,dr$. Upper left: field lines, and λ in units $\epsilon_0 V_1/24$. (Because it's hard to fit in 14 field lines, we've drawn one field line for each $V_1/12\,dr$ of field instead of each $V_1/24\,dr$.)

The preceding discrete example can be regarded as an approximation to the continuum problem of a square bar inside a square duct. Even though this discretization is a very coarse one, it shows the essential qualitative features of such continuum problems. Induced charge is concentrated at sharp corners of conductors, as in the cell with charge $+28$ in Figure 16.16. The field is also strongest (14) in this vicinity. The field and induced charge are small in recesses such as the upper right.

The examples we have done so far all involve systems that are completely surrounded by conductor 0, at potential V_0. It doesn't really matter whether this conductor extends to infinity and fills the whole universe, or is just a closed conducting container as in Figure 16.17. The solutions for the field and potential inside the container are the same either way: the equations for $V(c)$ inside the container involve $\rho(c)$ inside and V_0, but not $\rho(c)$ outside. The field, being a difference of $V(c)$'s, doesn't even depend on V_0. So

**charges outside a closed conducting box
have no effect on the electric fields inside**

Thus, a conducting box completely isolates its interior from the outside world. Strictly speaking, this is only true in electrostatics: an electromagnetic wave can penetrate a weakly conducting material. That's why the inside of a hot dog gets warm in a microwave oven, for example. We will study the propagation of waves in conducting media later: they get weaker as you look further into the conductor, at a

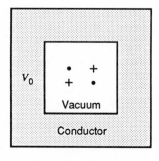

Figure 16.17 An electrostatic system enclosed in a conducting container.

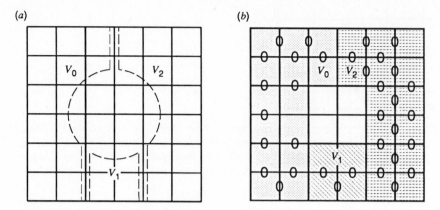

Figure 16.18 (*a*) A split conducting box. (*b*) The conducting region; conducting faces are marked by zeroes.

rate determined by the conductivity and the frequency. In high-conductivity materials (metals, for example) at radio frequencies or below, the fields are essentially zero further than a millimeter inside the material. So we can conclude that a conducting box thicker than this isolates its interior even for **non**-static fields. A container constructed for this purpose is called a "Faraday cage." Very sensitive electrical experiments are often conducted inside such a metal cage, to protect them from stray electric and magnetic fields from electric lights, motors, and the like. Conversely, devices such as electron tubes that **produce** strong electric fields are often enclosed in metal cans to protect nearby electronic components from these fields.

It is possible to work with problems in which the conducting box itself consists of more than one connected piece, as in Figure 16.18*a*. This is a discretization of a continuum problem in which there are three nearly touching pieces of conductor (dashed lines) held (by batteries, for example) at three different potentials V_0, V_1, and V_2. We have discretized the problem and marked the conducting faces with 0's in Figure 16.18*b*. The cells that can be reached by crossing conducting faces from each piece of conductor are shaded in three different ways; note that there is no conducting path from one piece to another. There is a finite number of nonconducting (vacuum) cells in the middle (and they are isolated by the conductor from any nonconducting cells outside the conductors) so this kind of problem can be done in exactly the same way as the previous problems.

■ **EXAMPLE 16.4**   Let us calculate the electrostatic potential and electric fields for a discrete approximation to the pseudo-2D split conducting duct shown in Figure 16.19. We will use a simple discretization in which there are only two nonconducting cells, *a* and *b*, as shown in Figure 16.20*a*. As before, the potentials V_a and V_b of these cells are determined from the Poisson equation for the nonconducting cells:

$$a: \qquad dr^2 \, \nabla^2 V = 3V_1 + V_b - 4V_a = 0$$

$$b: \qquad dr^2 \, \nabla^2 V = V_a - 4V_b = 0$$

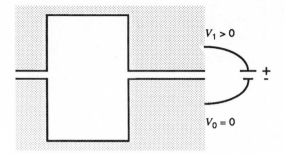

Figure 16.19 A split conducting duct, whose two sides are held at different potentials by a battery.

The solution is

$$V_a = \frac{4V_1}{5}, \qquad V_b = \frac{V_1}{5}$$

as shown in Figure 16.20b. The electric field is shown in Figure 16.20c. Gauss' law for the six boundary cells (those sharing a face with cells a or b) give the induced charge (Figure 16.20d).

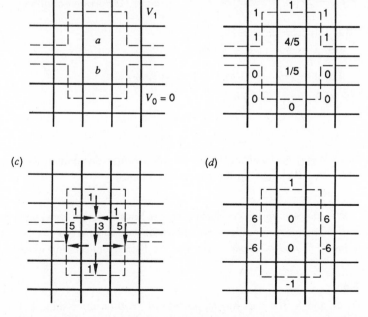

Figure 16.20 (a) Discretization of the split conducting duct. (b) The potential, in units of V_1. (c) The electric field, in units $V_1/5\,dr$. (d) Linear charge density λ, in units $\epsilon_0 V_1/5\,dr^2$.

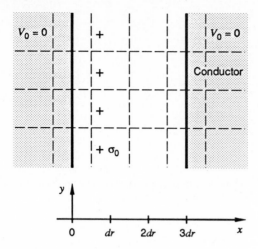

Figure 16.21 A pseudo-1D system.

■ **EXAMPLE 16.5** We can also calculate the potential from the Poisson equation in a pseudo-1D system such as that in Figure 16.21, in which the charge distribution $\rho(c)$ and the boundary conditions depend only on x. We assume there are charges only in the left-hand nonconducting cell (centered at $x = dr$); these charges are on a plane parallel to the yz-plane, and we may as well describe the charge distribution by the charge per unit area, $\sigma(c) = \rho(c)dr$. In the case shown, $\sigma(c)$ has only one nonzero value, which we can call σ_0. We will assume that V has the same symmetry, that is, depends only on x. Then, each cell has only two neighbors c' with different potentials (one to the left, one to the right) and from Eq. 16.4 we can get a pseudo-1D analog of Poisson's equation:

$$\sum_{c'} V(c') - 2V(c) = -\sigma(c)dr/\epsilon_0 \qquad \text{(pseudo-1D Poisson equation)} \qquad (16.10)$$

Writing this for the particular case in Figure 16.21,

$$\text{at } x = dr: \qquad 0 + V(2dr) - 2V(dr) = -\sigma_0 dr/\epsilon_0$$

$$\text{at } x = 2dr: \qquad V(dr) + 0 - 2V(2dr) = 0$$

whose solution for $V(dr)$ and $V(2dr)$ is

$$V(dr) = 2\sigma_0 dr/3\epsilon_0$$

$$V(2dr) = \sigma_0 dr/3\epsilon_0$$

Figure 16.22 shows V and the resulting field and total charge distribution (both the fixed charge σ_0 and the induced charge on the conducting planes).

■ **EXAMPLE 16.6** We can also let the conducting planes have different potentials, as shown in Figure 16.23. The boundary conditions are shown at the top (there is no fixed charge), and the solution for V, E, and σ below. Note that the field is uniform in this case.

▶   **Problem 16.6**

Compute V, E, and λ in a duct having potential $V_0 = 0$ on three sides, but $V_1 > 0$ on the fourth. Use four nonconducting cells as in Figure 16.24.   ◀

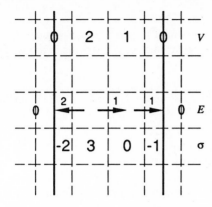

Figure 16.22 V in units $\sigma_0 dr/3\epsilon_0$, E in units $\sigma_0 dr/3\epsilon_0$, and σ in units $\sigma_0/3$. They are shown at different heights, but each, of course, is independent of height.

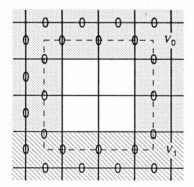

Figure 16.23   A quasi-1D problem with different potentials at the left and right, showing V in units of V_1, E in units of $V_1/3dr$, and σ in units $V_1/3\epsilon_0 dr$.

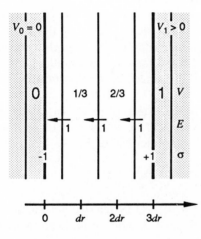

Figure 16.24   A duct with a different potential on one side.

▶   **Problem 16.7**

Compute V, E, and λ for the pseudo-2D system of Figure 16.18b, assuming $V_1 = 0$ and $V_2 = -V_0$. Does this system have any special symmetry? Give results in terms of V_0.   ◀

▶   **Problem 16.8**

Do the same, using $V_0 = 0$ and $V_2 = 3V_1$ (give results in terms of V_1). Does this have any symmetry? Would it if $V_2 = 2V_1$?   ◀

SECTION 16.5   THE METHOD OF RELAXATION

The method of relaxation is a way of calculating potentials from Laplace's or Poisson's equation that doesn't require solving systems of linear equations and is often convenient for computer calculation. We will discuss it first for the case of Laplace's equation, that is, when the charge density $\rho = 0$. The method is based on the fact that Laplace's equation requires the potential to be as smooth and featureless as possible, in a certain sense. Note that in each of the solutions of Laplace's equation in Section 16.4, the potential interpolated smoothly between the extreme values demanded by the boundary conditions. In the parallel-plate capacitor, for example (Figure 16.23), the potential changed at a uniform rate from one plate to the other.

This "smoothness" property can be expressed quantitatively by saying that the potential of each cell must be exactly the **average of its neighbors**. Solving Laplace's equation (Eq. 16.5 with $\rho = 0$) for $V(c)$ at the center cell of Figure 16.6a gives

$$V(c) = \sum_{c'} V(c')/6 \qquad\qquad (16.11)$$

This is the average of six neighbors; in a pseudo-2D system we get the average of four neighbors. A corollary of Eq. 16.11 is that **a solution of Laplace's equation cannot have a local maximum**: if it had a maximum at c, meaning $V(c) > V(c')$ for all neighbors c', then $V(c) > \sum V(c')/6$, contradicting Eq. 16.11. A similar argument shows there can be no minima.

If there is only one value of the potential on the boundaries, Eq. 16.11 is satisfied by having the potential the same everywhere; if the boundary conditions require some values to be different, each cell wants to be as much like its neighbors as possible.

To see how this tendency toward conformism helps us calculate the potential, let's look at how conformity is achieved in an analogous situation consisting of an array of humans instead of potentials in cells. A respect in which humans tend to conform is, of course, dress. Figure 16.25 shows a class of college students, chosen to be women because skirt length is easier to depict than, say, lapel width. Most of the students are conformists in dress, but a few (to form the left and right boundaries) dress independently of their neighbors. We will assume the class is infinite in the y-direction, so this is a pseudo-1D system in which skirt length is a function only of x. The right boundary consists of students (whom we shall call "rightists") who live at home with their parents and must wear skirts 8 inches below the knee, the left boundary consists of "leftists" who have their own apartments and wear skirts 8 inches above the knee, and the rest of the students are conformists and

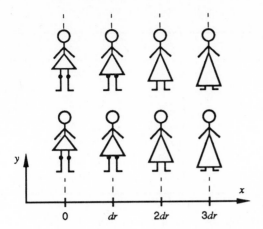

Figure 16.25   A pseudo-1D class.

wear skirts whose length is precisely the average of the skirt lengths of their four nearest neighbors. Thus, the skirt length (call it V) satisfies the pseudo-2D version of Laplace's equation: in the pseudo-1D system of Figure 16.25, this is equivalent to

$$V(c) = \sum_{c'} V(c')/2 \qquad (16.12)$$

where the sum is only over the two neighbors to the right and left.

Although the manner in which the static potential is attained in an electromagnetic system is rather complicated, in the fashion analogy we can see how it comes about. On the first day of class, the students don't know who they'll be sitting next to so they have arbitrarily chosen skirt lengths. To preserve the analogy with the parallel-plate capacitor (Example 16.6), we will measure skirt length V from the shortest (which is then $V = 0$), with the longest being 1.0 in arbitrary units. Suppose on the first day all of the students happen to choose to wear short skirts ($V = 0$), except, of course, the rightists at the right boundary, who are constrained to have $V = 1$. Some of the students between the boundaries will find, to their embarassment, that their skirt lengths differ from the average of their neighbors'. They will come to class on the second day with skirt lengths equal to the average of their neighbors' (first-day) lengths:

$$V^{\text{2nd day}}(c) = \sum_{c'} V^{\text{1st day}}(c')/2 \qquad (16.13)$$

The lengths are shown in Table 16.1. On the second day, the row of students at $x = dr$ find that their skirt length, although it is exactly the average of their neighbors' **previous** lengths, is now outmoded and must be changed for the third day. In general, the lengths on the nth day are

$$V^{n\text{th}}(c) = \sum_{c'} V^{(n-1)\text{th}}(c')/2 \qquad (16.14)$$

Table 16.1

x:	0	dr	$2\,dr$	$3\,dr$
V^{1st}:	0	0	0	1.00
V^{2nd}:	0	0	0.5	1.00
V^{3rd}:	0	0.25	0.5	1.00
V^{4th}:	0	0.25	0.625	1.00
V^{5th}:	0	0.312	0.625	1.00
V^{6th}:	0	0.312	0.656	1.00
Limit:	0	0.333	0.667	1.00

The table shows the result of applying this formula iteratively. By the sixth day, the skirt lengths satisfy Laplace's equation within 2%, which is probably imperceptible enough to allow the students to relax (hence the name of the method) and turn their attention to other matters. In the limit of an infinite number of iterations, this method converges to the exact answer computed in Example 16.6 ($V = 0, \frac{1}{3}, \frac{2}{3}$, and 1).

This same method will converge for any of the potential problems we have looked at, including those with nonzero ρ (i.e., involving Poisson's equation). The iteration equation for this case is obtained by solving Eq. 16.5 for $V(c)$. In three dimensions it is

$$V^n(c) = \frac{1}{6}\left[\sum_{c'} V^{n-1}(c') + \frac{\rho(c)\,dr^2}{\epsilon_0}\right] \tag{16.15}$$

(note that the superscript here is the iteration index, not an exponent) and for a pseudo-2D system

$$V^n(c) = \frac{1}{4}\left[\sum_{c'} V^{n-1}(c') + \frac{\lambda(c)}{\epsilon_0}\right] \tag{16.16}$$

(where the sum is now over four neighbors c') and for a pseudo-1D system (from Eq. 16.10)

$$V^n(c) = \frac{1}{2}\left[\sum_{c'} V^{n-1}(c') + \frac{\sigma(c)\,dr}{\epsilon_0}\right] \tag{16.17}$$

Hopefully, the fashion analogy gives some feeling for why the method works; another way of thinking about it is to think of the $(n-1)$th estimate V^{n-1} as differing from the exact answer by some error; then according to Eq. 16.15 only $1/6$ of this error is carried over to the next iteration. Of course, there are six neighbors, so it looks like we have to multiply the error by 6, but we are saved by the fact that often some of these are conducting boundary cells, at which V^{n-1} has **no** error. So on the average the error is multiplied by something less than $6/6$, and therefore gets smaller. This argument suggests that the method works best when most of the cells are next to boundary cells (note that in the examples of Section 16.3, **all** the cells were). This turns out to be true: if we subdivide a system into many cells, so most are not near a boundary, the method of relaxation converges only very slowly (see Problem 16.15).

■ **EXAMPLE 16.7**   Let's apply the relaxation method to the simplest pseudo-2D problem solved in Section 16.3, namely the two-cell system of Example 16.1. Choose the starting potential V^1 to be zero everywhere (of course, we could converge to the right answer faster if we made a more intelligent guess for V^1). Equation 16.16 then reads:

$$V_1^n \equiv V^n(c_1) = \tfrac{1}{4}[V_2^{n-1} + \lambda_0/\epsilon_0]$$

$$V_2^n \equiv V^n(c_2) = \tfrac{1}{4}[V_1^{n-1} + 0]$$

This gives for the second estimate V^2

$$V_1^2 = \tfrac{1}{4}[0 + \lambda_0/\epsilon_0] = 0.25\lambda_0/\epsilon_0$$

$$V_2^2 = \tfrac{1}{4}[0 + 0] = 0.0$$

Iterating again gives the third estimate

$$V_1^3 = \tfrac{1}{4}[0 + \lambda_0/\epsilon_0] = 0.250\lambda_0/\epsilon_0$$

$$V_2^3 = \tfrac{1}{4}[0.25\lambda_0/\epsilon_0 + 0] = 0.062\lambda_0/\epsilon_0$$

This is already not too far from the exact answer; the maximum error is 0.017. Iterating further gives Table 16.2. It takes four iterations to get three-figure accuracy in this case.

Table 16.2

Iteration $n =$	1	2	3	4	5	6	Exact
$V_1^n \epsilon_0/\lambda_0$	0.0	0.250	0.250	0.266	0.266	0.267	0.267...
$V_2^n \epsilon_0/\lambda_0$	0.0	0.000	0.062	0.062	0.066	0.066	0.067...
Maximum error	0.267	0.067	0.017	0.005	0.001	≤ 0.001	0

Many programs have been written to implement the relaxation method on a microcomputer. See, for example, the books by Koonin and by Gould and Tobochnik in the bibliography.

▶ **Problem 16.9**

Solve Example 16.2 by the method of relaxation. Iterate until two successive iterations are the same to within 0.001, and check your answer against the exact one.   ◀

▶ **Problem 16.10**

Do the same for Example 16.3 (the square bar inside a duct).   ◀

▶ **Problem 16.11**

(A model of a real capacitor, including fringing fields) Consider the three-conductor system shown in Figure 16.26, a model of a pseudo-2D capacitor whose plates have

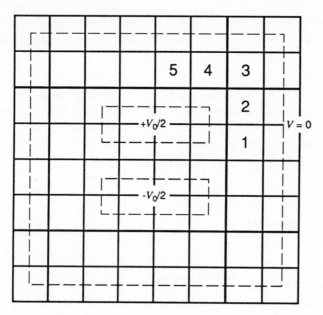

Figure 16.26 A three-conductor model of a capacitor, with labels for 5 inequivalent cells. (Dashed lines are continuum conductor boundaries.)

width 3 dr and are separated by dr. There are no nonconducting cells between the two plates, and only one layer of nonconducting cells between the plates and the outer boundary. There are only 5 independent potentials to solve for because of the symmetry. Set $V_0 = 1$ for calculational convenience.

a. Show that the relaxation equation at cell 5 can be written

$$V_5 = 3^{-1} \sum V(3 \text{ inequivalent neighbors}) = 3^{-1}(V_4 + 0 + 0.5)$$

b. Show that the equation at cell 1 can be written

$$V_1 = 5^{-1} \sum V(3 \text{ inequivalent neighbors}) = 5^{-1}(V_2 + 0 + 0.5)$$

[Hint: Write down the condition for horizontal-mirror symmetry of V carefully.]

c. Determine all five potentials to ± 0.001, and verify them using Laplace's equation.

d. Compute the electric field (show it on a diagram of the upper right quarter of the system). Sketch the field lines, using one line for each $0.1V_0\ dr$ of flux.

e. Compute the induced charge on each plate. ◀

▶ **Problem 16.12**

Write a computer program to do relaxation for a pseudo-1D system with zero charge density (i.e., using Laplace's equation). It should solicit as input the length of the system (in units of dr) and the potentials at the left and right. Use $V^1 = 0$ as a starting guess, and print out iterations $V^1, V^2, V^3, \ldots$. (After debugging, you might want to print out only every 10th iteration, or just the last one.) Stop when successive iterations differ by less than 10^{-4}. (Does that put you within 10^{-4} of the exact answer?) Hand in a listing of the program and the output for a system of six cells (four vacuum, two conducting boundary) using boundary conditions 0 and 1. (Do you need to store all the iterations at once? How many separate 6-element arrays $V(r)$ do you need? There is a variant of the relaxation method that uses only one, but you are to use Eq. 16.17 here.)   ◄

▶   **Problem 16.13**

    a. Write a computer program to compute the potential in an infinite 3D discrete system with a charge q in the cell at the origin. Use a box of length L, so $-\frac{1}{2}L \le x, y, z \le \frac{1}{2}L$, and calculate the potential for $L = dr$, $2\,dr$, and $3\,dr$. Estimate the $L \to \infty$ limit of $V(0)$ from the results for the largest box. Estimate the uncertainty from the apparent rate of convergence. Hint: You can do this with $V = 0$ at the boundary (i.e., a conducting box). But it will converge faster as $L \to \infty$ if you use the continuum formula $V = q/4\pi\epsilon_0 r$ to set the boundary values.

    b. Compute E at $0.5\,dr$ and $1.5\,dr$ from V, again estimating the uncertainty.   ◄

▶   **Problem 16.14**

Repeat Problem 16.13 using a pseudo-2D system with a linear charge density λ in one cell. What boundary condition should you use for most rapid convergence?

    Next compare to the results of MAXWELL with a current in which $\frac{1}{4}\lambda$ moves out to the boundary of the system during the first time interval. In an infinite system, do you expect the results to agree for finite dr, or only in the $dr \to 0$ limit?   ◄

▶   **Problem 16.15**

(Convergence Rate of the Relaxation Method) Consider a discrete pseudo-1D system with boundary cells centered at $x = 0$ and $x = L \equiv N\,dr$ (this could be a parallel-plate capacitor, or the skirt-length problem). Consider Laplace's equation with the boundary conditions $V(0) = 0$, $V(L) = 0$, and apply the relaxation method for the initial guess $V^0(x) = A \sin(2\pi x/L)$. Write an analytic formula for the nth iteration $V^n(x)$. Does it converge to the exact answer as $n \to \infty$? How many iterations are necessary to reduce the error by a factor of $e = 2.718\ldots$, for $L = 5\,dr$? For $L = 100\,dr$?   ◄

SECTION 16.6   EXISTENCE AND UNIQUENESS OF SOLUTIONS TO POISSON'S EQUATION

In this section, we will prove the following theorem about solutions to discrete boundary-value problems.

Existence and Uniqueness Theorem

Consider a discrete boundary value problem as defined in Section 16.4. We are given a finite region R and values $V(c)$ for the potential on its complement $\overline{R}$ (only the values on the boundary of R are relevant) as well as the charge density $\rho(c)$ for each cell $c \in R$. Then there is a **unique** potential $V(c)$ that satisfies the boundary conditions on $\overline{R}$ and satisfies Poisson's equation (Eqs. 16.4 or 16.5) at all cells in R. (Poisson's equation is satisfied on $\overline{R}$ too, if we use it to determine the charge density there.)

    PROOF: Outside the region R (i.e., in the conductors in Figure 16.27), $\rho(c)$ has not been specified, and we can choose it to be $-\epsilon_0\nabla^2 V(c)$ to satisfy Poisson's

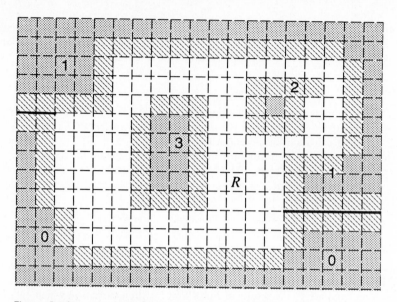

Figure 16.27 A possible 2D boundary value problem, showing the region R (white cells), its complement $\bar{R}$ (light or dark shading), and the boundary of $\bar{R}$ (light shading). (This case arose from a conducting-boundary problem, so V is constant on each of four connected pieces labeled 0, 1, 2, and 3.)

equation. We only need to find $V(c)$ for the cells in R, say $c_1, c_2, \ldots, c_m$, to satisfy the m Poisson equations

$$\epsilon_0 \nabla^2 V(c_i) = -\rho(c_i) \qquad (i = 1, 2, \ldots, m) \tag{16.18}$$

and prove these V's to be unique.

We will attack the "uniqueness" part first, by assuming the contrary and arriving at a contradiction. Suppose there could be two different solutions V and V' to Eq. 16.18. Substituting each into Eq. 16.18 and subtracting gives Laplace's equation

$$\nabla^2 W(c_i) = 0 \tag{16.19}$$

for the difference $W(c_i) = V(c_i) - V'(c_i)$. Also, the boundary conditions for V and V' imply that $W = 0$ on the boundary; we want to prove $W = 0$ everywhere.

We will first sketch the proof. It is based on the fact (discussed in Section 16.5) that a solution of Laplace's equation cannot have a maximum (except at a boundary), because each cell must be exactly the average of its neighbors. Either the m values $W(c_i)$ are all the same, or one must be higher than the others. The latter is ruled out by the absence of a maximum (there can't be one at the boundary either because $W = 0$ there), so they are all the same. Thus W is constant, and because it is zero at the boundaries, it must be zero everywhere. Another way of saying this is that a function that vanishes at boundaries but isn't constant must have a maximum or minimum (see Figure 16.28).

To make this argument precise, let's suppose one of the $W(c_i)$'s were nonzero and look for a contradiction. Suppose it's positive; if it's negative, we just need to

Figure 16.28 A 1D function vanishing at boundaries, which must have a maximum or minimum.

substitute $-W$ for W in the following. Let c be a cell of R in which W attains its maximum value: $W(c) \geq 0$ and $W(c) \geq W(c')$ for any $c' \in R$ (actually, since $W = 0$ on $\bar{R}$, this inequality is true for **all** c'). (We can be sure of finding such a c only because we assumed the system was finite: in an infinite system, W could increase forever and never have a maximum.) We must include the possibility $W(c) = W(c')$ because the maximum could be attained on a large subset of R. Let us pick a cell on the boundary of this "plateau," so c has at least one neighbor c' having smaller W: $W(c') < W(c)$. Then

$$\nabla^2 W(c) = (dr)^{-2} \sum_{c'} [W(c') - W(c)]$$

is strictly negative, because $W(c') - W(c) \leq 0$ for all neighbors c' and the strict inequality holds for at least one. But this contradicts Laplace's equation (Eq. 16.18), so our assumption that W was nonzero somewhere must have been wrong. W must have been zero, V must have been V', and the solution to Poisson's equation was unique.

That was the easy part. How do we show there is at least one solution V? The best way to do this is to write our Poisson equations (Eq. 16.18) in matrix form, as in Eq. 16.8. The general form is

$$(M) \begin{pmatrix} V(c_1) \\ V(c_2) \\ \vdots \\ V(c_m) \end{pmatrix} = \begin{pmatrix} r_1 \\ r_2 \\ \vdots \\ r_m \end{pmatrix} \tag{16.20}$$

where (M) is an $m \times m$ matrix (of integers in Section 16.3) and the right-hand side r_i of each equation has all the terms other than $V(c_1) \cdots V(c_m)$; this includes V's at the boundaries and charge density terms. Our proof of uniqueness amounted to showing that there is no nonzero column vector (W) such that $(M)(W) = 0$ (this is exactly Poisson's equation with zero boundary conditions). In the language of linear algebra, this is the same as saying (M) is **nonsingular**. (Recall that a matrix (M) is singular if and only if $(M)(W) = 0$ for some nonzero vector (W).) At this point, we must assume a basic theorem of linear algebra: every nonsingular square matrix is invertible. That is, there is a matrix $(M)^{-1}$ such that $(M)^{-1}(M) = (1)$, the identity $m \times m$ matrix. Multiplying Eq. 16.20 by this $(M)^{-1}$ gives

$$(M)^{-1}(M)(V) = (1)(V) = (V) = (M)^{-1}(r) \tag{16.21}$$

which is the desired solution.

Under certain circumstances, the theorem can be extended to infinite volumes (e.g., to free space in which there is no conducting box). However, one must be very careful in this case, as we have already seen in the case of the infinite charged plane:

there the field (and hence the potential) was **not** uniquely determined by the charge distribution. One can be fairly safe for charge distributions confined to a finite volume if one insists that the potential approach zero (or any fixed value) as the distance from the origin approaches infinity; this can be viewed as a sort of additional "boundary condition" at a "boundary at infinity." This is essentially what we did, for example, in determining the potential of a point charge in Section 15.1.

Pseudo-one- (or two-) dimensional problems technically have infinite extent in the directions of the two (or one) coordinates on which the fields do not depend. This does not, however, create a problem with the existence and uniqueness theorem, because the proof can be carried out in a one- (or two-) dimensional cross section that is finite. Thus, the potential in an infinite parallel-plate capacitor (Example 16.6) or an infinite duct (Example 16.1) is unique.

▶   **Problem 16.16**

It is not too hard to prove (formally, if not rigorously) the uniqueness of solutions to Poisson's equation in a continuum context (existence is much harder). The arguments given above work in the continuum case if one can show that a function $W(r)$ satisfying Laplace's equation can have no local maximum. Demonstrate this formally in two dimensions. (Hint: At a maximum, $dW/dx = dW/dy = 0$ and $d^2W/dx^2 < 0$, $d^2W/dy^2 < 0$. "Formally" here means that you are to assume the existence of all derivatives.)   ◀

SECTION 16.7   CONTINUUM CONDUCTING-BOUNDARY PROBLEMS

We have thus far looked only at discrete applications of Poisson's equation. In a discrete conducting-boundary problem we assumed a certain set of conducting faces and boundary values for the potential on each piece of bounding conductor. We then solved for the potential $V(c)$ in the cells of the vacuum region, given $\rho(c)$ there. We would like to define a continuum analog of this problem. Suppose that the conductor occupies a certain continuum region of space, such as that shaded in Figure 16.29a. The region is specified mathematically as a set of points $\mathbf{r} = (x, y, z)$, perhaps determined by a rule like

$$x^2 + y^2 + z^2 < r^2$$

(which gives a spherical region like the inner one in the figure). Such a continuum region, together with a specification of the potential on each connected piece of conductor (V_0 and V_1 in the figure) and the continuum charge distribution in the nonconducting regions, is called a **continuum conducting-boundary** problem.

Recall that a continuum charge density is a rule giving a discrete charge density (called a **discretization** of the continuum density) for each choice of lattice. In this spirit, we can use the continuum conducting region to determine a rule giving a discrete conducting region (i.e., a set of faces) for each lattice: the simplest such rule is to include a face if its center is in the continuum region. This has been done for a particular choice of lattice in Figure 16.29b. In most cases, it is obvious what value to use for the potential on each connected piece of the discrete conducting region:

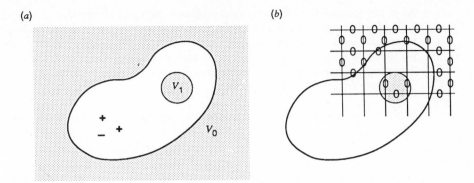

Figure 16.29 (*a*) A continuum conducting-boundary problem. (*b*) A discretization of part of the conducting region (conducting faces are marked by zeroes).

all the face centers are in the same continuum region and thus have the same potential.

If the cells are very large, or the continuum pieces of conductor very close together (see Figure 16.30*a*), then two continuum pieces can find themselves in the same discrete piece (i.e., they can "short out" as in Fig. 16.30*b*). We will handle this possibility by disallowing this lattice, and demanding one with a smaller *dr* or one more intelligently positioned, as in Figure 16.30*c*.

We have thus constructed a set of rules giving us a discrete conducting-boundary problem, i.e., a **discretization** of our continuum conducting-boundary problem, for each lattice. In addition, because of the existence and uniqueness theorem, each such discretization has a solution for the discrete potential. These do not necessarily have a limit, but if they do it is natural to regard it as the solution to the **continuum** problem. This is a continuum function $V_f(\mathbf{r})$ (the f stands for "continuum function"), defined formally by

$$V_f(\mathbf{r}) = \lim_{dr \to 0} V \text{ (cell containing } \mathbf{r}\text{)} \tag{16.22}$$

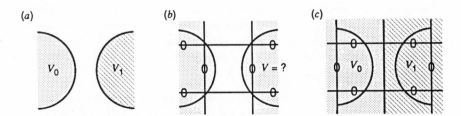

Figure 16.30 A system with close conductors (*a*), discretized incorrectly (*b*) and correctly (*c*).

As in our treatments of all continuum problems, we make no attempt to achieve the kind of mathematical rigor possible in treating discrete systems. In particular, we assume the limit is independent of the type of lattice (cubic, spherical, irregular), the placement of the cell centers, and so on. In reasonable cases, this is true. The problem of "shorting out" mentioned above disappears for small enough dr, unless the continuum conductors are infinitely close together.

We have given the above procedure for obtaining the solution V_f to a continuum conducting-boundary problem as a limit because that is the way most such problems are solved in practice; one cannot get an exact analytic expression for $V(\mathbf{r})$, but must approximate it by using a small dr. However, there are a few problems for which an exact answer can be obtained. Note that the continuum potential $V_f(\mathbf{r})$ must satisfy the continuum limit of Poisson's equation

$$\rho_f(\mathbf{r}) = -\epsilon_0 \nabla^2 V_f(\mathbf{r}) = -\epsilon_0 \left[d^2 V_f/dx^2 + d^2 V_f/dy^2 + d^2 V_f/dz^2 \right] \quad (16.23)$$

If we can find an exact analytic solution of Eq. 16.23 that satisfies the boundary conditions at the conductors, it must be the correct potential (we are assuming here that the existence and uniqueness theorem of Section 16.6 is valid in the continuum limit).

■ **EXAMPLE 16.8 PARALLEL-PLATE CAPACITOR**   In the pseudo-1D problem of a parallel-plate capacitor, the charge density is zero, and (if the plates are taken parallel to the yz-plane) V_f is only a function of x. Thus Poisson's equation reduces to the Laplace equation

$$0 = d^2 V_f(x)/dx^2 \quad (16.24)$$

The boundary conditions are at the conducting surfaces ($x = 0$ and $x = L$), say

$$V_f(0) = 0, \qquad V_f(L) = V_1 \quad (16.25)$$

We can find all possible solutions of the differential equation (Eq. 16.24) by integrating it with respect to x, getting

$$a = dV_f(x)/dx$$

for some constant a, and integrating again to get

$$ax + b = V_f(x) \quad (16.26)$$

for some other constant b. Requiring that Eq. 16.26 satisfy the boundary conditions ("bc's") (Eq. 16.25) gives $b = 0, aL = V_1$, so

$$V_f(x) = (V_1/L)x \quad (16.27)$$

This is an exact solution of Laplace's equation, which we obtained analytically from the continuum Laplace equation rather than as a limit of discrete solutions. Just to show that we could have done the latter instead, consider the discretization shown in Figure 16.31. For any such discretization, we can calculate the discrete potential $V(c)$ from the 1D discrete Laplace equation (Eq. 16.9):

$$\sum_c V(c') - 2V(c) = V(x + dr) + V(x - dr) - 2V(x) = 0 \quad (16.28)$$

where $V(x)$ denotes the potential $V(c)$ at the cell c centered at x. The boundary cells are at

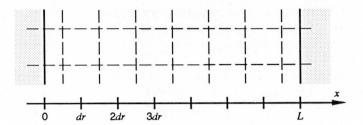

Figure 16.31 A discretization of the parallel-plate capacitor.

$x = 0$ and $x = L$, giving $V(0) = 0$ and $V(L) = 0$ as discrete bc's. These Laplace equations are $(L/dr) - 1$ linear equations in as many unknowns, and can easily be solved. In this special case, it turns out we can get the solution by evaluating the continuum potential V_f at the cell centers:

$$V(x) = V_f(x) = (V_1/L)x \qquad (16.29)$$

as you can see by substituting Eq. 16.29 into Eq. 16.28. We can now take the solutions (Eq. 16.29) of the discrete Laplace equations and take the $dr \rightarrow 0$ limit to get the continuum $V_f(x)$; in this case this is trivial because the solutions don't depend on dr.

The reason this example worked out so nicely is that we chose the cell positions so that the boundary of the conductor is at the **center** of the boundary cell (at the right as well as the left): this requires L to be an integer multiple of dr. This way the discrete boundary condition $V(c) = 0$ for the left boundary cell corresponds exactly to the continuum bc $V_f(x = 0) = 0$. We could have set up our discretization so that the conducting boundary was **not** at a cell center. For example, we could have put it at a face as in Figure 16.32. Then, the bc's would be $V(-\frac{1}{2}dr) = 0$, $V(L + \frac{1}{2}dr) = V_1$, and the solution of Laplace's equation would turn out to be

$$V(x) = \left[V_1/(L + dr)\right]\left(x + \tfrac{1}{2}dr\right) \qquad (16.30)$$

instead of Eq. 16.29. The limit as $dr \rightarrow 0$ is exactly the same, so either method would do to calculate the continuum potential V_f, but plainly Eq. 16.29 converges faster (in fact infinitely fast, because it doesn't depend on dr) than Eq. 16.30. This turns out to be true in general: although you can put your cells wherever you want,

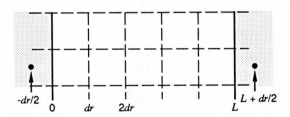

Figure 16.32 An alternate discretization for Example 16.8.

convergence to the right answer to a continuum problem is faster if you make sure the conducting boundaries correspond to cell centers. You may note that this rule was observed in all of the examples in Sections 16.3 through 16.5 that started with a continuum-like picture.

We can use the above example to illustrate the perils of assuming the uniqueness of solutions to Laplace's equation in an infinite system. If we removed the boundaries at $x = 0$ and $x = L$, Eq. 16.26 would be a solution for **any** choice of constants a and b, and therefore not unique.

From the discrete potential $V(x) = (V_1/L)x$ we can calculate the discrete electric field (Eq. 15.20)

$$E(x) = -\left[V\left(x + \tfrac{1}{2}dr\right) - V\left(x - \tfrac{1}{2}dr\right)\right]/dr$$
$$= -(V_1/L)\left[x + \tfrac{1}{2}dr - \left(x - \tfrac{1}{2}dr\right)\right]/dr = -V_1/L \tag{16.31}$$

We get the continuum field either by taking the limit of this or by using the continuum formula for the gradient:

$$\mathbf{E}_f(x) \equiv -\nabla V_f(x) \tag{16.32}$$

which has only an x-component, because V_f depends only on x (Eq. 16.27). This is the derivative

$$E_x(x) = -dV_f(x)/dx = -V_1/L \tag{16.33}$$

We can now calculate the charge induced on the conducting plates, from Gauss' law

$$\rho(c) = \epsilon_0 \nabla \cdot E(c) = \frac{\epsilon_0}{dr}\sum_f E(f) \tag{16.34}$$

where the sum is over the six faces f of the conducting boundary cell c. However, in our case, five of these are conducting faces and have $E = 0$. The only face with nonzero E is the one just outside the conducting surface and parallel to it (f in Figure 16.33). This corresponds to the component of the continuum field $\mathbf{E}_f$ normal to the surface, so we will denote it by E_n. We can remove the dr dependence from Eq. 16.34 by multiplying by dr; the left-hand side $\rho(c)\,dr$ then has units of charge per unit area. In terms of the area density $\sigma(c)$ defined by $\sigma(c)\,dr^2 = \rho(c)\,dr^3$, we

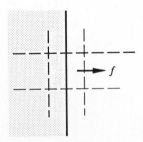

Figure 16.33 A cell c at a conducting surface; only one face f has nonzero $E(f)$.

get

$$\sigma(c) = \epsilon_0 E_n \qquad (16.35)$$

This is a particularly important formula, because it is independent of dr, and therefore true in the continuum limit $dr \to 0$. The continuum limit $\sigma_f(\mathbf{r})$ of the surface charge density is a rather peculiar function; it is a function of a continuous variable $\mathbf{r}$, but only makes sense when $\mathbf{r}$ is exactly on the conducting surface. Formally, we can define

$$\sigma_f(\mathbf{r}) = \lim_{dr \to 0} \sigma(c = \text{cell containing } \mathbf{r}) \qquad (16.36)$$

Then the continuum version of Eq. 16.35 is

$$\sigma_f(\mathbf{r}) = \epsilon_0 E_n(\mathbf{r}) \qquad (16.37)$$

where E_n is the component of the continuum vector field $\mathbf{E}_c$ along the **outward** normal to the surface (Figure 16.33).

In the pseudo-1D Example 16.8, Eq. 16.37 gives (using $E_n = E_x$ from Eq. 16.33)

$$\sigma_c(\mathbf{r}) = \epsilon_0(-V_1/L)$$

that is, a uniform negative charge density on the left-hand plate. Equations 16.35 and 16.37 give the correct charge density at the right-hand plate of Figure 16.31 as well, as long as we properly interpret the outward component E_n (the limit of $E(f)$ at a face f next to the right-hand plate) as $-E_x(x)$, because f is directed leftward. Thus,

$$\sigma_c(\mathbf{r}) = \epsilon_0(+V_1/L)$$

and the uniform induced charge on this plate is positive.

We have introduced the continuum limit using a flat surface in a pseudo-1D problem for the sake of simplicity. However, we can do it for a surface of any shape, by choosing a curvilinear coordinate system so that some coordinate is constant along the surface, and choosing cells centered at the surface. The arguments leading to Eqs. 16.35 to 16.37 work just as well for curved surfaces.

We can also say something about the components of the continuum electric field **tangent** (parallel) to the conducting surface. Such a component, which we will denote by E_t, is the limit of $E(f_t)$ at a sequence of faces f_t that are perpendicular to the conducting surface. However, the cells in front and in back of f_t are in the conductor and have the same potential (by our definition of a conductor). Thus, $E(f_t) = 0$ for **every** discretization, and the limit is certainly zero:

$$E_t = 0 \qquad \text{at a conducting surface} \qquad (16.38)$$

That is, the continuum field vector $\mathbf{E}_f$ is always perpendicular to a conducting surface. This fact is often useful in drawing field lines at peculiarly shaped conducting surfaces (see, for example, Figure 16.10).

■ **EXAMPLE 16.9 A CONTINUUM BOUNDARY VALUE PROBLEM**   Consider two concentric metal spheres, with vacuum between them, with radii $a < b$ as in Figure 16.34. The inner one is at potential $V_0 = 0$; the outer one is at potential V_1. (The wire grounding the inner one is routed through the fourth dimension via a space warp, in case you were wondering.) What is the potential everywhere, and how much charge is on each surface involved? There is a finite volume involved, so we are guaranteed a unique solution by the existence and uniqueness theorem. The nice thing about this theorem is that we don't have to be logical about how we look for a solution; as long as we check that it really satisfies Poisson's (in this case, really Laplace's) equation, it has to be the unique solution. We clearly should choose spherical coordinates because of the spherical symmetry, which also implies that $V(r, \theta, \phi)$ is really only a function of r (i.e., this is again a pseudo-1D problem). Laplace's equation for spherical coordinates (from the table of vector derivatives inside the back cover) becomes

$$r^{-2}\left(\frac{d}{dr}\right)\left[r^2\frac{dV(r)}{dr}\right] = 0 \qquad (16.39)$$

so $(d/dr)r^2(dV/dr) = 0$, that is, $r^2(dV/dr) = A$, a constant. Then, $dV(r)/dr = A/r^2$, so $V(r) = \int A\,dr/r^2 = -(A/r) + C$ where C is another constant. Our boundary conditions are

$$V(a) = 0 \qquad \text{(at the inner sphere)}$$

$$V(b) = V_1 \qquad \text{(at the outer sphere)}$$

and these two conditions determine the constants A, C:

$$V(a) = 0 = -(A/a) + C$$

(so $A = aC$)

$$V(b) = V_1 = -\frac{A}{b} + C$$

so

$$V_1 = \frac{-a}{b}C + C = C\frac{(b-a)}{b}$$

$$C = \frac{b}{(b-a)}V_1, \qquad A = \frac{ab}{(b-a)}V_1$$

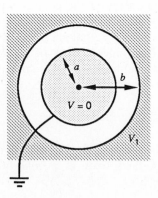

Figure 16.34   Concentric conducting spherical surfaces.

The solution is

$$V(r) = \frac{-V_1}{(b-a)}\left(\frac{ab}{r} - b\right)$$

You should check that the boundary conditions are satisfied. Now that we know V, we can determine the electric field by taking the gradient; in spherical coordinates

$$E_r = -dV(r)/dr = -\frac{V_1}{(b-a)}\frac{ab}{r^2}$$

At the inner sphere $(r = a)$ this is E_n (which conventionally means the component **outward** from the conductor), from which we can get the induced charge density:

$$\sigma_a = \epsilon_0 E_n = -\frac{\epsilon_0 V_1}{a}\frac{b}{(b-a)}$$

At the outer sphere, $E_n = -E_r$, so

$$\sigma_b = \frac{\epsilon_0 V_1}{b}\frac{a}{b-a}$$

Note that the **total** charge $4\pi a^2\sigma_a$ on the inner sphere is equal and opposite to the charge on the outer one.

▶   **Problem 16.17**

a. Suppose that the vacuum region between the plates of Figure 16.31 was filled with charge (fixed in position), with density ρ_0: $\rho_f(x) = \rho_0$ for $0 < x < L$. Compute (from the continuum Poisson equation) $V(x)$. Also give $E_x(x)$ and the induced charge σ on each plate.

b. Compute the Laplacian of the discrete potential field obtained by evaluating V at the cell centers of Figure 16.31. Does it exactly satisfy the discrete Poisson equation in this case?   ◄

▶   **Problem 16.18**

Consider the pseudo-2D problem of two concentric conducting cylinders of radii a and b $(a < b)$, at potentials $V(a) = 0$ and $V(b) = V_1$. (Use Figure 16.34 to visualize this, but imagine it to be a cross section of cylinders instead of spheres.) Compute $V(r)$ and $E(r)$ for $a \le r \le b$; compute as well the induced charge densities. What is the total charge per unit length (along the cylinder axis) on each conductor?   ◄

▶   **Problem 16.19**

Prove the continuum form of **Green's theorem**: for any two scalar functions V and W,

$$\int_R [V\nabla^2 W - W\nabla^2 V]\, d\tau = \int_S [V\nabla W - W\nabla V]\cdot d\mathbf{a}$$

where R is a continuum region and S is its boundary. [Hint: Apply the divergence theorem to $V\nabla W$ and $W\nabla V$.]   ◄

▶   **Problem 16.20**

Prove Green's theorem in the discrete form

$$\sum_{c\in R} \left[V(c)\nabla^2 W(c) - W(c)\nabla^2 V(c)\right] d\tau(c)$$
$$= \sum_{f\in S} \left[V(c_{\text{front}})(\text{grad }W)(f) - W(c_{\text{front}})(\text{grad }V)(f)\right] da(f)$$

where c_{front} is the cell in front of the face f on the boundary S of the region R (i.e., c_{front} is outside R). [Hint: Use the function $\text{sgn}(f, c)$, Eq. 15.41.] Is this theorem still true if you replace c_{front} by c_{back} (inside R)?   ◄

There is another interesting problem involving spherical conducting surfaces of an uncharged conducting shell. Let us assume there are some charges in the vacuum region **inside** the shell, but not outside. As we discussed in Section 16.3, any closed conducting container completely isolates any electrostatic charge inside it from the outside world, and vice versa. In this case, the potential outside the shell should be calculable **without** knowing the positions of the charges inside (which we assume **don't** have spherical symmetry). The problem of finding V outside does have spherical symmetry (V is constant on the outer surface of the conductor). It has a unique solution because of the existence and uniqueness theorem (Section 16.6) and we could calculate it from Laplace's equation. Alternatively, we can obtain it from Gauss' law for a sphere S of radius r (outside the conductor):

$$\int_S E \cdot da = q/\epsilon_0$$

$$(16.40)$$

$$E_r = q/4\pi r^2\epsilon_0$$

where q is the total charge on and inside the sphere. [You can then verify that the potential $V(r) = q/4\pi r\epsilon_0$ obtained by integrating E_r, which is the potential of a point charge, satisfies Laplace's equation (Eq. 16.39).] Our conclusion is that such a shell with an arbitrary (asymmetric) charge distribution inside looks from the outside **exactly** like a point charge (Figure 16.35).

If we think of this result in the context of Coulomb's law, whereby the field outside is the superposition of the (highly asymmetrical) fields due to the positive charge inside and the induced negative and positive charges on the conductor, it seems utterly incredible. One of our reasons for not trying to base the theory of electrostatics on Coulomb's law is that doing so would require thinking of the very simple field outside the shell as being the superposition of two very complicated fields, neither of which really exists. On the other hand, we have just seen that the result is quite clear in the context of the more fundamental Gauss' law or Poisson's equation. It is also clear from the even more fundamental point of view of creating the charge distribution by transient currents and solving Maxwell's equations for the

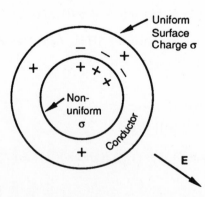

Figure 16.35 Radial electric field of a charge-containing spherical shell.

eventual static field. Suppose we created the inside charges at the inner surface of the shell and moved them toward the center leaving the "induced charges" behind, and created the positive induced charges on the outside of the shell by moving negative charges to infinity. The currents inside the shell would be asymmetrical, but those outside would be symmetrical. Then evolving the fields by Maxwell's equations would give electric and magnetic fields both inside and outside the sphere, but (if the charges are moved slowly so the magnetic field is small and wave propagation is not important) they would not penetrate very far into the conductor. Thus, the fields inside and outside would remain independent, and the one outside would remain symmetrical. The apparent dependence when we think in terms of Coulomb's law simply reflects the inappropriateness of Coulomb's law for this problem.

SECTION 16.8   CAPACITANCE

We have considered arrangements of conductors with specified potentials V_i ($i = 0, 1, \ldots$) such as that shown in Figure 16.36. We found that such a problem has a unique solution for the potentials $V(c)$ at cells in the nonconducting region, which determine the surface charge density. We can, of course, also calculate the total charge Q_i on (the surface of) the ith conductor:

$$Q_i = \sum_c \rho(c)\, d\tau(c) \tag{16.41}$$

where the sum is over the cells on the ith conductor. The charges Q_i therefore depend only on the assumed potentials V_i.

The special case of such a system in which there are **two** conductors is called a **capacitor**. You're already familiar with the parallel-plate capacitor. It consists of two conducting slabs whose plane surfaces are parallel to the yz-plane and separated by a distance L. We will assume that the capacitor has a lateral area (in the yz-plane) A, and that the lateral dimensions of the plates are much greater than L, so the lateral boundaries do not affect the field over most of the area A. The potential and

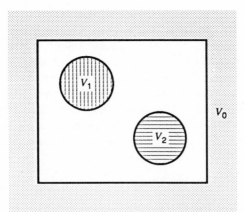

Figure 16.36 A general system of charged conductors.

charge density for this problem were calculated in Example 16.8, giving $\sigma = -\epsilon_0(V_1/L)$ for the $V = 0$ conductor and $\sigma = +\epsilon_0(V_1/L)$ for the $V = V_1$ conductor. The charge on the latter is therefore

$$Q_1 = \sigma A = \epsilon_0 V_1 A/L = CV_1 \qquad (16.42)$$

which is proportional to the potential with a proportionality constant

$$C = A\epsilon_0/L \qquad (16.43)$$

This proportionality is a very general property of capacitors (i.e., pairs of nearby conductors) independently of the shapes of the conductors. The proportionality constant $C = Q_1/V_1$ is called the **capacitance**; it is usually not expressible as simply as in Eq. 16.43, which holds only for parallel plates.

■ **EXAMPLE 16.10** Another simple example of a capacitor is the pair of concentric spheres of radii a and b for which we found V and σ in Example 16.9. We found a charge

$$Q_1 = 4\pi\epsilon_0 V_1/(a^{-1} - b^{-1})$$

on the inner sphere (and $Q_0 = -Q_1$ on the outer one). These two spheres can be regarded as a capacitor, with capacitance

$$C = Q_1/V_1 = 4\pi\epsilon_0/(a^{-1} - b^{-1}) \qquad (16.44)$$

In terms of the distance $d = b - a$ between them,

$$C = 4\pi\epsilon_0 a(a + d)/d$$

which approaches $4\pi\epsilon_0 a^2/d$ for $a \gg d$; this agrees with the parallel-plate formula with $A = 4\pi a^2$, as it should in this limit.

Note that if the outer conductor is much larger than the inner one ($b \gg a$), C becomes $4\pi\epsilon_0 a$, independently of the radius of the outer conductor. In fact, in this limit it is also independent of the **shape** of the outer conductor "at infinity," and may be regarded as a property of the inner conductor alone. In the limit of a "point conductor" ($a \to 0$) the capacitance vanishes because a point conductor cannot hold a charge. This fact is related to the divergent energy of a point charge.

■ **EXAMPLE 16.11** A similar case is that of an infinitely long coaxial cable, having two coaxial conducting cylindrical surfaces of radii a and b. (You can use Figure 16.34 to visualize this, if you think of it as a cross section of a cylinder instead of a sphere.) To discuss the charges on the cylinders, consider a section of length L (measured along the cylinder axis, perpendicularly to the paper), within which the inner conductor has charge Q_1 and the outer one Q_0. These must be equal and opposite ($Q_0 = -Q_1$) because the total charge must be zero within a cylindrical Gaussian surface embedded in the outer conductor. We can calculate the field between the two conductors by applying Gauss' law to a cylindrical Gaussian surface of radius r ($a < r < b$), on which the radial component of the field $E(r)$ is constant:

$$E(r)2\pi rL = Q_1/\epsilon_0$$

$$E(r) = Q_1/2\pi\epsilon_0 rL$$

The potential difference is obtained by integration:

$$V_0 - V_1 = -\int_a^b E(r)\,dr$$

$$= -(Q_1/2\pi\epsilon_0 L)\int_a^b r^{-1}\,dr = -(Q_1/2\pi\epsilon_0 L)\ln(b/a)$$

or (if $V_0 = 0$)

$$Q_1 = \frac{2\pi\epsilon_0 L V_1}{\ln(b/a)} = CV_1$$

where

$$C = \frac{2\pi\epsilon_0 L}{\ln(b/a)} \tag{16.45}$$

is the capacitance of a length L of cable.

The limiting case in which a is very small (compared to b), that is, a thin wire, is of great practical importance. We can use it as a model for a wire of radius a in a circuit whose other conductors are at distances of the order of b. The above result implies that the capacitance per unit length between the wire and its environment is small and approaches zero as $a \to 0$. This is the justification for the assumption that such connecting wires have no charge, which underlies most electrical circuit analysis.

▶ **Problem 16.21**

Calculate the capacitance of a wire of radius $a = 0.5$ mm and length 0.1 m, inside a cylinder of radius $b = 1$ cm. Express the result in picofarads (1 pf = 10^{-12} farad, 1 f = 1 coulomb/volt). ◀

SECTION 16.9   THE CAPACITANCE MATRIX

Our treatment above of the parallel-plate capacitor is hardly adequate for a real capacitor, because we ignored the effects of the edges of the plates. To understand a real capacitor, we must use finite plates (as, for example, in Figure 16.38) and enclose them in a very large box (whose radius we will later allow to become infinite). We now have a problem involving **three** conductors; it is not even clear what we mean by the "capacitance" of this arrangement.

Let us look at the problem of capacitance in the most general case of $n + 1$ conductors at potentials $V_0, V_1, \ldots, V_n$. (Figure 16.37 shows the case $n = 2$.) If we want to compute the total charges $Q_0, \ldots, Q_n$ on the conductors, we must first solve Laplace's equation $\nabla^2 V = 0$ for the potential everywhere (we are assuming the nonconducting region has no charge in it). Then the surface charge on the conductors is obtained by writing Poisson's equation $\nabla^2 V = -\rho/\epsilon_0$ for cells on the boundaries of the conductors. Although this may be difficult to do in a particular case, we know at least that all equations are **linear**. We have a linear matrix equation for the potentials in the nonconducting cells, which we wrote in Section 16.6 as

$$(M)(V) = (r) \tag{16.46}$$

where the column vector r involves the boundary values $V_0, \ldots, V_n$ (linearly). This can be solved for the column vector (V) in principle by inverting (M); the result for (V) depends **linearly** on (r) and hence on the boundary potentials. Plugging these linear expressions into $\rho = -\epsilon_0 \nabla^2 V$ at the boundary gives us ρ, and by summation Q_i $(i = 0, 1, \ldots, n)$, as linear functions of the boundary potentials. Let us denote the coefficients by C_{ij}:

$$Q_i = \sum_{j=0}^{n} C_{ij} V_j \tag{16.47}$$

The $(n + 1) \times (n + 1)$ matrix (C) is called the **capacitance matrix** of the system. It is easy to write for the two-conductor cases we have considered $(n = 1)$. For the infinite parallel-plate capacitor we obtained $Q_1 = CV_1$ on the plate, with $C = \epsilon_0 A/d$, assuming $V_0 = 0$. If we changed the reference potential so $V_0 \neq 0$, Q_1 would depend only on the difference:

$$Q_1 = C(V_1 - V_0) \tag{16.48}$$

The charge on the other plate is of course opposite:

$$Q_0 = C(V_0 - V_1) \tag{16.49}$$

These can be written in matrix form

$$\begin{pmatrix} Q_0 \\ Q_1 \end{pmatrix} = \begin{pmatrix} C & -C \\ -C & C \end{pmatrix} \begin{pmatrix} V_0 \\ V_1 \end{pmatrix} \tag{16.50}$$

The matrix will not always be so simple, but one thing we can learn from this matrix that is true in general is that the rows all add up to zero, as do the columns. The row constraint follows from the fact that only potential **differences** matter: If I change **all** the potentials from 0 to some V, the charges must remain zero. Thus,

$$Q_i = \sum_j C_{ij} V = 0$$

or

$$\sum_j C_{ij} = 0 \qquad \text{(row sum constraint)} \tag{16.51}$$

To see why the column sum is constrained, consider a Gaussian surface within the outer conductor V_0, enclosing the entire nonconducting volume (such as that shown in Figure 16.37). According to Gauss' law, the charge inside this surface must vanish:

$$0 = \sum_i Q_i = \sum_i \sum_j C_{ij} V_j = \sum_j V_j \left(\sum_i C_{ij} \right)$$

For this to vanish for **any** choice of V_j, the coefficient of each V_j must vanish:

$$\sum_i C_{ij} = 0 \qquad \text{(column sum constraint)} \tag{16.52}$$

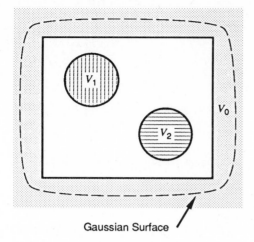

Gaussian Surface

Figure 16.37   A Gaussian surface within the outer conductor.

For any two-conductor capacitor ($n = 1$: infinite parallel plates, concentric spheres, and so on) these constraints impose the form we saw above. If we set $C \equiv C_{11}$, the constraints imply $C_{00} = C$, $C_{01} = C_{10} = -C$. The entire matrix is determined by the lower right entry C_{11}. In a general capacitance matrix

$$
\begin{pmatrix}
C_{00} & C_{01} & C_{02} & \cdots & C_{0n} \\
C_{10} & \begin{bmatrix} C_{11} & C_{12} & \cdots & C_{1n} \\
C_{20} & C_{21} & & \cdots & C_{2n} \\
C_{30} & & & & \\
\vdots & \vdots & & & \vdots \\
C_{n0} & C_{n1} & & \cdots & C_{nn} \end{bmatrix}
\end{pmatrix}
\tag{16.53}
$$

one need only specify the lower right $n \times n$ matrix involving conductors other than the 0th; the top row and left column are then determined by the constraints. One can think of the lower right $n \times n$ matrix as determining Q_i ($i = 1, \ldots, n$) from V_j via Eq. 16.47, if we always let $V_0 = 0$.

The capacitance matrix tells us how to determine the charges from the potentials. Suppose, however, that we wanted to do the reverse, that is, to determine the potentials from the charges. We cannot approach this simply by inverting the $(n + 1) \times (n + 1)$ capacitance matrix (C) appearing in Eq. 16.47, because a matrix whose rows are constrained to be linearly dependent is not invertible. This can also be seen from a physical point of view. The reference level of the potential is physically irrelevant, so we can never determine it from the charges. However, it turns out that **except** for this detail, the charges **do** uniquely determine the potentials. Mathematically, this means that the $n \times n$ lower right matrix **is** invertible. Thus, if we insist that $V_0 = 0$ and $\Sigma Q_i = 0$, then the Q_i's uniquely determine the V_i's (as well as vice versa). This can be proved by arguments similar to those involved in the uniqueness theorem, but we will not go through them here.

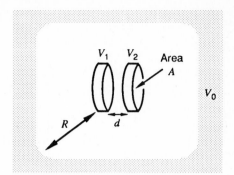

Figure 16.38   A model of a real capacitor.

Another interesting fact that can be similarly proved is the **reciprocity theorem**, which states that the capacitance matrix is symmetric:

$$C_{ij} = C_{ji} \tag{16.54}$$

This can also be proved by Green's theorem (Problem 16.22) or in a rather indirect way by arguments involving the electrostatic energy of the system of conductors (see p. 121 of the book by Reitz, Milford, and Christy cited in the bibliography).

■ **EXAMPLE 16.12 A MODEL OF A REAL CAPACITOR**   Consider a finite parallel-plate capacitor inside a large conducting box (approximately a distance R away, as in Figure 16.38). We want to approximately compute the capacitance matrix.

If $d \ll R$, most of the charge on each plate will be on the **inner** surface (facing the other plate). A little will be on the outside, though, and will induce charge on the surface of the large box across from it. This qualitative behavior is retained if we stretch the plates out to infinite area, while stretching the box out to stay a distance R away. The box has then become two parallel plates itself. This pseudo-1D problem (Figure 16.39) can now be solved exactly. The field between the left edge of the box and the left side of plate number

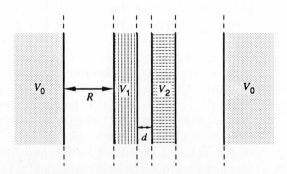

Figure 16.39   A pseudo-1D idealization of Figure 16.38.

1 is

$$E_x = -(V_1 - V_0)/R$$

just as for an isolated pair of parallel plates (this is the "leakage flux" per unit area, i.e., the flux corresponding to field lines that escape to the outer box rather than passing between the two capacitor plates). It determines the surface charge density $\sigma_{0,\text{left}} = -\epsilon_0(V_1 - V_0)/R$ on the left box surface, and also the surface density $\sigma_{1,\text{left}} = \epsilon_0 E_n = +\epsilon_0(V_1 - V_0)/R$ on the left side of plate 1 (note $E_n = -E_x$ there). The (much larger) field between the capacitor plates is $E_x = -(V_2 - V_1)/d$, which determines charges $\sigma_{2,\text{left}} = +\epsilon_0(V_2 - V_1)/d$, $\sigma_{1,\text{right}} = -\epsilon_0(V_2 - V_1)/d$. The charges $\sigma_{0,\text{right}} = \epsilon_0(V_0 - V_2)/R$, $\sigma_{2,\text{right}} = -\epsilon_0(V_0 - V_2)/R$ are determined similarly. If we now imagine the capacitor plates to have a large but finite area $A \gg R^2$, then even on the outer-box plates, which are still infinite, the induced charge will be confined approximately to the area A. So we can get the total charge by multiplying by A:

$$Q_0 = A(\sigma_{0,\text{left}} + \sigma_{0,\text{right}}) = \epsilon_0 A(2R^{-1}V_0 - R^{-1}V_1 - R^{-1}V_2)$$

$$Q_1 = \epsilon_0 A\left[-R^{-1}V_0 + (R^{-1} + d^{-1})V_1 - d^{-1}V_2\right]$$

$$Q_2 = \epsilon_0 A\left[-R^{-1}V_0 - d^{-1}V_1 + (R^{-1} + d^{-1})V_2\right]$$

so the capacitance matrix is

$$\begin{bmatrix} C_{00} & C_{01} & C_{02} \\ C_{10} & C_{11} & C_{12} \\ C_{20} & C_{21} & C_{22} \end{bmatrix} = \epsilon_0 A \begin{bmatrix} 2R^{-1} & -R^{-1} & -R^{-1} \\ -R^{-1} & R^{-1} + d^{-1} & -d^{-1} \\ -R^{-1} & -d^{-1} & R^{-1} + d^{-1} \end{bmatrix}$$

You can check that this satisfies the row and column sum constraints.

Although the above was an idealized example, every real capacitor in a real enclosure is characterized by a 3×3 capacitance matrix similar to this one. There is always some leakage flux, leading to leakage capacitances in the 0th row and column such as the R^{-1} terms above. When the capacitor is very far from other conductors (the $R \to \infty$ limit), the leakage capacitances C_{0i} and C_{i0} vanish, and the sum constraints force the matrix into the form

$$\begin{bmatrix} 0 & 0 & 0 \\ 0 & C & -C \\ 0 & -C & C \end{bmatrix}$$

which is determined by a single number C ($\epsilon_0 A/d$ in our example). It is only in this sense that a real capacitor can be characterized by a single capacitance C.

The only way to make the leakage flux vanish when there are other conductors within a finite distance is to enclose one conductor of the capacitor completely in the other, for example using concentric spheres (Figure 16.40). A real capacitor is often enclosed in a can that acts as one terminal, for precisely this reason; of course, it is necessary in practice to leave a hole in the can for the other lead wire.

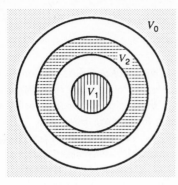

Figure 16.40   Shielded capacitor constructed of concentric spheres.

▶   **Problem 16.22**

Use Green's theorem (Problems 16.19 and 16.20) for a region R consisting of all space, to prove the reciprocity theorem. [Hint: Let V be the potential due to a charge on conductor i only and W be the potential due to a charge on j.]   ◀

▶   **Problem 16.23 A Continuation of Problem 16.11, a Discrete Model of a Real Capacitor**

a. Calculate the potential in the discrete system of Figure 16.26 for the case that the plates have the **same** potential $+ \frac{1}{2} V_0$, and the outer box has zero potential. [How is the symmetry different now? How does that affect the relaxation equation at cell 1?]

b. Calculate the charges Q_1 and Q_2 on the plates and Q_0 on the outer box (per length dr perpendicular to the paper) for this case.

c. Write the capacitance matrix equation $Q_i = \Sigma C_{ij} V_j$ for the case considered in parts **a** and **b**, and also (using the results of Problem 16.11) for the case of opposite plate potentials $\pm \frac{1}{2} V_0$. Solve these six equations for C_{01}, C_{02}, and C_{12} (which determine the capacitance matrix per dr). How different is C_{12} from the value $\epsilon_0 A / d = 3\epsilon_0 dr$ given by the infinite-plane formula? How different is C_{01} from the rough estimate $5\epsilon_0 dr/2$ suggested by the fact that a length $5\, dr$ of the perimeter of the plate is across from the box, at a distance $2\, dr$?   ◀

SECTION 16.10   EQUIVALENT LUMPED CAPACITANCE

Consider a system of conductors obtained by connecting ideal two-conductor capacitors (with no leakage flux, such as the concentric-sphere type) together by thin wires (which can have no charge, as we have seen). Such a system is depicted schematically in Figure 16.41. What is the capacitance matrix of this system? The terminal labeled 0 is connected to two plates in capacitors C' and C'', which together comprise a single conductor we will also label by 0. The total charge Q_0 on this conductor is the sum of charges on the plates of capacitors C' and C''

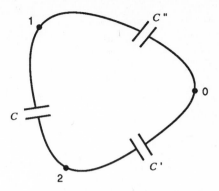

Figure 16.41   A system of three capacitors connected by wires.

connected to terminal 0:

$$Q_0 = C'(V_0 - V_2) + C''(V_0 - V_1) = (C' + C'')V_0 - C''V_1 - C'V_2$$

and similarly for the others

$$Q_1 = -C''V_0 + (C + C'')V_1 - CV_2$$

$$Q_2 = -C'V_0 - CV_1 + (C + C')V_2$$

We can then read off the capacitance matrix

$$\begin{bmatrix} C_{00} & C_{01} & C_{02} \\ C_{10} & C_{11} & C_{12} \\ C_{20} & C_{21} & C_{22} \end{bmatrix} = \begin{bmatrix} C' + C'' & -C'' & -C' \\ -C'' & C + C'' & -C \\ -C' & -C & C + C' \end{bmatrix} \quad (16.55)$$

Note that $-C_{01} = C''$ is the capacitance connecting terminals 0 and 1. It is not hard to see that this is true for any such network: if $i \neq j$,

$$-C_{ij} = \text{capacitance of capacitor connecting terminals } i \text{ and } j \quad (16.56)$$

The diagonal elements C_{ii} are then determined by the constraints (Eqs. 16.51 and 16.52) on row or column sums.

We have just seen how to trivially write down the capacitance matrix for any network of ideal capacitors. The relationship is actually more useful in the opposite direction: given **any** arrangement of conductors, we can find a network of ideal capacitors that is **equivalent** in the sense of having the same capacitance matrix (i.e., the same static electrical properties, or the same relationship between charge and voltage). This can be done by solving Poisson's equation for the given arrangement of conductors, computing the capacitance matrix, and then setting up a network having ideal capacitors $-C_{ij}$ in it (the C_{ij} will all be negative). An example with

(a)

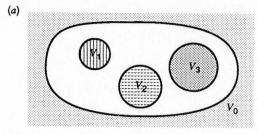

(b)

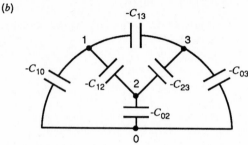

Figure 16.42 A system of three conductors (a) and its equivalent network of lumped capacitances (b).

(a)   (b)

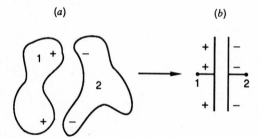

Figure 16.43 Two conductors interacting at many points (a) and the equivalent lumped capacitor (b).

$n = 3$ is shown in Figure 16.42. The capacitance $-C_{ij}$ is referred to as a "lumped" capacitance because all of the interactions between conductor i and conductor j, which might actually occur at many points spread over a large area, are combined into one "lump," that is, a single capacitor (Figure 16.43). The concept of lumped capacitance is very useful in electrical engineering applications; for example, the conductors in Figure 16.43 might be lead wires, transistor electrodes on an integrated-circuit chip, or the like, in an electronic device. It is very convenient to be able to treat these as lumped capacitances, because it allows one to use conventional circuit-theory methods (Kirchhoff's laws) instead of worrying explicitly about the electric fields in the regions between conductors.

CHAPTER SEVENTEEN

SPECIAL METHODS FOR
CALCULATING POTENTIALS

SECTION 17.1   THE IMAGE METHOD

The image method is a technique that can be used to calculate the electrostatic potential in a system of fixed charges with conducting surfaces of certain shapes, which include planes, spheres, and cylinders. The simplest example involves a conductor with an infinite planar surface (as shown in Figure 17.1; we can imagine the conducting material to fill the half-space below the plane). The simplest charge distribution is a point charge q, which we may place at a distance a above the surface. We want to find the potential everywhere above the plane; below the plane (i.e., within the conductor) it is a constant, to which we may as well assign it a value of zero.

   We have done conducting-boundary problems such as this (albeit discrete) by solving the discrete Poisson equation in Section 16.3; we expect V to be largest in the cells near q, and a total charge $-q$ to be induced on the conducting surface. Instead of trying to solve the continuum Poisson equation, however, we will use the image method.

   The image method is based on the uniqueness theorem for boundary-value problems: if we can find a potential V that satisfies Poisson's equation in the nonconducting region ($y > 0$) and the boundary condition ($V = 0$ at $y = 0$) then it **must** be the correct solution, no matter how we find it. We will find it by solving a completely different problem, shown in Figure 17.2. Here there is **no** conductor (the $y = 0$ plane is dotted in Figure 17.2 simply to remind you where the conductor used to be) but an extra charge $-q$ at $(0, -a, 0)$. It is called an **image charge** because it is where the image of $+q$ would be if there were a mirror at $y = 0$. This new problem is easy to solve because the potential of two point charges in empty space is given by Coulomb's law (Eq. 15.6). The potential at an arbitrary point $r = (x, y, z)$ is

$$V(\mathbf{r}) = [q/4\pi\epsilon_0][s_+^{-1} - s_-^{-1}] \tag{17.1}$$

where s_+ and s_- are the distances from $\mathbf{r}$ to the two charges, as shown in Figure 17.2. The position of $+q$ is $(0, a, 0)$, so

$$s_+ = x^2 + (y - a)^2 + z^2 \tag{17.2}$$

and similarly

$$s_- = x^2 + (y + a)^2 + z^2 \tag{17.3}$$

325

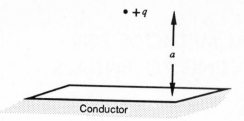

Figure 17.1 A point charge above a conducting plane.

We are particularly interested in the value of the potential at $y = 0$, where $s_+ = s_-$; here the contributions of $+q$ and $-q$ exactly cancel, yielding $V = 0$. However, this is exactly the boundary condition we wanted; we have found a potential that satisfies Poisson's equation for $y > 0$ in the presence of the charge $+q$, in which $V = 0$ for $y = 0$. (To be precise, we should define a new potential field that is equal to Eq. 17.1 for $y > 0$ and vanishes for $y < 0$.) It is therefore the only such potential, and must be the solution to the original conducting-boundary problem (Figure 17.1).

The potential defined by Eqs. 17.1 through 17.3 is called a (physical) dipole potential (as distinguished from the **point** dipole potential to be discussed in Section 17.3). The corresponding electric field lines are shown qualitatively in Figure 17.3. We can compute the electric field quantitatively by taking the gradient of V; for the y-component, for example,

$$E_y = -\frac{dV}{dy} = \frac{q}{4\pi\epsilon_0}\left[\frac{-ds_+^{-1}}{dy} + \frac{ds_-^{-1}}{dy}\right] \tag{17.4}$$

The easiest way to compute the derivative is by differentiating Eq. 17.2 before solving for s_+:

$$2s_+\,ds_+ = 2(y - a)\,dy \tag{17.5}$$

so

$$ds_+^{-1}/dy = -s_+^{-2}\,ds_+/dy = -s_+^{-3}(y - a) \tag{17.6}$$

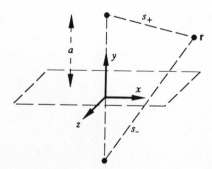

Figure 17.2 Two point charges (there are no conducting surfaces present).

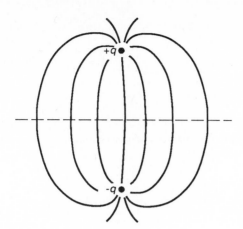

Figure 17.3 A sketch of the field lines of a dipole field.

and using a similar formula for ds_- we get

$$E_y = (q/4\pi\epsilon_0)\left[s_+^{-3}(y - a) - s_-^{-3}(y + a)\right] \qquad (17.7)$$

▶ **Problem 17.1**

Derive expressions like Eq. 17.7 for E_x and E_z. ◀

We can obtain the continuum density of induced charge at the conducting surface from **E** by using Eq. 16.37:

$$\sigma(\mathbf{r}) = \epsilon_0 E_n(\mathbf{r})$$

where E_n is the component normal to the conducting surface, in this case E_y. Since $y = 0$ at the surface, $s_+^2 = s_-^2 = x^2 + z^2 + a^2$, and Eq. (17.7) simplifies to give

$$\sigma(x, 0, z) = -(2aq/4\pi)(x^2 + z^2 + a^2)^{-3/2} \qquad (17.8)$$

The total induced charge can be computed as

$$Q = \int \sigma \, da = \int_{-\infty}^{\infty} dx \int_{-\infty}^{\infty} dz \, \sigma(x, 0, z) \qquad (17.9)$$

▶ **Problem 17.2**

Do the integral in Eq. 17.9 and show that the result is what we guessed earlier ($Q = -q$). Hint: It's easier in polar coordinates. ◀

Another quantity that is easy to calculate by the image method is the force on the charge. From the point of view of Coulomb's law, this force is exerted by the induced charge (Eq. 17.8) and could be calculated by integrating over the area element $dx \, dz$ the force due to its charge $\sigma(x, 0, z) \, dx \, dz$. However, it is much easier to observe that the fields in the vicinity of $+q$ are the same as in the image problem, so the force (being proportional to the field) must also be the same. If you make the mistake of thinking too hard about this, you will realize that in the continuum limit

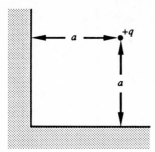

Figure 17.4 Two perpendicular conducting planes equidistant from a point charge.

the field near the charge diverges: for this procedure to make sense, we must **subtract** the Coulomb field of the charge $+q$ (which of course exerts no force on itself) from the total field. The remainder is well-behaved in the continuum limit. In fact, it is just the field of the point charge $-q$, and exerts a force

$$F_y = +qE_y = +q(-q/4\pi\epsilon_0)(2a)^{-2} = -q^2/16\pi\epsilon_0 a^2 \qquad (17.10)$$

downward on the charge $+q$.

▶ **Problem 17.3**

Compute the force (magnitude and direction) on a point charge q at a distance a from each of two grounded conducting planes that meet at a 90° angle (Figure 17.4). Hint: By judicious placement of three image charges, you can make $V = 0$ on both planes. ◀

▶ **Problem 17.4**

Suppose the planes in the preceding problem met at an angle θ (not necessarily 90°). For which values of θ can you still solve the problem by the image method? ◀

We did this calculation in the continuum limit, so we have had to gloss over some subtleties. In particular, the function (Eq. 17.1, cut off at $y = 0$) that we used to describe the continuum potential field is not differentiable at the position of $+q$ or at $y = 0$, and, therefore, doesn't satisfy the differential-equation form of Poisson's equation there. A more precise statement of the continuum existence and uniqueness theorem states that V must be twice-differentiable in the interior and continuous at the boundary, which fixes the $y = 0$ problem. The point charge is harder to deal with, however. The easiest way to do this consistently is simply to regard the continuum problem as a limit of **discrete** problems, each of which can be done without ambiguity. The resulting discrete potentials have a limit almost everywhere, which turns out to be Eq. 17.1. If we choose dr to be a submultiple of a, the image argument works for each discretization. The problem is then to solve the discrete Poisson equation for a discrete charge distribution having $\rho(c) = 0$ for all cells except the one centered at $(0, a, 0)$, for which $\rho(c) = q/dr^3$.

To solve this problem numerically, we need, of course, to restrict ourselves to a finite number of cells. We can do this by enclosing the system in a large box centered at the origin, and letting the box size approach infinity. This box must intersect the conducting plane, so its potential must be zero for consistency. Just as in the continuum case, we can solve this problem by comparing it to a corresponding problem without the conducting plane but with an image charge $-q$ in the cell at $(0, -a, 0)$. The mirror symmetry of this problem implies that $V = 0$ for the cells at $y = 0$. To show this rigorously, note that the reflected potential V' defined by $V'(x, -y, z) \equiv -V(x, y, z)$ satisfies Poisson's equation if V does (plug it in and see!) so by uniqueness, $V = V'$. At $y = 0$, this says $V = -V$, that is, $V = 0$. Thus, V is the solution to the original problem. When we take the continuum limit of this solution (letting the box size $\to \infty$ at the same time) we must get Coulomb's law, which gives us again Eq. 17.1.

It is clear from the way we solved the conducting-plane example that the image method can be used to solve problems involving conductors of any shape that can be produced as the equipotential of a simple charge distribution whose field is exactly known; in practice, this includes just point charges and line charges. Let us determine the equipotentials for a line dipole (Figure 17.5). The potential of a line charge is obtained by integrating the electric field (Example 16.11):

$$V = -(\lambda/2\pi\epsilon_0) \ln s + \text{constant} \tag{17.11}$$

where s is the distance from the line. Superposing two such terms for this case gives

$$V(x, y, z) = -(\lambda/2\pi\epsilon_0)\left\{\ln\left[x^2 + (y - a)^2\right]^{1/2} - \ln\left[x^2 + (y + a)^2\right]^{1/2}\right\} \tag{17.12}$$

(We have set the reference level of the potential so $V = 0$ at the origin.) Note that V is independent of z because the problem is pseudo-2D. Using the notation

$$M = \exp(2\pi\epsilon_0 V/\lambda) \tag{17.13}$$

this can be rewritten

$$\left[x^2 + (y + a)^2\right]/\left[x^2 + (y - a)^2\right] = M^2$$

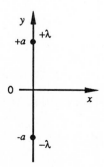

Figure 17.5 A line dipole: a pair of line charges $\pm\lambda$ (pointing out of the paper) at positions $y = \pm a$.

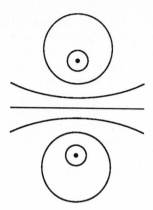

Figure 17.6   Equipotential surfaces for a line dipole.

or

$$x^2 + (y - y_0)^2 = 4a^2M^2/(M^2 - 1)^2 \qquad (17.14)$$

with

$$y_0 = a(M^2 + 1)/(M^2 - 1) \qquad (17.15)$$

This is, of course, the equation of a circle of radius

$$R = 2a|M/(M^2 - 1)| \qquad (17.16)$$

centered at y_0 (in 3D, a cylinder with its axis in the z-direction). For each choice of V, the set of points having that V is a cylinder; R and y_0 depend on V. These equipotentials are sketched in Figure 17.6. Note that for $V \to -\infty$ ($M \ll 1$), the circle is centered at $y_0 \cong -a$ as one might expect.

Suppose now that we wanted to know the potential in a system consisting of a cylindrical conductor of radius R and a line charge λ at a distance $b > R$ from its center (Figure 17.7a). We need to find an image charge inside the cylinder that gives an equipotential at this cylinder. That means we must find a (half the distance from the image to $+\lambda$) and M, satisfying Eq. 17.16 and

$$a - y_0 = b \qquad (17.17)$$

(from the geometry of Figure 17.7b), with y_0 (which is negative) given by Eq. 17.15. A little algebra gives the solution

$$a = (b^2 - R^2)/2b \qquad (17.18)$$

$$M = R/b \qquad (17.19)$$

Now we can calculate the potential on the cylinder by substituting this M into Eq. 17.13:

$$V = -\frac{\lambda}{2\pi\epsilon_0} \ln(b/R) \qquad (17.20)$$

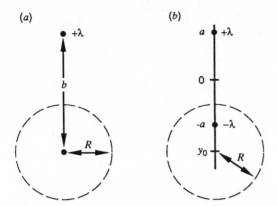

Figure 17.7   (*a*) A line charge near a cylindrical conductor. (*b*) The position inside the cylinder of the image charge.

We can determine the total charge per unit length on the cylinder by using a Gaussian surface just outside it: it has the same field there (and therefore the same charge $\epsilon_0 \int E\, da$) as the image charge $-\lambda$.

> Ordinarily in problems like this, we are allowed to choose the potential at a conducting surface arbitrarily. Why is the potential in this case imposed on us? Basically, because we have a system with an infinite amount of charge, and are insisting that the potential at infinity (in the plane of the paper) be finite (zero, actually). It is a peculiarity of line charges that if the net linear charge density of a group of nearby parallel line charges doesn't vanish, the potential at infinity diverges (because ln *r* does). So we have implicitly required the induced charge to be $-\lambda$, and this in turn imposes a value on the potential of the conducting cylinder.

The line-dipole image method we have used for this example can be used for a variety of problems involving line charges and/or cylindrical or planar conductors (note that the $V = 0$ equipotential is a plane). The other type of equipotential surface we can easily produce is a sphere. With a single point charge, all the equipotentials are, of course, spheres. But even with two point charges it turns out (see Problem 17.7) that there is a spherical equipotential, if the charges have different magnitudes. In this case, only **one** of the equipotentials is a sphere, namely the one with $V = 0$. (This contrasts with the line-dipole case, in which **all** equipotentials were cylinders.) One can produce other values of V, however, by adding a third point charge at the center of the sphere: this leaves the sphere an equipotential but changes the value of V. This allows us to solve any problem involving a point charge near a sphere at a given potential. The spherical image method is described in detail in Reitz, Milford and Christy, p. 65, and briefly in Griffiths, p. 110.

▶   **Problem 17.5**

A line charge ($\lambda = 6.28\ \mu c/m$) is at a distance $b = 2$ m from the axis of a conducting cylinder of radius $R = 1$ m. Compute numerically the distance of the image charge from the cylinder axis, and the potential of the cylinder.   ◀

▶ **Problem 17.6**

An infinite conducting cylinder of radius R with a given charge per unit length λ is centered at a distance $d > R$ from a grounded ($V = 0$) infinite plane, with its axis parallel to the plane.

a. It is possible to locate two line charges $\pm\lambda$ so their potential satisfies all the boundary conditions. Give their positions relative to the plane.

b. Find the capacitance per unit length of the cylinder-plane system. Evaluate it numerically for $R = 2 \times 10^{-3}$ m, $d = 20$ m (reasonable values for a transmission wire with a ground return).   ◀

▶ **Problem 17.7**

a. Consider a point charge Q at the origin, and another charge $-q$ at $(a, 0, 0)$ (Q and q are positive). Find the equation for a general surface of constant V, and show that for $V = 0$ the surface is a sphere. Give its radius and the coordinates of its center.

b. Compute the force on a charge q at a distance d from the center of a grounded sphere of radius R.   ◀

SECTION 17.2 SEPARATION OF VARIABLES

Consider the pseudo-2D electrostatic boundary-value problem shown in Figure 17.8: an infinitely long, infinitely deep slot of width d. The sides of the slot are

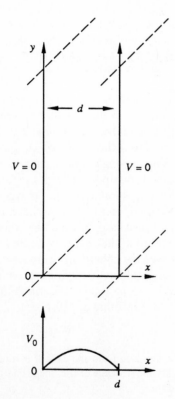

Figure 17.8 A rectangular slot in a conductor.

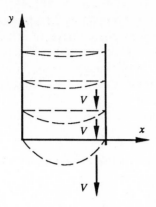

Figure 17.9 Plots of V as a function of x, for various choices of height y.

formed by grounded conductors, and the bottom (at $y = 0$) is somehow constrained to have a given potential $V_0(x)$, independently of z. We want to determine the continuum potential $V(x, y)$ everywhere within the slot. It is intuitively clear that the potential should approach zero for large y, because there we are far from the nonzero boundary condition but near the grounded side. It seems reasonable that the x-dependence might stay roughly the same as y increases, with just an overall amplitude factor decreasing, as depicted in Figure 17.9. Denoting the amplitude factor by $Y(y)$ and the x-dependent function (in this case V_0) in general by $X(x)$, the potential at any point would be given by

$$V(x, y) = X(x)Y(y) \tag{17.21}$$

Let us try out this potential, to see if we can make it satisfy Laplace's equation and the boundary conditions. Substituting it into Laplace's equation (Eq. 16.23) gives

$$0 = \partial^2 V/\partial x^2 + \partial^2 V/\partial y^2 = X(x)\,\partial^2 Y/\partial y^2 + Y(y)\,\partial^2 X/\partial x^2$$

$$0 = Y(y)^{-1}\,\partial^2 Y/\partial y^2 + X(x)^{-1}\,\partial^2 X/\partial x^2 \tag{17.22}$$

where we use ∂ instead of d to denote differentiation to emphasize that these are partial derivatives: V is a function of both x and y. However, the first term of this expression is a function of y alone. And if it varied with y, the second term would have to vary with y also (because they must add to zero), which it cannot because it is just a function of x! Thus the first term cannot vary at all, but must be a constant, which we will denote by C. Multiplying it by $Y(y)$ gives

$$d^2Y/dy^2 = CY(y) \tag{17.23}$$

in which we have replaced ∂ by d because it is now an **ordinary** (as opposed to partial) differential equation. You should recall that its solutions are exponentials

$$Y(y) = B\exp(\pm ky) \tag{17.24}$$

where B and k are constants. We can find the value of the constant k by substituting Eq. 17.24 into Eq. 17.23. The derivatives bring out two factors of k, and after dividing by $Y(y)$ (which we suppose to be nonzero) we have $k^2 = C$. If C is positive, k is real and $Y(y)$ is a growing or dying exponential, whereas if C is negative, k is imaginary and $Y(y)$ is oscillatory. In our case we want $Y(y)$ to be a dying exponential, so that it will approach zero properly as $y \rightarrow \infty$. Thus we choose $C > 0$. Defining k to be the positive square root of C

$$k \equiv C^{1/2} \tag{17.25}$$

Equation 17.24 is the solution to our differential equation. The equation for X is now

$$d^2X/dx^2 = -k^2X(x) \tag{17.26}$$

whose solutions are found by the same method to be $\exp(\pm ikx)$, where $i^2 = -1$. In our physical problem X must, of course, be real. Fortunately, it is easy to show that both the real and imaginary parts of the complex solutions we have found are also solutions; these are $\cos kx$ and $\sin kx$, which we may use as our two independent solutions. Our boundary condition demands $X(x) = 0$ at $x = 0$ and $x = d$. The former can be satisfied only by the sine (it forces the coefficient of the cosine to be zero) and the latter $[X(d) = 0]$ is satisfied by the sine only if $\sin kd = 0$. This requires kd to be an integer multiple of π, so

$$k = n\pi/d \qquad (n = \text{integer}) \tag{17.27}$$

The overall solution is then

$$V(x, y) = Be^{-ky} \sin kx \tag{17.28}$$

which satisfies the boundary conditions at $y = \infty$, $x = 0$, and $x = d$. The remaining boundary condition is at $y = 0$, and it requires

$$B \sin kx = V_0(x) \tag{17.29}$$

Thus, this method only works for certain special boundary conditions, for which V_0 is a sine function. One of these is shown in Figure 17.8, in which $V_0(x)$ is proportional to $\sin \pi x/d$; this is evidently the case $n = 1$. The decay length l for this case (the distance in the y-direction over which the potential decreases by a factor of e) is given by $kl = 1$, or $l = d/\pi$; this is illustrated in Figure 17.9. However, we can match other boundary conditions by superposing solutions like Eq. 17.28 for several different values of n. The most general solution is a superposition of all of them,

$$V(x, y) = \sum_k B_k e^{-ky} \sin kx \tag{17.30}$$

where the sum is over $k = \pi/d, 2\pi/d, 3\pi/d$, and so on. The boundary value of this function is

$$V(x, 0) = \sum_k B_k \sin kx \tag{17.31}$$

which is a Fourier series. One can prove (see for example the book by Boas cited in the bibliography) that **any** function (within reason) can be written as a Fourier series, so the method of separation of variables can in principle be used to solve this problem for any boundary potential $V_0(x)$. It is not easy to prove this, but if you are willing to accept it, it is quite easy to compute the necessary coefficients B_k for a given function V_0. The method relies on the **orthogonality** of the functions $\sin ky$, that is, the fact that

$$\int_0^d \sin(n\pi x/d)\sin(m\pi x/d) = (d/2)\delta_{mn} \tag{17.32}$$

where δ_{mn} is the Kronecker delta function (you can verify this yourself by doing the integral). If we multiply Eq. 17.31 by $\sin(m\pi x/d)$ and integrate, we get (using $k = n\pi/d$ so the allowed k's correspond to integer n)

$$\int_0^d V_0(x)\sin(m\pi x/d) = \sum_n B_k \sin(n\pi x/d)\sin(m\pi x/d) = B_{m\pi x/d}(d/2)$$

because only the term with $n = m$ is nonzero. Thus (denoting $m\pi x/d$ by k')

$$B_{k'} = (2/d)\int_0^d V_0(x)\sin(k'x) \tag{17.33}$$

■ **EXAMPLE 17.1**   As an example of the use of Eq. 17.33, suppose the bottom of the slot in Figure 17.8 consisted of a conductor at a constant potential V_0 (we are pretending here

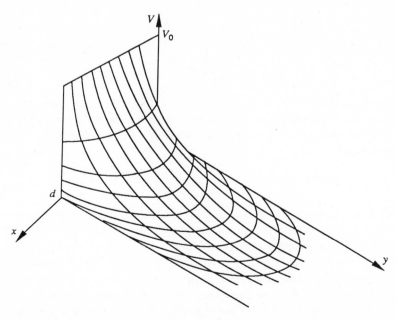

Figure 17.10   The exact solution (Eq. 17.35) for the case V_0 = constant on the slot bottom. (From *Electromagnetic Fields and Waves*, 2nd Ed., Paul Lorrain and Paul R. Corson, W. H. Freeman & Co., copyright (c) 1970.)

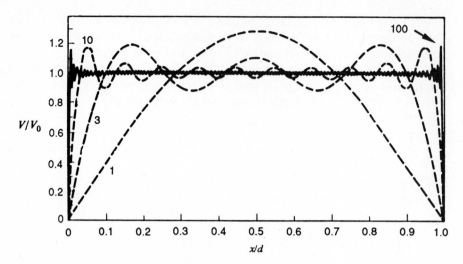

Figure 17.11 Convergence of partial sums in Eq. 17.35, for $y = 0$. Curves are labeled by the number of terms included. (From *Electromagnetic Fields and Waves,* 2nd Ed., Paul Lorrain and Paul R. Corson, W. H. Freeman & Co., copyright (c) 1970.)

that it could be insulated from the sides, in spite of the fact that it touches them — the potential is thus discontinuous at the corners). Then Eq. 17.33 gives

$$B_{k'} = (2V_0/d) \int_0^d \sin(k'x)\,dx = 4V_0/md \qquad \text{(odd } m)$$

$$= 0 \qquad \text{(even } m) \tag{17.34}$$

with the result that

$$V(x,y) = 4V_0 \sum (kd)^{-1} e^{-ky} \sin kx \tag{17.35}$$

where the sum is over $k = \pi/d, 3\pi/d, 5\pi/d, \ldots$. This infinite sum is the exact continuum potential, and is plotted in Figure 17.10. Various partial sums are drawn in Figure 17.11 to show how they converge to the infinite sum when $y = 0$. For large y the convergence is much more rapid.

■ **EXAMPLE 17.2** Suppose that instead of an infinitely deep slot we had a rectangular duct, whose width is d like that of the slot but that has a finite depth $2b$ (see Figure 17.12). The potential is again zero on the sides ($x = 0$ and $x = d$) but now is assumed to take the value $V_0(x)$ at the top boundary (at $y = b$) as well as at the bottom (now at $y = -b$). We expect the potential to drop as we move away from the bottom, as it did in the case of the slot, but now we expect it to rise again near the top, as shown in Figure 17.13. We again assume a product form (Eq. 17.21) and again use $C > 0$ so the solutions $X(x)$ are sines and cosines. This allows us to match the $V = 0$ boundary conditions at $x = 0$ and $x = d$ as in the previous example, by using only $\sin kx$ where $k = n\pi/d$. However, we can no longer use the $y \to \infty$ argument to rule out growing exponentials (the $+$ sign in Eq.

17.24), because y is bounded. Thus the most general Y is

$$Y(y) = Ae^{+ky} + Be^{-ky} \tag{17.36}$$

In the present case the mirror symmetry in the $y = 0$ plane (an operation that takes $y \rightarrow -y$) implies that $Y(y) = Y(-y)$, which requires $A = B$. Thus,

$$Y(y) = 2A \cosh ky \tag{17.37}$$

If the boundary condition is a single sine function

$$V_0(x) \equiv V_1 \sin(\pi x/d) \tag{17.38}$$

we can match it with a single product $X(x)Y(y)$,

$$V(x,y) = 2A \sin(\pi x/d) \cosh(\pi y/d) \tag{17.39}$$

The boundary conditions at $y = b$ and $y = -b$ are equivalent, and when solved for A

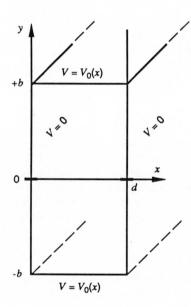

Figure 17.12 A pseudo-2D rectangular duct.

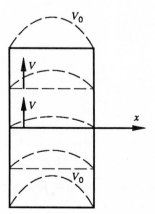

Figure 17.13 The potential at various heights in a rectangular duct.

give

$$V(x,y) = V_1[\cosh(\pi b/d)]^{-1} \sin(\pi x/d)\cosh(\pi y/d) \qquad (17.40)$$

which is sketched in Figure 17.13.

▶ **Problem 17.8**

Consider the exact solution (Example 17.2) for the rectangular duct with potentials $V(x, \pm b) = V_1 \sin(\pi x/d)$ at the top and bottom, and $V = 0$ at the left and right.

a. Numerically evaluate the exact series for the potential at the center of a square duct ($b = \frac{1}{2}d$), for $V_1 = 1$ volt.

b. Do the problem with the discrete methods of Section 16.4, using the discrete grid with $dr = \frac{1}{2}d$ shown in Figure 17.14. Evaluate the boundary conditions at cell centers, so there is just one unknown potential. What is the discretization error (the difference between the exact result and the discrete one), as a percentage of V_1?

c. Repeat, using a finer grid with $dr = \frac{1}{4}d$, again putting cell centers at the boundaries. This involves solving four equations in four unknowns. What is the error at the center now? Are the errors at the other three points larger?

d. If you have access to a computer program for solving this problem by the relaxation method (such as the one by Koonin reproduced on the IBM-PC program disk for this book), check parts (b) and (c) and give the result for $dr = \frac{1}{8}d$.   ◀

▶ **Problem 17.9**

Write a computer program to solve the previous problem by the relaxation method. It should solicit an integer N and solve the problem with cell size $dr = d/N$. It must set (and print out) the bc's, then start iterating and printing answers. (You might want to print only every fifth iteration or so, depending on N.) Debug the program until you get the right answers for the cases you solved by hand in the previous problem ($N = 2$ and $N = 4$). Hand in a printout for $N = 4$ and also give the potential at the center for $N = 8$, accurate to 10^{-4}. What is the discretization error at the center for $N = 8$?   ◀

▶ **Problem 17.10**

Consider a modification of the rectangular-duct problem in which the potential is zero on the bottom (taken at $y = 0$) as well as on the left and right, and has nonzero values $V_0 \sin(\pi x/d)$ only on the top ($y = +b$).

a. Compute an exact analytic solution for $V(x, y)$.

b. Compute the exact value at the center of a square duct, and compare to the result of a discrete calculation using the grid of Figure 17.14.

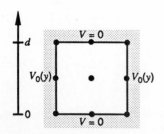

Figure 17.14   A discretized square duct (the dots are cell centers).

    c. Modify the program of Problem 17.9 for this case, and give results for $N = 4$ and $N = 8$.

    d. Can you recover the answers to Problem 17.9 from these results? How and why? ◄

In these cases the continuum method is probably easier than the discrete one if high accuracy is desired. However, keep in mind that these are very special problems, chosen precisely because they are easy to do analytically. In any real situation, the problem we have posed would only be an approximation anyway, so there wouldn't be much point in solving it with high accuracy. The great advantage of the discrete method is that you can **always** do it, without learning a lot of special techniques, and you can tailor it to any particular real problem as accurately as needed. The relaxation method is very slow for fine grids (large N); the number of iterations needed to reduce the error by a factor of e is roughly $2N^2/\pi^2$ (this can be shown by eigenvalue methods to be found in numerical analysis texts—see also Problem 16.15). Other computer methods (e.g., solving simultaneous equations, as we did by hand in Section 16.3) are much faster but are harder to program and require more memory.

    We can also use the method of separation of variables in curvilinear coordinates. We will illustrate this for the case of spherical coordinates, using for simplicity a system with **azimuthal symmetry**, that is, one in which the potential is $V(r, \theta)$, independent of ϕ. In regions of space where there is no charge, the potential must satisfy Laplace's equation; in spherical coordinates this is (from the inside back cover)

$$0 = \nabla^2 V(r, \theta)$$

$$= \frac{1}{r^2} \frac{d}{dr} r^2 \frac{dV}{dr} + \frac{1}{r^2 \sin\theta} \frac{d}{d\theta}\left[\sin\theta \frac{dV}{d\theta}\right] \tag{17.41}$$

where we have left off the term involving $d/d\phi$ because V is independent of ϕ.

    In analogy with the rectangular-coordinates case, we assume V can be written as a product

$$V(r, \theta) = R(r)\Theta(\theta) \tag{17.42}$$

We derive differential equations for R and Θ by substituting Eq. 17.42 into Eq. 17.41:

$$0 = \frac{\Theta(\theta)}{r^2} \frac{d}{dr}\left[r^2 \frac{dR(r)}{dr}\right] + \frac{R(r)}{r^2 \sin\theta} \frac{d}{d\theta}\left[\sin\theta \frac{d\Theta(\theta)}{d\theta}\right] \tag{17.43}$$

Assuming R and Θ are nonzero, we can multiply by $r^2/R(r)\Theta(\theta)$ to give

$$0 = \frac{1}{R(r)} \frac{d}{dr}\left(r^2 \frac{dR}{dr}\right) + \frac{1}{\Theta(\theta) \sin\theta} \frac{d}{d\theta}\left[\sin\theta \frac{d\Theta(\theta)}{d\theta}\right] \tag{17.44}$$

Each term now involves only r or only θ, and exactly as before, each must be constant; suppose the first term is A. Then the equation for R is

$$\frac{d}{dr} r^2 \frac{dR}{dr} = AR(r) \tag{17.45}$$

Although this cannot be directly integrated, we can make the reasonable guess that R has a power law form

$$R(r) = r^l \tag{17.46}$$

and verify by substitution that this is a solution if

$$A = l(l + 1) \tag{17.47}$$

In general, a second-order differential equation has two linearly independent solutions and we have found both, because for each A Eq. 17.47 has two solutions for l. Turning to the equation for Θ,

$$\frac{d}{d\theta} \left[\sin\theta \frac{d\Theta(\theta)}{d\theta} \right] = -l(l + 1) \sin\theta \, \Theta(\theta) \tag{17.48}$$

we can eliminate the trigonometric functions by the substitution $x = \cos\theta$. Using

$$\frac{d\Theta(\theta)}{d\theta} = \frac{d\Theta}{dx} \frac{dx}{d\theta} = -\frac{d\Theta(x)}{dx} \sin\theta \tag{17.49}$$

we get

$$\frac{d}{dx} (1 - x^2) \frac{d\Theta(x)}{dx} = -l(l + 1)\Theta(x) \tag{17.50}$$

This is called **Legendre's equation**. Some of the solutions are not hard to find, because they are polynomials. For each integer $l \geq 0$, there is a **Legendre polynomial** of order l that satisfies Eq. 17.50 and is denoted by $P_l(x)$. A general method for calculating them is described in Problem 17.11. The first few are

$$P_0(x) = 1, \qquad P_1(x) = x$$
$$P_2(x) = \tfrac{1}{2}(3x^2 - 1), \qquad P_3(x) = \tfrac{1}{2}(5x^3 - 3x) \tag{17.51}$$

as you can verify by substituting them into Legendre's equation.

▶ **Problem 17.11**

  a. Compute the first three Legendre polynomials P_0, P_1, and P_2 from scratch, by assuming $P_l(x) = c_0 + c_1 x + \cdots + c_l x^l$, substituting into Eq. 17.50, equating coefficients of like powers of x, and solving for the c's. [First choose c_l arbitrarily, and then normalize so $P_l(1) = 1$.]

  b. For $l > 2$, the easiest way to do this is to derive a general formula for c_n by substituting the formal sum $\Sigma c_n x^n$ into Eq. 17.50, equating like powers of x (you will need to change dummy variables in some terms), and obtaining an equation for each c_n in terms of higher c_n's. This is called a **recursion relation**. For $l = 2$, for example, it is

$$[6 - n(n + 1)]c_n = -(n + 2)(n + 1)c_{n+2}$$

Find the recursion relation for general l, and compute P_2, P_3, and P_4, normalized. Assume $c_n = 0$ for $n > l$, and choose c_l arbitrarily. (Note that the recursion relation allows any c_l, and requires $c_{l-1} = 0$.)   ◄

▶   **Problem 17.12**

Write a computer program to calculate the c_i's for any l, using the method of the previous problem. Check that it gives the correct results for $l \le 3$. Hand in a program listing and the normalized results for $l = 10$ and $l = 20$.   ◄

You might recall that a second-order differential equation generally has **two** solutions; the second solution of Legendre's equation is denoted Q_l (the **associated Legendre function**) and turns out not to be allowable because it diverges at $x = \pm 1$ (i.e., at $\theta = 0$ and π) and we usually have a boundary condition that demands that the potential be finite there. There are also solutions for non-integer l, which are again ruled out because of divergences.

Now we have a solution of the Laplace equation for each integer l:

$$V(r, \theta) = R(r)\Theta(\theta) = r^l P_l(\cos \theta) \tag{17.52}$$

Note, however, that we can pair the non-negative and negative l's, associating the negative $l' = -l - 1$ with each non-negative l. The differential equations for R and Θ are the same for l and l' because $l'(l' + 1) = l(l + 1)$. Both r^l and r^{-l-1} are solutions to the radial equation (Eq. 17.45), and the angular equation has the single solution P_l. It is customary to use only non-negative l, and associate with l the two solutions $r^l P_l(\cos \theta)$ and $r^{-l-1}P_l(\cos \theta)$. Thus a general linear combination of these solutions is

$$V(r, \theta) = \sum_{l=0}^{\infty} \left[A_l r^l P_l(\cos \theta) + B_l r^{-l-1}P_l(\cos \theta) \right] \tag{17.53}$$

Although we have not proved it, this set of solutions is **complete** in the sense that any solution of the Laplace equation can be expressed as an infinite sum of this form.

■ **EXAMPLE 17.3**   Consider the electrostatic system shown in Figure 17.15, in which the potential $V(a, \theta)$ is known on the surface of a sphere of radius a, and there is no charge outside the sphere (there must be charge inside, of course, to produce a nonzero potential). The sphere is in infinite space and we also assume $V \to 0$ as $r \to \infty$. The figure is drawn for the case $V(a, \theta) = V_1 \cos \theta$ for some constant V_1. We want to calculate the potential everywhere outside the sphere. We will find it by assuming it has the form of Eq. 17.53 and deducing the values of A_l and B_l (it is even all right to guess, because the uniqueness theorem guarantees that any solution we find that satisfies the bc's is the correct one.) None of the r^l terms approaches zero as $r \to \infty$ (indeed all but r^0 diverge) and, because they are linearly independent, these divergences cannot cancel. Thus the

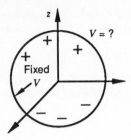

Figure 17.15 A spherical surface with a known potential. The $+$ and $-$ signs indicate the sign of V.

coefficients A_l of all these terms must vanish. The remaining terms must satisfy the bc at $r = a$, which we assume to be $V_1 \cos\theta$:

$$\sum_{l=0}^{\infty} B_l a^{-l-1} P_l(\cos\theta) = V(a,\theta) = V_1 P_1(\cos\theta) \tag{17.54}$$

The Legendre polynomials are a linearly independent set of functions of θ, so the coefficients of each must match:

$$B_1 a^{-1-1} = V_1$$

$$B_l = 0 \qquad \text{for } l \neq 1$$

Thus $B_1 = a^2 V_1$, and our final result is

$$V(r,\theta) = V_1(a/r)^2 \cos\theta \tag{17.55}$$

You should check that this satisfies all the bc's and Laplace's equation.

What if $V(a,\theta)$ in the preceding example had not happened to already be a Legendre polynomial? Then finding the B_l's from Eq. 17.53 would amount to expanding $V(a,\theta)$ in a series of Legendre polynomials in $\cos\theta$. Fortunately, this can be done for any function $V(a,\theta)$; the Legendre polynomials are a complete set of functions on the interval $[-1,1]$. They also happen to be **orthogonal**, that is,

$$\int_0^{\pi} P_l(\cos\theta) P_{l'}(\cos\theta) \sin\theta \, d\theta = 2(2l+1)^{-1}\delta_{ll'} \tag{17.56}$$

where $\delta_{ll'}$ is the Kronecker delta function. Thus we can calculate B_l in general by multiplying Eq. 17.54 by $P_{l'}(\cos\theta)\sin\theta \, d\theta$ and integrating:

$$B_l = 2^{-1}(2l+1)a^{l+1}\int_0^{\pi} V(a,\theta) P_l(\cos\theta) \sin\theta \, d\theta \tag{17.57}$$

You can show that this reduces to the above result when $V(a,\theta) = V_1 \cos\theta$.

▶ **Problem 17.13**

Two conducting hemispheres of radius a are insulated from each other and held at potentials $\pm V_0$ (Figure 17.16). There is no charge outside the sphere. Compute the potential for $r > a$ (you will get an infinite series). Evaluate the coefficients explicitly for $l \leq 3$, and calculate the potential directly above the sphere (at $\theta = 0$) for $r = 1.5a$ and $r = 3a$. Estimate the error in each case. ◀

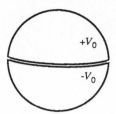

Figure 17.16 A split conducting sphere.

▶   **Problem 17.14**

Find V **inside** the (hollow) sphere of Example 17.3, assuming now that all charges are outside. [Hint: Imagine that there is a "boundary condition" that V not diverge at $r = 0$, which forces many coefficients to vanish.]   ◀

■ **EXAMPLE 17.4 A CONDUCTING SPHERE IN A UNIFORM FIELD**   Consider a conducting sphere of radius a in a uniform electric field $\mathbf{E}_0$. (You can think of the uniform field as having been created by capacitor plates far away.) Find the potential everywhere outside the sphere.

Although we assume the **total** charge on the sphere vanishes, we expect to find positive and negative charge density induced on the surface, which distorts the field near it. As in Example 17.3, because the potential satisfies Laplace's equation with azimuthal symmetry it must be expressible in the form of Eq. 17.53. The boundary condition at $r = \infty$ is that $V(r, \theta)$ must approach the potential $-E_0 z = -E_0 r \cos \theta$ of the uniform field (we have chosen the arbitrary additive constant so that $V = 0$ at $z = 0$ for large x or y). In this limit $r^{-l-1} \to 0$, so we have only the A_l terms:

$$V(r, \theta) = \sum_{l=0}^{\infty} A_l r^l P_l(\cos \theta) \to -E_0 r P_l(\cos \theta)$$

This implies that the coefficients match because the Legendre polynomials are linearly independent.

$$A_1 r^1 \to -E_0 r$$

$$A_l r^l \to 0 \qquad \text{for } l \neq 1$$

The former expression requires

$$A_1 = -E_0$$

and the latter requires $A_l = 0$ for $l \neq 1$. To determine the B_1's, we must use the boundary condition at $r = a$. This states that the potential there is a constant which, because of the horizontal mirror symmetry, must be zero:

$$\sum_{l=0}^{\infty} \left[A_l a^l + B_l a^{-l-1} \right] P_l(\cos \theta) = 0$$

so

$$A_l a^l + B_l a^{-l-1} = 0$$

This gives $B_l = 0$ for $l \neq 1$, and

$$B_1 = a^3 E_0$$

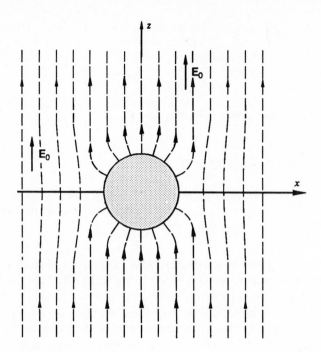

Figure 17.17 Distortion of the electric field lines near a conducting sphere when the field is uniform far away.

Thus, the exact result is

$$V(r, \theta) = -E_0(r - a^3/r^2)\cos\theta = 0 \qquad (17.58)$$

which you can verify to satisfy all the boundary conditions. From this we can obtain the electric field

$$E_r = -dV/dr = +E_0(1 + 2a^3/r^3)\cos\theta$$

$$(17.59)$$

$$E_\theta = -r^{-1}dV/d\theta = E_0(1 - a^3/r^3)\sin\theta$$

This field is shown in Figure 17.17.

▶ **Problem 17.15**

A spherical shell of radius a has a uniform positive surface charge $+\sigma_0$ on the upper hemisphere and a uniform negative charge $-\sigma_0$ on the lower one. Find the potential inside and outside the sphere, giving the coefficients of all the Legendre polynomials for $l \leq 6$. ◀

▶ **Problem 17.16**

a. Consider an electrostatic system with **cylindrical** symmetry (in which the potential is $V(r, \theta)$ in cylindrical coordinates, independent of z). Use the method of separation of variables to determine the most general solution.

b. Solve the cylindrical analog of Example 17.4, that is, find the field outside a cylindrical conducting surface when the field $\mathbf{E}_0$ far away is uniform and perpendicular to the cylinder axis (use Figure 17.17 to visualize this, thinking of the circle as a cross section of a cylinder perpendicular to the paper rather than of a sphere). Find the surface charge density on the conductor.

c. Compute the potential inside and outside a cylindrical surface of radius a that has a fixed surface charge density $\sigma(\phi, z) = \sigma_0 \cos 3\phi$. ◄

SECTION 17.3 THE MULTIPOLE EXPANSION

The potential $V(\mathbf{R})$ of the electrostatic field of an arbitrary charge distribution $\rho(\mathbf{r})$ such as that in Figure 17.18 can be determined from Coulomb's law, Eq. 15.8:

$$V(\mathbf{R}) = (4\pi\epsilon_0)^{-1} \int (\rho/s)\, d\tau \qquad (17.60)$$

This integral must be calculated separately for every $\mathbf{R}$, because the s's are completely different for each $\mathbf{R}$. It turns out that if $\mathbf{R}$ is very far away from the charge distribution, the potential changes only very slowly as we move $\mathbf{R}$. Furthermore, the potential far away does not depend on the fine details of the charge distribution, but only on certain gross properties we will refer to as the "multipole moments."

To see why this is so, look at the factor $1/s$ in the integral when $R \gg a$ (a is the size of the charge distribution, as indicated in Figure 17.18.) Pick an origin in or near the charge distribution, and denote the vector position of the volume element $d\tau$ by $\mathbf{r}$. Then

$$s^2 = (\mathbf{R} - \mathbf{r}) \cdot (\mathbf{R} - \mathbf{r}) = R^2 - 2\mathbf{R} \cdot \mathbf{r} + r^2$$
$$= R^2\left(1 - 2\mathbf{R} \cdot \mathbf{r}/R^2 + r^2/R^2\right)$$

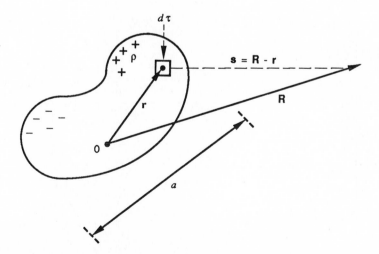

Figure 17.18 The charge distribution of size a, showing the vector positions of the field point $\mathbf{R}$ and the source point $\mathbf{r}$.

so

$$s^{-1} = (s^2)^{-1/2} = R^{-1}[1 - 2\mathbf{R}\cdot\mathbf{r}/R^2 + r^2/R^2]^{-1/2} \qquad (17.61)$$

But $r \cong a \ll R$, so the terms added to 1 in the bracket are small. We expand in a power series in this small number, $x = -2\mathbf{R}\cdot\mathbf{r}/R^2 + r^2/R^2$. Use the binomial theorem

$$(1 + x)^n = 1 + [n/1]x + [n(n-1)/1\cdot 2]x^2$$
$$+ [n(n-1)(n-2)/1\cdot 2\cdot 3]x^3 + \cdots \qquad (17.62)$$

with $n = -\frac{1}{2}$:

$$(1 + x)^{-1/2} = 1 - \tfrac{1}{2}x + \frac{\left(-\frac{1}{2}\right)\left(-\frac{3}{2}\right)}{2}x^2 + \cdots$$

so

$$s^{-1}$$

$$= R^{-1}\left[1 + \frac{\mathbf{R}\cdot\mathbf{r}}{R^2} - \frac{r^2}{2R^2} + \frac{3}{8}\left(4\frac{(\mathbf{R}\cdot\mathbf{r})(\mathbf{R}\cdot\mathbf{r})}{R^4} - 4\frac{(\mathbf{R}\cdot\mathbf{r})r^2}{R^4} + \frac{r^4}{R^4}\right) + \cdots\right]$$

$$(17.63)$$

To see the significance of these terms, let us apply them to the case of a "physical dipole": a positive and a negative charge separated by a distance a (Figure 17.19). If we put the negative charge at the origin, the potential at $\mathbf{R}$ is

$$V(\mathbf{R}) = (4\pi\epsilon_0)^{-1}[-q/R + q/s]$$

so (using Eq. 17.63 for q/s with $\mathbf{r} = \mathbf{a}$)

$$4\pi\epsilon_0 V(\mathbf{R}) = \frac{q}{R}\left[-1 + 1 + \frac{\mathbf{R}\cdot\mathbf{a}}{R^2} - \frac{1}{2}\frac{a^2}{R^2} + \frac{3}{2}\frac{(\mathbf{R}\cdot\mathbf{a})^2}{R^4} + O(a^3/R^3)\right] \qquad (17.64)$$

$$= q\frac{\hat{\mathbf{R}}\cdot\mathbf{a}}{R^2} + \frac{1}{2}\frac{q}{R^3}\left(-a^2 + 3(\hat{\mathbf{R}}\cdot\mathbf{a})^2\right) + \cdots$$

Figure 17.19 A physical dipole.

where $O(a^3/R^3)$ represents terms of third and higher order in the small number a/R, which are much smaller than the terms we have written out. We write the numerators in terms of the unit vector $\hat{\mathbf{R}} = \mathbf{R}/R$ in the direction of the field point $\mathbf{R}$ to make clear the dependence on the magnitude R: the first term decreases as $1/R^2$. In fact, we can separate its dependence on the source distribution from its dependence on $\mathbf{R}$:

$$V(\mathbf{R}) = \frac{(q\mathbf{a}) \cdot \hat{\mathbf{R}}}{4\pi\epsilon_0 R^2} + \cdots = \frac{\mathbf{p} \cdot \hat{\mathbf{R}}}{4\pi\epsilon_0 R^2} + \cdots$$

where we have defined the **dipole moment**

$$\mathbf{p} = q\mathbf{a} \tag{17.65}$$

to contain the entire dependence on the size and separation of the charges. Note that this term in the field depends only on the product of the charge and the separation.

This first term is clearly a good approximation to the potential for large R (later terms go as R^{-3} or higher powers). When is it a good approximation for **all** R? If we let q get large and a small, so their product p remains constant, the later terms (which all have extra factors of a) become arbitrarily small. This limit ($q \rightarrow \infty$, $a \rightarrow 0$) is called the **point dipole** and its potential is **exactly**

$$V(\mathbf{R}) = \mathbf{p} \cdot \hat{\mathbf{R}}/4\pi\epsilon_0 R^2 \qquad (\text{the "dipole potential"}) \tag{17.66}$$

(a)

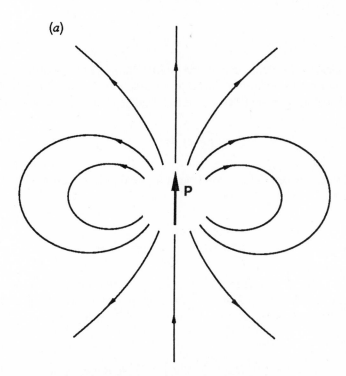

Figure 17.20 The field lines near a point dipole **p**.

We can express this in spherical coordinates (R, θ) if we take $\mathbf{p}$ along the z axis:

$$V(R, \theta) = p \cos \theta / 4\pi\epsilon_0 R^2 \qquad (17.67)$$

The electric field $\mathbf{E}(\mathbf{R}) = -\text{grad}\, V(\mathbf{R})$ associated with a dipole potential looks something like Figure 17.20. The field of a physical dipole (Figure 17.3) looks like this except within a distance of order a of the charges.

▶ **Problem 17.17**

Calculate the components E_R and E_θ of the dipole field.   ◀

Returning to the general case (Eq. 17.60) and using Eq. 17.63,

$$4\pi\epsilon_0 V(\mathbf{R}) = \int \rho \frac{d\tau}{s}$$

$$= \frac{1}{R} \int \rho(\mathbf{r})\, d\tau \left[1 + \frac{\hat{\mathbf{R}} \cdot \mathbf{r}}{R} + \frac{1}{2R^4}\left(-r^2 R^2 + 3(\mathbf{R} \cdot \mathbf{r})^2\right) + O(R^{-3}) \right]$$

$$= \frac{1}{R}\left[\int \rho(\mathbf{r})\, d\tau \right] + \frac{\hat{\mathbf{R}}}{R^2} \cdot \left[\int \rho(\mathbf{r})\mathbf{r}\, d\tau \right] \qquad (17.68)$$

$$+ \frac{1}{2R^5} \int \rho(\mathbf{r})\left(-r^2 R^2 + 3(\mathbf{R} \cdot \mathbf{r})^2\right) d\tau$$

The first term is clearly the field of a single point charge (a "monopole") at the origin. It is the dominant term for large R: when one is far from a charge distribution, the distribution resembles a point charge Q at the origin, where

$$Q = \int \rho(\mathbf{r})\, d\tau \qquad (17.69)$$

is just the total charge present. The second term is the dipole potential (Eq. 17.66), with dipole moment

$$\mathbf{p} = \int \rho(\mathbf{r})\mathbf{r}\, d\tau \qquad (17.70)$$

and it dominates if $Q = 0$.

We would like to separate out the R-dependence of the third term in a similar way. Unfortunately, $\mathbf{R}$ is well buried in a dot product, which we will have to expand explicitly in terms of components R_i ($i = x$, y, or z so $R_i = R_x$, R_y, or R_z) and r_i ($= r_x$, r_y, or r_z):

$$(\mathbf{R} \cdot \mathbf{r})^2 = \left(\sum_i R_i r_i \right)\left(\sum_j R_j r_j \right) = \sum_{i,j} R_i R_j r_i r_j \qquad (17.71)$$

so the third ("quadrupole") term is

$$\int \rho(\mathbf{r}) \, d\tau \left(-r^2 R^2 + 3 \sum_{i,j} R_i R_j r_i r_j \right) / 2R^5 \tag{17.72}$$

and we can factor the R's out of the integral. It is convenient to express R^2 in terms of components as well:

$$R^2 = \sum_i R_i R_i = \sum_{i,j} R_i R_j \delta_{ij} \tag{17.73}$$

(δ_{ij} is the Kronecker delta function, 1 if $i = j$ and 0 if $i \neq j$) so the term can be written

$$R^{-5} \sum_{i,j} R_i R_j \int \rho(r) \, d\tau \left(-\tfrac{1}{2} r^2 \delta_{ij} + \tfrac{3}{2} r_i r_j \right) = R^{-5} \sum_{i,j} R_i R_j Q_{ij} \tag{17.74}$$

where Q_{ij} is a 3×3 matrix (the "quadrupole moment matrix" or "quadrupole moment tensor") defined by

$$Q_{ij} = \int \rho(r) \, d\tau \left(-\tfrac{1}{2} r^2 \delta_{ij} + \tfrac{3}{2} r_i r_j \right) \tag{17.75}$$

Like the monopole moment (i.e., the charge Q) and the dipole moment $\mathbf{p}$, the quadrupole moment depends only on the charge distribution and not on $\mathbf{R}$. Notice that these moment terms in the potential decrease with R as successively higher powers R^{-1}, R^{-2}, R^{-3}. In general, the 2^l-pole term goes as R^{-l-1}, where $l = 0$ (1-pole = monopole) $l = 1$ (2-pole = dipole), $l = 2$ (4-pole = quadrupole), $l = 3$ (8-pole = octupole), and so on. The practical usefulness of the multipole concept, however, lies in cases in which only one or two terms are important, so it is not worth worrying too much about the general multipole.

Our final result is that the potential of a charge distribution is given by

$$V(\mathbf{R}) = \frac{Q}{4\pi\epsilon_0 R} + \frac{\mathbf{p} \cdot \hat{\mathbf{R}}}{4\pi\epsilon_0 R^2} + \frac{1}{4\pi\epsilon_0 R^3} \sum \hat{R}_i \hat{R}_j Q_{ij} + O(R^{-4}) \tag{17.76}$$

where the multipole moments Q, $\mathbf{p}$, and Q_{ij} are computed from Eqs. 17.69, 17.70, and 17.75. Alternatively, if we happen to know the potential we can work backwards and deduce the moments from Eq. 17.76. For example, in Example 17.4 we found the potential of a conducting sphere in a uniform field: the total potential was the potential of the original uniform field plus a dipole potential. Comparing coefficients, we can identify the dipole moment of a conducting sphere of radius a in a field $\mathbf{E}_0$:

$$\mathbf{p} = 4\pi\epsilon_0 \mathbf{E}_0 a^3 \tag{17.77}$$

In this case the multipole expansion (Eq. 17.76) has only one term; the quadrupole and higher moments are exactly zero. The potential due to the sphere is exactly the same as that of a point dipole at its center. This peculiarity is due to the spherical shape of the conductor, and is not true in general.

There are two ways to compute the quadrupole or dipole moments of a collection of point charges q_α ($\alpha = 1, 2, \ldots, n$) at positions $\mathbf{a}_\alpha$. One is to write the continuum charge density $\rho(\mathbf{r})$ for this distribution (i.e., a sum of Dirac delta functions) and use Eq. 17.75. A simpler way is to write the potential as the sum of those of the charges individually:

$$V(\mathbf{r}) = \sum_\alpha q_\alpha / 4\pi\epsilon_0 s_\alpha \qquad (17.78)$$

and expand each $s_\alpha^{-1} = |\mathbf{R} - \mathbf{a}_\alpha|^{-1}$ in powers of R^{-1} exactly as we did in the case of a smooth density, using Eq. 17.63. The result is

$$Q = \sum_\alpha q_\alpha$$

$$\mathbf{p} = \sum_\alpha q_\alpha \mathbf{a}_\alpha \qquad (17.79)$$

and

$$Q_{ij} = \sum_\alpha q_\alpha \left(-\tfrac{1}{2} a_\alpha^2 \delta_{ij} + \tfrac{3}{2} a_{\alpha i} a_{\alpha j} \right) \qquad (17.80)$$

Here $a_\alpha^2 \equiv |\mathbf{a}_\alpha|^2$, the square of the distance from the αth charge to the origin, and $a_{\alpha i}$ is the i-component of the vector a_α, that is, if $\alpha = 1$ and $i = x$ it is a_{1x}.

It is clear that the monopole moment is independent of the choice of origin. If we compute the dipole moment of a collection of point charges with respect to a new origin at $\mathbf{b}$, however, it becomes

$$\mathbf{p}' = \sum_\alpha q_\alpha(\mathbf{a}_\alpha - \mathbf{b}) = \sum_\alpha q_\alpha a_\alpha - \mathbf{b} \sum_\alpha q_\alpha$$

$$= \mathbf{p} - \mathbf{b}Q \qquad (17.81)$$

This depends on the new origin, **unless** it happens that $Q = 0$. Thus, we must be very careful in talking about the dipole moment of a distribution that has a net charge; ordinarily we simply don't. One can talk about (and measure) the dipole moment of a water molecule (H_2O), for example, but it doesn't make sense to try to measure the dipole moment of an OH^- ion.

This general principle extends to the higher 2^l-poles: the 2^l-pole moment is independent of the choice of origin if and only if the $2^{l-1}, 2^{l-2}, \ldots$, 1-pole moments all vanish. We can talk unambiguously about the quadrupole moment of CO_2 (a linear molecule that has $\mathbf{p} = 0$) but not of H_2O.

■ **EXAMPLE 17.5**   Consider four charges at the corners of a square of side b (Figure 17.21). The signs are chosen so the two sides (left and right) of the square form dipoles whose

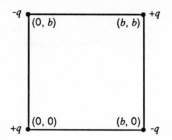

Figure 17.21 A quadrupole composed of four point charges.

moments cancel; thus the quadrupole moment will be independent of the origin. It is

$$Q_{xx} = \sum_\alpha q_\alpha \left(-\tfrac{1}{2} a_\alpha^2 + \tfrac{3}{2} a_{\alpha x}^2 \right)$$

$$= \underset{\uparrow}{0} - q\underset{\uparrow}{\left(-\tfrac{1}{2} b^2 \right)} - q\underset{\uparrow}{\left(-\tfrac{1}{2} b^2 + \tfrac{3}{2} b^2 \right)} + q\underset{\uparrow}{\left(-\tfrac{1}{2} 2b^2 + \tfrac{3}{2} b^2 \right)} = 0$$

[from $\mathbf{a} = (0,0) \qquad (0,b) \qquad\qquad (b,0) \qquad\qquad\qquad (b,b)$]

$Q_{yy} = 0 \qquad$ similarly

$$Q_{zz} = \sum_\alpha q_\alpha (\tfrac{1}{2} a_\alpha^2 + \tfrac{3}{2} a_{\alpha z}^2)$$

$$= 0 - q(-\tfrac{1}{2} b^2) - q(-\tfrac{1}{2} b^2) + q(-b^2) = 0$$

$$Q_{xy} = \sum_\alpha q_\alpha(-\tfrac{1}{2} a_\alpha^2 \cdot 0 + \tfrac{3}{2} a_{\alpha x} a_{\alpha y}) = q(\tfrac{3}{2} b^2)$$

$Q_{yx} =$ same (Q_{ij} is symmetric)

$Q_{xz} = Q_{yz} = 0$

so

$$Q_{ij} = \begin{vmatrix} 0 & \tfrac{3}{2} qb^2 & 0 \\ \tfrac{3}{2} qb^2 & 0 & 0 \\ 0 & 0 & 0 \end{vmatrix}$$

The name "quadrupole" has its origin in the fact that the easiest way to construct a distribution having a nonzero quadrupole moment but $\mathbf{p} = Q = 0$ is to superpose two dipoles with opposite $\mathbf{p}$'s as in the previous example, for a total of four charges. Similarly, you can make an octupole by superposing two quadrupoles with equal and opposite Q_{ij}'s.

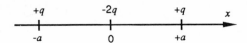

Figure 17.22   A linear quadrupole.

▶   **Problem 17.18**

a. Compute the quadrupole moment Q_{ij} of three point charges equally spaced along a line (Figure 17.22).

b. Compute the potential due to these charges to order $1/R^3$ (i.e., the leading nonzero term in the multipole expansion) at a point $\mathbf{R} = (x, 0, 0)$ on the x-axis, starting with the Q_{ij} you obtained in part **a**.

c. Compute the potential far away along the x-axis directly [i.e., expand $4\pi\epsilon_0 V = q/(x-a) - 2q/x + q/(x+a)$ for large x, to order $(a/x)^3$]. Compare to part **b**.   ◀

SECTION 17.4   POLARIZABILITY OF ATOMS AND MOLECULES

We have already mentioned that some molecules (e.g., H_2O, HCl) have a dipole moment. This can be understood by imagining the atoms to have nonzero net charge, and considering them as point charges as in Figure 17.23. This sort of a moment is referred to as a **permanent** dipole moment (because it exists whether or not an electric field is applied to the molecule).

A molecule or atom **without** a permanent dipole moment (e.g., CH_4, Ne) can acquire one ("become polarized") in an electric field. Such a moment is called an **induced** dipole moment. Essentially, it arises because every molecule or atom has positive and negative charges in it (nuclei and electrons) even if normally the positive and negative charge distributions are centered at the same point (as in Figure 17.24a) so there is no dipole moment. An electric field (to the right, say) will tend to push the positive nucleus to the right and the electrons to the left (Figure 17.24b). The field of a spherical negative charge distribution is the same as that of a point charge at its center (recall that this follows from Gauss' law), so the field of this distorted atom is the same as that of two point charges (Figure 17.24c) separated by a distance a, and so is its dipole moment, as we could verify from the definition (Eq. 17.70).

It is very difficult to calculate the induced dipole moment quantitatively for a real atom, but we can understand its general behavior from the simple model suggested by Figure 17.24. This model regards the electron cloud as a rigid sphere with **uniform** charge density. (The fact that real electron clouds are denser near the center doesn't change the qualitative results; see Problem 17.19.) When we turn on

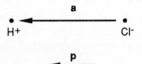

Figure 17.23   A molecule with a dipole moment.

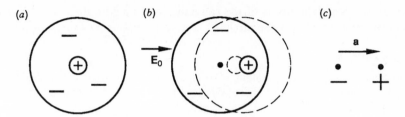

Figure 17.24 (a) A schematic view of an atom. (b) The effect of an electric field. Solid lines are displaced charges, dashed lines are initial positions. (c) The equivalent dipole.

an electric field $\mathbf{E}_0$ the nucleus will move to the right until the force $+Ze\mathbf{E}_0$ to the right exerted by the electric field $\mathbf{E}_0$ is exactly cancelled by the force of attraction of the electron cloud, $+Ze\mathbf{E}_e$. This stable equilibrium occurs when the magnitudes of the fields are equal, $E_e = E_0$. We can get E_e by applying Gauss' law to a small sphere of radius a, concentric with the electron cloud (which has radius R, say):

$$E_e \times 4\pi a^2 = q/\epsilon_0$$

where q is the charge inside the small sphere. This charge is the volume multiplied by the (uniform) density $-Ze/(4\pi R^3/3)$:

$$q = (4\pi/3)a^3\big[-Ze/(4\pi/3)R^3\big] = -Zea^3/R^3$$

so

$$E_e = \big(-Ze/4\pi a^2\epsilon_0\big)a^3/R^3 = -\big(Ze/4\pi\epsilon_0\big)a/R^3$$

and the condition for equilibrium is satisfied if

$$E_0 = \big(Ze/4\pi\epsilon_0\big)a/R^3$$

or

$$a = 4\pi\epsilon_0 R^3 E_0/Ze$$

Then the dipole moment induced is

$$p = (Ze)a = 4\pi\epsilon_0 R^3 E_0$$

(Oddly enough, this is exactly the same as the moment of a conducting sphere, Eq. 17.77.) The most important thing to notice about this result is that it is **linear** in the electric field E_0. This is true very generally (not just in this model). The proportionality constant is called the **polarizability** α:

$$\alpha \equiv p/E_0 \qquad\qquad (17.82)$$

For our simple model,

$$\alpha = 4\pi\epsilon_o R^3 \qquad\qquad (17.83)$$

In general, the polarizability is of the same order of magnitude as $\epsilon_0 \times$ the volume of the atom or molecule (in the model it is exactly three times this).

▶ **Problem 17.19**

For the case of the hydrogen atom ($Z = 1$), a more realistic model than the above is one with a nonuniform negative charge density $\rho(r) = -\rho_0 e^{-2r/R}$ (the positive nucleus is still a point charge).

a. What must ρ_0 be for the atom to be neutral?

b. Compute the restoring force on the nucleus if it is displaced by a small distance $x \ll R$. [Hint: Only $\rho(r)$ for $r < x$ contributes, and $\rho(r) \cong -\rho_0$ here.]

c. Compute the polarizability α. How does it differ from the uniform-sphere model? ◀

CHAPTER EIGHTEEN

FIELDS IN DIELECTRIC MATERIALS

SECTION 18.1 THE POLARIZATION FIELD

We know how to calculate electrostatic fields in systems of conductors and fixed charges in a vacuum. We would like to be able to do this in systems containing insulating materials (called "dielectrics" in this context), such as that in Figure 18.1. In insulating materials, the electrons are not free to leave their molecules (as they are in a metal), but this does not mean that dielectrics can be treated as though they were vacuum. Electric fields due to nearby charges can shift charge around within molecules, as we saw in Section 17.4 for an isolated molecule in a uniform field $\mathbf{E}_0$. This means that there is a nonzero charge density in a dielectric when it is subjected to an electric field $\mathbf{E}$. Each molecule develops an induced dipole moment; we may visualize it as a positive and negative charge held together by a spring that is stretched by the electric forces due to $\mathbf{E}$.

In principle, we know how to calculate the electrostatic field in a system such as that in Figure 18.1: we just solve Poisson's equation $\nabla^2 V = -\rho/\epsilon_0$ for V. However, the charge density on the right-hand side must include the charges that have shifted within the dielectric; this is called the **bound** charge. The density field of the bound charge can easily be visualized discretely by dividing the dielectric into cells, as in Figure 18.2. Before the material is polarized, the charge density in each cell is, of course, zero. Polarizing a material affects this discrete charge distribution only when it carries charge from one cell to another. This happens only within molecules that straddle a face. Count the charges transferred in this way across the face f, and define

$$P(f) \equiv \text{net charge transferred}/da(f) \tag{18.1}$$

the charge transferred per unit area. This face vector field is called the **polarization vector field**, or just the **polarization**. Note that the definition of the polarization is very similar to that of the electric current density $j(f)$ in a conductor; the only difference is that the charge in a conductor can flow continuously, so we must consider the charge per unit area **per unit time** instead of just per unit area.

We would like to use the polarization $P(f)$ to determine the charge density of the polarized material; this is the bound charge density, denoted ρ_b. We can find the total charge that flows out of a cell c while it is being polarized by adding up the charges going out each face. This amounts to taking the divergence of P:

$$\text{charge flowing out of } c = \sum_f P(f)\, da(f)$$
$$= \text{div}\, P(c)\, d\tau(c) \tag{18.2}$$

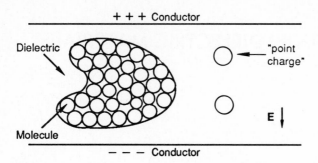

Figure 18.1 A system containing dielectrics as well as conductors and fixed charges.

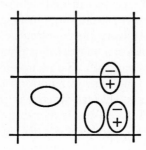

Figure 18.2 A polarized dielectric material with a discrete lattice.

For the upper right-hand cell in Figure 18.2, for example, this sum has a positive contribution from the bottom face. We can now compute the charge $\rho_b(c)\,d\tau(c)$ in cell c, because the net charge density was zero before the material was polarized:

$$(\text{initial charge}) - (\text{charge flowing out}) = (\text{final charge})$$

$$0 - \operatorname{div} P(c)\,d\tau(c) = \rho_b(c)\,d\tau(c)$$

$$\rho_b(c) = -\operatorname{div} P(c) \tag{18.3}$$

Thus the bound charge density is very simply related to the polarization vector.

▶ **Problem 18.1**

a. Figure 18.3a shows a pseudo-2D system, a long bar of dielectric, with its discrete polarization field $P(f)$ (in units $\mu\mathrm{coul/m}^2$). Compute the discrete density of bound charge $\rho_b(c)$, assuming $dr = 1$ m, and show it in a diagram.

b. Do the same for Figure 18.3b (a dielectric cylinder).

c. Same for Figure 18.3c.   ◀

These problems are highly artificial in that we can never know P without first calculating the electric field E; we will do this in Section 18.3.

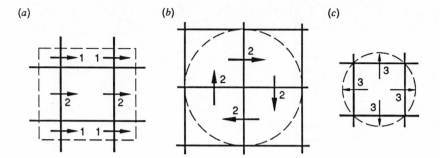

Figure 18.3 Cross sections of (*a*) a long bar of dielectric (dotted line) with a discretization (solid line) superposed. (*b*) A dielectric cylinder with an azimuthal polarization field. (*c*) A dielectric cylinder with a radial field.

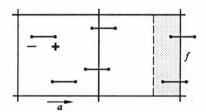

Figure 18.4 A system of randomly placed identical dipoles.

SECTION 18.2 DIPOLE MOMENT PER UNIT VOLUME

There are several ways of defining the polarization vector P that are equivalent in the continuum limit. We have defined it as charge displaced per unit area; this turns out to be equivalent to the dipole moment per unit volume. We will show this for a system consisting of many molecular dipoles, each with charges $\pm q$ separated by **a** as in Figure 18.4. Suppose there are many of these in each cell, on the average n dipoles per unit volume. What is the polarization vector **P**? We must count the dipoles that straddle a given face, say the one labeled f in the figure. These are exactly the ones whose left end (the negative charge) lies in the shaded region, of volume $a\,dr^2$. The number of these (on the average) is $na\,dr^2$, so the total charge carried across f when the material is polarized (q per dipole) is $qna\,dr^2$:

$$P(f) \equiv qna\,dr^2/dr^2 = qna \qquad (18.4)$$

Of course $p = qa$ is the dipole moment of each dipole, so the dipole moment per unit volume is

$$pn = qna \qquad (18.5)$$

Thus $P(f)$ is equal to the dipole moment per unit volume. This proof becomes

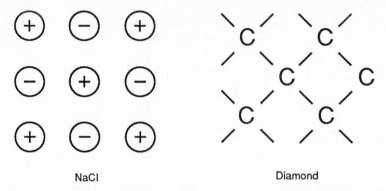

NaCl   Diamond

Figure 18.5   Schematic crystal structures of sodium chloride and diamond.

wordier, but is not essentially altered, if we consider the general case in which **p** does not lie along a coordinate axis and not all dipoles have the same **p**.

> Most electromagnetism texts begin by defining **P** as the dipole moment per unit volume. The derivation of the basic properties of **P** is very awkward when it is defined this way, however. Simply showing that $\rho_b = -\text{div}\, P$ (Eq. 18.3) requires a rather abstract argument involving integration by parts, as you can verify by looking at almost any text. Our approach leads to the essential results by enormously simpler arguments; the only thing that requires some work to show is that **P** is the dipole moment per unit volume, but this is not needed for most applications. The present approach also makes clear the physical meaning of bound charge density (as pieces of dipoles straddling faces), which can be very obscure in the traditional treatment (students frequently get the impression that the bound charge density is "fictitious," which is of course untrue). Another advantage of our approach is that it applies to systems that are not unambiguously separable into individual dipoles. The identification of individual dipoles and the specification of their dipole moments is easy enough for H_2O or CH_4. But what are the dipoles in a crystal of Na^+Cl^- (Figure 18.5)? Which way do they point? Also, in a covalently bonded crystal such as diamond, it is not clear what the "molecules" are, and how to define their dipole moments when the electron distributions shift in response to an electric field. In each of these cases, the notion of counting the charge pulled past a face by the field makes perfect sense (although actually calculating it correctly involves quantum mechanics).

> Lest a reader think it is revisionist to use the displaced-charge definition, the following quotations from the 1864 paper in which James Clerk Maxwell first wrote down the "Maxwell equations" are offered:

> *In a dielectric under the action of electromotive force, we may conceive that the electricity in each molecule is so displaced that one side is rendered positively and the other negatively electrical, but that the electricity remains entirely connected with the molecule, and does not pass from one molecule to another. The effect of this action on the whole dielectric mass is to produce a general displacement of electricity in a certain direction . . . . Electrical displacement consists in the opposite electrification of sides of a molecule or particle of a body which may or may not be accompanied with transmission through the body. Let the quantity of electricity which would appear on the faces dy . dz of an element dx dy dz cut from the body be f dy dz, then f is the component of electric displacement parallel to x.*

> (MAXWELL, TRANSACTIONS OF THE ROYAL SOCIETY, VOL. CLV, 1864, PP. 531 AND 554).

Maxwell's approach is essentially equivalent to ours. However, one should note for completeness that Maxwell's "electric displacement" f was not exactly the same as our P_x; it included a term related to the displacement of the "aether," and actually corresponds to our D_x (Section 18.3).

SECTION 18.3   GAUSS' LAW IN A DIELECTRIC: THE *D* FIELD

Our definition in Section 18.1 of the polarization **P** doesn't tell us how to calculate it, or how to calculate the electric field **E** in a system (such as that in Figure 18.1) containing dielectrics. To do this (using Poisson's equation, say), we would need to know the total charge density ρ. This includes the bound charge in the dielectric, which we calculated in Eq. 18.3, as well as any other charges in the system. These other charges are called "free charges" and include charges on the conductors and any small charged objects (such as the "point charge" $+q$ in Figure 18.1) whose charges are attached to the objects but "free" in the sense that we can move the objects around at will. We will denote the free charge density by ρ_{free}, so the total charge density is

$$\rho(c) = \rho_f(c) + \rho_b(c) \tag{18.6}$$

We will use the subscript f as an abbreviation for "free"; we avoid confusion with a face label f by always using "free" as a subscript and a face label as an argument. We have also used the subscript f to denote a continuous function; from now on we will rely on context, for example whether or not the argument is continuous, to distinguish continuous functions from discrete fields.

In principle, we can calculate the electrostatic field E from the Poisson equation:

$$-\epsilon_0 \nabla^2 V = \rho = \rho_f + \rho_b$$

Written in terms of E, this is simply Gauss' law:

$$\epsilon_0 \operatorname{div} E = \rho_f + \rho_b = \rho_f - \operatorname{div} P$$

Unfortunately, we can't use this in the usual way to calculate E, because the P on the right-hand side depends on E: the forces that polarize the molecules are exerted by E. However, we can move all the E-dependence to the left-hand side:

$$\operatorname{div}(\epsilon_0 E + P) = \rho_f \tag{18.7}$$

This is similar to Gauss' law, but with the role of E played by

$$D(E) \equiv \epsilon_0 E + P(E) \tag{18.8}$$

and the free charge instead of the total charge on the right-hand side:

$$\operatorname{div} D = \rho_f \quad \text{(Gauss' law for } D) \tag{18.9}$$

Having ρ_f instead of $\rho_f + \rho_b$ on the right-hand side is convenient because ρ_f is known. We can now try to solve for D. (Actually, the part of ρ_f representing induced charge on conductors isn't known initially, but we know how to handle that by requiring the potential to be constant on the conductor, and calculating the surface charge at the end.) This equation may be very difficult to solve, if P depends on E in a complicated way. Fortunately, the polarization $P(f)$ at a face f is frequently just proportional to $E(f)$:

$$P(f) = \epsilon_0 \chi(f) E(f) \tag{18.10}$$

for some constant χ (the Greek letter chi) called the **dielectric susceptibility** (the ϵ_0 is included to make χ dimensionless). In the continuum limit, the vectors are proportional:

$$\mathbf{P(r)} = \epsilon_0 \chi(\mathbf{r}) \mathbf{E(r)} \tag{18.11}$$

A medium in which Eq. 18.11 holds is called **linear** and **isotropic**. ["Isotropic" means that all directions in space are equivalent, which implies that **P** is parallel to **E** because there is no other preferred direction. "Linear" implies further that the magnitudes are proportional.] In this text we will assume that all dielectric media are linear and isotropic.

In a linear, isotropic medium, **D** is given by

$$\mathbf{D} = \epsilon_0 \mathbf{E} + \mathbf{P} = \epsilon_0 E + \epsilon_0 \chi E = \epsilon_0 (1 + \chi) E = \epsilon E \tag{18.12}$$

where

$$\epsilon \equiv \epsilon_0 (1 + \chi) \tag{18.13}$$

is called the **permittivity** of the dielectric. We now see why ϵ_0 was condemned to drag a subscript along through all these chapters: it is a special case of ϵ, namely the permittivity of the vacuum. The dimensionless ratio of the permittivity of a medium to that of the vacuum

$$K = \epsilon/\epsilon_0 = 1 + \chi \tag{18.14}$$

is called the **dielectric constant**.

In general, Gauss' law is not sufficient to solve for D, just as it was usually insufficient to solve for E in the absence of dielectrics (we needed curl $E = 0$ as well). However, just as in the case of Gauss' law for E, there is a class of highly symmetrical problems for which it is easy to solve. If the free charge distribution has spherical, cylindrical, or planar symmetry, we can assume D is uniform over a sphere, cylinder, or plane respectively. This allows us to factor out D in the integral form of Gauss' law:

$$\int_S \mathbf{D} \cdot d\mathbf{a} = q_{\text{free}} \text{ (the free charge inside a surface } S) \tag{18.15}$$

and solve for D.

■ **EXAMPLE 18.1** Consider a free point charge $+q$ embedded in a dielectric that is either infinitely large or a sphere with its center at q. Due to the spherical symmetry, **D** must be radial and have the same magnitude everywhere on a sphere S of radius r (Figure 18.6).

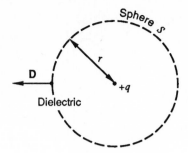

Figure 18.6 A spherical Gaussian surface around a point charge imbedded in a dielectric.

In terms of the magnitude of *D*, the integral is

$$\int_S \mathbf{D} \cdot d\mathbf{a} = 4\pi r^2 D$$

so Gauss' law (Eq. 18.15) can be solved for *D*

$$D = q/4\pi r^2 \tag{18.16}$$

If the dielectric is linear, we can also compute the electric field

$$E = D/\epsilon = q/4\pi\epsilon r^2 = q/4\pi K\epsilon_0 r^2 \tag{18.17}$$

(but note that Eq. 18.16 for *D* is correct even in a **nonlinear** dielectric).

Equation 18.17 is the generalization of Coulomb's law to dielectrics. The electric field due to q is decreased by a factor $1/K$ from its value in a vacuum: the field is said to be partially "screened" by the dielectric. This terminology arose from the fact that a charge q surrounded by grounded **conducting** material (such as a metal box) would be **entirely** screened (a charge $-q$ would be induced on the inside of the box and there would be no field outside the box). A box made of metal window screen would be almost as effective, and would allow visual inspection of the charge inside the sphere. The same idea is used to confine the fields in a microwave oven by putting a screen over the window.

To see physically why the electric field of a free charge is smaller in a dielectric than in a vacuum, we must look at the bound charge near it. A positive free charge $+q$ attracts the negative charges in the dielectric and repels the positive ones (Figure 18.7). There is an outward shift of charge past the faces shown (pieces of spheres, if we use a spherical coordinate system), so the polarization field (Eq. 18.1) is directed outward. (This is also clear from the dipole-moment-per-unit-volume definition in Section 18.2—the dipole moment of each dielectric molecule points outward.) There is a negative bound charge left in the central cell, given quantitatively by Eq. 18.3; in a linear, isotropic material, it is proportional to the total charge ρ

$$\rho_b(c) = -\operatorname{div} P(c) = -\chi\epsilon_0 \operatorname{div} E(c) = -\chi\rho(c) \tag{18.18}$$

The free charge is also proportional to ρ, by Gauss' law (Eq. 18.9)

$$\rho_f(c) = \operatorname{div} D = K\epsilon_0 \operatorname{div} E = K\rho(c) \tag{18.19}$$

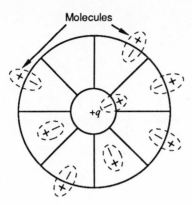

Figure 18.7 A positive charge embedded in a dielectric, showing the polarization of the dielectric.

The bound and free charge must add up to the total, as we can check from Eqs. 18.14, 18.18, and 18.19:

$$\rho_f(c) + \rho_b(c) = K\rho(c) - \chi\rho(c)$$

$$= (1 + \chi)\rho(c) - \chi\rho(c) = \rho(c) \tag{18.20}$$

These proportionalities are not restricted to the simple case of Figure 18.7; they hold for any charge distribution in a uniform, linear, isotropic dielectric. The important point is that the total charge (which determines E by Gauss' law) is **less** than the free charge alone; some of the free charge gets canceled out ("screened") by bound charge.

Another example of screening is afforded by a parallel-plate capacitor filled with dielectric (Figure 18.8). The free charge is now on the bottom and top metal plates; we describe it by the area density $\sigma_f(c) = \rho_f(c)\, dr$ because it resides only on the surface, but it obeys the same proportionality as in Eq. 18.19:

$$\sigma_f = K\sigma \tag{18.21}$$

(It may not be obvious that the derivation of Eq. 18.19 works in the presence of

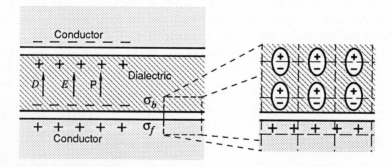

Figure 18.8 A dielectric-filled capacitor, with detail of the region near the lower plate.

conductors; we will discuss this in detail in Chapter 19.) The bound charge (Eq. 18.18) is

$$\sigma_b = -\chi\sigma \tag{18.22}$$

Its physical origin is indicated in the detail of Figure 18.8. The boundary cells have a net negative bound charge due to molecules straddling the face. In the other dielectric cells, the extra negative charge at the top of each cell is canceled by positive charge at the bottom of the cell.

We can calculate the fields by Gauss' law because of the high symmetry of the system, just as we did with no dielectric in Section 14.4. Using Gauss' law for a pillbox at the lower surface (Figure 14.21) gives as before

$$E = \sigma/\epsilon_0 \tag{18.23}$$

But note that this is smaller than the field σ_f/ϵ_0 that σ_f would produce in the absence of dielectric: the field is partly screened out. We could alternatively use Gauss' law for D (Eq. 18.15) in the same way to get

$$D = \sigma_f \tag{18.24}$$

which is equivalent to Eq. 18.23 because $D = \epsilon E$.

Recalling that the capacitance C is the (free) charge divided by the potential difference $V = Ed$, the capacitance per unit area is

$$C/\text{area} = \sigma_f/V = \epsilon/d \tag{18.25}$$

compared to ϵ_0/d for a vacuum-filled capacitor. [This is very nearly the same as an air-filled one; $\epsilon(\text{air}) \cong 1.0006\epsilon_0$.] That is, inserting the dielectric **increases** the capacitance by a factor of $K = \epsilon/\epsilon_0$. For polyethylene, for example, $K = 2.3$.

Dielectric screening is an extremely important phenomenon in the real world. For example, NaCl dissociates and dissolves in water, in spite of the strong attraction of Na^+ for Cl^-, only because the field of each ion is strongly screened by the water ($K \cong 80$). A more esoteric example involves the screening of the charge on an electron by polarization of the vacuum. The theory of quantum electrodynamics holds that quantum fluctuations of the vacuum produce "virtual" electron–positron pairs that can have dipole moments; the vacuum therefore acts as a dielectric medium. The charge $q_{\text{total}} = -e$ that we observe on an electron is related to its actual ("bare") charge q_{free} by

$$-e = q_{\text{free}}/K$$

where K (and therefore the bare charge) is believed to be infinite.

▶ **Problem 18.2**

A cylindrical conductor has charge per unit length λ, and is embedded in a dielectric of permittivity ϵ. Find the electric field at a distance r from the cylinder axis (assume $r > R$, the cylinder radius).   ◀

CHAPTER NINETEEN

BOUNDARY VALUE PROBLEMS
IN DIELECTRICS

SECTION 19.1 POISSON'S EQUATION IN A DIELECTRIC

So far, we have looked only at dielectric problems in which symmetry alone ensured that curl $E = 0$, so that D could be computed from Gauss' law alone. In most cases (such as that in Figure 18.1), of course, we also need to use curl $E = 0$. We showed in Chapter 15 and Chapter 16 that instead of solving Gauss' law and curl $E = 0$ simultaneously, we could equivalently define a potential V so $E = -\operatorname{grad} V$ (this ensures curl $E = 0$) and just solve Gauss' law (which in terms of V is called Poisson's equation) for V. This latter procedure works in a dielectric if we use Gauss' law for D (Eq. 18.9) and $D = \epsilon E = \epsilon \operatorname{grad} V$:

$$-\operatorname{div}(\epsilon \operatorname{grad} V) = \rho_{\text{free}} \tag{19.1}$$

This "dielectric Poisson equation" differs from the usual vacuum one only by a factor ϵ/ϵ_0, and can easily be solved by the usual methods whenever ϵ is uniform. In particular, V satisfies Laplace's equation wherever there is no free charge. However, ϵ is never uniform over all space, and we must be careful in applying Eq. 19.1 where ϵ changes.

It is simplest to do this in a discrete calculation. D is, of course, a face vector field because E and P are. The coefficient ϵ will be different in different dielectric materials and we denote its value at a particular face by $\epsilon(f)$. Thus,

$$D(f) = \epsilon(f)E(f) = \epsilon_0 E(f) + P(f) \tag{19.2}$$

We can give a general prescription for solving a general problem (e.g., Figure 18.1) with conductors, dielectrics, and isolated charges (which may be in the vacuum or the dielectric regions):

1. Solve the dielectric Poisson equation (Eq. 19.1) for $V(c)$, subject to boundary conditions on conductors.
2. Extract the fields:

$$E(f) = -[\operatorname{grad} V](f)$$

$$D(f) = \epsilon(f)E(f)$$

Explicitly, the discrete dielectric Poisson equation is

$$\rho_f(c) = -d\tau(c)^{-1} \sum_{f \text{ of } c} da(f)\epsilon(f) \operatorname{grad} V(f) \tag{19.3}$$

which in cartesian coordinates is

$$\rho_f(c) = dr^{-3} \sum_{f} dr^2 \epsilon(f)\, dr^{-1}[V(\text{front}) - V(\text{back})] \tag{19.4}$$

where the sum is over the six faces of the cell c; $V(\text{front})$ is the potential of a neighboring cell and $V(\text{back})$ is $V(c)$ itself, as in Figure 16.6. If $\epsilon \neq 1$, this Poisson equation is not the same as the one for which we proved, in Section 16.6, the existence and uniqueness of solutions. However, the proof can be straightforwardly generalized if $\epsilon(f) > 0$ at every face.

■ **EXAMPLE 19.1**   Consider the pseudo-one-dimensional system shown in Figure 19.1a, which is a discretization of the parallel-plate capacitor problem of Section 18.3. The potential and fields depend only on z, so there is only one unknown potential V_2, at the cells whose centers have $z = dr$. We have indicated $\epsilon(f)$ at the three inequivalent faces f in the nonconducting region, assuming a dielectric constant $K = 3$. (Note that $\epsilon(f)$ is irrelevant at a conducting face, because $E(f) = 0$ there. In electrostatics we may assume $P = D = 0$ in a conductor.) To write the Poisson equation, it is easiest to first determine E in terms of V:

$$E(f) = dr^{-1}[V(\text{back}) - V(\text{front})] \tag{19.5}$$

as shown in Figure 19.1b. We show $D(f) = \epsilon(f)E(f)$ in Figure 19.1c. Writing Gauss' law with this D gives us an equation that is equivalent to Poisson's equation for the

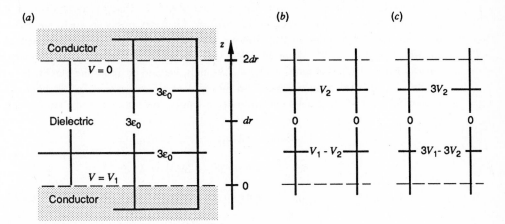

Figure 19.1   (a) A dielectric-filled parallel-plate capacitor. (b) The electric field. The quantity shown at each face f is $E(f)dr$. (c) $D(f)dr/\epsilon_0$, for each face f.

nonconducting cell,

$$\mathrm{div}D = dr^{-1}[3V_2 - (3V_1 - 3V_2)] = \rho_{free} = 0$$

whose solution is

$$V_2 = V_1/2$$

From V we can obtain the electric field from Eq. 19.5, which gives

$$E = V_1/2\,dr$$

at both faces ($z = 0.5\,dr$ and $1.5\,dr$). This is the same answer we would have obtained without the dielectric (and in fact it is the exact continuum answer, because we chose the conducting surfaces to lie at cell centers). The other fields are

$$D = K\epsilon_0 V_1/2\,dr = 3\epsilon_0 V_1/2\,dr$$

$$P = D - \epsilon_0 E = 2\epsilon_0 V_1/2\,dr$$

pointing upward at these same faces. The divergences of E, D, and P (i.e., ρ/ϵ_0, ρ_{free}, and $-\rho_{bound}$) are zero at cells in the dielectric (the contributions from the upper and lower faces cancel). They are only nonzero at the boundary cells of the conductors, where each divergence has only one nonzero term. At the surface of the bottom plate,

$$\rho(c) = \epsilon_0 dr^{-1} \sum E(f) = \epsilon_0 V_1/2\,dr$$

$$\rho_{free}(c) = dr^{-1} \sum D(f) = 3\epsilon_0 V_1/2\,dr$$

(this is induced charge on the conductor), and

$$\rho_{bound}(c) = -dr^{-1} \sum P(f) = -2\epsilon_0 V_1/2\,dr$$

(on the dielectric). These results are in exact agreement with the ones we got in Section 18.3 by a less general method.

■ **EXAMPLE 19.2**   As a less trivial example, we will work out in detail the pseudo-2D discrete dielectric problem shown in Figure 19.2a. It can be thought of as a discretization of the problem of a long square metal duct, $4\,dr$ on a side, with a concentric square $2\,dr \times 2\,dr$ dielectric bar down the middle that has a line of free charge down its axis with λ_0 coulombs per meter. The dielectric constant is $K = 3$. Thus the permittivity is $\epsilon = 3\epsilon_0$ inside the dielectric and $\epsilon = \epsilon_0$ in the vacuum; for faces partly in each, we will take an average, $\epsilon = 2\epsilon_0$ (Figure 19.2b). There are only three unknown potentials because of the square symmetry, say V_1, V_2, and V_3. (We take the potential V_0 of the outer conductor to be zero — Figure 19.2c.) In terms of these potentials, we can compute the electric field (Figure 19.3a). (Fields are taken positive if to the right or up, as usual.) The D field is then as shown in Figure 19.3b. So we can write down Gauss' law (i.e., Poisson's equation) for each cell:

$$\text{Center cell } (V_1): \qquad \mathrm{div}D = \rho_f, \text{ or}$$

$$dr^{-1}[4\epsilon_0 dr^{-1}(3V_1 - 3V_2)] = \lambda_0 dr^{-2}$$

$$V_2\text{-cell:} \qquad dr^{-1}\epsilon_0 dr^{-1}[-(3V_1 - 3V_2) + (2V_2 - 2V_3) - (2V_3 - 2V_2) + V_2]$$

$$= \rho_f = 0$$

$$V_3\text{-cell:} \qquad dr^{-1}\epsilon_0 dr^{-1}[2(2V_3 - 2V_2) + 2V_3] = 0$$

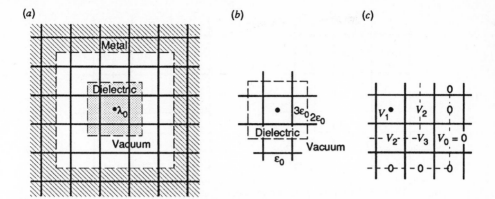

Figure 19.2 (*a*) A cross section of a metal duct with a concentric dielectric bar (the surfaces of the metal and the dielectric are shown as dashed lines). The solid lines show the discrete lattice we will use to compute the fields. (*b*) The dielectric constant $\epsilon(f)$ for each face touching the dielectric. (*c*) The labeling of the potentials in the lower right quarter of the system (the other three quarters are equivalent).

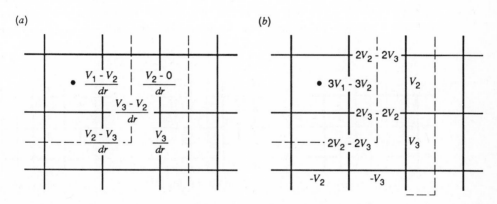

Figure 19.3 (*a*) The electric field $E(f)$, in terms of the potentials. (*b*) The D field, in units of ϵ_0/dr.

These are three equations for three unknowns, which reduce to

$$12V_1 - 12V_2 = \lambda_0/\epsilon_0$$

$$8V_2 - 4V_3 - 3V_1 = 0$$

$$6V_3 - 4V_2 = 0$$

Solve the first and third equations for V_1 and V_3 in terms of V_2, and substitute in the second equation, giving

$$\tfrac{7}{3}V_2 - \tfrac{3}{12}\lambda_0/\epsilon_0 = 0$$

so

$$V_2 = \tfrac{9}{84}\lambda_0/\epsilon_0, \qquad V_1 = \tfrac{16}{84}\lambda_0/\epsilon_0, \qquad V_3 = \tfrac{6}{84}\lambda_0/\epsilon_0$$

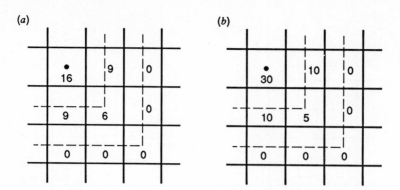

Figure 19.4 (a) V for the dielectric system, in units $\lambda_0/84\epsilon_0$. (b) V for the case of no dielectric, in units $\lambda_0/80\epsilon_0$.

This potential field is shown in Figure 19.4a. Without the dielectric, the answers would have been those in Figure 19.4b. The greatest change is inside the dielectric. We can now easily compute $E = -\text{grad}V$, $D = \epsilon E$, and the polarization $P = D - \epsilon_0 E$ (Figure 19.5). Figure 19.6 shows the bound charge density $\rho_b = -\text{div}P$ and the free charge density $\rho_f = \text{div}D$, from which we can obtain the charge induced on the conducting box $(-6-9-6 = -21$ on each side, for a total of -84, which is equal and opposite to

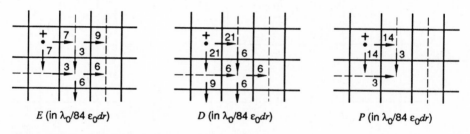

Figure 19.5 E, D, and P for the dielectric system.

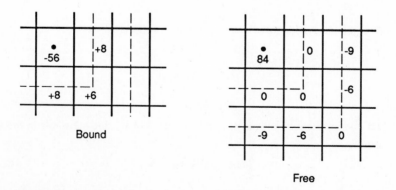

Figure 19.6 The bound and free charge densities, in units $\lambda_0/84\,dr^2$.

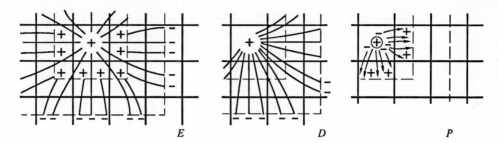

Figure 19.7 The electric field lines and the lines of D and P. To make the number of E lines — which must be proportional to $E(f)$ — tractable, we have used one line for each $(3/84)\lambda_0/\epsilon_0\,dr$ in $E(f)$. The lines begin at charges — one for each $(3/84)\lambda_0$ of total linear charge density. For the P and D fields, we have drawn one line for each $(3/84)\lambda_0/dr$.

the charge at the center). The fact that this ρ_f agrees with what we originally assumed in the nonconducting cells (0 and 84) is a good check on our algebra.

It is interesting to use these results to sketch the lines of E, as in Figure 19.7. Figure 19.7 also shows the D-lines; these are the same as the E-lines in the vacuum ($D = \epsilon_0 E$) but start and stop only at **free** charge (i.e., div$D = \rho_f$). The P-lines begin and end at **bound** charges, and exist only in the dielectric.

▶   **Problem 19.1**

Consider a pseudo-2D system (Figure 19.8) consisting of a $2\,dr \times 2\,dr$ square bar of dielectric having $K = 3$, halfway between two infinite plane conductors at $y = \pm 2\,dr$.

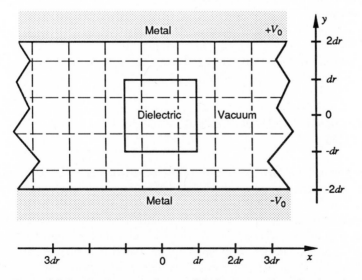

Figure 19.8 Cross section of a parallel-plate capacitor with a long bar of dielectric inside.

The conductors are at potentials $V = \pm V_0$, respectively. The boundary condition at $x \to \pm \infty$ (far from the dielectric) is that the potential approaches that of a parallel-plate capacitor. To make the calculation finite, assume this is achieved already at $x = \pm 3\, dr$: $V(\pm 3\, dr, y) = V_0 y/2\, dr$. (One could check the reasonableness of this by assuming it farther out instead and seeing how the answer changes. The error turns out to be of order 10^{-2}.) Use the discretization shown, assigning the proper ϵ to each face. For faces that are half in the dielectric, use the average $\epsilon = 2\epsilon_0$.

a. Exploit the mirror symmetries $V(x, -y) = -V(x, y)$ and $V(-x, y) = +V(x, y)$ to show there are only three different potentials to solve for in this discretization.

b. Solve for V and show it on a diagram.

c. Find E, D, P, ρ_{free}, and ρ everywhere, and indicate on diagrams. (As a check, verify that ρ_{free} resides only on conductors.) Roughly sketch the lines of **E**. ◄

▶ **Problem 19.2**

(Discrete formula for the dipole moment of a dielectric.) Given $P(f)$ and its discrete divergence $\rho_b(c) = -\text{div}\, P(f)$, there are two plausible ways to get the total dipole moment vector **p** of a piece of dielectric:

(1) **P** can be thought of as dipole moment per unit volume, so

$$p_x = \sum_f P(f)\, dr^3 \qquad (\text{sum over } f\text{'s normal to } x\text{-axis})$$

$$p_y = \sum_f P(f)\, dr^3 \qquad (\text{sum over } f\text{'s normal to } y)$$

and similarly for p_z.

(2) The continuum formula, Eq. 17.70, gives $p_x = \int x \rho_b(\mathbf{r})\, d\tau$, which suggests

$$p_x = \sum_c c_x \rho_b(c)\, dr^3$$

where c_x is the x-coordinate of the center of cell c, and similar formulas for p_y and p_z.

Show that these two methods give the same answer (exactly, not just as $dr \to 0$). Hint: Prove it first for a polarization field $P(f)$ that is nonzero only at **one** face, and appeal to linearity. ◄

▶ **Problem 19.3**

Use the results of Problems 19.1 and 19.2 to compute the dipole moment per unit length of the bar in Figure 19.8. ◄

SECTION 19.2 BOUNDARY CONDITIONS AT DIELECTRIC INTERFACES

In this section, we will set up continuum problems involving dielectrics and derive the proper boundary conditions on the fields. Let us begin by briefly reviewing how we set up and solved continuum problems without dielectrics in Section 16.7. A continuum conducting-boundary problem in electrostatistics is specified by giving the shapes of, and potentials on, conducting surfaces and giving the continuum charge density. For any lattice with spacing dr, we can define a corresponding discrete problem, in which we can solve Poisson's equation for the potential and find the fields and induced surface charges on the conductors. We defined the

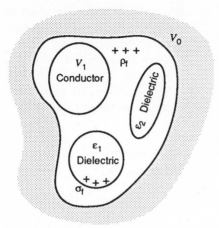

Figure 19.9 A general electrostatics problem involving dielectrics.

solution to the continuum problem as the $dr \to 0$ limit of these discrete solutions. This limit was a continuous function $V(\mathbf{r})$, which (except at conducting boundaries and where ρ is singular) satisfied the continuum Poisson equation (with $\nabla^2 = d^2/dx^2 + d^2/dy^2 + d^2/dz^2$) and had the correct values at the conducting boundaries. In special cases in which we could solve the Poisson differential equation exactly, we could thereby avoid having to discretize the problem and take the limit. And from the continuum potential V we could calculate the continuum function $E(\mathbf{r}) = -\operatorname{grad} V(\mathbf{r})$ and the continuum surface charge density $\sigma(\mathbf{r})$.

Consider now the corresponding problem including dielectrics (Figure 19.9). It is specified by giving, in addition to the shapes of the conductors, the shapes and permittivities ($\epsilon_1, \epsilon_2, \ldots$) of the pieces of dielectric, and the continuum free charge density everywhere except on the conductors (where the potential is given instead). The free charge can reside in the vacuum or in the dielectrics, and can include in general a volume density $\rho_{\text{free}}(\mathbf{r})$, surface densities $\sigma_{\text{free}}(\mathbf{r})$, line charges $\lambda_{\text{free}}(\mathbf{r})$, and free point charges q_{free}. (See Appendix C for a discussion of what these functions mean.)

As before, we will define the solution to the continuum dielectric conducting-boundary problem as the limit of the solutions to discretizations of the problem (such as those solved in Section 19.1) as $dr \to 0$. Exactly how we handle ambiguities such as determining $\epsilon(f)$ for a face f that straddles an interface should be irrelevant in the limit. Evidently, the resulting continuum potential satisfies the continuum Poisson equation for a dielectric [$\operatorname{div}(\epsilon \operatorname{grad} V) = -\rho$] in each region, because this is the limit of the discrete Poisson equation. (We will omit the subscript we have previously used to distinguish continuum from discrete fields, relying on the context or the continuum argument $\mathbf{r}$ to indicate this.) We might hope that (as in the nondielectric case) we could find the continuum solution by solving this differential equation for V, requiring it to have the correct values at the conducting boundaries. However, this does not work at a dielectric–dielectric boundary: we cannot demand that the continuum Poisson equation be satisfied there because the derivatives appearing in it do not exist. To see this, note that if E is continuous, then, because ϵ is obviously discontinuous across a dielectric–vacuum or dielectric–dielectric interface, so is $D = \epsilon E$. If D is continuous, then E isn't; they can't both be continuous.

So fields in dielectric problems must be allowed to be **piecewise continuous** functions. That is, they are continuous within each homogeneous region (dielectric ϵ_1, vacuum, dielectric ϵ_2, and so on) but may be discontinuous across interfaces. There is no analogous argument implying that V must be discontinuous and it turns out to be consistent to assume that it is continuous. However, because it has discontinuous derivatives, if we want to represent it by analytic functions we must use a different function in each region. Thus, we will represent V by several functions $V_1, V_2, \ldots$, such that

$$V(\mathbf{r}) = \begin{cases} V_1(\mathbf{r}) \text{ if } \mathbf{r} \text{ is in region 1} \\ V_2(\mathbf{r}) \text{ if } \mathbf{r} \text{ is in region 2} \\ \text{and so on} \end{cases} \tag{19.6}$$

To ensure that V is continuous, we must require $V_i(\mathbf{r}) = V_j(\mathbf{r})$ at each point $\mathbf{r}$ of the interface between regions i and j. Although it is continuous everywhere, it is only piecewise differentiable (i.e., only in the interior of each dielectric region). Within region i, we get the electric field by differentiating V_i:

$$\mathbf{E}_i(\mathbf{r}) = -\operatorname{grad} V_i(\mathbf{r}) \tag{19.7}$$

so that

$$\mathbf{E}(\mathbf{r}) = \begin{cases} \mathbf{E}_1(\mathbf{r}) \text{ if } \mathbf{r} \text{ in 1st region} \\ \mathbf{E}_2(\mathbf{r}) \text{ if } \mathbf{r} \text{ in 2nd region} \\ \text{and so on} \end{cases} \tag{19.8}$$

These $\mathbf{E}_i$ of course determine the corresponding $\mathbf{D}_i(\mathbf{r}) = \epsilon_i \mathbf{E}_i(\mathbf{r})$.

Now the Poisson equation at an interior cell of region i has the continuum limit

$$\epsilon_i \nabla^2 V_i(\mathbf{r}) = \rho_f(\mathbf{r}) \tag{19.9}$$

so we must require V_i to satisfy this partial differential equation in the interior of each region. To completely determine V, however, we need to use the Poisson equation (or Gauss' law) at the cells on the **interface**. To be specific, let us choose a coordinate system such that the interface is a surface of constant value of some coordinate (if the surface is a sphere, for example, this coordinate would be r) and put a layer of cell centers at the interface, as in Figure 19.10a. Consider a cell on the interface (Fig. 19.10b). We will assume cartesian coordinates here for simplicity; the result would be the same in any coordinate system. Then $da = dr^2$, $d\tau = dr^3$, and Gauss' law for this cell is

$$dr^{-1}[D(f_1) - D(f_2) + D(f_3) - D(f_4)] = \rho_{\text{free}}(c) \tag{19.10}$$

(There are two more faces f_5 and f_6 parallel to the paper that do not appear in this cross section; these can be treated exactly like f_3 and f_4, and we will not include them explicitly.) One must be careful in computing the continuum fields at the point $\mathbf{r}$ as limits of discrete fields. The limit that defines $\mathbf{D}(\mathbf{r})$ elsewhere (the limit of $D(f)$

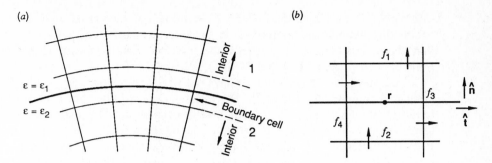

Figure 19.10 (*a*) Cross section of a dielectric interface, with an appropriately chosen discrete cell lattice. (*b*) A cell centered at a point **r** of the interface, with four of its faces.

as $dr \to 0$ for a sequence of f's approaching **r**) doesn't exist here. If we look at D at f_1, which approaches **r** from region 1, we will find $D(f_1) \to D_{1n}(\mathbf{r})$, the component of $\mathbf{D}_1$ normal to the interface (because this is the limit for any **r** above the interface). On the other hand, $D(f_2) \to D_{2n}(\mathbf{r})$, because f_2 approaches from region 2. These limits of D can be different, because $\mathbf{D}(\mathbf{r})$ can be discontinuous. It is less obvious what the limit of $D(f_3)$ is; at a point **r** above the interface it is $D_{1t}(\mathbf{r})$, the tangential component of $\mathbf{D}_1$, and below the interface it is $D_{2t}(\mathbf{r})$. The same is true of $D(f_4)$. At the interface, each turns out to be the average $\frac{1}{2}[D_{1t}(\mathbf{r}) + D_{2t}(\mathbf{r})]$, although we will not prove this here. Actually it doesn't matter what the limit is, as long as it is the same for $D(f_3)$ and $D(f_4)$, so they cancel from Eq. 19.10. Multiplying that equation by dr and taking the limit yields

$$D_{1n}(\mathbf{r}) - D_{2n}(\mathbf{r}) = \lim \rho_{\text{free}}(c)\, dr \qquad (19.11)$$

If the continuum free charge density is given by a (nonsingular) continuum volume charge $\rho_{\text{free}}(\mathbf{r})$, the factor of dr on the right-hand side causes it to vanish in the limit; Eq. 19.11 then states that the normal components of $\mathbf{D}_1$ and $\mathbf{D}_2$ are equal at the interface (D_n is continuous across the interface). A more general case is that in which the continuum free charge density includes a surface charge density $\sigma_{\text{free}}(\mathbf{r})$ at the interface (one could create such a surface charge at a dielectric–vacuum interface by rubbing the dielectric, for example). Then the charge in the cell is $\rho_{\text{free}}(\mathbf{r})\, d\tau + \sigma_{\text{free}}(\mathbf{r})\, da$ for small dr (see Appendix C for details), and the right-hand side of Eq. 19.11 is $\lim \rho_{\text{free}}(\mathbf{r})\, dr + \lim \sigma_{\text{free}}(\mathbf{r})$. The surface charge term survives the limiting process, and

$$D_{1n}(\mathbf{r}) - D_{2n}(\mathbf{r}) = \sigma_{\text{free}}(\mathbf{r}) \qquad (19.12)$$

Note that while σ_{free} is often zero, a surface density of **bound** charge is usually to be expected (recall we found one at a dielectric–conductor interface in the capacitor problem).

We now have written continuum equations equivalent to the discrete Gauss' law at cells in the interior of dielectrics (Eq. 19.9) and at dielectric interfaces (Eq. 19.12). The limit of the existence and uniqueness theorem suggests that there is a unique solution to these equations. That is, if we can find continuum functions V_i

satisfying Poisson's equation in each dielectric region, having the right values at the conducting boundaries, continuous at the interfaces between dielectrics, and such that the corresponding D_{in}'s satisfy the interface condition (Eq. 19.12), then the potential (Eq. 19.6) determined by the V_i's is the correct one.

You may have noticed that we have not written any boundary condition for the **tangential** component of **E** or **D**. There is such a condition, but it is equivalent to requiring the continuity of V and therefore redundant. To derive the condition, consider a point **r** moving in a direction $\hat{\mathbf{t}}$ along the interface. The rate of change of $V_1(\mathbf{r})$ (with respect to distance) is exactly the component of its gradient along the motion, that is, $-E_{1t}(\mathbf{r})$. Similarly, the rate of change of $V_2(\mathbf{r})$ is $-E_{2t}(\mathbf{r})$. These V_i's are the same at the interface (i.e., V is continuous), so we have there

$$E_{1t}(\mathbf{r}) = E_{2t}(\mathbf{r}) \tag{19.13}$$

This condition can also be derived directly from curl $E = 0$, by an argument similar to the one we used for the boundary condition on D_n.

▶　　**Problem 19.4**

Write the discrete equation curl $E(e) = 0$ for the edge e indicated in Figure 19.11. The faces touching it are f_1, f_2, f_3, and f_4. Express $E(f_2)$, $E(f_3)$, and $E(f_4)$ in terms of $\mathbf{E}_1$ or $\mathbf{E}_2$ in the $dr \to 0$ limit (all faces approach **r**). Assume $E(f_1) \to \frac{1}{2}[E_{1t}(\mathbf{r}) + E_{2t}(\mathbf{r})]$ by analogy with the D_n case. Show that the curl equation reduces to Eq. 19.13.　　　◀

Once the D and E fields have been computed in a continuum problem, one can of course calculate $\mathbf{P} = \mathbf{D} - \epsilon_0 \mathbf{E}$ on both sides of an interface. Like **D**, it can be discontinuous across the interface; indeed, if one side is vacuum, $\mathbf{P} = 0$ on that side, whereas it will not, in general, vanish at the surface of the dielectric. We can calculate the density of bound charge in the interiors of regions 1 and 2 from $\rho_b = -\nabla \cdot \mathbf{P}$, the continuum limit of Eq. 18.3. It turns out that there is also a

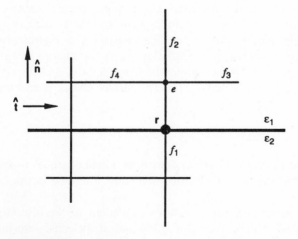

Figure 19.11　An edge e (normal to the paper) next to a dielectric interface, and the four faces touching it.

surface density of bound charge. To see this, consider a cell c at the interface, such as the one centered at $\mathbf{r}$ in Figure 19.10b, where Eq. 18.3 gives

$$\rho_b(c) = -\operatorname{div} P$$

$$= -\left[P(f_1) - P(f_2) + P(f_3) - P(f_4)\right]/dr$$

$$dr\,\rho_b(c) = -P(f_1) + P(f_2) - P(f_3) + P(f_4) \tag{19.14}$$

We now take the $dr \to 0$ limit. If we allow a continuum volume density $\rho_b(\mathbf{r})$ and a surface density $\sigma_b(\mathbf{r})$, the discrete $\rho_b(c)$ is given for small dr by (see Appendix C)

$$\rho_b(c) = \rho_b(\mathbf{r}) + \sigma_b(\mathbf{r})/dr$$

When we multiply by dr, the term in $\rho_b(\mathbf{r})$ vanishes in the $dr \to 0$ limit, leaving the surface density of bound charge σ_b. On the right-hand side of Eq. 19.14, the f_3 and f_4 terms cancel as $dr \to 0$ (both involve the same average of P_{1t} and P_{2t}, and $f_3 \to f_4$). The contributions from the faces f_5 and f_6 parallel to the plane of the paper and not shown in Figure 19.10b also cancel (this is why we ignored them in Eq. 19.14). However, $P(f_1) \to P_{1n}(\mathbf{r})$ and $P(f_2) \to P_{2n}(\mathbf{r})$ as $f_1, f_2 \to \mathbf{r}$, and these limits may not be the same. So the limit of Eq. 19.14 is

$$\sigma_b(\mathbf{r}) = P_{2n}(\mathbf{r}) - P_{1n}(\mathbf{r}) \tag{19.15}$$

and there is in general a nonzero surface density of bound charge. Evidently, the first term yields the charge on dielectric 2, and the second term the charge on dielectric 1. In the common case that region 1 is a vacuum ($\epsilon_1 = \epsilon_0$) as in Figure 19.12, we can denote $\mathbf{P}_2$ by $\mathbf{P}$ (there is no $\mathbf{P}_1$) and

$$\sigma_b = P_n \tag{19.16}$$

That is, any dielectric surface has a charge density given by the outward component of the polarization vector. This is in addition to the continuum volume charge density $\rho_b = -\nabla \cdot \mathbf{P}$; the divergence doesn't make sense at the surface because $\mathbf{P}$ is discontinuous. This complication doesn't arise in the discrete formulation, because there $\nabla \cdot \mathbf{P}$ does make sense at the surface, and gives the correct surface charge.

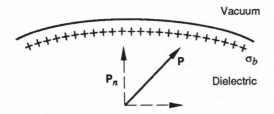

Figure 19.12 Vacuum–dielectric interface, showing the normal component of the polarization vector.

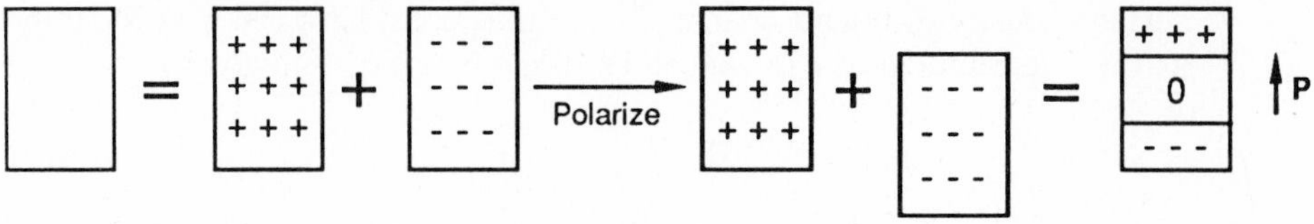

Figure 19.13 How a uniform polarization produces a surface density of bound charge.

There is a simple way to visualize the surface bound charge density in the common case that the polarization is uniform (i.e., the charge is shifted by the same amount and in the same direction throughout the dielectric). Think of the unpolarized dielectric as a superposition of positive and negative charge (nuclei and electrons, if you like) that move in opposite directions when the material is polarized. Then they still cancel in the interior, giving zero net charge (note that $\nabla \cdot \mathbf{P} = 0$ if $\mathbf{P}$ is uniform, so $\rho_b = 0$) but there is a thin layer of uncanceled positive charge at one end and negative charge at the other, as shown in Figure 19.13.

The absence of bound charge in the interior is a quite general property of linear, isotropic dielectrics. Because P is proportional to D, $\rho_b = -\operatorname{div} P$ is proportional to $\rho_{\text{free}} = \operatorname{div} D$. Ordinarily there is no **free** charge inside a dielectric, so there can be no **bound** charge either.

SECTION 19.3　SOLVING CONTINUUM CONDUCTING-BOUNDARY PROBLEMS WITH DIELECTRICS

In this section we will solve three sample conducting-boundary problems. Let us first summarize the rules we arrived at in the previous section. To compute the fields in a continuum boundary-value problem involving dielectrics (such as that in Figure 19.9), we must find a piecewise twice-differentiable function $V(\mathbf{r})$, composed of twice-differentiable pieces $V_1, V_2, \ldots$, which give its value in the various dielectric ($\epsilon = \epsilon_1, \epsilon_2, \ldots$) or vacuum ($\epsilon = \epsilon_0$) regions. Each piece must satisfy Poisson's equation in its own region

$$\epsilon_i \nabla^2 V_i(\mathbf{r}) = \rho_{\text{free}}(\mathbf{r}) \tag{19.17}$$

and at the interfaces between dielectrics (such as that between regions 1 and 2 in Figure 19.14) they must satisfy the boundary conditions

$$D_{1n} - D_{2n} = \sigma_{\text{free}} \tag{19.18}$$

$$V_1 = V_2 \tag{19.19}$$

We can consider a dielectric–vacuum interface such as those in Figure 19.9 as a special case: the vacuum can be considered a dielectric with $\epsilon = 1$. At dielectric–conductor interfaces, we require only the correct prespecified potential. It turns out, however, that Eq. 19.18 is still correct at a dielectric–conductor interface (you can check that the derivation in Section 19.2 still works in this case). Since $\mathbf{D} = 0$ in the conductor, it can be used to calculate the free surface charge induced on the conductor.

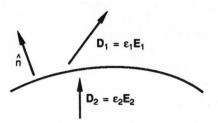

Figure 19.14 A boundary between continuum dielectrics.

There is a useful mnemonic device for remembering Eq. 19.18. Recall that it arose from Gauss' law, which can be written

$$\int_S \mathbf{D} \cdot d\mathbf{a} = q_{\text{free}}$$

where q_{free} is the total free charge inside a surface S. Choose for S a "Gaussian pillbox," a very short cylinder with cross-sectional area A, with one circular face in each dielectric (Figure 19.15). Then if A is small enough so $\mathbf{D}$ doesn't vary over the flat surfaces, and the height is small enough that we can ignore the cylindrical surface,

$$\int \mathbf{D} \cdot d\mathbf{a} = AD_{1n} - AD_{2n}$$

and because

$$q_{\text{free}} = A\sigma_{\text{free}}$$

our boundary condition (Eq. 19.18) follows. We refer to this as a mnemonic device rather than a proof because we have ignored questions of how the height and the area of the pillbox should vanish in the limit (the former has to vanish faster) but it is nonetheless very useful. There is a similar mnemonic device for showing the continuity of the tangential electric field (Eq. 19.13), which is equivalent to Eq. 19.19. Write the curl equation $\nabla \times \mathbf{E} = 0$ in its integral form using Stokes' law:

$$\int \mathbf{E} \cdot d\mathbf{l} = 0$$

and apply this to the closed rectangular path shown in Figure 19.16. The height of the rectangle is allowed to approach zero faster than the length L along the surface,

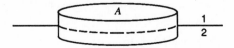

Figure 19.15 A Gaussian pillbox for calculating the free surface charge at a dielectric interface.

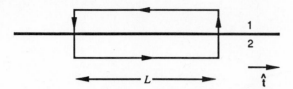

Figure 19.16 A path cutting a dielectric interface, for deriving the boundary condition on transverse electric field.

so the line integral is

$$E_{2t}L - E_{1t}L = 0$$

so that

$$E_{2t} = E_{1t}$$

Once $V(\mathbf{r})$ has been found, we can get $\mathbf{E} = -\nabla V$ and $\mathbf{D} = \epsilon \mathbf{E}$, $\mathbf{P} = \chi \epsilon_0 \mathbf{E}$ in each region. There is no bound charge in the interior of the dielectrics, but there is a surface bound charge density on each dielectric that can be obtained from the outward normal component of the polarization vector

$$\sigma_b(\mathbf{r}) = P_n \tag{19.20}$$

Equation 19.20 is almost obvious from the definition of P (the amount of charge per unit area transferred across a surface); for a derivation see Eq. 19.16.

■ **EXAMPLE 19.3** The simplest example of the use of these rules is the dielectric-filled infinite parallel-plate capacitor for which we have calculated the fields twice before (once using the continuum Gauss law and once as a discrete boundary value problem). To introduce a dielectric interface we will generalize it to the system shown in Figure 19.17, which has **two** dielectric slabs, with equal thickness $\frac{1}{2}d$ and permittivities ϵ_2 and ϵ_3. (We will use the labels 0 and 1 to refer to the conducting plates.) This is a pseudo-1D problem, so the potential depends only on z and Poisson's equation in region 2 is

$$d^2 V_2(z)/dz^2 = 0$$

But this is just Laplace's equation (there is no free charge in the dielectric), for which we determined the most general solution by integration in Section 16.7:

$$V_2(z) = a_2 z + b_2 \tag{19.21}$$

Similarly, the most general solution in region 3 is

$$V_3(z) = a_3 z + b_3 \tag{19.22}$$

The four coefficients a_2, b_2, a_3, and b_3 must be determined by the boundary conditions. Fortunately there are also four of these: V and D_n are continuous across the $2-3$ interface, and V has the correct values at the $0-2$ and $3-1$ interfaces. Continuity of V at $z = \frac{1}{2}d$ implies

$$a_2 d/2 + b_2 = a_3 d/2 + b_3 \tag{19.23}$$

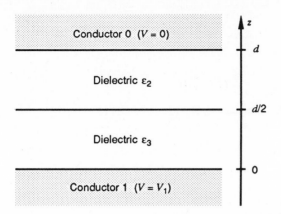

Figure 19.17 A parallel-plate capacitor with an inhomogeneous dielectric.

The D-fields have only z-components,

$$D_{2z}(z) = -\epsilon_2 dV_2/dz = -\epsilon_2 a_2$$

$$D_{3z}(z) = -\epsilon_3 a_3$$

the continuity of which requires

$$\epsilon_2 a_2 = \epsilon_3 a_3 \tag{19.24}$$

Finally, at the $z = 0$ conducting boundary

$$V_3(z) = b_3 = V_1 \tag{19.25}$$

and at the $z = d$ boundary

$$V_2(z) = a_2 d + b_2 = 0 \tag{19.26}$$

Solving Eqs. 19.23 through 19.26 and substituting the a's and b's into Eqs. 19.21 and 19.22 for V gives

$$V_2(z) = 2\epsilon_3 V_1(d - z)/d(\epsilon_2 + \epsilon_3) \tag{19.27}$$

$$V_3(z) = V_1 - 2\epsilon_2 V_1 z/d(\epsilon_2 + \epsilon_3) \tag{19.28}$$

which is the exact solution. The potential and the D and E fields are plotted in Figure 19.18 for the case $\epsilon_3 = 3\epsilon_2$; note that E is discontinuous at the interface.

■ **EXAMPLE 19.4 A DIELECTRIC SPHERE IN A FIELD**   As another example of the application of these rules, consider a dielectric sphere in an otherwise uniform electric field. There are two regions of uniform permittivity in this problem, the inside and the outside of the sphere. The potential, therefore, has two continuous pieces, V_{in} and V_{out}, which determine the fields $\mathbf{E}_{in} = -\nabla V_{in}$ and $\mathbf{D}_{in} = \epsilon \mathbf{E}_{in}$, and $\mathbf{E}_{out} = -\nabla V_{out}$ and $\mathbf{D}_{out} = \epsilon_0 \mathbf{E}_{out}$. The functions V_{in} and V_{out} may have completely different analytic forms, as long as they satisfy our boundary conditions at the interface. In this case, there is another boundary condition at infinity, that $\mathbf{E}_{out} \rightarrow \mathbf{E}_0$, some fixed uniform field. There is no free charge anywhere, so Poisson's equation reduces to Laplace's equation $\nabla^2 V = 0$ both inside and outside the dielectric. In spherical coordinates we know the most general function that is independent of ϕ and

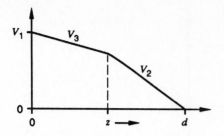

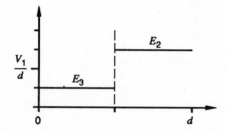

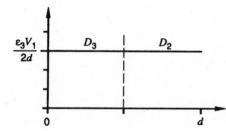

Figure 19.18 The V, E, and D fields for the inhomogeneous capacitor of Figure 19.17.

satisfies Laplace's equation (Section 17.2), which we can write here as

$$V_{out}(r, \theta) = \sum_l A_l^{out} r^l P_l(\cos \theta) + \sum_l B_l^{out} r^{-l-1} P_l(\cos \theta)$$

where A_l^{out} and B_l^{out} are constants. The form of V is the same inside, but the constants may be different:

$$V_{in}(r, \theta) = \sum_l A_l^{in} r^l P_l(\cos \theta) + \sum_l B_l^{in} r^{-l-1} P_l(\cos \theta)$$

We must determine the A's and B's from the boundary conditions. If we take the z-axis along the field $\mathbf{E}_0$, then as $r \to \infty$

$$V_{out}(r, \theta) \to -E_0 r \cos \theta \qquad \text{(uniform field)}$$

$$= -E_0 r P_1(\cos \theta)$$

Matching coefficients of each P_l gives

$$A_l^{out} = 0 \qquad (l \neq 1)$$

$$A_1^{out} = -E_0$$

At $r = 0$, the "boundary condition" is that V_{in} must not diverge, so

$$B_l^{in} = 0$$

The boundary condition at $r = R$ requires the continuity of V:

$$V_{out}(R, \theta) = V_{in}(R, \theta)$$

or

$$-E_0 R P_1(\cos \theta) + \sum_l B_l^{out} R^{-l-1} P_l(\cos \theta) = \sum_l A_l^{in} R^l P_l(\cos \theta)$$

Equating coefficients of each P_l (we can now omit the superscripts from A^{in} and B^{out} without ambiguity; we will not need A^{out} or B^{in} again):

$$-E_0 R + B_1 R^{-2} = A_1 R \qquad (l = 1)$$

$$B_l R^{-l-1} = A_l R^l \qquad (l \neq 1)$$

This gives only one equation for each two unknowns; we need the other boundary condition (Eq. 19.18) at $r = R$;

$$D_{out, n} = D_{in, n}$$

$$-\epsilon_0 dV_{out}/dr = -\epsilon \, dV_{in}/dr$$

$$-E_0 P_1(\cos \theta) + \sum_l B_l(-l-1)R^{-l-2}P_l(\cos \theta)$$

$$= (\epsilon/\epsilon_0)\sum_l A_l l R^{l-1} P_l(\cos \theta)$$

Equating coefficients of P_l,

$$-E_0 + B_1(-2)R^{-3} = KA_1 \qquad (l = 1)$$

$$B_l(-l-1)R^{-l-2} = KA_l l R^{l-1} \qquad (l \neq 1)$$

We can eliminate the A's now:

$$K(-E_0 R + B_1 R^{-2}) = -E_0 R - 2B_1 R^{-2}$$

$$B_l(-l-1)R^{-l-2} = KlB_l R^{-l-2} \qquad (l \neq 1)$$

or

$$B_1(K + 2) = E_0(K - 1)R^3$$

$$B_l(+l+1+Kl) = 0 \qquad (l \neq 1)$$

Since $l \geq 0$ and $K > 1$, this means $B_l = 0$ for $l \neq 1$ and

$$B_1 = \frac{K - 1}{K + 2} E_0 R^3$$

so

$$A_1 = -E_0 + B_1 R^{-3} = \frac{-3E_0}{(K + 2)}$$

The final results for the potentials are

$$V_{in}(r,\theta) = -\frac{3}{K+2}E_0 r\cos\theta \tag{19.29}$$

$$V_{out}(r,\theta) = -E_0 r\cos\theta + \frac{K-1}{K+2}E_0\frac{R^3}{r^2}\cos\theta \tag{19.30}$$

Outside, this is the sum of the original faraway uniform field and a **dipole** term ($\propto r^{-2}\cos\theta$) due to the induced dipole moment of the sphere [you can check that the dipole moment is $\mathbf{p} = 4\pi\epsilon_0 R^3 \mathbf{E}_0(K-1)/(K+2)$]. Surprisingly, **inside** the sphere the electric field is **uniform**

$$\mathbf{E}_{in} = \frac{3}{(K+2)}\mathbf{E}_0 \tag{19.31}$$

as is the polarization

$$\mathbf{P} = (K-1)\epsilon_0\mathbf{E}_{in} = \frac{3(K-1)}{(K-2)}\mathbf{E}_0 \tag{19.32}$$

As a consistency check, note that if $K = 1$ (i.e., the "dielectric" is a vacuum) the field is uniformly $\mathbf{E}_0$ everywhere and $\mathbf{P} = 0$. If $K > 1$, the field inside the sphere is **less** than E_0. However, the D field

$$\mathbf{D}_{in} = \frac{3K}{(K+2)}\epsilon_0\mathbf{E}_0 \tag{19.33}$$

is **larger** inside than out because $3K/(K+2) > 1$. To sketch the fields, it is easiest to start with D, whose lines never begin or end ($\rho_f = 0$). They are parallel and equally spaced far away, and also parallel but closer together inside the sphere. Outside, the lines must be distorted near the sphere to connect up with those inside (Fig. 19.19a shows the $K = 2$ case). The electric field lines (Fig. 19.19b) are identical to those of D outside, but there are

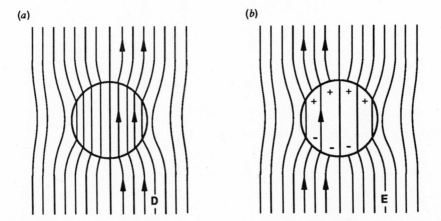

(a)   (b)

Figure 19.19 (a) Distortion of the D field lines in and near a dielectric sphere (b) The E field.

Figure 19.20 The bound charge at the surface of a dielectric sphere in a uniform field.

fewer inside. Some of the E-lines start and end on induced bound surface charges,

$$\sigma_b(R,\theta) = P_n = 3\left(\frac{K-1}{K+2}\right)\epsilon_0 E_0 \cos\theta \tag{19.34}$$

(where we have used $\mathbf{E}_0 \cdot \hat{\mathbf{n}} = E_0 \cos\theta$). The bound charge can be visualized as displaced positively and negatively charged spheres because P is uniform (Figure 19.20).

■ **EXAMPLE 19.5** Consider a free charge q at the center of a small spherical cavity of radius R in an infinite dielectric of dielectric constant K (Figure 19.21). We could treat this as a boundary-value problem, but it has enough symmetry to be solvable by Gauss' law so we will save some time by doing it that way. In Section 18.3, we computed the fields E and D in the dielectric for the limit $R \to 0$ (a point charge embedded in a uniform dielectric) by applying Gauss' law. The same argument gives $E(r)$ and $D(r)$ in the $R > 0$ case, for $r > R$:

$$\mathbf{E}(r) = (q/4\pi K\epsilon_0 r^2)\hat{\mathbf{r}} \tag{19.35}$$

Then we can calculate

$$\mathbf{P}(r) = (K-1)\epsilon_0\mathbf{E}(r) = \frac{(K-1)}{K}\frac{q}{4\pi r^2}\hat{\mathbf{r}} \tag{19.36}$$

and see explicitly that there is no volume density of bound charge $[\nabla \cdot (\hat{\mathbf{r}}/r^2) = 0]$.

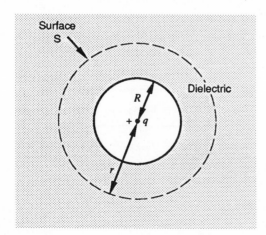

Figure 19.21 A free charge in a spherical cavity in a dielectric material.

However, there **is** a surface charge on the inside surface of the dielectric. To calculate it, we must note that the outward normal from the dielectric is in the $-\hat{r}$ direction, so

$$\sigma_b = P_n = \mathbf{P} \cdot (-\hat{r}) = -\frac{(K-1)}{K}\frac{q}{4\pi R^2} \tag{19.37}$$

is the charge per unit area on the cavity surface. The total bound charge is

$$q_b = 4\pi R^2 \sigma_b = -\frac{(K-1)}{K}q \tag{19.38}$$

This bound charge partly cancels the free charge, in the sense that the total charge on and inside the spherical cavity is

$$q_{total} = q + q_b = \frac{q}{K} \tag{19.39}$$

which is **smaller** than the free charge q. This is the same result we obtained for the case $R = 0$; the free charge is partially screened by the dielectric.

▶ **Problem 19.5**

Rework Example 19.3 for the case that the thicknesses of the two dielectric slabs of permittivity ϵ_2 and ϵ_3 in Figure 19.17 are d_2 and d_3 respectively (so $d_2 + d_3 = d$, the separation of the conducting plates).

a. Calculate V, E, and D in each slab.

b. Check that your result reduces to Eqs. 19.27 and 19.28 for the special case $d_2 = d_3 = \frac{1}{2}d$.

c. Plot V, E, and D vs z for the case $d_2 = \frac{1}{4}d$, $d_3 = \frac{3}{4}d$, $\epsilon_3 = 3\epsilon_0$, $\epsilon_2 = \epsilon_0$.

d. Calculate the free and bound surface charge densities at each interface for the case considered in **c**.

e. Derive from Gauss' law for E an equation for the discontinuity of E_n across an interface, in terms of the surface density of **total** charge σ. Verify that it is satisfied in the case considered in **c**. ◀

▶ **Problem 19.6**

a. Compute the capacitance of a parallel-plate capacitor of area A and plate separation d, which is half-filled with dielectric of dielectric constant K as shown in Figure 19.22a. Hint: Try a potential that is the same as though the plates were infinite. If it

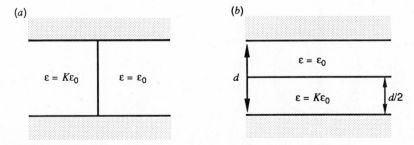

Figure 19.22 A parallel-plate capacitor filled with an inhomogeneous dielectric with a vertical interface (a) and a horizontal interface (b).

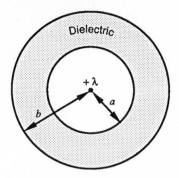

Figure 19.23 A cylindrical dielectric pipe with a line charge on its axis.

satisfies all the boundary conditions, the uniqueness theorem guarantees that it is correct.

b. Compare this to the capacitance of another half-filled capacitor whose dielectric is in the form of a slab of thickness $\frac{1}{2}d$ (Figure 19.22*b*). You may use the results of Example 19.3. ◄

▶ **Problem 19.7**

Consider a line charge (linear charge density λ) on the axis of a dielectric pipe of inner radius a, outer radius b, and dielectric constant K, as in Figure 19.23. Compute the fields D, E, and P in all three regions ($r < a$, $a < r < b$, $r > b$) and the bound charge on each surface. Verify that the total bound charge is zero. ◄

▶ **Problem 19.8**

A sphere of radius R is filled with a dielectric material of dielectric constant K. There is a uniform **free** charge density ρ throughout the material. Find the potential at the center, relative to infinity. ◄

▶ **Problem 19.9**

Consider a rectangular block of dielectric ($a \times b \times c$) having the continuum polarization field

$$\mathbf{P}(x, y, z) = \left(Kx, Ly^2, 0 \right)$$

a. Compute the bound charge density.

b. Compute the bound surface charge density on each face.

c. Compute the total bound charge by integrating the results of **a** and **b** and adding. ◄

▶ **Problem 19.10**

A coaxial cable consists of a cylindrical conductor of radius a surrounded by a cylindrical insulating layer of dielectric constant K_2 from a to b, another layer of dielectric constant K_3 from b to c, and an outer conductor of inner radius c (its outer radius is irrelevant). Compute the capacitance per unit length. ◄

▶ **Problem 19.11**

a. Compute the $\mathbf{D}$ and $\mathbf{E}$ fields at a distance r from a point charge q embedded at the center of a **graded** dielectric sphere of radius b, with dielectric constant K varying

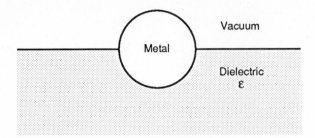

Figure 19.24 A conducting sphere embedded in a planar dielectric surface.

radially according to

$$K(r) = \begin{cases} b/a & \text{for } r < a \\ b/r & \text{for } a < r < b \\ 1 & \text{for } b < r \end{cases}$$

b. Compute the bound charge (this may be a point charge, a line charge, a surface charge density, a volume charge density, or any combination of these).

c. Compute the total bound charge explicitly, by adding the charges found in **b.**   ◀

▶ **Problem 19.12**

A thin spherical shell of radius R has a uniform (free) surface charge σ_0 on the top hemisphere (for $\theta < \frac{1}{2}\pi$ in spherical coordinates) and surface charge $-\sigma_0$ on the bottom $(\theta > \frac{1}{2}\pi)$. Compute the potential inside and outside the sphere in terms of Legendre polynomials. Give a formula for the coefficient of P_l for each l, and calculate them explicitly for $l = 0$, 2, and 4. (You need to treat the inside and the outside of the sphere as different regions, even though they have the same ϵ, to allow for a surface density of free charge and its accompanying discontinuity in D_n.)   ◀

▶ **Problem 19.13**

A conducting sphere of radius R is halfway embedded in dielectric which fills a half-space (Figure 19.24). In terms of the total charge Q on the sphere, compute the potential everywhere. Hint: The potential depends only on the distance from the sphere's center.   ◀

▶ **Problem 19.14**

a. Consider a dielectric cylinder with dielectric constant $K = 3$, in a uniform field E_0 **perpendicular** to its axis. Compute E and P everywhere. Sketch the lines of E and P. Hint: The most general solution of Laplace's equation in cylindrical coordinates that is independent of z is

$$V(r, \phi) = A_0 + B_0 \ln r$$

$$+ \sum_{n=1}^{\infty} (A_n \sin n\phi + B_n \cos n\phi)(C_n r^n + D_n r^{-n}) \qquad (19.40)$$

b. Compute the dipole moment per unit length of the cylinder.

c. Compare the answer to **b** with your answer to the discrete Problem 19.3, by evaluating it for a cylinder of cross-sectional area $4\,dr^2$ in a field $-V_0/2\,dr$.   ◀

SECTION 19.4   ENERGY IN A DIELECTRIC

In this section we will generalize the formula we got in Chapter 13 for the energy of a system containing electric fields. We started with a formula (Eq. 13.5) for the work done by an electromagnetic system on moving charges described by a current density $\mathbf{j}$:

$$dW/dt = \int \mathbf{j} \cdot \mathbf{E}\, d\tau \tag{19.41}$$

This work should be reflected in a decrease in the total energy $U_E(t)$ of the electric-field system:

$$dW/dt = -dU_E/dt \tag{19.42}$$

Using Maxwell's equation to relate $\mathbf{j}$ to $d\mathbf{E}/dt$ (and assuming that the charges are moved slowly, so the magnetic field is negligible), we found

$$dW/dt = -(d/dt)\int \tfrac{1}{2}\epsilon_0 E^2\, d\tau \tag{19.43}$$

so it was consistent to choose

$$U_E = \int \tfrac{1}{2}\epsilon_0 E^2\, d\tau \tag{19.44}$$

All this is perfectly valid in the presence of dielectrics. However, in a dielectric other forms of energy are present. In particular, to distort an atom or molecule by polarizing it requires some energy, which is stored as potential energy in the molecule. We will refer to this as "elastic" potential energy. In thinking about this energy, it is useful to model the atom by two point charges, which could represent the centers of electronic (negative) and nuclear (positive) charge, connected by an elastic spring of spring constant k. For this model the elastic potential energy is $\tfrac{1}{2}ka^2$, where a is the distance between the charges.

Much of the usefulness of the concept of energy lies in the fact that the mechanical work we do on a system can be calculated as the change in its energy (see Example 13.2 for one such calculation). This requires that we know its **total** energy, which in the case of a dielectric includes the **elastic** energy. There is no way to calculate the elastic energy directly, but we can get the **total** (elastic plus electric) energy by a modification of the above argument. Note that the work (Eq. 19.41) done by the electric field includes work on both free and bound charge. Defining j_f to be the current of free charge (e.g., currents in wires or due to the motion of free charged particles) and j_b to be the current of bound charge (due to motions of charges inside the molecules of dielectric), we can divide the total work into two terms, $\int \mathbf{j}_f \cdot \mathbf{E}\, d\tau$ and $\int \mathbf{j}_b \cdot \mathbf{E}\, d\tau$. If we consider a system consisting of the electric field and the molecules of the dielectric, the integral involving $\mathbf{j}_b$ (the work done by the field on the charges in the dielectric) represents energy transfer within the system. The work done by the system on the rest of the universe includes only work done on **free** charge, which we will denote by W_f. So we should replace the $\mathbf{j}$ in Eq.

19.41 by the current of free charge:

$$dW_f/dt = \int \mathbf{j}_f \cdot \mathbf{E} \, d\tau \qquad (19.45)$$

The Maxwell equation we used to relate $\mathbf{j}$ to $d\mathbf{E}/dt$ is more complicated when expressed in terms of $\mathbf{j}_f$:

$$d\mathbf{E}/dt = -\mathbf{j}/\epsilon_0 = -\mathbf{j}_f/\epsilon_0 - \mathbf{j}_b/\epsilon_0 \qquad (19.46)$$

However, the extra term $\mathbf{j}_b$ is defined in a discrete system by the fact that $j_b(f, t') \, da(f) \, dt$ is the amount of bound charge crossing the face f during the time interval around t', where t' is a half-integer multiple of dt and refers to the interval from $t' - \frac{1}{2}dt$ to $t' + \frac{1}{2}dt$. If we add these numbers for all time intervals up to a time t (a multiple of dt), we get exactly the **total** charge that has crossed since the fields were first turned on. However, this total (divided by da) is just what we have defined as the polarization field $P(f, t)$! Thus,

$$P(f, t) = \sum_{t' < t} j_b(f, t') \, dt \qquad (19.47)$$

and taking the discrete time derivative gives exactly

$$\frac{d}{dt} P(f, t') = j_b(f, t') \qquad (19.48)$$

Substituting Eq. 19.48 into Maxwell's equation (Eq. 19.46) and moving dP/dt to the left-hand side yields the simple result

$$dD/dt = -j_f \qquad (19.49)$$

Finally, we can substitute this "Maxwell equation for D" (Eq. 19.49) into Eq. 19.45 to obtain

$$dW_f/dt = -\int (d\mathbf{D}/dt) \cdot \mathbf{E} \, d\tau \qquad (19.50)$$

This is as far as we can go, unless the system is linear. If it is, $D(f) = \epsilon(f)E(f)$, so

$$d(\mathbf{D} \cdot \mathbf{E})/dt = 2\epsilon \mathbf{E} \cdot (d\mathbf{E}/dt) = 2\mathbf{E} \cdot (d\mathbf{D}/dt) \qquad (19.51)$$

and

$$dW_f/dt = -(d/dt) \int \tfrac{1}{2} \mathbf{D} \cdot \mathbf{E} \, d\tau \qquad (19.52)$$

Thus it is consistent to define

$$U_D = \int \tfrac{1}{2} \mathbf{D} \cdot \mathbf{E} \, d\tau \qquad (19.53)$$

as the energy of the dielectric system. This is our final result. It is easy to remember because it differs from the old Eq. 19.44 only by the by-now familiar replacement of ϵ_0 by ϵ. Although we have derived this result only in the continuum limit for brevity, if you go through the steps in Section 13.2 replacing E by D as we have done here, you will see that it works exactly for a discrete system as well.

The above derivation is valid only when the fields change so slowly that the magnetic field B can be taken to be zero. For a derivation of a formula for the energy of materials in the presence of both electric and magnetic fields, see Section 23.11.

■ **EXAMPLE 19.6** Let us do the analog of Example 13.2 (the force on a plate of a parallel-plate capacitor) for a capacitor that has a dielectric material between its plates. To allow the plates to move, suppose that the dielectric is a liquid (oil, for example), in which the plates are immersed. We again denote the total charge by Q and the plate separation by d, and assume that the fields are the same as those in an infinite capacitor, though the plate area A is actually finite. We have $Q/A = \sigma_{\text{free}} = D = \epsilon E$, volume $= Ad$, so

$$U_D = \tfrac{1}{2}ED(Ad) = Q^2 d/2A\epsilon$$

If F is the upward force on the lower plate, the work the system does when the lower plate is moved up by Δd is $F\Delta d$, which must be the decrease in energy

$$U_D(d) - U_D(d - \Delta d) = Q^2 \Delta d/2A\epsilon$$

Thus

$$F = Q^2/2A\epsilon \tag{19.54}$$

Comparing this with Example 13.2, we see that the dielectric decreases the force by a factor of the dielectric constant K.

■ **EXAMPLE 19.7** In the previous example, we assumed the charge Q did not change when the plate was moved. This implies that the capacitor was not connected to a battery, because

$$Q = CV = (\epsilon_0 A/d)V \tag{19.55}$$

would change if d changed while V remained constant. Let us now suppose instead that the plates **are** connected to the terminals of a battery, so the potentials at the top and bottom plates are maintained at $+V$ and 0 respectively. Then we must rewrite the energy U_D in terms of V (using Eq. 19.55) to see how it changes when d changes:

$$U_D = V^2 A\epsilon/2d$$

Now when d is decreased, the energy U_D **increases**! If we calculate the force as we did before, we will find that the plates **repel** each other. This is not right, of course; the force doesn't suddenly reverse itself when we disconnect the battery. The problem is that we have introduced a new object (the battery) into the problem without accounting for its energy. When d decreases by Δd, the charge Q increases by (differentiating Eq. 19.55)

$$\Delta Q = \left(\epsilon_0 AV/d^2\right)\Delta d$$

This extra charge is put onto the plates by the battery, which must do work $V\Delta Q$ in the process. The work $F\Delta d$ the dielectric system can do is now not just its decrease in potential energy, but also includes the work done by the battery

$$F\Delta d = U_D(d) - U_D(d - \Delta d) + V\left(\epsilon_0 AV/d^2\right)\Delta d$$

$$= -\left(\epsilon_0 AV^2/2d^2\right)\Delta d + \left(\epsilon_0 AV^2/d^2\right)\Delta d$$

$$= +\left(\epsilon_0 AV^2/2d^2\right)\Delta d$$

(to first order in Δd) which gives exactly the same value we got before for the force.

■ **EXAMPLE 19.8 THE FORCE ON A DIELECTRIC SLAB**   Figure 19.25 shows a dielectric slab partially inserted into a parallel-plate capacitor. We would like to calculate the electrostatic **force** on the slab. We can guess the direction of this force from the sketches of the induced charges on the conductors and the bound charges on the dielectric shown in the figure. The bound charge on the dielectric just outside the capacitor feels a force pulling it **into** the capacitor, because of the attraction of the induced charge on the conductor. (Note that there is no force exerted at the other end of the dielectric slab, **inside** the capacitor, because the D and E and E fields there have no horizontal components, as you found in Problem 19.6a.)

To calculate the force quantitatively, we must calculate the total electrostatic energy of the system as a function of the length x of the part of the dielectric remaining in the capacitor, and differentiate with respect to x. It is easiest to do this at constant potential difference V (i.e., assume the capacitor is connected to a battery) because then the fields change very little with x. In fact, the electric field doesn't change at all. Between the plates it is $E = V/d$, both in the dielectric and in the vacuum (Problem 19.6a). The fringing field near the left sides of the capacitor plates is independent of x, as long as the left end of the dielectric is far away ($L - x \gg d$). The fringing field on the right is similarly independent of x. The only effect of increasing x by Δx is that a volume $w\,d\Delta x$ (w is the width of the system perpendicular to the paper) between the plates, formerly occupied by vacuum, becomes occupied by dielectric, so its D changes from $\epsilon_0 V/d$ to $K\epsilon_0 V/d$. Its contribution to the energy (Eq. 19.53) therefore changes by

$$\Delta U_E = \tfrac{1}{2}(K - 1)(\epsilon_0 V/d)(V/d)w\,d\Delta x = \tfrac{1}{2}(K - 1)\epsilon_0 V^2 w\Delta x/d$$

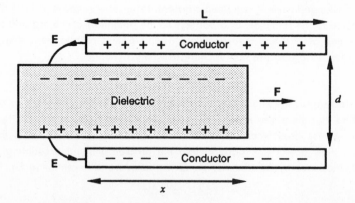

Figure 19.25   A cross section of a parallel-plate capacitor with a dielectric slab partially inserted.

To get the force, we must compute the change in energy of the **whole** system, including the battery as well as the capacitor. The change in the battery energy is $-V\Delta Q$, where ΔQ is the change in the free charge on the plates. The charge changes only in the area $w\Delta x$ of the plate that is newly adjacent to dielectric; its surface charge density $\sigma_f = D_n$ changes by $K\epsilon_0 V/d - \epsilon_0 V/d$. Thus $\Delta Q = w\Delta x \epsilon_0 V(K-1)/d$, and the change in battery energy is

$$\Delta U_{batt} = -(K-1)\epsilon_0 V^2 w\Delta x/d$$

The total energy change is

$$\Delta U = (\tfrac{1}{2}-1)(K-1)\epsilon_0 V^2 w\Delta x/d = -\tfrac{1}{2}(K-1)\epsilon_0 V^2 w\Delta x/d$$

The negative of this (the energy lost by the system) must be the work $F\Delta x$ done by the system, so

$$F = \tfrac{1}{2}(K-1)\epsilon_0 V^2 w/d$$

(Note that if we had forgotten the contribution of the battery, we would have gotten the same magnitude but the wrong sign.)

▶ **Problem 19.15**

Calculate the capacitance of the partially filled capacitor in Figure 19.25. ◀

▶ **Problem 19.16**

Calculate the force on the dielectric slab in Example 19.8 assuming Q (instead of V) is held constant, that is, the battery is disconnected. Neglect fringing fields. ◀

▶ **Problem 19.17**

In Section 14.4 we derived Gauss' law for E from Maxwell's equation for dE/dt. Use a similar method to derive Gauss' law for D from Maxwell's equation for dD/dt (Eq. 19.49). ◀

SECTION 19.5 MICROSCOPIC ESTIMATES OF SUSCEPTIBILITY

Consider a dielectric material made up of molecules with a known polarizability α, so that in an electric field $\mathbf{E}$ the dipole moment of each molecule is $\mathbf{p} = \alpha\mathbf{E}$. If the density (number of molecules per unit volume) is n, the polarization field can be computed as the dipole moment per unit volume (Section 18.2):

$$\mathbf{P} = n\mathbf{p} = n\alpha\mathbf{E} \tag{19.56}$$

Comparing this to Eq. 18.11 defining the dielectric susceptibility χ, we see that

$$\chi = n\alpha/\epsilon_0 \tag{19.57}$$

Equation 19.57 is a very accurate estimate of the susceptibility when the density n is low, for example in a gas. In solids and liquids, however, an effect we have left out becomes important. The susceptibility is supposed to relate the **macroscopic** electric field to the macroscopic polarization field; these fields are defined for large faces, much larger than molecular dimensions. (If we assume the macroscopic fields are smooth, it doesn't matter whether we average the fields over a volume or a face, as long as it is much larger than the molecules.) The notion of the dipole moment of a molecule, however, only makes sense when the fields are defined on a lattice with dr much **smaller** than the molecule: the $\mathbf{E}$ in $\mathbf{p} = \alpha\mathbf{E}$ must be the field exactly **at** the position of the molecule, not an average over a large volume containing it. This is called the **local field** and denoted $\mathbf{E}_{\mathrm{loc}}$, so that

$$\mathbf{p} = \alpha\mathbf{E}_{\mathrm{loc}} \qquad (19.58)$$

The local field at a molecule does not include the field of the molecule itself, but does include the fields of its neighbors. If the neighbors are few and far between, as in a gas, we can ignore their fields, $\mathbf{E}_{\mathrm{loc}}$ is essentially the macroscopic field, and the calculation leading to Eq. 19.57 is correct. If not, we must try to calculate the difference, called the **local field correction**.

To this end, consider the block of dielectric shown in Figure 19.26. We have put it between two capacitor plates to provide a smooth (in fact uniform) macroscopic field, which we will call $\mathbf{E}_{\mathrm{macro}}$. It is simply the potential difference divided by the distance between the plates, the same field we have calculated before using macroscopic dielectric methods. We will calculate the local field as the sum of contributions from each of the types of charges in the system:

$$\mathbf{E}_{\mathrm{loc}} = \mathbf{E}_{\mathrm{ext}} + \mathbf{E}_{\mathrm{depol}} + \mathbf{E}_{\mathrm{sphere}} + \mathbf{E}_{\mathrm{nbrs}} \qquad (19.59)$$

The first two terms are uniform fields: $\mathbf{E}_{\mathrm{ext}}$ is the field of the "external" charges on the capacitor plates and $\mathbf{E}_{\mathrm{depol}}$ is the field of the bound surface charge on the outside of the dielectric, next to the plates (it is called the "depolarization field" because it partly cancels $\mathbf{E}_{\mathrm{ext}}$, decreasing the local field so the dipole moment of each molecule is less than it would be if the rest of the dielectric were absent). These first two terms are the only ones that would be present in a macroscopic calculation, so that

$$\mathbf{E}_{\mathrm{macro}} = \mathbf{E}_{\mathrm{ext}} + \mathbf{E}_{\mathrm{depol}} \qquad (19.60)$$

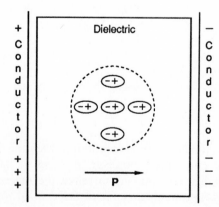

Figure 19.26 A rectangular block of dielectric subjected to the electric field of a capacitor. The molecule at which we want to calculate the local field is shown explicitly, as are its neighbors in a sphere of radius R about it.

The last two terms of Eq. 19.59 are due to the rest of the bound charge. In a macroscopic calculation they would be exactly zero; there is no net bound charge density in the interior because the polarization is uniform. However, if we look at the polarization on a microscopic scale (using dr smaller than molecular dimensions) it is not uniform, and there are variations in the bound charge density (though it averages to zero over many cells). The electric field due to this bound charge density also vanishes on the average, but may fluctuate locally. We calculate it by mentally dividing the molecules into two groups, those in a sphere of radius R (much larger than the spacings between molecules) and those outside this sphere. The purpose of this is to allow the molecules outside the sphere to be treated as a continuum dielectric medium. We have already included the charge on its outside surface and noted that its bound charge density vanishes in the interior. That leaves the inside (spherical) surface of this region. The surface is not real, in that we have not physically removed the molecules inside it to make a cavity in the dielectric, but we can still use the usual methods to compute its bound charge density. Its surface charge is

$$\sigma_b = \mathbf{p} \cdot \hat{\mathbf{n}} = -P\cos\theta \qquad (19.61)$$

as shown in Figure 19.27. The field due to this charge can be calculated from Coulomb's law (Problem 19.18) and is exactly

$$\mathbf{E}_{\text{sphere}} = \mathbf{P}/3\epsilon_0 \qquad (19.62)$$

at the center (in fact, everywhere inside).

The last contribution to the total local field, $\mathbf{E}_{\text{nbrs}}$, is due to the molecules **inside** the sphere, and is the hardest to calculate, because our macroscopic methods are of no use and we don't know exactly where the molecules are. If we make the simple assumption that they lie on the vertices of a cubic lattice, one can show (see for example the book by Jackson cited in the bibliography) that the sum is exactly zero. Figure 19.28 shows qualitatively how this cancellation occurs. It turns out that the vanishing of $\mathbf{E}_{\text{nbrs}}$ is quite common: it occurs for any crystal lattice of cubic symmetry and (on the average) for liquids and gases. Although there exist anisotropic materials in which $\mathbf{E}_{\text{nbrs}}$ is nonzero, we will consider here only the case that it vanishes.

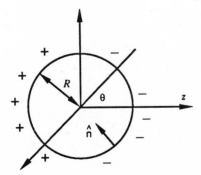

Figure 19.27 The charge on the inside surface of an imaginary spherical cavity in the dielectric.

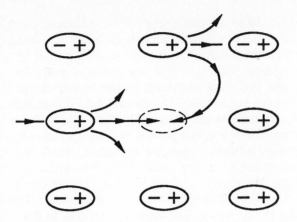

Figure 19.28 A cubic lattice of molecular dipoles. A few selected E-field lines of the dipole fields of neighbors of the central (dashed line) molecule are included, showing their tendency to cancel. Note that it would not make sense to draw lines of D on this diagram; D is meaningful only on a macroscopic scale.

Finally, we can write the total local field from Eqs. 19.59, 19.60, and 19.62:

$$\mathbf{E}_{loc} = \mathbf{E}_{macro} + \mathbf{P}/3\epsilon_0 \tag{19.63}$$

Thus the field at a molecule is **larger** than the average field $\mathbf{E}_{macro}$; it follows of course that the field between molecules is on the average smaller. Substituting $\mathbf{E}_{loc}$ for $\mathbf{E}$ in Eq. 19.56 gives

$$\mathbf{P} = n\alpha(\mathbf{E}_{macro} + \mathbf{P}/3\epsilon_0) \tag{19.64}$$

which can easily be solved for $\mathbf{P}$:

$$\mathbf{P} = \frac{n\alpha\mathbf{E}_{macro}}{1 - n\alpha/3\epsilon_0} \tag{19.65}$$

The proportionality constant here gives us the dielectric susceptibility

$$\chi = \frac{n\alpha/\epsilon_0}{1 - n\alpha/3\epsilon_0} \tag{19.66}$$

In the limit $n \to 0$, this returns us to the low-density expression, Eq. 19.57. From Eq. 19.66 we can calculate ϵ and K; by solving for α in terms of K we can obtain an equivalent equation

$$\alpha = \frac{K - 1}{K + 2}\frac{3\epsilon_0}{n} \tag{19.67}$$

called the **Clausius—Mossotti equation.**

▶ **Problem 19.18**

Show from Coulomb's law (Section 14.3) that the field due to a surface charge density $\sigma = -P\cos\theta$ on a spherical surface (Figure 19.27) has magnitude $P/3\epsilon_0$ at the center of the sphere and points in the z-direction. ◀

▶ **Problem 19.19**

Consider Figure 19.27 as a boundary value problem, and find the potential everywhere. Show that it is uniform inside the sphere. ◀

Magnetostatics and Induction

CHAPTER TWENTY

MAGNETOSTATICS

SECTION 20.1 THE GENERAL EQUATIONS OF MAGNETOSTATICS

In electrostatics we learned to calculate the fields of a charge distribution a long time after it stops changing, that is, becomes "static." Magnetostatics is the study of static **current** distributions, meaning those in which the current density $\mathbf{j}$ is independent of time. If we allow a completely arbitrary current distribution, we will in general have a changing charge density according to the continuity equation $d\rho/dt = -\nabla \cdot \mathbf{j}$, and our system is not really "static." So we will restrict the current distribution to be one for which

$$\nabla \cdot \mathbf{j} = 0 \tag{20.2}$$

When such a current distribution is first turned on, there are, of course, changing electric and magnetic fields. However, after a long time they settle down to "static" fields that are independent of the details of how the current distribution was turned on. Whereas in electrostatics the magnetic field approaches zero after a long time, in magnetostatics it will turn out to be the electric field that disappears; the static field is solely a magnetic field (hence the name, "magnetostatics"). The most general problem of magnetostatics is that of calculating the static magnetic field for an arbitrary time-independent divergenceless current distribution.

The static fields must, of course, satisfy Maxwell's equations. Everything is time-independent, so we can leave off the d/dt terms and Maxwell's equations become

$$\nabla \times \mathbf{E} = 0 \qquad \nabla \cdot \mathbf{E} = \frac{\rho}{\epsilon_0} = 0$$

$$c^2\nabla \times \mathbf{B} = \frac{\mathbf{j}}{\epsilon_0} \qquad \nabla \cdot \mathbf{B} = 0 \tag{20.3}$$

Note that these equations can be solved for $\mathbf{E}$ and $\mathbf{B}$ independently; they are decoupled. In electrostatics, $\mathbf{j} = 0$ so the lower two equations have the solution $\mathbf{B} = 0$: there is no B field in the static limit. We merely solved the upper two equations for $\mathbf{E}$. In magnetostatics, the opposite is true: the $\mathbf{E}$ equations have the solution $\mathbf{E} = 0$ (the E-field goes away after a long time) and we need to solve only

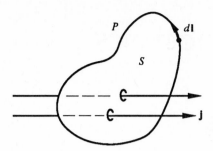

Figure 20.1   A surface S and its bounding path P, showing current passing through S and a length element $d\mathbf{l}$ along the path.

the **B** equations

$$\nabla \times \mathbf{B} = \mu_0 \mathbf{j} \quad \text{(Ampère's law)}, \quad \nabla \cdot \mathbf{B} = 0 \qquad (20.4)$$

We have denoted the constant $1/(c^2\epsilon_0)$ by μ_0; it is called the magnetic permeability of space. Its numerical value is $4\pi \times 10^{-7}$ newtons/ampere^2; we will see shortly why it is such a peculiarly round number. These equations do not determine **B** uniquely (for example, adding a constant to **B** gives another solution). Just as in electrostatics, we need some additional information related to how the current was turned on, which is usually supplied in the form of a boundary condition, such as **B** = 0 at infinity.

SECTION 20.2   SYSTEMS WITH HIGH SYMMETRY: USING AMPÈRE'S LAW

Before attacking the general problem of computing magnetostatic fields, let us observe that there are a few problems of very high symmetry that can be solved almost trivially using Ampère's law, just as a few problems (such as that of a spherical charge distribution) can be solved easily using Gauss' law in electrostatics. We will need the integral form of Ampère's law, obtained by integrating the microscopic form $\nabla \times \mathbf{B} = \mu_0 \mathbf{j}$ over a surface S bounded by a closed path P (Figure 20.1):

$$\int \mathbf{B} \cdot d\mathbf{l} = \mu_0 I \qquad (20.5)$$

where the path integral is over the path P and I is the integral of the current density over the surface S, that is, the total current through the surface. Ampère's law is directly useful in calculating **B** only if the tangential component of **B** is constant along the curve, so it can be factored out of the path integral:

$$\int \mathbf{B} \cdot d\mathbf{l} = B \int dl = Bl$$

where l is the total length of the path P. Then

$$B = \frac{\mu_0 I}{l} \qquad (20.6)$$

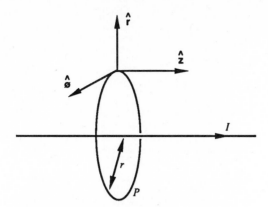

Figure 20.2   An amperian loop P around a straight wire carrying a current I.

■ **EXAMPLE 20.1 THE FIELD OF A LONG WIRE**   One problem that can be solved with Ampère's law is that of a current flowing in an infinite straight wire. Take the path P to be a circle of radius r about the wire, in a plane perpendicular to the wire (Figure 20.2). Then the vector $d\mathbf{l}$ points along the $\hat{\phi}$ unit vector, so $\mathbf{B} \cdot d\mathbf{l}$ is just $B_{\phi}\, dl$, and B_{ϕ} is the same everywhere on the circle because of the cylindrical symmetry. Let's assume for the moment that it's the only nonzero component and just refer to it as B. Ampère's law then becomes

$$2\pi r B = \mu_0 I$$

where I, the current through the surface S, is just the current in the straight wire, r is the radius at which we are calculating the field, and $2\pi r$ is the length of the path P. Solving for B, we get the magnetic field at a distance r from an infinite straight wire:

$$B = \frac{\mu_0 I}{2\pi r} \tag{20.7}$$

It is clear from the derivation that it doesn't matter how tightly the current density is concentrated near the axis of the cylinder, as long as it is all inside the radius r. This same formula gives the magnetic field of **any** cylindrically symmetric axial current distribution.

Strictly speaking, there is no such thing in the real world as an infinite straight wire. However, if the wire is straight for a distance much larger than r in both directions, it turns out that the static field is very close to that given by our formula for an infinite wire.

■ **EXAMPLE 20.2 THE FORCE BETWEEN PARALLEL WIRES**   We can use our result for the magnetic field of a long wire to calculate the force between two long parallel wires. Recall from Chapter 12 that the force on a length dl of wire carrying current I in a magnetic field **B** involves the vector cross product of **B** and $d\mathbf{l}$:

$$\mathbf{F} = I\mathbf{B} \times d\mathbf{l} \tag{20.8}$$

In the present case, the **B** field of the left-hand wire carrying current I is perpendicular to the $d\mathbf{l}$ of the other wire carrying current I' (see Figure 20.3) so the force is just the product

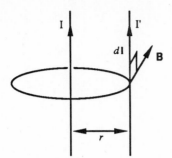

Figure 20.3 Two straight parallel wires carrying currents I and I' respectively, in the same direction.

of magnitudes $IB\,dl$. Using our result for B, the force per unit length is

$$\frac{F}{dl} = \mu_0 I \frac{I'}{2\pi r} \tag{20.9}$$

between two parallel wires carrying currents I and I' and separated by a distance r. The force is attractive if the currents flow in the same direction, and repulsive if they flow in opposite directions.

This sort of experiment turns out to be a very accurate way to relate electrical to mechanical units. Note that current is the only nonmechanical quantity in Eq. 20.9. Rather than define the ampere from the coulomb (which would have to be defined through some less accurate experiment) and determine μ_0 from this experiment, we follow international convention and define μ_0 arbitrarily, and use this experiment to define the ampere. So μ_0 is defined to be **exactly** $4\pi \times 10^{-7}$ N/a^2, a value which is chosen so that the ampere is that current which, when carried in each of two parallel wires 1 m apart, causes a magnetic force between them of 2×10^{-7} N/m. Thus the coulomb, which logically is the more fundamental unit, is in practice defined in terms of the ampere (1 c $= 1$ a $\cdot$ s).

▶ **Problem 20.1**

    a. Consider a square loop of side a with current I', placed a distance a from a long straight wire carrying a current I, as in Figure 20.4a. Calculate the net force on the square loop.

    b. Calculate the force on an isosceles right triangle near a long wire, as shown in Figure 20.4b.　　　　　　　　　　　　　　　　　　　　　　　　◀

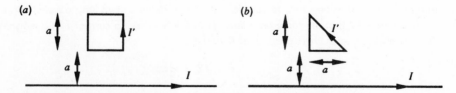

Figure 20.4 (a) A square loop near a long current-carrying wire. (b) A triangular loop near a wire.

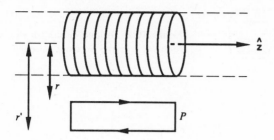

Figure 20.5   A cylindrical solenoid (above). Below it is an amperian loop P for showing that the axial magnetic field is independent of r.

■ **EXAMPLE 20.3 A LONG SOLENOID**   Another problem that can be solved using only Ampère's law and symmetry is that of the infinitely long cylindrical solenoid (Figure 20.5), that is, a wire wound around a cylinder and carrying a current I. Imagine the wires are so close together that we can regard the current as being uniformly spread over the surface of the cylinder. That is, we effectively have an azimuthal surface current density $K = nI$ where n is the number of turns of wire per unit distance axially; K has units amperes/meter. Ampère's law forbids an azimuthal B because of the azimuthal symmetry; $2\pi r B_\phi$ is proportional to the current through a circle in the plane of a wire loop, which is zero when the current is azimuthal. (In the real solenoid this axial current is exactly I, but this is negligible compared to the azimuthal current, which is many times I.) In addition, $\nabla \cdot \mathbf{B} = 0$ forbids a radial B. Thus B must be axial in this case. (Alternatively, this could be shown directly from Maxwell's equations by the methods described in the next section.) We can apply Ampère's law to relate B at two distances r and r' from the axis (both outside the solenoid, as shown). Ampère's integral law for the rectangular path P shown in Figure 20.5 gives

$$B_z(r)l - B_z(r')l = 0$$

(l is the length of P in the z-direction). So B_z is independent of r. It's physically reasonable that $\mathbf{B}$ would not extend to $r = \infty$, so $\mathbf{B}$ must be zero **everywhere** outside the cylinder (we could verify this also by solving Maxwell's equations; see Section 20.3). Now let us move r **inside** the cylinder, as in Figure 20.6. There is now a current Kl through P:

$$B_z(r)l - B_z(r')l = \mu_0 Kl$$

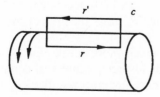

Figure 20.6   An amperian loop for calculating the axial magnetic field inside a solenoid.

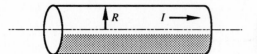

Figure 20.7 A long cylindrical wire of radius R carrying an axial current I.

and, because $B_z(r') = 0$,

$$B_z = \mu_0 K \qquad\qquad (20.10)$$

inside the solenoid. This result actually holds (and can be computed in the same way) for a solenoid with any cross-sectional shape (not necessarily circular) as long as the cross section and the surface current density K are independent of z. It can be verified for various discrete current distributions using program MAXWELL.

▶ **Problem 20.2**

    a. Compute the magnetic field both inside and outside the solid wire shown in Figure 20.7, assuming that the current density is uniform throughout the wire.

    b. Compute the magnetic field again, this time assuming that the current flows only in a thin layer on the surface.

    c. Repeat the calculation for an inhomogeneous volume current density $j_z(r)$ that is proportional to r^2 inside the wire and zero outside the wire. ◀

▶ **Problem 20.3**

Consider two long coaxial solenoids. The inner one has radius R_1 and n_1 turns per unit length, and the outer one radius R_2 and n_2 turns. Each turn carries a current I, but the current directions are opposite in the inner and outer solenoids. Ignoring any axial current, compute the magnetic field everywhere (inside both solenoids, between them, and outside both). ◀

SECTION 20.3 PROVING SYMMETRY PROPERTIES OF MAGNETIC FIELDS

Let's return to the question of why the radial (r) and axial (z) components of **B** must vanish in the magnetostatic field of a long straight wire (Figure 20.2). One can argue from translational symmetry along the axis and rotational symmetry about it that B_r must be uniform over the surface of a cylinder of radius r around the axis (Figure 20.8). However, the integral of B_r over this surface is given by the integral form of $\nabla \cdot \mathbf{B} = 0$ (an analog for **B** of Gauss' law for **E**):

$$\int \mathbf{B} \cdot da = 0 \qquad \text{(over any closed surface)} \qquad (20.11)$$

To close the surface we need to include the integral of B_z over the ends of the cylinder. Translational symmetry implies that B_z must be independent of z, so the integrals over the two ends cancel each other. Thus B_r times the area of the cylinder is zero, and $B_r = 0$.

The question of why B_z must vanish is a much more subtle one. Actually a uniform axial field is perfectly consistent with the divergence and curl equations we are solving for **B**. In fact, it is easy to show from Ampère's law that B_z **must** be

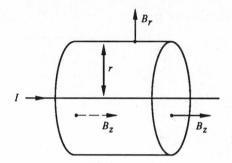

Figure 20.8 A gaussian surface for determining the radial magnetic field (zero) of a long straight wire.

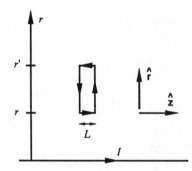

Figure 20.9 An amperian loop for showing that B_z is constant.

uniform, out to infinite r. If we just evaluate the integral in Eq. 20.5 over the loop in Figure 20.9, using $B_r = 0$, we obtain $LB_z(r) - LB_z(r') = 0$. If you are willing to dismiss such a uniform B_z as ridiculous, you may skip the rest of this section. If you aren't so willing, we will have to use the methods described in Chapter 8 for determining the symmetry of solutions to dynamic equations to see why an axial B will not arise under the action of Maxwell's equations. We will do this using discrete variables. The first step is to choose a coordinate system that preserves the symmetry of the problem, namely the cylindrical coordinate system shown in Figure 20.10. The cell lattice shown could have been produced by attacking the system with an apple corer. It differs from that in Figure 7.5 because it has cell centers rather than vertices along the axis. The lattice has a cylindrical cell of radius $\frac{1}{2}dr$ at the center, surrounded by cylindrical shells of thickness dr sliced lengthwise into cells at

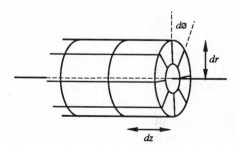

Figure 20.10 A discrete lattice in a cylindrical coordinate system.

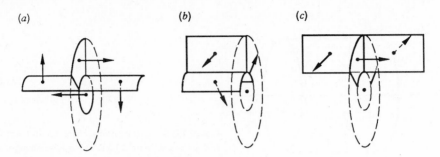

Figure 20.11 "Paddlewheel" for computing curl E at (a) an azimuthal edge, (b) an axial edge (because it adjoins a cell on the axis, it has only three vanes), and (c) a radial edge.

angles differing by $d\phi$ and sliced crosswise at intervals of dz. Suppose we just turned the current on suddenly at $t = 0$; all fields are zero for $t < 0$ and there are discrete current densities $j(f, \frac{1}{2}dt)$ at the circular faces (normal to $\hat{z}$) centered at $r = 0$. Maxwell's equation for dE/dt has a j term and therefore gives $E(f, dt) \neq 0$ on these same faces. To then compute $B(e, 1.5\,dt)$ on the edges e, we use $dB/dt = -\text{curl } E$. We defined the curl of an edge field for this coordinate system in Chapter 7; by associating each face with an edge of the dual lattice we can extend the definition to face fields (see Section 20.5 for details). Except for weighting factors, curl E is a sum of E's on faces forming a "paddlewheel" around the edge e. For an azimuthal edge the paddlewheel is the strange-looking object in Figure 20.11a. It has only one nonzero contribution, that from $E(f)$ at the circular face. Thus the azimuthal $B(f, 1.5\,dt)$ is nonzero. For an axial edge (Figure 20.11b), none of the E's is nonzero, so the axial B stays zero. Figure 20.11c shows a radial edge; all E's are zero and the radial B stays zero.

We can now calculate $E(f, 2\,dt)$ from j and curl B; the curl at an axial face is nonzero and the axial E-field spreads farther from the axis. By looking at the non-axial faces it is not hard to see that curl B vanishes there. Thus, E stays axial. The next B's (at $t = 2.5\,dt$) can then be seen to remain azimuthal, and so on. This must remain true at all times, and is therefore true for the static fields (the limit $t \to \infty$), and we have explicitly justified our earlier assumption that B is azimuthal.

We can also prove these symmetry properties from the continuum Maxwell equations (though less rigorously, because in some situations the fields are not differentiable). Assume an initial current density distribution in which the current points along the axis and is cylindrically symmetric: $j_z(r)$ is independent of ϕ and z, and $j_r = j_\phi = 0$. Then Maxwell's equation $d\mathbf{E}/dt = c^2 \nabla \times \mathbf{B} - j/\epsilon_0$ gives $E_z(r) \neq 0$ at later times, but $E_r = E_\phi = 0$ (as long as $\nabla \times \mathbf{B}$ doesn't have other components: we'll check that next). The evolution of $\mathbf{B}$ is governed by $d\mathbf{B}/dt = -\nabla \times \mathbf{E}$, which (from the continuum formula for the curl in cylindrical coordinates) has only a ϕ component, $-dE_z(r)/dr$. This gives our desired result, that $\mathbf{B}$ itself has only a ϕ component. We can then go back to check the consistency of our assumption that $\nabla \times \mathbf{B}$ has only a z-component. It turns out to be $r^{-1}d(rB_\phi)/dr$, and the other components are indeed zero.

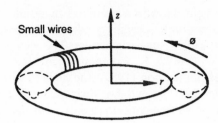

Figure 20.12 A toroidal solenoid. The wires are so small and close together that the ϕ-component of the current is negligible.

▶ **Problem 20.4**

Consider a toroidal solenoid of arbitrary cross section, as in Figure 20.12. Assume that the current distribution $j(f)$ in cylindrical coordinates has these azimuthal symmetry properties: $j_\phi = 0$, j_r and j_z are independent of ϕ.

a. Show from Maxwell's equations (discrete or continuous) that the electric and magnetic fields resulting from turning this current on have related symmetry properties. Specify what these are.

b. Using the symmetry you found for B, calculate from Ampère's law the exact magnetostatic field everywhere. Assume the current flows only on the surface of the toroid, with total current NI (N is the number of wires, I the current in each). ◀

▶ **Problem 20.5**

Consider the fields very near a flat (ribbon-shaped) wire in the xy-plane, in which a current flows in the x-direction. At a field point (x, y, z) with z much smaller than the distance to the edge of the ribbon, this problem can be idealized as an infinitely thin planar current sheet (with current per unit width K) of infinite extent in the x and y directions. Assume that the symmetry of $\mathbf{B}$ is such that its only nonzero component is $B_y(z)$, which depends only on z and is an odd function of z.

a. Use Maxwell's equation for $d\mathbf{E}/dt$ to determine the corresponding symmetry for $\mathbf{E}$.

b. Show that this symmetry for $\mathbf{E}$ gives a $d\mathbf{B}/dt$ (via Maxwell's equation) which is consistent with the original assumption. (Thus the evolving fields must have these symmetries at all times, assuming they vanish when the current is turned on.)

c. Does $\mathbf{E} \to 0$ after a long time in this ideal system? Do you think it would for the original problem (the wire ribbon)? ◀

SECTION 20.4 BOUNDARY VALUE PROBLEMS IN MAGNETOSTATICS

Having essentially exhausted the set of problems that can be conveniently done with Ampère's law alone, we consider the case of a general static current distribution, in which we must solve the system of equations $\nabla \times \mathbf{B} = \mu_0 \mathbf{j}$ and $\nabla \cdot \mathbf{B} = 0$. As in electrostatics, this is much easier to think about if we can confine the fields to a finite volume. We did this for the electric field by enclosing our system in a conducting box, defining a conductor to have $\mathbf{E} = 0$ in its interior. Unfortunately, the static magnetic field $\mathbf{B}$ in a conductor need not vanish, so we cannot use a

conductor to confine a static magnetic field. However, there are materials that do
not allow a magnetic field in their interiors under certain conditions (in particular,
at low enough temperatures). These are called **superconductors**. The element lead,
for example, is a superconductor below $T = 7.2$ K. Compounds are known with
much higher transition temperatures; at publication (1989), the highest was 125 K
but rising rapidly. Superconductors are best known for their related ability to
conduct current without resistance, but we will be concerned only with their ability
to expel magnetic fields. This expulsion is called the Meissner effect. We will define
a **perfect superconductor** as a material in which the magnetic field is always zero. It
doesn't matter if you don't know anything about the microscopic properties of
superconductors (indeed, you probably don't know much about the microscopic
properties of conductors that make $\mathbf{E} = 0$ statically, but that doesn't prevent you
from solving electrostatic boundary value problems). Real superconductors behave
this way to an excellent approximation if the magnetic field isn't too large; actually
the magnetic field penetrates a distance called the London penetration depth, which
is typically of the order of a micrometer and can safely be ignored in macroscopic
problems.

  We will therefore begin our discussion of general magnetostatic problems by
considering some examples confined to perfectly superconducting boxes, so that the
fields are confined to a finite volume. We can do without the superconducting boxes
only by using a system with infinite volume. We will do some continuum problems
of this sort in Section 20.8.

  Consider a current density distribution j inside an otherwise evacuated superconducting box, as shown in Figure 20.13. As in electrostatics, it turns out that the
best way to discretize this problem is to put cell centers on the boundary, as shown
in Figure 20.13b. The edges e that are entirely in the superconductor have
$B(e) = 0$, by definition. We will assume $B(e)$ also vanishes for an edge with one
end v in the superconductor. (This actually follows from div $B = 0$ whenever the
other edges touching v are entirely in the superconductor, as in Figure 20.13b.) We
will refer to the edges with at least one end in the superconductor as "superconduct-

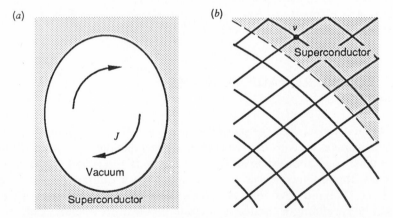

Figure 20.13 (a) A cavity in a superconducting material, in which magnetic fields can be confined
(b) The superconductor – vacuum interface, with a discrete lattice.

ing edges." We are given a current $j(f)$, defined for each face f that is entirely in the nonsuperconducting region. There may be currents in the superconductor as well ("supercurrents") but we will not need to know what they are; their role is similar to that of the induced charges on conducting boundaries in electrostatics. Our problem is to calculate $B(e)$ at the nonsuperconducting edges e, from the equations curl $B = \mu_0 j(f)$ (at nonsuperconducting faces), div $B = 0$, and the boundary condition $B = 0$. In a cubic discrete lattice, these equations have the explicit form

$$\sum_e B(e) = \mu_0 j(f) \, dr \tag{20.12}$$

$$\sum_e B(e) = 0 \tag{20.13}$$

where the sums are over the edges of the face f and the edges diverging from a vertex v, respectively.

■ **EXAMPLE 20.4**   To see how this scheme really works, let's solve a simple magnetostatics problem. Consider a rectangular $2dr \times 2dr \times dr$ cavity in a superconductor with a straight wire passing through it carrying a current I, as in Figure 20.14. (If you're worried about what happens where the wire enters the superconductor, imagine that it is soldered to it, and a small battery is soldered in series in the wire to drive the current, which returns through the superconductor.) Putting in cells of length dr with centers on the boundaries, we see that there are only four nonsuperconducting vertices and four nonsuperconducting edges, as shown in Figure 20.15. There is only one nonsuperconducting face, with current density $j(f) = I/dr^2$. Label the fields as shown in Figure 20.15b. The div $B = 0$ equation (Eq. 20.13) at the vertex v is $dr^{-1}[B(e_2) - B(e_1)] = 0$, implying $B(e_2) = B(e_1)$. Writing the corresponding equation for each vertex in turn as we move around the square, we see that all the B's are the same. We will call their common value B. Then Ampère's law states

$$\text{curl } B = dr^{-1}[4B] = \mu_0 I/dr^2$$

$$B = \mu_0 I/4dr \tag{20.14}$$

which is the solution to the problem.

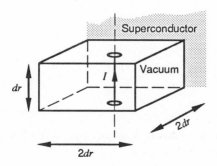

Figure 20.14   A cavity in a superconductor with a current-carrying wire through it.

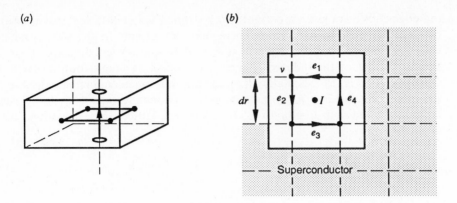

Figure 20.15   The four nonsuperconducting edges at which the magnetic field must be calculated. (*a*) Side view. (*b*) top view, showing the edge labels and the rest of the lattice.

■ **EXAMPLE 20.5**   We could give the system in the previous example translational symmetry in the vertical direction by using an infinitely long superconducting square duct with a wire down the axis as shown in Figure 20.16. The magnetic fields on the vertical edges can be shown to vanish just as the axial field did in the continuum case of a long wire (by turning on I and solving Maxwell's equations). So the problem of computing the fields on the edges of each horizontal square is independent of the other squares, and leads again to Eq. 20.14. We can think of this as a discretization of the problem of a long wire in a cylindrical hole in a superconductor. This can be solved exactly by Ampère's law, as it was in Section 20.2 in empty space. The result here is the same (Eq. 20.7); the superconductor

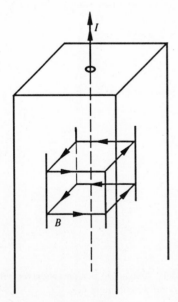

Figure 20.16   A square superconducting duct with a current-carrying wire down its axis.

makes no difference in this case. At a distance $r = \frac{1}{2} dr$ from the wire, the continuum field is

$$B = \mu_0 I / 2\pi(\tfrac{1}{2} dr) = \mu_0 I / \pi \, dr$$

which differs from the discrete result (Eq. 20.14) by a factor of $4/\pi = 1.27$. Considering the coarseness of the discretization, the discrete result is quite good.

■ **EXAMPLE 20.6** Let's solve another magnetostatics problem, in which the current closes on itself, rather than extending to infinity or exiting through the superconducting boundary. Consider a loop of wire carrying a current I in a cylindrical superconducting box, as shown in Figure 20.17a. Draw a discrete lattice with a vertical edge at the center, with cell length dr such that the only edges in the nonsuperconducting region are the 13 shown in Figure 20.17b. The current is azimuthal, and each face has the same current density $j = I/dr^2$, directed as shown. Clearly the problem has $90°$ rotational symmetry — the four squares in the paddlewheel are equivalent. Thus, the magnetic fields at the three other top edges are the same as $B(e_2)$, and the fields at the other three outer vertical edges are $B(e_3)$. Evidently the top edge e_2 is also equivalent to the edge e_4 directly below it, because of the mirror symmetry in a horizontal plane through the center (the $z = 0$ plane, if z is the vertical coordinate). It is tempting to use this symmetry to conclude that B is the same (and has the same direction) at e_2 and e_4. This is actually false — the directions are opposite, as we have suggested by drawing e_4 directed to the left. To see what the actual symmetry is, we must again use the procedure from Chapter 8 and compute the fields from Maxwell's equations as we turn on the current. The j term in $d\mathbf{E}/dt$ gives us E-fields at the faces shown in Figure 20.17b, and then $d\mathbf{B}/dt = -\text{curl } \mathbf{E}$ gives us positive $B(e_1)$, $B(e_2)$, $B(e_3)$, and $B(e_4)$, with $B(e_2) = B(e_4)$.

To prove that this symmetry holds at all times, we must write down the condition for z-reflection symmetry at an arbitrary face. The symmetry of the initial currents can be described by $j_z = 0$ and $\mathbf{j}(x, y, z) = \mathbf{j}(x, y, -z)$ (for any face at x, y, z). This gives for E (using $d\mathbf{E}/dt = -\mathbf{j}/\epsilon_0$ and ignoring curl B for now, because $B = 0$ initially) the

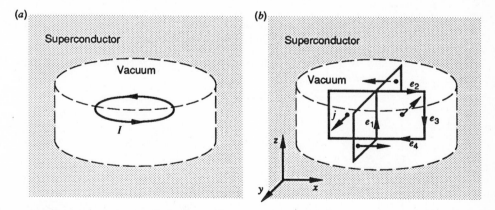

Figure 20.17 (a) A current-carrying loop in a superconducting cavity. (b) A discrete lattice, showing the nonsuperconducting edges.

same symmetry: $E_x(x, y, z) = +E_x(x, y, -z)$, $E_y(x, y, z) = +E_y(x, y, -z)$, and $E_z(x, y, z) = 0$. Then the symmetry of curl E (and therefore B) can be seen to be $B_x(x, y, z) = -B_x(x, y, -z)$, $B_y(x, y, z) = -B_y(x, y, -z)$, and $B_z(x, y, z) = +B_z(x, y, -z)$ (these are most easily deduced from the continuum curl equations). The resulting curl B is consistent with the symmetry written above for E, except that we can't be sure its z-component vanishes; Maxwell's equations imply only that $E_z(x, y, z) = -E_z(x, y, -z)$. However, this still gives the symmetry written above for B. These relations can be shown by mathematical induction to be true at every time, and therefore true of the magnetostatic limit. Geometrically, this means that we have to reverse the B's on horizontal edges when we reflect them; in particular, $B(e_4) = B(e_2)$.

So far we have shown there are really only three distinct B's in this problem: B_1, B_2, and B_3 (we will refer to $B(e_1)$ as B_1 for short). The divergence equations they must satisfy are:

$$\text{top center vertex:} \qquad 4B_2 - B_1 = 0$$

$$\text{top outer vertex:} \qquad B_3 - B_2 = 0$$

which imply $B_3 = B_2 = \frac{1}{4}B_1$. Ampère's law for each face is

$$dr^{-1}(B_1 + 2B_2 + B_3) = \mu_0 i$$

or

$$[1 + 2 \times \tfrac{1}{4} + \tfrac{1}{4}]B_1 = \mu_0 i$$

so

$$B_1 = \tfrac{4}{7}\mu_0 i \, dr$$

$$B_2 = B_3 = \tfrac{1}{7}\mu_0 i \, dr$$

The final B field is shown in Figure 20.18. We may regard this calculation as an approximation to the magnetic field of a circular current loop of radius R ($\cong \frac{1}{2}dr$) in free space, which can be improved by decreasing the cell size and enlarging the box. This continuum problem can **not** be solved exactly in terms of simple functions, so the discrete method is a useful and practical way of approaching it.

▶ **Problem 20.6**

Do this same example using a slightly larger box that contains the two additional edges e_5 and e_6 shown in Figure 20.19 (and their equivalents under symmetry).

a. Show by a symmetry argument that the oriented edges labeled "5" must have the same value of B.

b. By how much does this change B_1? ◀

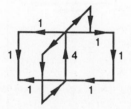

Figure 20.18 B, in units of $\mu_0 i \, dr/7 = \mu_0 I/7 dr$.

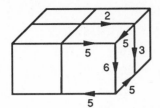

Figure 20.19 The nonsuperconducting edges in a larger superconducting cavity than that in Figure 20.17b. Numbers are edge labels.

■ **EXAMPLE 20.7 A SOLENOID IN A SUPERCONDUCTING DUCT** The current loop system in Example 20.6, like the one involving the straight wire (Example 20.4), can be given translational symmetry in the z-direction by using an infinitely tall cylindrical box and having identical current densities on all faces above and below the ones we used in Figure 20.17b. Translational symmetry then implies that B is in the same direction at e_2 and at e_4. However, we concluded before from reflection symmetry that it had to be in the **opposite** direction. Thus, it is zero, the only number that is equal to minus itself. So in this problem only B_1 and B_3 are nonzero. The divergence equations are trivially satisfied, and Ampère's law says

$$dr^{-1}(B_1 + B_3) = \mu_0 i$$

We have one equation for two variables; the fields are not uniquely determined! This difficulty is related to the fact that the superconducting region is not simply connected (a **simply connected** region is one in which any closed curve can be smoothly shrunk to a point without leaving the region, as a curve looping around our cylinder cannot). We can make the region simply connected by imagining the cavity to be capped at the top and bottom very far from where we are looking. The only effect of the capping is that the integral form of $\nabla \cdot \mathbf{B} = 0$ applied to a region that contains one of the caps and whose surface cuts the cylinder through e_1 and e_3 implies that $0 = \int \mathbf{B} \cdot d\mathbf{a} = dr^2(4B_3 - B_1)$ (that is, there is no magnetic flux through the cylinder). Now we can solve for B_1 and B_3:

$$B_1 = \tfrac{4}{5}\mu_0 i \, dr, \qquad B_3 = \tfrac{1}{5}\mu_0 i \, dr$$

You may recognize this example as a discrete approximation to the field of an infinitely long solenoid. The exact field inside such a solenoid (Eq. 20.10) is $\mu_0 i \, dr$; our answer differs from this only by 20% and would approach this as $dr \to 0$ for fixed solenoid size if we let the radius of the cylindrical box become large. The magnetic flux $\int \mathbf{B} \cdot d\mathbf{a}$ outside the solenoid would spread itself out over an infinite area, so the field (flux per unit area) would approach zero outside the solenoid.

▶ **Problem 20.7**

The related continuum problem of a cylindrical solenoid of radius a with surface current density K, inside a concentric cylindrical superconducting cavity of radius R, can be solved exactly. Assume the cavity is capped far away, so the total flux $\int \mathbf{B} \cdot d\mathbf{a}$ across any cross section is zero. Use Ampère's law to compute the fields inside and outside the solenoid, and show that they have the correct limits as $R \to \infty$. ◀

SECTION 20.5 BOUNDARY VALUE PROBLEMS IN CURVILINEAR COORDINATES

The two continuum problems for which we have solved discretizations both have cylindrical symmetry. We began by solving them in cartesian coordinates because

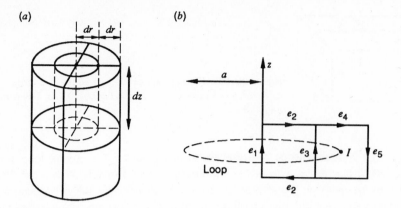

Figure 20.20 (*a*) A discrete lattice in cylindrical coordinates. The figure is drawn for the case $d\phi = \pi/2$; the final result is independent of $d\phi$. (*b*) Cross section in a plane containing the axis of the loop.

the discrete equations are simpler then, but obviously that is quite inefficient. It fails to exploit the exact rotational symmetry of the problem; when we use a small dr, fields at various points on a circle about the axis will be calculated separately, even though they are really equivalent. So let us rework the current-loop problem (Example 20.6) using a **cylindrical** coordinate system, as in Figure 20.20*a*. The symmetry of the problem is such that $B(e)$ is zero on the azimuthal edges ($B_\phi = 0$), so we can show all the relevant edges in a cross section through the axis, Figure 20.20*b*. Cell size $dr = \frac{2}{3}a$ and $dz = dr$ (a is the loop radius) were chosen so as to yield a finer resolution for the fields inside the loop (two edges instead of one). To write the vector derivatives we must use the general formulas given in Chapter 7. The curl is the simplest. In general, it is the line integral per unit area:

$$[\text{curl } B](f) = da(f)^{-1} \sum_e B(e)\, dl(e) \tag{20.15}$$

where the sum is over the four edges of the face f and $da = dr\, dz = dr^2$ for all faces shown (and the curl is zero for those not shown). For the face next to the axis, Ampère's law is

$$\text{curl } B = dr^{-2}(B_1 + 2B_2 - B_3)\, dr = \mu_0 j = 0 \tag{20.16}$$

[as before, B_1 means $B(e_1)$]. For the face crossed by the current,

$$dr^{-2}(B_3 + 2B_4 + B_5)\, dr = \mu_0 I/dr^2 \tag{20.17}$$

We must now use the divergence equation div $B = 0$. We defined the divergence in curvilinear coordinates (Chapter 7) at cell centers, as a sum of B's at the faces of the cell. Unfortunately B is not **defined** at faces, but at edges. Recall, however, that the edges of a lattice are centered at the same points as the faces of the dual lattice.

Let us denote by f' the face of the dual lattice with the same center as an edge e of the original lattice. The natural way to compute the divergence of B is to define a face field on the dual lattice (which we can also call B without ambiguity) by $B(f') = B(e)$. Then the divergence of B at a cell center c' of the dual lattice is defined by

$$[\operatorname{div} B](c') = d\tau(c')^{-1} \sum_{f'} B(f') \, da(f') \qquad (20.18)$$

where the sum is over the faces f' of the cell c', which is centered at a vertex of the original lattice. We can write Eq. 20.18 in terms of the original lattice as

$$[\operatorname{div} B](v) = d\tau(v)^{-1} \sum_{e} B(e) \, da(e) \qquad (20.19)$$

where $d\tau(v)$ is the volume of a cell of the dual lattice centered on the vertex v of the original lattice, and $da(e)$ is the area of the face dual to e.

The dual lattice is shown for the present case in Figure 20.21. To visualize it, note that each axis is divided exactly between the divisions of the original lattice. Radial divisions are at $r = 0.5 \, dr, 1.5 \, dr, \ldots$, instead of $r = dr, 2 \, dr, \ldots$. Axial divisions are at odd-half-integer instead of integer multiples of dz, and angular divisions are at $\frac{1}{4}\pi, \frac{3}{4}\pi, \frac{5}{4}\pi, \ldots$, instead of $0, \frac{1}{2}\pi, \pi, \frac{3}{2}\pi, \ldots$. The circular face f_1' of the dual lattice (Figure 20.22) has the same center as the e_1 in Figure 20.20b, f_2' has the same center as e_2, and so on. The areas of the faces are easily computed to

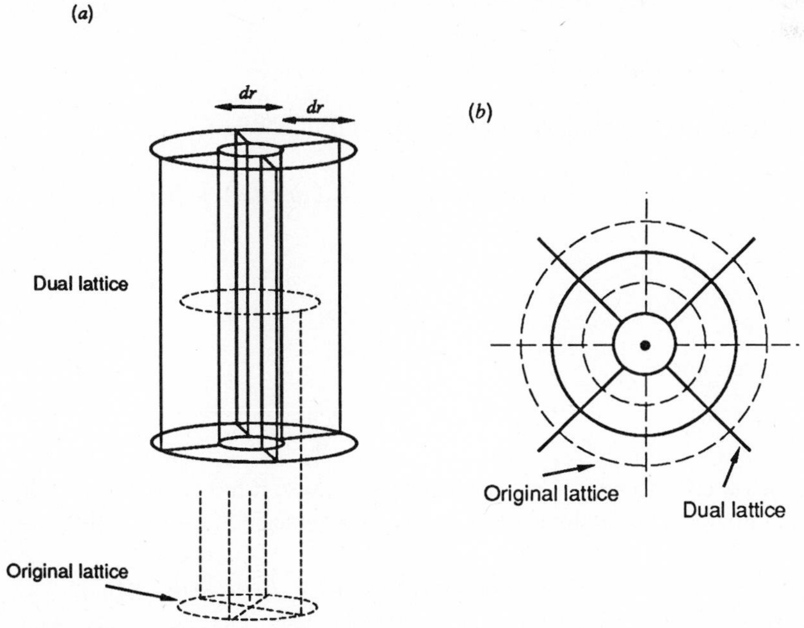

Figure 20.21   The dual lattice (solid lines) of the lattice in Figure 20.20a (indicated by dashed lines). (a) Oblique view (b) Top view.

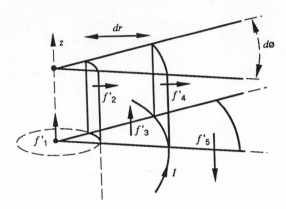

Figure 20.22 Detail of the faces of the dual lattice which are dual to the edges shown in Figure 20.20*b*.

be

$$da(f_5') = \tfrac{1}{2} d\phi \left[(2.5\, dr)^2 - (1.5\, dr)^2 \right] = 2\, dr^2\, d\phi$$

$$da(f_3') = dr^2\, d\phi$$

$$da(f_4') = 1.5\, dr^2\, d\phi$$

$$da(f_2') = 0.5\, dr^2\, d\phi$$

$$da(f_1') = 0.25\, dr^2\, \pi$$

Therefore the divergence equation for the cell c' above f_3' is

$$0 = d\tau(c')^{-1} \left[-B_2(0.5\, dr^2\, d\phi) - B_3(dr^2\, d\phi) + B_4(1.5\, dr^2\, d\phi) \right]$$

or

$$0 = -0.5B_2 - B_3 + 1.5B_4 \tag{20.20}$$

For the cell above f_5',

$$0 = -1.5B_4 + 2B_5 \tag{20.21}$$

The central cell (above f_1') is a special case because it has more than six faces. There are $2\pi/d\phi$ faces equivalent to f_2', so the divergence equation for this cell is

$$0 = B_2(0.5\, dr^2\, d\phi)(2\pi/d\phi) - B_1(0.25\, dr^2\, \pi)$$

or

$$0 = B_2 - 0.25B_1 \tag{20.22}$$

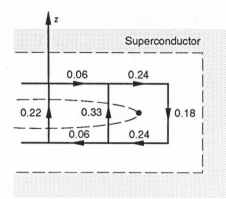

Figure 20.23   The B field, in units of $\mu_0 I/dr$.

Note that these divergence equations have the same general form we would have obtained by naively applying the cartesian-coordinates formula to Figure 20.20b, except for weighting factors.

We now have five equations (Eqs. 20.16, 20.17, 20.20, 20.21, and 20.22) in five unknowns, which can immediately be reduced to three because the last two imply

$$B_2 = 0.25B_1, \qquad B_5 = 0.75B_4 \qquad\qquad (20.23)$$

Eliminating B_2, B_5, B_3, and B_4 successively gives the solution

$$B_1 = \frac{48}{215}\frac{\mu_0 I}{dr}, \qquad B_2 = \frac{12}{215}\frac{\mu_0 I}{dr}$$

$$B_3 = \frac{72}{215}\frac{\mu_0 I}{dr}, \qquad B_4 = \frac{52}{215}\frac{\mu_0 I}{dr}$$

$$B_5 = \frac{39}{215}\frac{\mu_0 I}{dr}$$

This field is shown in Figure 20.23. This solution gives a higher-resolution picture of the field inside the loop (the field is largest near the loop) than did the solution (Figure 20.18) with just one edge inside the loop. (If you want to make a quantitative comparison, remember that the present dr is about one-third of the old one.)

▶   **Problem 20.8**

Compute the magnetic field of a long straight wire carrying a current I along the axis of a cylindrical superconducting box of radius $2\,dr$, using the discrete cylindrical coordinate system of Figures 20.21 and 20.22 (so that the dual lattice is as in Figure 20.20). Show that your answer is the same as the continuum one (Eq. 20.7).   ◀

SECTION 20.6   THE MAGNETIC VECTOR POTENTIAL

You may have noticed that problems in magnetostatics tend to be harder to visualize and to solve than problems involving the same number of cells in electrostatics. This is because in magnetostatics we have been solving for a **vector** field (the magnetic field **B**) rather than a scalar field (the potential V), and this involves three variables per cell rather than one. Of course, we would have to deal with a vector field (**E**) in electrostatics also, if we hadn't found a way to express **E** in terms of a scalar potential: $\mathbf{E} = -\nabla V$. Can we find an analogous potential formulation for the magnetic field? We proved V existed because $\nabla \times \mathbf{E} = 0$; unfortunately this is not true of **B**. [It is true that $\nabla \times \mathbf{B} = 0$ in a region having no currents, and a "magnetic scalar potential" whose gradient is **B** can be defined in a simply connected region of this type; see for example the book by Reitz, Milford, and Christy cited in the bibliography. But this is of limited usefulness; we want to be able to treat regions with currents.] However, we do know that $\nabla \cdot \mathbf{B} = 0$. One can prove that this implies the existence of a vector field **A** so that

$$\mathbf{B} = \nabla \times \mathbf{A} \qquad (20.24)$$

The converse is actually quite easy to show: if there is an **A** whose curl is **B**, then $\nabla \cdot \mathbf{B} = 0$ because of the identity $\nabla \cdot \nabla \times \mathbf{A} = 0$ (true for any **A**, as we proved in Section 14.4).

We will prove that **A** exists whenever $\nabla \cdot \mathbf{B} = 0$ by simply writing down a formula for it. This is made easier by the fact that **A** is highly nonunique. For example, we will see in Eq. 20.30 that we can add any gradient to **A** without changing $\nabla \times \mathbf{A}$. (The electrostatic scalar potential V was also not unique, of course, but only to the extent that we could add a constant. The vector potential is **much** more nonunique.) Taking the continuum case first, we want **A** to satisfy Eq. 20.24. Explicitly,

$$B_x = \frac{\partial A_z}{\partial y} - \frac{\partial A_y}{\partial z} \qquad (20.24a)$$

$$B_y = \frac{\partial A_x}{\partial z} - \frac{\partial A_z}{\partial x} \qquad (20.24b)$$

$$B_z = \frac{\partial A_y}{\partial x} - \frac{\partial A_x}{\partial y} \qquad (20.24c)$$

We could solve the last equation for A_y if A_x were zero, by integrating with respect to x:

$$A_y(x, y, z) = \int_0^x B_z(x', y, z) \, dx' + A_y(0, y, z) \qquad (20.25)$$

So let's just assume

$$A_x(x, y, z) = 0 \qquad (20.26)$$

and see where it gets us. Let's also assume the integration constant $A_y(0, y, z)$ is zero, that is, $A_y = 0$ on the $x = 0$ plane. Then Eq. 20.24a can be solved for A_z on this plane

$$A_z(0, y, z) = \int_0^y B_x(0, y', z)\, dy' + A_z(0, 0, z)$$

Again assuming that the integration constant $A_z(0, 0, z)$ is zero, we can obtain A_z away from this plane by integrating Eq. 20.24b with respect to x:

$$A_z(x, y, z) = -\int_0^x B_y(x', y, z)\, dx' + A_z(0, y, z)$$

$$\text{(20.27)}$$

$$= -\int_0^x B_y(x', y, z)\, dx' + \int_0^y B_x(0, y', z)\, dy'$$

Then Eqs. 20.25 through 20.27 give a possible vector potential $\mathbf{A}$. Obviously there are many other ways to choose $\mathbf{A}$; we could have started by insisting $A_y = 0$ instead of $A_x = 0$, for example.

▶ **Problem 20.9**
   Check Eqs. 20.25 through 20.27 by substituting them into Eq. 20.24.   ◀

▶ **Problem 20.10**
   Compute the vector potential given by Eqs. 20.25 through 20.27 for the special case of a uniform magnetic field having $B_x = B_y = 0$, $B_z = B_0$, where B_0 is a constant.   ◀

▶ **Problem 20.11**
   Construct an exact discrete analog of the above procedure for computing $\mathbf{A}$. That is, given an arbitrary edge vector field B satisfying $\operatorname{div} B = 0$, show how to construct a particular face vector field A satisfying $B = \operatorname{curl} A$. Hints: Take the origin at a cell center. Assume $A(f) = 0$ for faces normal to the x-axis (in terms of components, $A_x = 0$). You can let $A_y = 0$ for faces with centers on the $x = 0$ plane and $A_z = 0$ for faces centered on the z-axis. Show that the x-component of $B = \operatorname{curl} A$ uniquely determines A_z on the rest of the $x = 0$ plane. (It may help to do this first in an explicit case such as the next problem.) Then the y-component determines A_z everywhere, and the z-component determines A_y. You then have to verify that the x-component of the curl equation is also satisfied everywhere.   ◀

▶ **Problem 20.12**
   Carry out the procedure of Problem 20.11 explicitly for the discrete magnetic field shown in Figure 20.24 ($B_z = 0$ everywhere, and B_x and B_y are nonzero only in the plane of the figure). Show A on a similar figure. Verify that $\operatorname{curl} A = B$ at each edge.   ◀

Often there is a simpler choice for $\mathbf{A}$ than Eqs. 20.25 through 20.27, as in the following example.

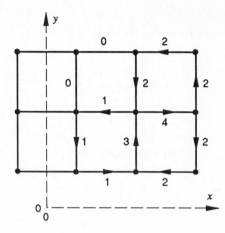

Figure 20.24 The magnetic field B, in the plane $z = dr/2$.

■ **EXAMPLE 20.8 THE VECTOR POTENTIAL OF A LONG WIRE**   Find a vector potential for the magnetic field of a long straight wire. In cylindrical coordinates, this field is $B_r = B_z = 0$, $B_\phi = \mu_0 I / 2\pi r$. Thus, the only nonzero component of curl A should be the azimuthal one, which is given by the continuum formula

$$(\nabla \times A)_\phi = \frac{dA_r}{dz} - \frac{dA_z}{dr}$$

First try to find a solution having only an r-component, so we can ignore the second term. This gives

$$\frac{dA_r}{dz} = B_\phi = \frac{\mu_0 I}{2\pi r}$$

which integrates to

$$A_r = \frac{\mu_0 I z}{2\pi r} \qquad (A_z = 0) \tag{20.28}$$

(taking the integration constant to be zero). This is a possible vector potential for the given magnetic field. However, the nonuniqueness is clear: we could have assumed instead that **A** has only a z component, so only the second term in the curl is nonzero. Integrating with respect to r would then have yielded

$$A'_z = -(\mu_0 I / 2\pi) \ln r \qquad (A'_r = 0) \tag{20.29}$$

Once we have a **particular** potential A we can get the set of **all** possible potentials A' for which curl $A' = B$ by adding an arbitrary gradient

$$A'(f) = A(f) + \nabla \Lambda(c) \tag{20.30}$$

where Λ is any cell scalar field. To show this, we must show that for any A' satisfying curl $A' = B$, there is a Λ so that A' is given by Eq. 20.30. But for any such A', curl$(A' - A) = 0$ and therefore $A' - A$ is a gradient for the same reason the electrostatic field E was in Chapter 15. This transformation from A to A' is

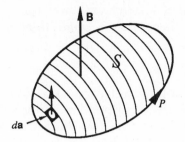

Figure 20.25   A surface S and its boundary path P, in a magnetic field.

called a **gauge transformation**, and Λ is called a **gauge function**. The particular gauge function that gives Eq. 20.29 from Eq. 20.28 is $\Lambda = (\mu_0 I/2\pi) z \ln r$.

▶   **Problem 20.13**

    a. Find a continuum vector potential for the uniform continuum magnetic field $B_x = B_y = 0$, $B_z = B_0$, different from the one you found in Problem 20.10.

    b. Show that their difference can be expressed as a gradient.   ◀

There is a useful connection between the magnetic vector potential and the magnetic flux $\Phi(S) = \int \mathbf{B} \cdot d\mathbf{a}$ through a surface S bounded by a path P, as in Figure 20.25. Stokes's theorem states that $\int (\nabla \times \mathbf{A}) \cdot d\mathbf{a} = \int \mathbf{A} \cdot d\mathbf{l}$, so the flux is

$$\Phi(S) = \int \mathbf{A} \cdot d\mathbf{l} \qquad\qquad (20.31)$$

the line integral of $\mathbf{A}$ around the perimeter path of S. This sometimes makes it possible to solve problems involving flux without actually calculating the B field.

▶   **Problem 20.14**

Compute the magnetic flux through a square of side a, a distance a from a long straight wire carrying a current I (see Figure 20.4a). Compute the flux

    a. from $\int \mathbf{A} \cdot d\mathbf{l}$, and

    b. from $\int \mathbf{B} \cdot d\mathbf{a}$. Are the answers the same?   ◀

▶   **Problem 20.15**

    a. Verify that $\mathbf{A} = -\frac{1}{2}\mathbf{r} \times \mathbf{B}$ is a possible vector potential for a **uniform** field $\mathbf{B}$.

    b. Calculate the components of this $\mathbf{A}$ for the $\mathbf{B}$ of Problem 20.10.

    c. Show that the difference between this vector potential and the one you computed in Problem 20.10 is a gradient (i.e., give a scalar function of which it is the gradient).   ◀

SECTION 20.7   THE POISSON EQUATION IN MAGNETOSTATICS

What good does the vector potential do us? To a certain extent it can simplify magnetostatic calculations. It does not reduce the equations to scalar equations, as in electrostatics, but it does reduce two equations ($\nabla \times \mathbf{B} = \mu_0 j$, $\nabla \cdot \mathbf{B} = 0$) for $\mathbf{B}$

to **one** equation for **A**. This is because $\mathbf{B} = \nabla \times \mathbf{A}$ satisfies $\nabla \cdot \mathbf{A} = 0$ automatically for any **A**. The only requirement we must impose on **A** comes from expressing Ampère's law in terms of **A**:

$$\nabla \times \nabla \times \mathbf{A} = \mu_0 \mathbf{j} \tag{20.32}$$

It is very awkward to work with this double curl; the vector second derivative we are familiar with is the Laplacian ∇^2. Can we express Ampère's law in terms of $\nabla^2 \mathbf{A}$? So far, this expression has no meaning for a vector field **A**. When the Laplacian operator ∇^2 acts on a scalar like V, it means $\nabla \cdot \nabla V$. However, you can't take the gradient of a vector field. We do know, though, that $\nabla^2 V(c)$ can also be thought of geometrically in terms of the six neighbors c' of the cell c, displaced by $\pm dr$ in the x, y, or z direction as in Figure 16.6.

$$\nabla^2 V(c) = dr^{-2} \sum_{c'} [V(c') - V(c)]$$

We can generalize this notion to a face vector field A. A face f also has six neighbors f', parallel faces displaced by $\pm dr$ in each direction, as shown in Figure 20.26. We will **define** the Laplacian $\nabla^2 A$ in cartesian coordinates by

$$\nabla^2 A(f) = dr^{-2} \sum_{f'} [A(f') - A(f)]$$

$$= dr^{-2} \left[\sum_{f'} A(f') - 6A(f) \right] \tag{20.33}$$

However, remember that the gradient ∇A still has no meaning. In the continuum limit, it is easy to show that our definition corresponds to

$$\nabla^2 A_x = \frac{d^2}{dx^2} A_x + \frac{d^2}{dy^2} A_x + \frac{d^2}{dz^2} A_x \tag{20.34}$$

and similarly for A_y and A_z.

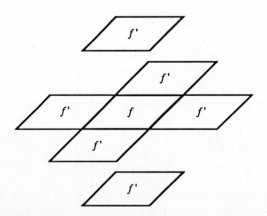

Figure 20.26   The six "neighbors" f' of a face f in a cartesian lattice.

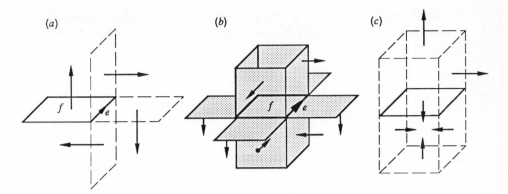

Figure 20.27 (a) The four faces involved in computing a curl at e, (b) the faces involved in computing the left-hand side of Eq. 20.35, (c) those involved in the first term on the right-hand side of Eq. 20.35.

Using this definition of $\nabla^2 A$, we can prove an identity

$$\nabla \times \nabla \times A = \nabla(\nabla \cdot A) - \nabla^2 A \qquad (20.35)$$

To prove it, simply express both sides of Eq. 20.35 (evaluated at a face f) in terms of $A(f)$'s at nearby faces. They turn out to be the same. To see this, start by thinking about [curl A](e) at an edge e of f as a sum of $A(f')$'s in a "paddlewheel" (Figure 20.27a). The curl of curl A involves adding these up for all four edges of f, which yields $A(f')$ at all the directed faces f' shaded in Figure 20.27b (and $A(f)$ itself, four times). On the right-hand side of the equation, [div A](c) at the cell c above f gives the sum of the faces of the top cube in Figure 20.27c, and subtracting the cell below f (to get the gradient) gives all the faces in Figure 20.27c [with a coefficient -2 for $A(f)$]. Subtracting $\nabla^2 A$ (Figure 20.26) changes the coefficient of $A(f)$ to $-2 - (-6) = +4$, cancels the top and bottom horizontal faces, and adds four "wings" along the edges of f, giving exactly Figure 20.27b (i.e., the left-hand side $\nabla \times \nabla \times A$).

▶ **Problem 20.16**

Verify that Eq. 20.35 is correct in the continuum limit by expressing both sides in terms of partial derivatives. ◀

▶ **Problem 20.17**

a. Prove the "BAC–CAB" rule: for any three vectors **A**, **B**, and **C**

$$\mathbf{A} \times (\mathbf{B} \times \mathbf{C}) = \mathbf{B}(\mathbf{A} \cdot \mathbf{C}) - \mathbf{C}(\mathbf{A} \cdot \mathbf{B})$$

by expressing both sides in terms of components of **A**, **B**, and **C**.

b. Show that Eq. 20.35 can be formally obtained by substituting the del vector $\nabla \equiv (\partial/\partial x, \partial/\partial y, \partial/\partial z)$ into the BAC–CAB rule. ◀

Note that we have defined $\nabla^2 A$ **only** in cartesian coordinates; the procedure we have used does *not* generalize to curvilinear coordinates (nor can we apply the continuum formulas for ∇^2 to $\mathbf{A}$ in curvilinear coordinates, as we did in Eq. 20.34 in cartesian coordinates). If you want to use $\nabla^2\mathbf{A}$ in curvilinear coordinates, you must **define** it by Eq. (20.35):

$$\nabla^2\mathbf{A} \equiv \nabla(\nabla \cdot \mathbf{A}) - \nabla \times \nabla \times \mathbf{A} \qquad (20.36)$$

which makes sense in any coordinate system.

We can now express Ampère's law for $\mathbf{A}$ (Eq. 20.32) in terms of the Laplacian via Eq. 20.35:

$$\nabla(\nabla \cdot \mathbf{A}) - \nabla^2\mathbf{A} = \mu_0\mathbf{j}$$

We would like to make this look like Poisson's equation by getting rid of $\nabla \cdot \mathbf{A}$. We can do this by using our freedom to change $\mathbf{A}$ by a gradient (Eq. 20.30), giving a new $\mathbf{A}'$ such that $\nabla \cdot \mathbf{A}' = 0$; then (using $\mathbf{A}'$ instead of $\mathbf{A}$ in the above equation)

$$-\nabla^2\mathbf{A}' = \mu_0 j \qquad (20.37)$$

The choice of potential so that $\nabla \cdot \mathbf{A}' = 0$ is called **transverse gauge**.

> The proof that we can find such an $\mathbf{A}'$ goes as follows. Take the $\mathbf{A}$ we constructed at the beginning of this section, satisfying $\mathbf{B} = \nabla \times \mathbf{A}$, but not necessarily $\nabla \cdot \mathbf{A} = 0$. Consider $\mathbf{A}' \equiv \mathbf{A} + \nabla\Lambda$; then $\nabla \cdot \mathbf{A}' = 0$ means $\nabla \cdot \mathbf{A} + \nabla \cdot \nabla\Lambda = 0$, that is, Λ satisfies the Poisson equation $\nabla^2\Lambda = -\nabla \cdot \mathbf{A}$. However, the existence and uniqueness theorem of Chapter 16 tells us we can always solve this (the solution isn't unique here because we haven't insisted on boundary conditions). This gives the desired $\mathbf{A}'$.

We have now shown that any magnetostatics problem can be solved using the magnetic vector potential, by solving the Poisson equation (Eq. 20.37). The magnetic field is then obtained by taking the curl of $\mathbf{A}$. However, in practice there is a fly in the ointment: we don't know the boundary condition on $\mathbf{A}$. Many different boundary conditions will work ($\mathbf{A}$ is not unique), but unfortunately some will not—they give $\nabla \cdot \mathbf{A} \neq 0$, so the Poisson equation we solved is not equivalent to Ampère's law. Fortunately, the boundary condition $\mathbf{A} = 0$ happens to work in many cases of interest. If we solve Poisson's equation for $\mathbf{A}$ and find $\nabla \cdot \mathbf{A} = 0$, we know we have such a case, and we have found the correct solution.

> Actually, we need verify $\nabla \cdot \mathbf{A} = 0$ only on the boundary; this implies $\nabla \cdot \mathbf{A} = 0$ everywhere, for the following reason: The divergence of Eq. 20.37 is
>
> $$\nabla \cdot (\nabla^2\mathbf{A}) = \mu_0\nabla \cdot \mathbf{j} = 0$$
>
> (using Eq. 20.2). If we could interchange ∇^2 with $\nabla \cdot$, this would imply that $\nabla \cdot \mathbf{A}$ obeys Laplace's equation with a zero boundary condition ($\nabla \cdot \mathbf{A} = 0$ is assumed at the boundary), and is therefore zero. It is left as an exercise for the reader to show that $\nabla \cdot$ and ∇^2 do indeed commute, using the discrete and/or continuum equations. (In the continuum case it requires assuming the equality of crossed partial derivatives, $d^2\mathbf{A}/dx\,dy = d^2\mathbf{A}/dy\,dx$.) Therefore $\nabla \cdot \mathbf{A} = 0$ everywhere, and we have found the correct vector potential in transverse gauge.

■ **EXAMPLE 20.9**   One case in which the boundary condition $A = 0$ works is the current-loop problem (Example 20.6) for which we have already calculated B. We will demonstrate how the solution by vector potential works by solving it again. Suppose we allow the vector potential $A(f)$ to be nonzero at the nonsuperconducting faces shown in Figure 20.17b, which form a paddlewheel. However, these are equivalent by symmetry; call their common vector potential A. Let us write Poisson's equation for the face f:

$$-\nabla^2 A(f) = -dr^{-2}\left[\sum_{f'} A(f') - 6A(f)\right] = -dr^{-2}[-A - 6A] = \mu_0 i$$

The sum is over the neighbors f' of f, that is, the six nearest parallel faces, of which only one has nonzero A, due to the $A = 0$ boundary condition (nonparallel faces are not considered neighbors here). This equation can be immediately solved for A:

$$A = \tfrac{1}{7}\mu_0 i \, dr^2$$

We can check the transverse gauge condition $[\text{div} A](c) = 0$ at each cell c. In this case it holds by symmetry — there is an outward and an inward A, and they cancel in $\text{div} A = dr^{-1}\sum A(f)$. Now compute $B = \text{curl } A$: at the top or right edge (e_3 or e_2 in Figure 20.17b) there is only one A in the paddlewheel:

$$B_3 = B_2 = dr^{-1}\sum_f A(f) = \frac{A}{dr} = \tfrac{1}{7}\mu_0 i \, dr$$

whereas at the axis (e_1) all four faces in the paddlewheel have nonzero A:

$$B_1 = \frac{4A}{dr} = \tfrac{1}{7}4\mu_0 i \, dr$$

This is exactly the same field we calculated in Section 20.4 by solving three coupled equations; this time there was only one.

▶   **Problem 20.18**

Consider the superconducting cavity of Figure 20.28. The magnetic vector potential is zero except on the faces shown. There is a loop of current near the outer boundary,

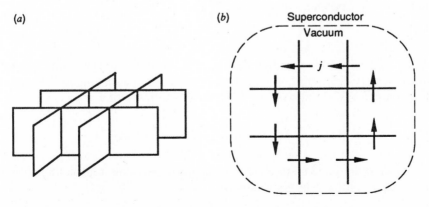

Figure 20.28   The nonsuperconducting faces in a cavity in a superconductor: (a) Oblique view, (b) Top view.

giving a discrete current density $j(f)$ as shown in the top view (all the nonzero $j(f)$'s are equal—call them j). (This can be thought of as a finer discretization of the system of Figure 20.17.)

a. Use the symmetry to reduce the number of independent A's to one or two. [Hint: In principle you should use Maxwell's equations when the current is turned on to determine the symmetry of E and B, and hence that of A. In fact, however, you can always choose A to have the same symmetry as E (note that curl $dA/dt = -$curl E), and this determines the symmetry of B. Therefore, you need not consider the symmetry of B explicitly at all.]

b. Write down the Poisson equation and solve it for A.

c. Verify that $\nabla \cdot A = 0$ everywhere.

d. Compute $B = \nabla \times A$.

e. Verify that $\nabla \times B = \mu_0 j(f)$ at each inequivalent face, and $\nabla \cdot B = 0$ at each vertex.   ◄

Another case in which $A = 0$ is an adequate boundary condition is that of a finite current distribution in infinite space; this will be critical in Sections 20.8 and 20.9. We will show in Section 20.9 that $A = 0$ works for this case, by explicitly determining A far away from the current, and showing $\nabla \cdot A = 0$ there (i.e., "on the boundary"). Further applications of the magnetic vector potential may be found in many electromagnetism texts, for example, that by Griffiths cited in the bibliography.

SECTION 20.8   THE LAW OF BIOT AND SAVART

Consider the continuum problem of solving the Poisson equation for $\mathbf{A}$ (Eq. 20.37) in infinite space, with the boundary condition that $A \to 0$ at infinity. (We showed in Section 20.7 that this boundary condition works as long as the resulting $\mathbf{A}$ satisfies $\nabla \cdot \mathbf{A} = 0$ on the boundary; we will verify this later.) Although this is a vector equation (the potential formulation did not reduce it to a scalar, as in electrostatics) the components are **decoupled**: the equation for A_x does not involve A_y, and so on. This makes it as easy to solve as a scalar equation. In fact, each component is equivalent to Poisson's equation for V

$$-\nabla^2 V = \rho/\epsilon_0 \tag{20.38}$$

which we solved in Chapter 15 using a Green's function method (in effect, superposing Coulomb's law potentials for each charge making up the distribution). The solution (Eq. 15.8) was, for arbitrary position vector $\mathbf{r}$

$$V(\mathbf{r}) = (4\pi\epsilon_0)^{-1} \int \rho(\mathbf{r}') \, d\tau'/|\mathbf{r}' - \mathbf{r}| \tag{20.39}$$

We can use the same method here. In fact, we need only replace the symbols V by A_x, ρ by j_x, and ϵ_0^{-1} by μ_0 to obtain

$$A_x(\mathbf{r}) = (\mu_0/4\pi) \int j_x(\mathbf{r}') \, d\tau'/|\mathbf{r}' - \mathbf{r}|$$

Since we can do this for each component, we have the vector equation

$$\mathbf{A}(\mathbf{r}) = (\mu_0/4\pi) \int \mathbf{j}(\mathbf{r}')\, d\tau'/|\mathbf{r}' - \mathbf{r}| \qquad (20.40)$$

We can now compute the magnetic field by taking the curl:

$$\mathbf{B}(\mathbf{r}) = (\mu_0/4\pi)\nabla_r \times \int \mathbf{j}(\mathbf{r}')\, d\tau'/|\mathbf{r}' - \mathbf{r}| \qquad (20.41)$$

The subscript "r" on the ∇, is to remind the reader that it involves differentiation with respect to $\mathbf{r}$, rather than $\mathbf{r}'$. The importance of keeping this straight in manipulating expressions of this sort cannot be overemphasized. The integration is over $\mathbf{r}'$, which varies independently of $\mathbf{r}$, so it doesn't matter whether we integrate before or after differentiating; the ∇ can be brought inside the integral. It can also be brought past the $\mathbf{j}(\mathbf{r}')$, which doesn't depend on $\mathbf{r}$. However, $\mathbf{j}$ is the vector into which ∇ is being crossed. To avoid confusion, let us separate the $\mathbf{r}'$-dependence of $\mathbf{j}$ from its vector character by writing it in terms of unit vectors

$$\mathbf{j}(\mathbf{r}') = \sum_i j_i(\mathbf{r}')\hat{\mathbf{i}} \qquad (20.42)$$

where the sum is over the three coordinate directions, $i = x$, y, or z, and $\hat{\mathbf{i}}$ is a unit vector in the appropriate direction. Thus, Eq. 20.41 becomes

$$\mathbf{B}(\mathbf{r}) = (\mu_0/4\pi)\sum_i \int j_i(\mathbf{r}')\left[\nabla_r \times (\hat{\mathbf{i}}/|\mathbf{r}' - \mathbf{r}|)\right] d\tau' \qquad (20.43)$$

We can now evaluate the curl explicitly. Let us denote the vector separation between the source and field points by $\mathbf{s} = \mathbf{r} - \mathbf{r}'$. Note that any derivative with respect to a component of $\mathbf{r}$ is the same as a derivative with respect to that component of $\mathbf{s}$. (Use the chain rule of differentiation, and note that $\mathbf{r}'$ is just a constant in this context). So we can replace ∇_r by ∇_s. It turns out there is a simple formula for the expression in brackets,

$$\nabla_s \times \frac{\hat{\mathbf{i}}}{|s|} = \frac{\hat{\mathbf{i}} \times \mathbf{s}}{|s|^3} \qquad (20.44)$$

To prove Eq. 20.44, look at it component by component. Suppose for concreteness $i = x$, and look at the y-components

$$\frac{\partial}{\partial s_z}\left(\frac{\hat{\mathbf{x}}}{s}\right)_x - \frac{\partial}{\partial s_x}\left(\frac{\hat{\mathbf{x}}}{s}\right)_z \overset{?}{=} \frac{0 \cdot s_x - 1 \cdot s_z}{s^3}$$

or

$$\frac{\partial}{\partial s_z}\frac{1}{s} \overset{?}{=} -\frac{s_z}{s^3} \qquad (20.45)$$

But this follows directly from a very useful formula for differentiating the magnitude of a vector with respect to a component:

$$\frac{ds}{ds_z} = \frac{s_z}{s} \tag{20.46}$$

which you can prove by differentiating $s^2 = s_x^2 + s_y^2 + s_z^2$. Equation 20.45 then follows from the chain rule: $(d/ds_z)s^{-1} = -s^{-2}(ds/ds_z)$. We can make this same argument for the z-component of Eq. 20.44, and the x-component is trivial. The same proof we have used for $i = x$ works for $i = y$ or z, so we have proved Eq. 20.44. We can now use it to transform Eq. 20.43 to

$$\mathbf{B}(\mathbf{r}) = \frac{\mu_0}{4\pi} \sum_i \int j_i(\mathbf{r}') \frac{\hat{\mathbf{i}} \times \mathbf{s}}{s^3} \, d\tau'$$

$$= \frac{\mu_0}{4\pi} \int \frac{\mathbf{j}(\mathbf{r}') \times \mathbf{s}}{s^3} \, d\tau' \tag{20.47}$$

Defining a unit vector $\hat{\mathbf{s}}$ by

$$\hat{\mathbf{s}} = \frac{\mathbf{s}}{s} \tag{20.48}$$

we can express the magnetic field as

$$\mathbf{B}(\mathbf{r}) = \frac{\mu_0}{4\pi} \int \frac{\mathbf{j}(\mathbf{r}') \times \hat{\mathbf{s}}}{s^2} \, d\tau' \tag{20.49}$$

This is the **Biot–Savart law**. For the special case of a current confined to a wire, we can replace $\mathbf{j}(\mathbf{r}') \, d\tau'$ by $I \, d\mathbf{l}$, where $d\mathbf{l}$ is a length element along the wire. It is a vector, and points tangent to the wire. This replacement may be justified by looking at a face normal to the wire as in Figure 20.29, and noting that $j = I/dr^2$, $d\tau' = dr^3$, and $dl = dr$. Then the Biot–Savart law becomes

$$\mathbf{B}(\mathbf{r}) = \frac{\mu_0 I}{4\pi} \int \frac{d\mathbf{l} \times \hat{\mathbf{s}}}{s^2} \tag{20.50}$$

This is the form in which the law was first worked out, from experiments on currents in wires, by Ampère. There is another law named for Ampère, so this one

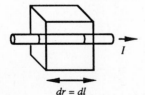

Figure 20.29   Sketch illustrating the equivalence of $\mathbf{j}\,d\tau$ and $I\,d\mathbf{l}$.

has become known as the law of Biot and Savart, who had earlier measured B for the special case of a long straight wire.

> A cautionary remark about the Biot–Savart law is in order. It is not unique, as one might guess from the fact that it is derived from the nonunique vector potential. It is not hard to see, for example, that we could add a constant times $d\mathbf{l}$ to the integrand of Eq. 20.50 without changing $B(\mathbf{r})$, because $\int d\mathbf{l} = 0$ around a closed loop. This is related to the fact that the integrand is in some sense being regarded as the "field of a point current," whereas, of course, there is no such thing as a "point current."

The Biot–Savart law may be used to calculate the magnetostatic field of any current distribution (though care must be taken if the current extends to infinity, as for an infinite current sheet, because the derivation assumed it didn't). We can calculate the fields we have already deduced from Ampère's law for the long straight wire and the long solenoid, though this is much more tedious with the Biot–Savart law. However, some problems can be solved easily with the Biot–Savart law that cannot be solved with Ampère's law.

■ **EXAMPLE 20.10** We will determine the field on the axis of a circular wire loop of radius a carrying a current I, as shown in Figure 20.30. For a given length element $d\mathbf{l}$, the integrand of Eq. 20.50 (call it $d\mathbf{B}$) points at an angle θ to the axis as shown. As we integrate around the circle, this vector traces out a cone, and the horizontal components of $d\mathbf{B}$ cancel. The vertical components dB_z integrate to

$$B_z = \frac{\mu_0 I}{4\pi} \int \frac{dl \cos \theta}{s^2}$$

because $d\mathbf{l}$ and $\hat{\mathbf{s}}$ are perpendicular. However, $\int dl = 2\pi a$, the circumference, and

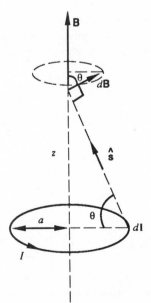

Figure 20.30 Calculating the field of a circular loop of wire, at a point above the center.

$\cos \theta = a/s$, so

$$B = \frac{\mu_0 I}{4\pi} \frac{\cos \theta}{s^2} 2\pi a$$

(20.51)

$$= \frac{\mu_0 I}{2} a^2 (a^2 + z^2)^{-3/2}$$

Computing B off the axis is not so easy — it involves elliptic integrals.

■ **EXAMPLE 20.11** Compute the Biot–Savart integral (Eq. 20.50) for the segment of straight wire shown in Figure 20.31, carrying current I. (Note that we did not call this the **field** of this segment. It is not possible to have a static system with current only in a bounded segment of the wire, because charge would then build up at the end. It therefore makes no sense to ask what the resulting magnetostatic field would be. The integral is useful only as a building block for larger circuits, as in the next example.) The cross product $d\mathbf{l} \times \hat{\mathbf{s}}$ points at the viewer in Figure 20.31 and has magnitude $dl \sin(\theta + \frac{1}{2}\pi) = dl \cos \theta$. The integral over the segment turns out to be easier to calculate in terms of the angle θ (which runs from θ_1 to θ_2) than in terms of the length l along the wire. They are related by $l = r \tan \theta$, so $dl = r\, d\theta / \cos^2 \theta$. In addition, $s = r/\cos \theta$, so the integral becomes

$$B(r) = \frac{\mu_0 I}{4\pi} \int_{\theta_1}^{\theta_2} \frac{r\, d\theta}{\cos \theta} \frac{\cos^2 \theta}{r^2}$$

(20.52)

$$= \frac{\mu_0 I}{4\pi r} \int_{\theta_1}^{\theta_2} \cos \theta\, d\theta = \frac{\mu_0 I}{4\pi r} (\sin \theta_2 - \sin \theta_1)$$

The direction of **B** is toward the viewer. As a simple check, we can look at the special case $\theta_1 = -\frac{1}{2}\pi$, $\theta_2 = \frac{1}{2}\pi$, which **is** an allowable static current (an infinite wire). Our result reduces to $\mu_0 I/2\pi r$, which agrees with the result we obtained much more simply from Ampère's law.

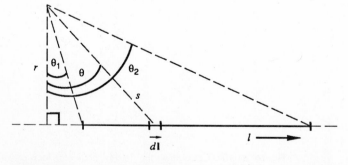

Figure 20.31   A segment (solid line) of a straight wire, with a length element $d\mathbf{l}$.

■ **EXAMPLE 20.12 THE FIELD OF A SQUARE LOOP**   Find the magnetostatic field at the center of a square loop of wire, of length a on each side, carrying a current I. We can break the integral in Eq. 20.50 into four terms, one over each side of the square. We can apply the result (Eq. 20.52) of the previous example, where in this case $\theta_1 = -\frac{1}{4}\pi$, $\theta_2 = \frac{1}{4}\pi$, $r = \frac{1}{2}a$. This gives a contribution to **B** of $2^{1/2}(\mu_0 I/2\pi a)$, normal to the square. Each of the other three sides makes an identical contribution, giving a total B of $2^{3/2}\mu_0 I/\pi a$.

▶   **Problem 20.19**

Find the magnetostatic field a distance z above the center of the square in Example 20.12. Check that your answer is correct for $z = 0$.   ◀

▶   **Problem 20.20**

In Example 20.3 we determined the magnetic field inside an infinite solenoid carrying an azimuthal surface current (current per unit length) K to be axial, with magnitude $B = \mu_0 K$, independently of the radius R. Compute the field inside the solenoid of **finite** length shown in Figure 20.32. Give your answer in terms of K, R, θ_1, and θ_2. Check that your answer reduces to the correct answer for an infinite solenoid (what are θ_1 and θ_2 in that case?).   ◀

▶   **Problem 20.21**

Compute the magnetic field vector **B** at the point **r** when a current I flows in each of the circuits shown in Figure 20.33.   ◀

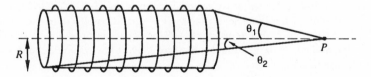

Figure 20.32   A finite solenoid, the radii of whose circular ends subtend angles θ_1 and θ_2 at the field point **r**.

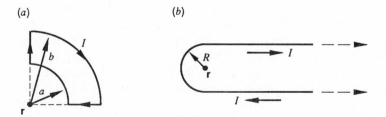

Figure 20.33   Current loops for Problem 20.21.

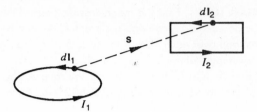

Figure 20.34 Two arbitrary current-carrying wire loops, following paths P_1 and P_2 and carrying currents I_1 and I_2 respectively.

▶   **Problem 20.22**

Compute the total force (magnitude and direction) on a long straight wire, due to a square loop of side a, a distance a away, as shown in Figure 20.4a. Each carries a current I, and the current in the side of the square closest to the wire is parallel to that in the wire.   ◀

▶   **Problem 20.23**

Show that the force on the wire loop carrying current I_2 in Figure 20.34 can be written as a double path integral:

$$\mathbf{F} = -\frac{\mu_0}{4\pi} \int_{P_1} \int_{P_2} (\mathbf{dl}_1 \cdot \mathbf{dl}_2) \frac{\hat{\mathbf{s}}}{s^2} \qquad (20.53)$$

[Hint: Start with Eq. 20.50 (the Biot–Savart law) and the Lorentz force. Note that $\mathbf{dl}_2 \cdot \hat{\mathbf{s}}$ is ds.] Show from this result that the magnetostatic forces on loops 1 and 2 obey Newton's third law.   ◀

Most texts on electricity and magnetism approach magnetostatics from a different angle, starting with the Biot–Savart law, which is postulated as being experimentally verified (for steady currents). From this, the static $\mathbf{B}(\mathbf{r})$ can be calculated for any current distribution $\mathbf{j}(\mathbf{r})$. One can prove $\nabla \cdot \mathbf{B} = 0$ and Ampère's law ($\nabla \times \mathbf{B} = \mu_0 \mathbf{j}$) from the Biot–Savart law (although some messy vector calculus is involved, comparable to that in our above derivation of the Biot–Savart law). However, Maxwell's dynamic version of Ampère's law (with the $d\mathbf{E}/dt$ term) must still be separately postulated as being experimentally verified. Thus, postulating the Biot–Savart law was redundant, which is why I do not use this approach.

SECTION 20.9   MULTIPOLE EXPANSION OF MAGNETOSTATIC FIELDS

Recall that in electrostatics, we found that all charge distributions look similar from far away. If there is a nonzero net charge, its field far away is like that of a point charge; if not it is like that of a dipole. These fields turned out to be successive terms in an expansion in powers of $1/r$, where r is the distance from the field point to some origin near the charge distribution. We can do a very similar thing with a current distribution such as that shown in Figure 20.35. Suppose the current distribution is confined to a region outlined by the dashed curve, and the vector $\mathbf{r}'$ labels an arbitrary point within the current distribution. The magnetostatic vector potential at $\mathbf{r}$ is then given by Eq. 20.40. The factor $|\mathbf{r} - \mathbf{r}'|^{-1}$ can be expanded in powers of the small quantity r'/r, using the binomial theorem exactly as in

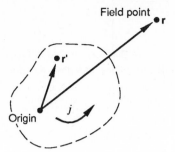

Figure 20.35   An arbitrary current distribution with current density **j**, showing a field point **r**.

electrostatics:

$$|r - r'|^{-1} = r^{-1}\left[1 - \frac{2r \cdot r'}{r^2} + \frac{r'^2}{r^2}\right]^{-1/2}$$

$$= r^{-1}\left[1 + \frac{r \cdot r'}{r^2} + O\left(\frac{r'}{r}\right)^{-2}\right]$$

(The symbol $O(x^n)$ means "terms of order x^n.") Then Eq. 20.40 becomes

$$\mathbf{A(r)} = \left(\frac{\mu_0}{4\pi r}\right)\int \mathbf{j(r')}\, d\tau'$$

$$+ \left(\frac{\mu_0}{4\pi r^2}\right)\int \mathbf{j(r')}\frac{\mathbf{r} \cdot \mathbf{r'}}{r}\, d\tau' + O(r^{-3})$$

(20.54)

because **r** is independent of the integration variable **r'** and can therefore be taken out of the integral. The first term is proportional to $1/r$ and is analogous the monopole (point charge) term in electrostatics. We might expect it to be zero because Maxwell's equations do not allow magnetic charge. We can show this explicitly by writing one component (say the x-component) of the volume integral as a triple integral, with volume element $dx'\, dy'\, dz'$, and doing the x'-integration last:

$$\int \mathbf{j}_x(r')\, d\tau' = \int dx' \int\int dy'\, dz'\, j_x(x', y', z')$$

(20.55)

The integral over y' and z' is evidently a surface integral $\int \mathbf{j} \cdot d\mathbf{a}$, which is the net current through a plane $x' = $ constant. However, there can be no net current through this plane or else charge would pile up on one side. The integral over x' therefore vanishes, and so does the monopole term in Eq. 20.54.

▶   **Problem 20.24**

Show formally that Eq. 20.55 vanishes by using the integral form of $\nabla \cdot \mathbf{j} = 0$; over what surface must you integrate **j**?   ◀

Figure 20.36 The path used for the path integral defining the magnetic vector potential of a wire loop; it follows the loop.

We are left with the second (dipole) term. Defining a unit vector in the direction of $\mathbf{r}$,

$$\hat{\mathbf{r}} = \mathbf{r}/|\mathbf{r}|$$

we can write the dipole term as

$$A(\mathbf{r}) = \frac{\mu_0}{4\pi r^2} \int \mathbf{j}(\mathbf{r}')(\mathbf{r}' \cdot \hat{\mathbf{r}})\, d\tau' \tag{20.56}$$

For the special case of a current in a loop of wire (Figure 20.36) this is the path integral

$$A(\mathbf{r}) = \frac{\mu_0}{4\pi r^2} I \int (\mathbf{r}' \cdot \hat{\mathbf{r}})\, d\mathbf{l} \tag{20.57}$$

As in the case of the electric dipole moment, we want to express this in terms of an integral involving $\mathbf{r}'$ only, that is, factor $\hat{\mathbf{r}}$ out of the integral. To do this, we must write out the dot product explicitly in terms of components: the i-component of A is

$$A_i = \frac{\mu_0}{4\pi r^2} I \int \sum_j r_j' \hat{r}_j\, dl_i \tag{20.58}$$

$$= \frac{\mu_0}{4\pi r^2} \sum_j \hat{r}_j \left[I \int r_j'\, dl_i \right] = \frac{\mu_0}{4\pi r^2} \sum_j \hat{r}_j T_{ji}$$

where

$$T_{ji} = I \int r_j'\, dl_i \tag{20.59}$$

is a two-index tensor, which we may call the magnetic dipole moment tensor. It is the magnetostatic analog of the electric dipole moment vector, in that it uniquely determines the magnetic field far away from the current loop. This tensor has an interesting property, which is a result of the fact that it is an integral over a closed loop. Rewriting dl_i as dr_i' to emphasize that it is the change in r_i' across a path length element,

$$T_{ji} + T_{ij} = I \int (r_j'\, dr_i' + r_i'\, dr_j')$$

$$= I \int d(r_j' r_i') = 0 \tag{20.60}$$

because the total change in any function around a closed loop is zero. Thus $T_{ji} = -T_{ij}$: the tensor is **antisymmetric**. We will now show that any antisymmetric tensor is equivalent to a **vector**, in a certain sense. The diagonal matrix elements of an antisymmetric tensor vanish ($T_{ii} = -T_{ii}$ implies $T_{ii} = 0$) and the six off-diagonal elements are equal in pairs; there are only three independent elements, which we may call m_x, m_y, and m_z and write

$$T = \begin{pmatrix} T_{xx} & T_{xy} & T_{xz} \\ T_{yx} & T_{yy} & T_{yz} \\ T_{zx} & T_{zy} & T_{zz} \end{pmatrix} = \begin{pmatrix} 0 & m_z & -m_y \\ -m_z & 0 & m_x \\ m_y & -m_x & 0 \end{pmatrix} \qquad (20.61)$$

The indices and signs are chosen so that

$$m_x = T_{yz} \qquad (20.62)$$

and the other components are given by cyclic permutations of the indices ($x \to y \to z \to x$) in this formula. Usually one describes the loop by giving the m's instead of the T's (it's more concise); they can be thought of as comprising a vector $\mathbf{m} = (m_x, m_y, m_z)$, called the **magnetic dipole moment vector** (or just the **magnetic moment**) of the loop. From Eq. 20.59, we can write m_x as

$$m_x = I \int r_y' \, dl_z$$

or, because of the antisymmetry, as

$$m_x = -I \int r_z' \, dl_y$$

The mean of these two formulas is particularly convenient because it is expressible as a cross product:

$$m_x = \tfrac{1}{2} I \int \left(r_y' \, dl_z - r_z' \, dl_y \right) = \tfrac{1}{2} I \int (\mathbf{r}' \times d\mathbf{l})_x$$

so that the magnetic moment vector can be expressed as

$$\mathbf{m} = \tfrac{1}{2} I \int \mathbf{r}' \times d\mathbf{l} \qquad (20.63)$$

An exactly analogous treatment of a volume current distribution $\mathbf{j}(\mathbf{r})$ leads to

$$\mathbf{m} = \tfrac{1}{2} \int \mathbf{r}' \times \mathbf{j}'(\mathbf{r}') \, d\tau' \qquad (20.64)$$

(Proving T_{ij} antisymmetric in this case requires the use of $\nabla \cdot \mathbf{j} = 0$.)

In terms of **m**, Eq. 20.58 for **A** becomes

$$\frac{4\pi r^2}{\mu_0} A_x = \sum_j \hat{\mathbf{r}}_j T_{jx} = 0 + \hat{\mathbf{r}}_y T_{yx} + \hat{\mathbf{r}}_z T_{zx}$$

$$= -\hat{\mathbf{r}}_y m_z + \hat{\mathbf{r}}_z m_y = (\mathbf{m} \times \hat{\mathbf{r}})_x$$

The formulas for A_y and A_z are similar, giving

$$\mathbf{A} = \left(\frac{\mu_0}{4\pi r^2}\right)\mathbf{m} \times \hat{\mathbf{r}} \tag{20.65}$$

For a circular loop of radius R, $\mathbf{r}'$ and $d\mathbf{l}$ are perpendicular so their cross product is simple, and **m** is easy to evaluate: $|\mathbf{r}' \times d\mathbf{l}| = R\,dl$, so

$$m = \tfrac{1}{2}IR \int dl = \tfrac{1}{2}IR(2\pi R) = \pi R^2 I$$

which is the area of the loop times the current. We can show that this relation to the area is general, as long as the loop lies in a plane so that $\hat{\mathbf{r}} \times d\mathbf{l}$ is always along the same normal unit vector $\hat{\mathbf{n}}$. In terms of the components dl_r and dl_ϕ of $d\mathbf{l}$ parallel and perpendicular to the radius vector,

$$\tfrac{1}{2}|\mathbf{r}' \times d\mathbf{l}| = \tfrac{1}{2}r'\,dl_\phi = da$$

which is the area of the narrow triangle in Figure 20.37. Then

$$\mathbf{m} = I \int \hat{\mathbf{n}}\,da = Ia\hat{\mathbf{n}} \tag{20.66}$$

where a is the total area of the loop.

For completeness, we should fulfill the promise made at the beginning of Section 20.8 and check that the vector potential (Eq. 20.65) of an arbitrary bounded current distribution satisfies $\nabla \cdot \mathbf{A} = 0$ at the boundary, that is, in the limit $r \to \infty$. The condition is satisfied because $1/r^2 \to 0$ there.

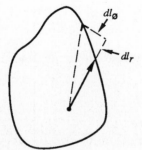

Figure 20.37 The components of the path element $d\mathbf{l}$ for an arbitrary planar loop.

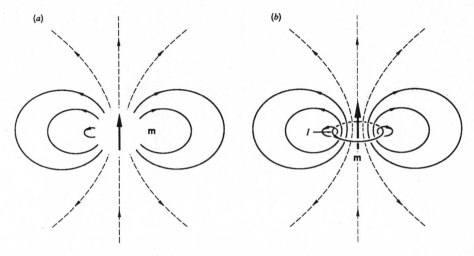

Figure 20.38   Magnetic field of (*a*) a point dipole, (*b*) a physical dipole.

Just as we defined a "point" electric dipole by taking the limit in which the separation of two charges approaches zero while the dipole moment remains constant, we may define a **point magnetic dipole** by letting $a \to 0$ while Ia remains constant in Eq. 20.66. The magnetic vector potential of this point dipole is given (**exactly**, not approximately as for a finite-sized dipole) by Eq. 20.65. We can evaluate this explicitly for a point dipole at the origin pointing in the z-direction. At the point whose spherical coordinates are (r, θ, ϕ) the potential is

$$\mathbf{A}(r, \theta, \phi) = \left(\frac{\mu_0}{4\pi r^2} \right) m \sin \theta \, \hat{\phi} \tag{20.67}$$

The magnetic field is therefore

$$\mathbf{B} = \nabla \times \mathbf{A} = \left(\frac{\mu_0}{4\pi r^3} \right) m (2 \cos \theta \, \hat{\mathbf{r}} + \sin \theta \, \hat{\theta}) \tag{20.68}$$

This field is sketched in Figure 20.38*a*; you may notice that it is proportional to the **electric** field of an **electric** point dipole.

As a mnemonic device for keeping track of signs, remember that the magnetic moment of a loop points along the magnetic field **inside** the loop, as shown in Figure 20.38*b* (although the magnetic field is in the opposite direction just outside the loop).

■ **EXAMPLE 20.13**   Compute the magnetic moment of the circuit shown in Figure 20.39*a*, which carries a current *I*. One way to do this is to evaluate the integral (Eq. 20.63) directly. It is easier, however, to note that it is equivalent to a circuit composed of two **square** loops of side *R*. The square loops each have an extra segment of wire along the edge joining them, but it is easy to see that the contributions to the integral from these

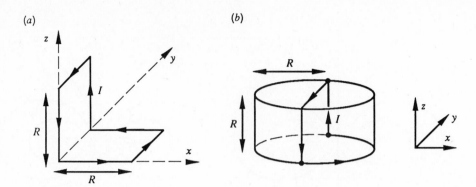

Figure 20.39 Nonplanar current loops, consisting of (*a*) squares and (*b*) a semicircle and three straight lines.

cancel (the currents are in opposite directions). The magnetic moments of these square loops can be exactly calculated from Eq. 20.66, and are $IR^2\hat{\mathbf{x}}$ and $IR^2\hat{\mathbf{z}}$ respectively. The total **m** is the vector sum of these, which points between the *x* and *z* axes.

▶ **Problem 20.25**

A circuit carrying a current I goes halfway around a cylinder of radius and height R (Figure 20.39*b*), up the side, across the top, and back down. Compute the magnetic moment vector.   ◀

▶ **Problem 20.26**

a. Compute the vector magnetic moment of the loop whose field we computed in Example 20.10, and show that Eq. 20.68 for the magnetic field on the axis reduces to the field we obtained in the example, in the appropriate limit.

b. Do the same for the field above the center of the square in Problem 20.19.   ◀

▶ **Problem 20.27**

Show that Eq. 20.63 for the magnetic moment of a current loop is independent of the choice of origin (i.e., that it is unchanged when $\mathbf{r}'$ is replaced by $\mathbf{r}' - \mathbf{b}$).   ◀

▶ **Problem 20.28**

Consider a current loop that is **not** confined to a plane. Show that Eq. 20.66 generalizes to

$$\mathbf{m} = I\mathbf{a} \tag{20.69}$$

where a_x is the area of the projection of the loop onto the *yz*-plane, a_y the area in the *xz*-plane, and a_z the area in the *xy*-plane.   ◀

CHAPTER TWENTY-ONE

IMPERFECT CONDUCTORS

SECTION 21.1   OHM'S LAW

The conductors we considered in electrostatics were perfect ones, in the sense that the charges in them are so free to move that whenever an electric field starts to appear, they move instantaneously to counteract it and keep $\mathbf{E} = 0$. This, of course, is an idealization—real charges require some time to move. One's first guess (based on Newton's second law $\mathbf{F} = m\mathbf{a}$) might be that they should have an acceleration proportional to the electric field (which is, after all, the force per unit charge). However, this is based on a picture in which the charges are isolated and free to move independently of each other and of the atoms and ions in the conducting material. This is quite unrealistic; not only are they packed rather close together, but the electrons and ions in a typical metallic conductor are constantly in random thermal motion. The fields they experience in their random motion are enormous compared to the few newtons per coulomb that we typically apply to a conductor. Our applied fields cause only a tiny perturbation of the motions of the electrons in a metal; they cause the electrons to go a fraction of a percent faster or slower for the brief instant before the next collision changes the velocity drastically. The only reason the effects of an applied field are detectable at all is that they are systematic: the accelerations, and therefore the perturbations in the displacements, are along the direction of the applied electric field. The current due to the random motions, on the other hand, averages to zero. The result is that in most materials, for fields that are not unattainably large, the macroscopic current density is found experimentally to be proportional to the static electric field:

$$\mathbf{j} = \sigma \mathbf{E} \qquad \text{(Ohm's law)} \qquad (21.1)$$

where the proportionality constant σ is called the **conductivity**. It should be clear that this equation cannot be true for currents on a very small scale; on the atomic scale (or even a few times this scale) the current fluctuates wildly and must be treated statistically. The study of such fluctuations is part of a very large field of physics called statistical mechanics, which is, unfortunately, beyond the scope of this text. Our Eq. 21.1 can only make sense when the current is averaged over a region containing many atoms. That is, the discrete fields are related by

$$j(f) = \sigma(f)E(f) \qquad (21.2)$$

for faces of a cell lattice whose spacing dr is many times the atomic size. On this large scale, the random currents average to zero and we can see the effects of the average field $E(f)$.

> You may notice that this treatment of Ohm's law is slightly inconsistent, in the following sense. We have implicitly assumed that the electric field **E** is a static one. (Otherwise, we would have to worry about the time delay between the application of a field and the motion of the charges; this is very short but is sometimes important.) However, in previous discussions of electrostatics we have always assumed that there is no current, in fact, that it had stopped a long time before. Now we have a current that must exist as long as the electric field does. Fortunately, it turns out that we never really **used** the assumption $\mathbf{j} = 0$ in electrostatics. For the electric field to become constant we need a constant charge density ρ (since $\rho = \epsilon_0 \nabla \cdot \mathbf{E}$), which requires

$$\nabla \cdot \mathbf{j} = \frac{-d\rho}{dt} = 0 \tag{21.3}$$

> However, we may allow a nonzero divergenceless $\mathbf{j}$ and still use the methods of electrostatics. Of course then there is a magnetic field too; we should perhaps just call this "statics."

> We should also mention here the existence of very important systems that do not obey Ohm's law. Among these are superconducting materials, which we discussed briefly in Section 20.4. Other "non-ohmic" systems are semiconductor devices such as *pn* junctions and transistors: although semiconductors obey Ohm's law on a large distance and time scale when electric fields are small, the distance scales important in semiconductor devices are too short (of order 10^{-6} m), and the fields too large (of order 10^6 v/m), for Ohm's law to be valid. The important consequence of this is that these are **nonlinear** devices: the current in a *pn* junction diode is not a linear function of the voltage across it. The microscopic motions of electrons must be studied in more detail to understand the functioning of such devices.

What kind of problems involving currents can we solve with statics? Strictly speaking, a current cannot be maintained forever in an isolated static system. The electromagnetic fields would continually be doing work ($\mathbf{j} \cdot \mathbf{E}$, as in Chapter 13) and would eventually run out of energy. We must feed energy in from outside to drive the current. This can be done, conceptually at least, by using perfect-conductor electrodes maintained at different potentials, as in Figure 21.1. We have shown a system completely enclosed by a perfect conductor, so the region in which we want

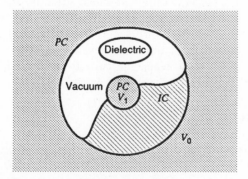

Figure 21.1 A general boundary-value problem containing imperfect conductors (labeled IC) as well as perfect ones (PC).

to calculate the fields is finite. Such a problem can be solved by a modification of the techniques we have used in electrostatics. In a discrete problem, one assigns a conductivity $\sigma(f)$ to each face (zero in nonconductors) and imposes $\nabla \cdot \mathbf{j} = 0$ as well as Poisson's equation. All the properties we previously attributed to "perfect" conductors can be recovered by setting $\sigma(f) = \infty$ in these conductors: we insist that $\mathbf{j}$ be finite, so Ohm's law implies $\mathbf{E} = 0$. The new $\nabla \cdot \mathbf{j} = 0$ condition is trivially satisfied in nonconductors. In a uniform conductor it says $\rho = \epsilon\nabla \cdot \mathbf{E} = (\epsilon/\sigma)\nabla \cdot \mathbf{j} = 0$; that is, there is no charge in the interior of a uniform conductor. There may be charge at the surfaces of imperfect as well as perfect conductors; this introduces one unknown variable for each imperfect-conductor boundary cell, but we have one additional non-trivial equation there too ($\nabla \cdot \mathbf{j} = 0$). The whole problem still has a unique solution, although it is harder to prove in this case than it was for Poisson's equation by itself.

In a continuum problem, we must use different functions for the potentials in each material. The potential must satisfy Poisson's equation in the imperfect conductor as well as in the vacuum and the dielectric. The boundary conditions on the fields can be deduced using the same types of argument we used in discussing interfaces between dielectrics. In fact we need not use different rules for dielectric and conducting materials; we can consider them special cases of a general material in which both the conductivity and the dielectric susceptibility can be nonzero. We get the same conditions as before (E_t and D_n continuous, with the continuity of E_t being equivalent to the continuity of the potential). In addition, at each boundary there is a condition that requires the conservation of current: the normal component j_n must be continuous across the boundary. At a conductor–nonconductor boundary, for example, this implies $E_n = 0$ in the conductor (though the component E_t along the interface can be nonzero if the conductor is imperfect). At such an interface Gauss' law implies a surface charge $\epsilon_0 E_n$, where E_n is the outward normal component of the field on the nonconductor side.

We will content ourselves here with doing a couple of very simple continuum examples.

■ **EXAMPLE 21.1** Consider an infinite slab of imperfect conductor between two perfectly conducting planes separated by a distance d and a potential difference V_0 (Figure 21.2). The solution to Laplace's equation for this case is exactly as it was for a vacuum or a dielectric because the boundary conditions are the same. The potential is

$$V(z) = \frac{-V_0 z}{d} \tag{21.4}$$

The only new thing is the current, which is uniform and points upward:

$$j_z = \sigma E_z = \frac{\sigma V_0}{d} \tag{21.5}$$

■ **EXAMPLE 21.2** A more practical problem involves a **finite** block of imperfect conductor subjected to a potential difference V_0, as in Figure 21.3a. This looks like it might be much harder; we can no longer assume from the symmetry of the problem that V will be independent of x and y. Let's begin by setting the potentials in both the vacuum and the

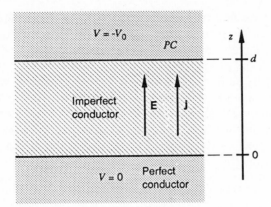

Figure 21.2 Two parallel infinitely conducting planes with a slab of imperfect conductor between them.

conductor equal to the obviously over-simplified Eq. 21.4, to see how the boundary conditions are violated. It gives $-V_0$ and 0 correctly at the top and bottom, and $E_n = 0$ at the side boundaries. But that's all the conditions there are: Eq. 21.4 must be the unique solution! This solution works for any block shape, as long as it has the same cross section at every height z. (Incidentally, it also works for such a block of dielectric.)

Because the cross-sectional area A in the previous example was finite, we can compute the total **current** passing through the block:

$$I = \int \mathbf{j} \cdot d\mathbf{a} = j_z A = \left(\frac{\sigma A}{d} \right) V_0$$

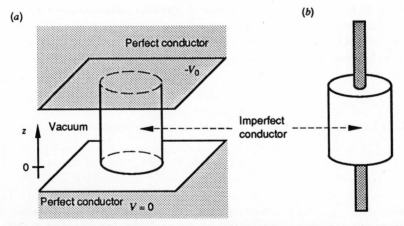

Figure 21.3 (a) Two perfectly conducting planes with a finite piece of imperfect conductor connecting them, in the shape of a cylinder (not necessarily circular) with cross-sectional area A. (b) A more realistic model of a resistor, with wire leads.

(using Eq. 21.5). Solving for V_0, we get the macroscopic form of Ohm's law,

$$V_0 = IR \tag{21.6}$$

where

$$R = \frac{d}{\sigma A} \tag{21.7}$$

is called the **resistance** of the block. In this context, the block is referred to as a **resistor**. Note that V_0 is the "voltage drop" as we cross the resistor from bottom to top (along the current flow); this was the reason for the minus sign on V_0 in Figure 21.2.

The units of resistance are (from Eq. 21.6) volts/ampere. This is called an **ohm** and denoted by a capital omega (Ω). Instead of the conductivity, one sometimes refers to its reciprocal, the **resistivity**, usually denoted ρ (but not to be confused with charge density):

$$\rho \equiv \sigma^{-1} \tag{21.8}$$

so that $R = \rho d / A$. The SI units of resistivity are then ohm–meters (Ωm), and those of conductivity are $\Omega^{-1} m^{-1}$ (Ω^{-1} is sometimes written "mho").

The circuit (Figure 21.3a) in which we have proved the macroscopic Ohm's law and the resistance formula (Eq. 21.7) to hold is not one we meet frequently in practice. We are more likely, for example, to want to attach the resistor to wire electrodes, as in Figure 21.3b, than to infinite planes. To find the resistance in this case, we would need to solve Poisson's equation within the vacuum and imperfect conductor regions. We would find Eq. 21.6 with R given by Eq. 21.7 to be only approximately true; the necessity to funnel the current into a small lead wire decreases the current slightly. However, we would **still** find I to be proportional to V_0. That is, Ohm's law (Eq. 21.6) would still be true, with a different value for the resistance. The macroscopic Ohm's law is a very general consequence of the linearity of the equations of electrostatics (among which we now include the microscopic Ohm's law). It is obeyed by a resistor of **any** shape, even if the conductivity is nonuniform. The value of R will depend on the exact configuration of infinite-conductivity electrodes to which the resistor is attached, but it can be calculated for any configuration by solving for the potential using a discrete lattice and letting $dr \to 0$. For the configuration in Figure 21.3b, Eq. 21.7 is an excellent approximation as long as the height d is much larger than the lateral dimensions of the resistor.

■ **EXAMPLE 21.3**   Figure 21.4 shows two long perfectly conducting concentric cylinders of length L and radii a and b respectively, separated by an imperfect conductor of conductivity σ and permittivity ϵ. Find the resistance between them.

We must apply a potential difference between the cylinders and calculate the potential everywhere. We could approach this as a boundary value problem, but if we ignore the region near the ends, the symmetry allows us to determine D (and therefore E and j) from Gauss' law. The result (Problem 18.2) is proportional to $1/r$; integrating gives the potential $V(r) = (\lambda/2\pi\epsilon)\ln r$, where λ is the free charge per unit length on the inner

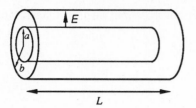

Figure 21.4   A resistor with two cylindrical inner and outer electrodes.

cylinder. The potential difference is therefore

$$\Delta V = \left(\frac{\lambda}{2\pi\epsilon}\right) \ln\left(\frac{b}{a}\right)$$

The total current passing through a cylindrical surface of radius r $(a < r < b)$ is

$$I = \int \mathbf{i} \cdot d\mathbf{a} = \sigma \int \mathbf{E} \cdot d\mathbf{a} = \sigma\epsilon^{-1} \int \mathbf{D} \cdot d\mathbf{a} = \sigma\epsilon^{-1}\lambda L$$

where λL is the total free charge on the inner cylinder. Thus ΔV and I are proportional, consistently with the linearity argument above, and the proportionality constant is

$$R = \frac{\Delta V}{I} = (2\pi\sigma L)^{-1} \ln\left(\frac{b}{a}\right)$$

▶   **Problem 21.1**

Consider a resistor consisting of a spherical shell of imperfectly conducting material with conductivity σ and permittivity ϵ, bounded inside and outside by perfectly conducting concentric spherical electrodes (Figure 21.5a).

a. Find the resistance between the electrodes.

b. The result of part **(a)** should approach a limit as $b \rightarrow \infty$, for fixed a. Using this fact, can you guess the resistance between two solid metal spheres (Figure 21.5b), each of

(a)   (b)

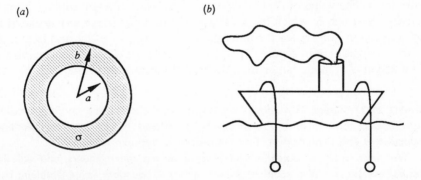

Figure 21.5   (a) A resistor in the shape of a spherical shell. (b) Two spherical electrodes imbedded in an imperfectly conducting medium (this can be used to measure the conductivity of sea water). From *Electrodynamics*, D. J. Griffiths, Prentice-Hall, 1981.

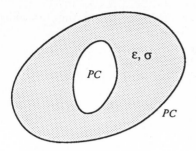

Figure 21.6 Two electrodes separated by an imperfect conductor. To make the imperfect conductor finite, one electrode is drawn to entirely enclose the other, but this is not necessary.

radius a, imbedded in a dielectric and separated by a distance $2b \gg a$? (Hint: One way is to think of this as involving three resistors in series.) ◄

▶ **Problem 21.2**

Figure 21.6 shows two perfectly conducting electrodes imbedded in an imperfectly conducting material. The system can be thought of either as a resistor or a capacitor. Show that the resistance and the capacitance are related by

$$RC = \frac{\epsilon}{\sigma}$$

where σ is the conductivity and ϵ is the permittivity. ◄

▶ **Problem 21.3**

Two long perfectly conducting cylinders, each of radius R, are parallel to each other and their centers are separated by a distance $2d$. They are imbedded in an infinite imperfect conductor of conductivity σ. Find the resistance between them. You may use the results of Section 17.1. ◄

SECTION 21.2 OHM'S LAW FOR NONSTATIC FIELDS

In a nonstatic situation, both the current and the electric field in a conductor are functions of time: in a discrete lattice, we can write them as $j(f, t)$ and $E(f, t)$. Clearly Ohm's law in the simple form $j(f, t) = \sigma(f)E(f, t)$ cannot be valid for time-varying fields; for one thing, E is defined at integer multiples of dt, while j is defined at half-integer multiples. Underlying this technical difficulty is the physical fact that the influence of the field on the current cannot occur instantaneously. The most general linear relation between $j(f, t)$ and $E(f, t)$ would express $j(f, t)$ as a linear combination of **all** previous electric fields, at all positions. Usually the fields are smooth enough (as a function of position) that it is sufficient to use E at the same face f, but in dealing with high-frequency (i.e., rapidly varying) fields it is necessary to include all previous times. It is found that $j(f, t)$ depends on the values of the electric field at times differing from t by amounts of the order of a **relaxation time** τ, which can be as small as a picosecond, but is much larger in certain materials. We will assume here that the electric field, even when it varies with time, varies on a time scale much longer than this relaxation time. Then the electric fields involved in this linear expression for j are essentially the same, and can be replaced by a single term involving the most recent time at which E is defined:

$$j(f, t) = \sigma(f)E\left(f, t - \tfrac{1}{2}dt\right) \tag{21.9}$$

In the continuum limit, this becomes

$$\mathbf{j}(\mathbf{r}, t) = \sigma(\mathbf{r})\mathbf{E}(\mathbf{r}, t) \tag{21.10}$$

In our discussions of Maxwell's equation up to now we have regarded the current density $\mathbf{j}$ (the free current density, in a dielectric) as a given quantity, which we know before we begin a calculation. In the presence of conducting materials, this is, of course, not true. In general, we can think of the free current as being composed of an externally imposed part $\mathbf{j}_{ext}$ and a conduction current $\mathbf{j}_{cond}$, which is given by Eq. 21.10, so the total discrete current density at a time t at a face f is

$$j(f, t) = j_{ext}(f, t) + j_{cond}(f, t) + j_{bound}(f, t) \tag{21.11}$$

Substituting this into the Maxwell equation for dE/dt, we can move j_{bound} to the left side to change dE/dt to dD/dt just as in Section 19.4, obtaining

$$dD(f, t)/dt = \epsilon_0 c^2 [\operatorname{curl} B](f, t) - j_{ext}(f, t) - \sigma(f)E\left(f, t - \tfrac{1}{2}dt\right) \tag{21.12}$$

This equation is used in program MAXWELL, when the boundaries are assumed conducting in order to absorb radiant energy and allow relaxation to a static field.

■ **EXAMPLE 21.4 XEROGRAPHY**   Consider a pseudo-1D system consisting of a metal plate upon which a thin layer of a photoconductive semiconductor (typically an amorphous selenium compound) has been deposited, as shown in Figure 21.7. Before it is exposed to light, the semiconductor has negligible conductivity, so it can be treated as an insulating dielectric with permittivity ϵ. A free charge density s_0 (we can't denote it by σ here because we are using that for conductivity) is sprayed onto the surface by a spark discharge. The dielectric field rapidly achieves its static value, which is the same as that of a dielectric-filled capacitor,

$$D_0 = s_0 \tag{21.13}$$

(downward). Thus $E_0 = s_0/\epsilon$. An image is then projected onto the surface. In the dark

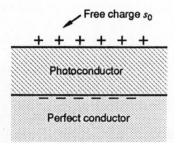

Figure 21.7   Schematic diagram of the surface of the photosensitive drum in a xerographic copying machine.

parts of the image, the semiconductor remains insulating, and there is no rapid change in the surface charge density. In the light parts, however, the conductivity changes from essentially zero to some value σ. Thus there is a downward conduction current σE, and Maxwell's equation for D (Eq. 21.12) reads in the continuum limit

$$\frac{dD(t)}{dt} = -i_{cond}(t) = -\sigma E(t) = -\left(\frac{\sigma}{\epsilon}\right)D(t) \tag{21.14}$$

independently of position z within the semiconductor. This differential equation has an exponential solution; matching the initial condition gives

$$D(t) = D_0 \exp\left(\frac{-\sigma t}{\epsilon}\right) \tag{21.15}$$

Thus the field (and therefore the charge density) decays by a factor e^{-1} in a characteristic time (the "time constant")

$$\tau = \epsilon/\sigma = \rho\epsilon \tag{21.16}$$

If the resistivity $\rho \approx 10^5\ \Omega$m, for example, $\tau \approx 1\ \mu s$. After a time long compared with τ, the plate (actually it's usually bent into a cylindrical drum) is brought near a supply of low-melting-point colored plastic beads ("toner") that are attracted to the charged parts of the plate. The drum then rolls over a sheet of paper, to which the toner particles are transferred, and they are then melted onto the paper.

▶   **Problem 21.4**

In order that the charge not leak off before the image can be transferred to the paper, one wants the time constant τ in the non-illuminated parts of the drum to be at least 1 second. What requirement does this place on the "dark resistivity" of the semiconductor? ◀

▶   **Problem 21.5**

Consider an infinite imperfect conductor with an arbitrary initial electric field $\mathbf{E}(\mathbf{r}, 0) = \mathbf{E}_0(\mathbf{r})$ (and the corresponding current and charge density, of course). Compute its evolution $\mathbf{E}(\mathbf{r}, t)$ exactly for all t. ◀

Needless to say, the above is an oversimplification of the actual xerography process. Th conductivity, which is produced when the light creates electron – hole pairs in the semicor ductor, does not remain uniform when these electrons and holes begin to drift in respons to an electric field. However, the essential features of the phenomenon can be understoo using the approximate method we have described. For more details and references, se *Physics Today*, vol. 39, no. 5, p. 47 (May, 1986).

■ **EXAMPLE 21.5 CONDUCTION EFFECTS IN EMP**   In Section 12.5 we calculated the electro-magnetic pulse (EMP) produced by a nuclear explosion. We left out one important effect, due to the conductivity of the ionized atmosphere behind the slab of gamma rays. To take this into account, we must use Eq. 21.12 to update the electric field. Ionization doesn't change the dielectric constant (1.0) of air significantly, so we can use $\epsilon = \epsilon_0$, giving

$$d\mathbf{E}/dt = c^2 \operatorname{curl} B - \epsilon_0^{-1} i_{Compton}(f, t) - \epsilon_0^{-1} \sigma \mathbf{E} \tag{21.17}$$

Qualitatively, it is clear that the conductivity term slows the growth of the electric field. In fact, if it cancels the Compton current term, the pulse will not grow at all (it will propagate according to the source-free Maxwell equations). This occurs when

$$E = i_{Compton}/\sigma$$

This is called the **saturation field**; the field can approach this value, but not exceed it. We can get an approximate value for the saturation field by estimating $\sigma = 10^{-6} \, \Omega^{-1} m^{-1}$, so the saturation field is 0.02 a m$^{-2}/10^{-6} \, \Omega^{-1} m^{-1} = 2 \times 10^4$ v/m. This is about 1/4 the value we obtained before by ignoring the conductivity. The numbers (for very early times) shown in Figure 12.14 are still essentially correct; the electric field doesn't saturate until $t \approx 25 dt$.

SECTION 21.3   EMF

We will sometimes want to apply the macroscopic Ohm's law to situations that are not static, in which an electrostatic potential cannot be defined. So it is useful to note that the macroscopic Ohm's law can be expressed directly in terms of the electric field. The potential difference V_0 between the electrodes across a resistor (say in Figure 21.2 or Figure 21.3) can be written

$$V_0 = \int_0^d E_z \, dz = \int E \cdot dl$$

as a path integral of the electric field over a path going from the bottom to the top electrode. This relation is true very generally for a current path (any combination of conducting materials, for example wires) between two perfectly conducting electrodes; because $\nabla \times \mathbf{E} = 0$ it doesn't matter which path we choose for the path integral. Thus the general form of Ohm's law is

$$\int E \cdot dl = IR \qquad (21.18)$$

This path integral is called the **electromotive force** or **EMF** along the path between the two perfectly conducting electrodes, and is sometimes denoted by a script capital E ($\mathscr{E}$). Equation 21.18 is more general than Eq. 21.6 involving V_0, only because it is valid in certain nonstatic situations (for example, for electromagnetic induction in a solenoid as discussed in Chapter 22) in which the fields are not constant. It must be used with care, however, because it **does** require that the **electric** field not be rapidly varying. (For example, if E were turned on suddenly in only the top half of the resistor in Figure 21.3, there would be current only in the top, and the I on the right-hand side of Eq. 21.18 would not even be uniquely defined; it's zero in a part of the circuit, and nonzero in another.)

We will even use Ohm's law (Eq. 21.18) for a **closed** circuit of wire, for which the two ends of the path are at the same point. The electrostatic potentials at the ends are the same, so there can never be any electrostatic potential difference V_0. However, it is quite possible to have a nonzero EMF in this circuit, in a nonstatic situation.

CHAPTER TWENTY-TWO

INDUCED ELECTRIC FIELDS

SECTION 22.1 PHYSICAL ORIGIN OF INDUCED ELECTRIC FIELDS

We now return to the study of systems having both electric and magnetic fields. In particular, we want to look at **electromagnetic induction**, meaning the association between electric fields and changing magnetic fields. The electric field is often said to be **induced** by the change in the magnetic field. This terminology is sometimes misleading in that it tends to obscure the actual cause-and-effect relationships. Rather than belabor the terminology, however, let us look at the phenomenon.

The example of electromagnetic induction that is easiest to analyze is the infinitely long solenoid transformer. This consists of two solenoidal coils of wire, of radii a and b, called the **primary** and **secondary** coils respectively. A cross section of such a transformer is shown in Figure 22.1. What is observed experimentally is that when a current I flows in the primary, a current flows, at least momentarily, in the secondary solenoid. We know Maxwell's equations, so we can analyze exactly what is going on here. It is easier mathematically if we flatten the solenoids and let them be infinitely tall as well as infinitely long (see Figure 22.2a, in which the height h is assumed to become infinite). The qualitative behavior will be the same as for the cylindrical solenoid. Suppose all the fields are zero for $t \leq 0$, and we turn on a surface current K (current per unit length) at $t = 0$. Until the fields propagate to the secondary solenoid, this is exactly the same problem as that of an isolated current sheet turned on at $t = 0$. The exact solution to this continuum problem is shown in Figure 22.2b. A wave moves out at the speed of light c, with uniform E

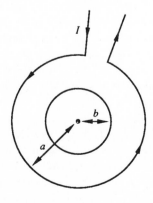

Figure 22.1 Cross section of a solenoid transformer, whose cylinder axis is perpendicular to the paper.

449

(a)   (b)

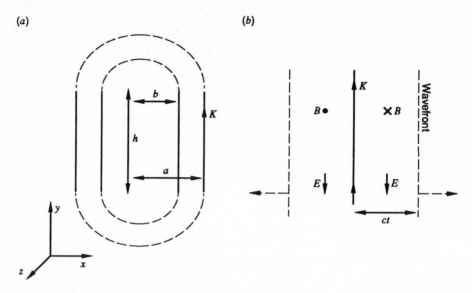

Figure 22.2 (*a*) A planar solenoid transformer. (*b*) The wave produced by turning on an infinite planar current *K*.

and *B* fields behind it given by

$$B = \frac{\mu_0 K}{2}, \qquad E = \frac{K}{2\epsilon_0 c} \tag{22.1}$$

with **E** antiparallel to the current, and **B** away from the viewer to the right of the current sheet and toward the viewer to the left. Both fields are of course zero beyond the wave front; because of the discontinuity at the wavefront this bears some resemblance to a shock wave.

We can easily simulate this wave using a program such as MAXWELL to solve the discrete Maxwell equations for some *dr* and *dt*. We will not, in general, see the perfectly sharp wavefront characteristic of the continuum limit *dr* → 0, *dt* → 0, but one can get a clear general picture of the outward motion using finite *dr* and *dt*. It happens that there is a special choice of *dr* and *dt* (corresponding to the delta-pulses discussed in Chapter 2) for which the wavefront remains sharp even for finite *dr* and *dt*: this is *dr* = *c dt*, for which the front moves out one cell in each interval *dt*. As you may recall from Chapter 2, for such a large *dt* there is an unstable "corrugation" mode, so one has to be careful that one's initial conditions or sources (currents) do not create such a mode. A choice of current that works is shown in Figure 22.3. There is a plane of cell boundaries at the center of the thin current sheet (the *x* = 0 plane) so half the current is on each side. Then the current density is *j* = *K*/2 *dr* on each of two layers of cells, at *x* = ± ½*dr*. The evolution of the *E* and *B* fields according to the discrete Maxwell equations is indicated in Figure 22.3. You can check the numbers for yourself by using the discrete Maxwell equations for *dE/dt* and *dB/dt* (Eqs. 9.31 and 9.32) to update *E* and *B*. From the series of snapshots in Figure 22.3 it is clear that the wavefront (farthest-right nonzero field)

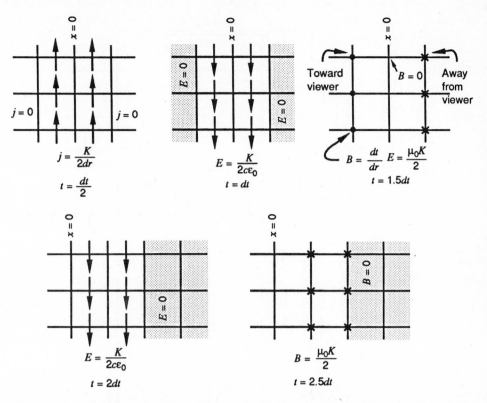

Figure 22.3   Evolution of the electric and magnetic fields of a discretization of an infinite planar current.

moves ahead by a distance dr during each time interval dt (i.e., moves at the speed $dr/dt = c$). Taking the continuum limit is particularly easy in this case, because E and B behind the wavefront are independent of dr and dt; this proves the result in Eq. 22.1.

> We could also solve this problem with a **single** column of current-carrying faces; we would need to use a smaller dt to avoid instability, and the wavefront would not be perfectly sharp, but the qualitative results and the continuum limit would be the same.

Let us return now to the "planar solenoids" of Figure 22.2, using our result (Eq. 22.1) for the fields emanating from the surface current K. Initially, waves move outward from each side of the primary, as shown in Figure 22.4a. Nothing happens at the secondary solenoid until $t = (a - b)/c$ (we could have predicted that, of course, because no effect can propagate faster then the speed of light). After that, there is an electric field $E = K/2\epsilon_0 c$ along the secondary (Figure 22.4b), which causes a current to flow in the secondary. For simplicity assume the secondary has a fairly high resistance, so this current is small compared to the primary current and we can ignore its effect on the fields. Then the wavefront goes on past the secondary, encountering the wavefront from the left side of the primary solenoid when it

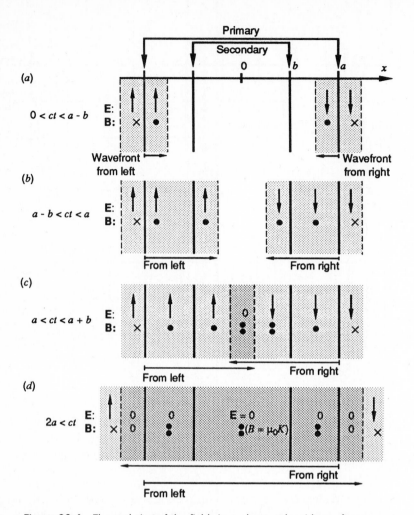

Figure 22.4 The evolution of the fields in a planar solenoid transformer.

reaches the center $x = 0$, at time a/c. Because Maxwell's equations are linear, we may just superpose the fields of the two waves in the region where they overlap; neither influences the propagation of the other. The electric fields from the two sides just cancel in the overlap region, but the magnetic fields add, giving $B = \mu_0 K$, as shown in Figure 22.4c. For $t > 2a$ (Figure 22.4d), the magnetic field is a uniform $\mu_0 K$ inside the solenoid (this agrees with our result from magnetostatics!). At each point $x > a$ outside the solenoid, the B and the E fields both become zero at $t = (x + a)/c$; the static magnetic field is zero outside, as predicted by magnetostatics. Our interest, however, is in the electric field felt by the secondary solenoid. It is zero except between $t = (a - b)/c$ (when the first wavefront passes it and turns on E) and $t = (a + b)/c$ (when the second wavefront passes and turns off E); during this interval of length $2b/c$ there is a field $E = \frac{1}{2}\mu_0 cK$. To a human this all appears to be instantaneous; if $b = 0.1$ m, the duration of the electric field pulse is 7×10^{-10} seconds. Afterwards, as long as the current K remains constant, there is

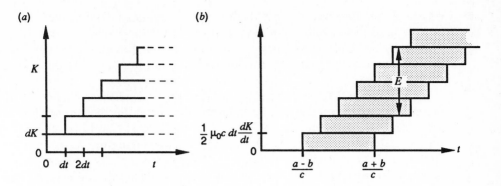

Figure 22.5 (*a*) A linearly increasing current, decomposed into flat pulses. (*b*) Electric field (height of shaded region) at the secondary solenoid.

no E field. If we want the electric field to continue, we have to continue turning on current. Suppose we turn on an additional current dK in each time interval dt, that is, on the average the current K is changing at a uniform rate dK/dt. Figure 22.5*a* shows this linearly increasing current as a superposition of small currents, each of magnitude dK, turned on at successively later times. This allows us to determine the electric field without solving Maxwell's equations over again; we know the E field produced by each current step ($E = \frac{1}{2}\mu_0 c\,dK$), and we can just superpose them. The electric field at any time is just the sum of the pulses present at that time, as shown in Figure 22.5*b*. E is just the vertical thickness of the "pile of bricks" in the figure. The figure has been drawn for $dt \cong 2b/4c$, but the result we will get is independent of dt. Initially E rises linearly, but after a short transient (specifically, after $t = (a + b)/c$), there are always a constant number n of pulses present ($n = 4$ in the figure). A given E pulse lasts for a time $2b/c$, and n pulses must start during this time for n pulses to be present at the end of the given pulse. They start at intervals of dt, so $n\,dt = 2b/c$. Thus, $n = 2b/c\,dt$, and the electric field is

$$E = \tfrac{1}{2}\mu_0 c\,dt\,\frac{dK}{dt}\,\frac{2b}{c\,dt} = \mu_0 b\,\frac{dK}{dt} \qquad (22.2)$$

For a given rate of change dK/dt, this is independent of the value of dt, as advertised. This new result for a smoothly increasing current is more useful than our previous one for a current suddenly turned because it is hard to achieve sudden changes in current in a real solenoid.

SECTION 22.2 FARADAY'S FLUX LAW

We have seen that in the particular case of a planar solenoid, if we are willing to ignore transient effects over short times, the electric field induced by a changing current is proportional to the rate of change of the current, and is given by Eq. 22.2. Let us now forget about the details of the derivation and ask if there is an easier way to obtain this result. Clearly the Maxwell equation analysis we have done would be very hard to generalize to a more realistic transformer shape, even to an infinite cylindrical solenoid. Now that we have a physical understanding of what happens, however, it turns out that we can re-derive the result in a way that generalizes more easily.

First, notice that we don't really need to know the electric field everywhere, to predict the experimentally measured effect (the current in the secondary solenoid). The current depends only on the integral of the E field (the EMF) through Ohm's law (Eq. 21.18) (as long as the electric field isn't changing too fast: remember that Ohm's law was derived only in statics!). The result we have obtained for E in the planar solenoid implies an EMF around the secondary solenoid (assuming a finite height h for this purpose, as in Figure 22.2)

$$\int \mathbf{E} \cdot d\mathbf{l} = 2\mu_0 hb \frac{dK}{dt} \tag{22.3}$$

(We include the integrals only over the vertical parts of the solenoid in Figure 22.2 because $h \gg b$ and a.) This EMF is all we need to know to calculate the current in the secondary solenoid from Ohm's law.

How could we calculate the EMF more easily? The key observation to make is that one of Maxwell's equations gives a direct relation between the EMF and changes in the magnetic field: if we integrate

$$\frac{d\mathbf{B}}{dt} = -\nabla \times \mathbf{E} \tag{22.4}$$

over the surface S in Figure 22.6 we get

$$\int \frac{d\mathbf{B}}{dt} \cdot d\mathbf{a} = -\int \nabla \times \mathbf{E}$$

Then by Stokes's theorem

$$\frac{d}{dt} \int \mathbf{B} \cdot d\mathbf{a} = -\int \mathbf{E} \cdot d\mathbf{l} \tag{22.5}$$

where the path integral is around the path P. This result is the integral form of

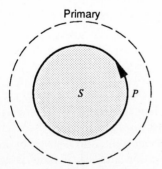

Figure 22.6 A surface S whose boundary P follows the secondary solenoid.

Faraday's law, and is usually written

$$\text{EMF} = -\frac{d\Phi}{dt} \qquad \text{(Faraday's flux law)} \qquad (22.6)$$

It gives the EMF around the secondary solenoid in terms of the magnetic flux $\Phi \equiv \int \mathbf{B} \cdot d\mathbf{a}$ through it. The presence of a wire along the path P is actually irrelevant; Faraday's law gives the EMF around any path. Faraday's law is useful for calculating the EMF only if we can calculate the magnetic field as a function of time. This is very difficult if the current generating the field is rapidly changing. However, if it changes smoothly (as in our uniformly increasing current example, and as in many practical situations) the field attains its static value for a particular current before the current changes significantly. Then we can use the methods of magnetostatics to calculate the field at each time; such fields are sometimes called "quasistatic."

The upshot of this is that there is an approximate but relatively easy way to calculate the EMF around a secondary coil induced by a slowly changing current in a primary coil:

1. Use magnetostatics to calculate the magnetic field of a current I in the primary coil; it will be proportional to I.

2. Calculate the flux of this field through the secondary coil; it will also be proportional to I.

3. Using Faraday's flux law, the time derivative of this flux gives the EMF (it will be proportional to dI/dt).

Let us apply these rules to our example of a planar solenoid. In this case, the static field for a given current in the primary solenoid can be calculated very easily: the field in a solenoid with surface current K is exactly $B = \mu_0 K$. The flux through the secondary (whose area is $2bh$) is $2bh\mu_0 K$. The EMF is therefore $2bh\mu_0(dK/dt)$. This is precisely the result (Eq. 22.3) we obtained directly from Maxwell's equations.

Historically, Faraday determined his flux law from experimental observations of EMF's in changing magnetic fields long before Maxwell realized the existence of the waves that cause these EMF's. It is possible to treat Faraday's law as an experimental fact without discussing waves at all. However, the EMF then appears to arise via a mysterious action-at-a-distance due to $d\mathbf{B}/dt$. In fact, there is nothing mysterious about it at all; it is the electric field of a wave that is emitted by the primary coil when its current is increased.

■ **EXAMPLE 22.1 A SOLENOID TRANSFORMER** As a slightly more practical example, let us consider a transformer of two concentric cylindrical solenoids with the outer (primary, of radius a) one longer than the inner one (secondary, of radius b), as shown in Figure 22.7. We want to calculate the EMF in the secondary, given the current $I(t)$ in the primary. If the primary coil has N_p turns in a length L, the current per unit length is $K = N_p I/L$. Step 1 of the above method again gives $B = \mu_0 K$ by Ampère's law (Eq. 20.5), or $B = \mu_0 N_p I/L$. Note that this requires assuming that the field at the secondary coil is the same as that of an infinitely long primary, that is, the secondary coil stays far from the ends of the primary;

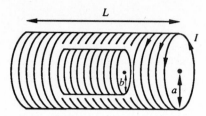

Figure 22.7 A cylindrical solenoid transformer.

this is why we had to take the primary to be longer. Then the flux through the secondary (step 2) is $\Phi = B \times \text{area} = (\mu_0 N_p I/L)\pi b^2$. Faraday's flux law then gives an EMF of

$$-\frac{\pi\mu_0 N_p b^2}{L}\frac{dI}{dt}$$

This is the EMF around each turn of the secondary. If there are N_s turns, the total EMF is N_s times larger (you can think of the individual turns as being connected in series, that is, end to end):

$$\text{EMF(total)} = -\frac{\pi\mu_0 N_p N_s b^2}{L}\frac{dI}{dt} \tag{22.7}$$

If the secondary circuit has a resistance R, the current I_s in it is therefore

$$I_s = -\frac{\pi\mu_0 N_p N_s b^2}{RL}\frac{dI}{dt} \tag{22.8}$$

▶   **Problem 22.1**

A current is turned on in a long straight wire at a rate dI/dt. Compute the EMF around a square loop of wire in the plane of the straight wire, of side a and at a distance a from the straight wire (Figure 20.4a—see Problem 20.14).   ◀

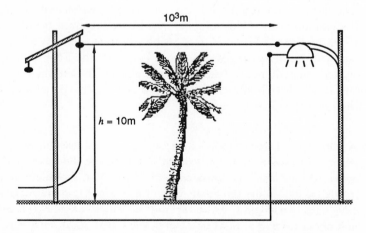

Figure 22.8 A simplified street lamp circuit consisting of a 1 km horizontal wire at height $h = 10$ m and a ground return.

▶   **Problem 22.2**

The EMP pulse described in Section 12.5 and Example 21.5 is incident vertically on the street lamp circuit shown in Figure 22.8. The pulse travels vertically downward, and has electric field amplitude $E = 2 \times 10^4$ v/m and magnetic field amplitude E/c. The magnetic field rises from zero to this value in a time h/c; assume a linear rise. Calculate the induced EMF in the circuit. A number of street lamp failures were observed in Honolulu at the time of the 1962 high-altitude Starfish test. Assuming that the EMF there had the same order of magnitude (though the test was not conducted directly over Honolulu), is this EMF a plausible explanation?   ◀

SECTION 22.3   LENZ'S LAW

The sign of the EMF (and therefore of the induced current) is always correctly given by the Faraday flux law (Eq. 22.6). However, determining it requires keeping track of the sign conventions for the primary current, the magnetic field, and the secondary current, and correctly using the right-hand rules that relate them. There is a rule of thumb called Lenz's law which can be used to check the final sign:

Lenz's law

The magnetic field produced by an induced current **opposes** the magnetic field change that induced the current.

This can be thought of as a sort of inertia of magnetic fields: **B** does not want to change, and when it is forced to change, it induces a current that tries to change it back. Lenz's law can be restated in terms of **current** in a form that is more useful in cases such as Example 22.1:

An induced current points oppositely to the current change which induced it.

Thus in Figure 22.7 an increasing counterclockwise I in the primary coil induces a clockwise EMF and I_s in the secondary. Using the same sign convention (say counterclockwise = positive) for both I and I_s, they must be of opposite sign, in agreement with Eq. 22.8.

▶   **Problem 22.3**

A metal ring is placed coaxially with a solenoid, near one end, and a current is turned on in the solenoid. Indicate in a sketch the direction of the current induced in the ring and the direction of the net force on the ring.   ◀

SECTION 22.4   FLUX CHANGES DUE TO MOTION

Our examples of induced electric fields have thus far always involved fields that change because currents in coils do. However, it is clear that Faraday's flux law (having been derived directly from Maxwell's equations) is true **independently** of the reason for the change in the magnetic field. Instead of being due to a change in the current, it could be because of a motion of the primary coil, as in Figure 22.9a. A thin rectangular loop of wire is threaded between the wires of the solenoid, with a meter attached to measure the EMF around it. The loop remains stationary as the

(a)   (b)

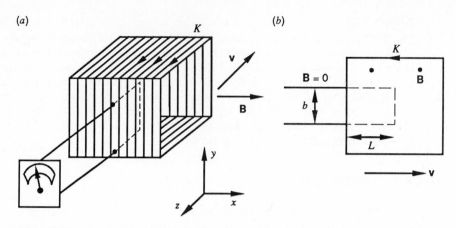

Figure 22.9 (a) A long square solenoid with a surface current density K, moving in the $-z$ direction with a speed v. (b) A cross-sectional view in the plane of the probe loop.

solenoid is pulled away. We know the magnetostatic field is $\mu_0 K$ inside the solenoid and zero outside. Therefore at a point on the front ($+z$) surface of the solenoid it is changing (from $\mu_0 K$ to 0) as the solenoid passes the point in question. It can be seen from Figure 22.9 that the flux through the loop is $\Phi = BbL = \mu_0 KbL$. Only L is changing, decreasing at the rate $dL/dt = -v$. Faraday's flux law then gives

$$\text{EMF} = \mu_0 Kbv \qquad (22.9)$$

In principle we could analyse the fields of the moving solenoid directly with Maxwell's equations. Changes in the current (due to motion of the solenoid) must produce electric fields that turn off the magnetic field in some regions, turn it on in others, and then recede to infinity. However, this would be quite difficult (see Problem 22.11): this is a case in which Faraday's law is a **much** easier way to calculate the EMF.

Of course we could as easily (probably more easily, in fact) move the loop in Figure 22.9 instead of the solenoid, leaving the solenoid stationary. If we believe in the fundamental notion of Galilean relativity (that the laws of physics should be the same in any uniformly moving coordinate system, for example one moving with the solenoid) we must expect the measured current to be the same when the solenoid is stationary. However, this current cannot be driven by an electric field, because $\mathbf{E} = 0$ everywhere in this magnetostatic situation. We do in fact measure the same current in such an experiment, so some force other than the electric force $q\mathbf{E}$ must have pushed the charges around. A little thought will convince you that this must have been a Lorentz force $\mathbf{F} = q\mathbf{v} \times \mathbf{B}$; there **is** a $\mathbf{B}$ field and the charges in the loop are moving with a velocity $\mathbf{v}$ (equal and opposite to the velocity of the solenoid in the other coordinate system). If the EMF is to include **all** forces, we should generalize it to

$$\text{EMF} = \int (\mathbf{E} + \mathbf{v} \times \mathbf{B}) \cdot d\mathbf{l} \qquad (22.10)$$

where **v** is the vector velocity of the wire. This reduces to our previous form in all the cases we considered before, because **v** was zero. The contribution of the second term is called the **motional EMF**, as distinguished from the first term involving the induced **E** field, which is called the **induced** EMF. We can generalize the microscopic Ohm's law (Eq. 21.1) to moving media by replacing $q\mathbf{E}$ by the total force on a charge, $q(\mathbf{E} + \mathbf{v} \times \mathbf{B})$:

$$\mathbf{j} = \sigma(\mathbf{E} + \mathbf{v} \times \mathbf{B}) \qquad (22.11)$$

This reduces to the previous form when $\mathbf{v} = 0$, and agrees with experiment for nonzero **v** as well. The corresponding generalization of the macroscopic Ohm's law (Eq. 21.18) is

$$\text{EMF} = IR \qquad (22.12)$$

where the EMF is given by Eq. 22.10.

In our example, $\mathbf{v} \times \mathbf{B}$ is in the y-direction (Figure 22.9), which is along $d\mathbf{l}$ only in the vertical part of the loop, so the path integral in Eq. 22.10 giving the EMF around the moving loop is vBb. This is identical with the prediction (Eq. 22.9) of the Faraday flux law for the case in which the **solenoid** moved, in agreement with Galilean relativity.

Let us return to Faraday's flux law, $\text{EMF} = -d\Phi/dt$. We derived it for the induced EMF in a stationary loop, and we modified it to a form that works (at least in one case) for a moving loop. Actually, we can prove that it works for any moving loop, as follows. The EMF and Φ can be measured in an inertial (constant-velocity) coordinate system moving with the loop; in this system we know Faraday's law to be valid. But the value of the EMF in this system will be the same as in the laboratory system, if we use Eq. 22.10 for the EMF; this is proportional to the total force on a charge, and forces are the same in all inertial frames. (We assume the velocity is much less than the speed of light here, so we don't have relativistic effects.) Assuming the magnetic flux is also the same, the flux law is seen to be valid in the laboratory frame as well as in the moving frame. So Faraday's flux law has a wider range of validity than we thought: it gives the total EMF in a loop of wire when the flux through the loop changes due to changes in magnetic field (from turning on or moving a coil or permanent magnet) **or** due to the motion of the loop itself. One can even show it works for flux changes due to changing the shape of the wire (bending it); a proof that applies to this case is given on p. 254 of Griffiths' book cited in the bibliography.

▶   **Problem 22.4**

A circular loop of wire of radius r, initially horizontal, is rotated by an angle π about a horizontal axis in a vertical magnetic field **B** (see Figure 22.10). The conductivity of the wire is σ and its cross-sectional area is A $(A \ll r^2)$. Compute the **charge** passing a point on the wire $(Q = \int I\,dt)$.   ◀

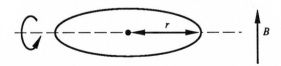

Figure 22.10   A circular loop normal to a magnetic field, which rotates about an axis perpendicular to the field.

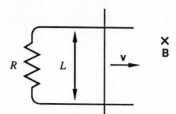

Figure 22.11 A sliding-wire circuit in a magnetic field
of magnitude B pointing into the page.

▶   **Problem 22.5**

Figure 22.11 shows a U-shaped wire with a resistor of resistance R. The circuit is
completed by a straight wire of mass m that slides frictionlessly along the two arms of
the U.

a. What is the current in the resistor when the sliding wire moves to the right at a speed
v? In which direction does it flow?

b. What is the (vector) force on the sliding wire?

c. If the initial speed of the sliding wire is v_0 and no external force is applied, what is its
speed after a time t?

d. After a long time $v(t)$ approaches zero. Where did the kinetic energy of the moving
wire go? Show that an equal amount of energy appeared somewhere else.   ◀

▶   **Problem 22.6**

Consider a square loop of wire of side s, coplanar with a long straight wire carrying a
current I, such as that shown in Figure 20.4a. Suppose that the loop is moving away
from the straight wire, with one side remaining parallel to it. Calculate the EMF in the
loop, in terms of the speed v and the instantaneous distance a between the long wire
and the closest side of the loop. In which direction does current flow around the loop?   ◀

▶   **Problem 22.7**

Consider the **Faraday disc** shown in Figure 22.12. Compute the EMF in the circuit, in
terms of the angular velocity ω of the disc and the radius R. (Hint: The force on each
charge carrier in the disc is the same as though all of the disc were cut away except for a

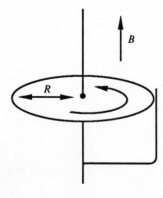

Figure 22.12 A metal disk rotating in a plane perpendicular
to a uniform magnetic field, with a sliding contact on the
perimeter to complete the circuit.

small wire from the center to the sliding contact.) Does current flow inward or outward in the disk?   ◄

▶ **Problem 22.8**

Prove that the flux Φ through a path P, which is defined as $\int \mathbf{B} \cdot d\mathbf{a}$ over a surface S bounded by P, is independent of the choice of S.   ◄

▶ **Problem 22.9**

An AC generator consists of a loop of area A, rotating about an axis in its own plane, in a magnetic field B perpendicular to the axis. Compute the time-dependent EMF in the loop in terms of B, A, and the frequency and phase of the loop's rotation.   ◄

▶ **Problem 22.10**

A square wire loop of side s, resistance R, and mass m has its top side in a uniform magnetic field B normal to the loop, and its bottom side outside the field.

a. If it moves downward with a velocity v, what is the magnetic force (magnitude and direction) on it?

b. If it falls from rest ($v = 0$ initially), calculate its velocity as a function of time, assuming that the top remains in the field.

c. Calculate the terminal velocity if $R = 10^{-4}$ Ω, $s = 0.1$ m, $B = 1$ T, and $m = 0.01$ kg.   ◄

▶ **Problem 22.11**

Faraday's flux law for the moving-solenoid problem (Figure 22.9) tells us only the total EMF around the wire loop when the solenoid is moving uniformly. It does not tell us the effect of a particular part of the motion (say that during dt, over a distance $v\,dt$) on the force on the current carriers at a particular place on the loop. Analyse this effect using Maxwell's equations, in a simplified model in which the height of the solenoid becomes infinite (it looks like the primary solenoid in Figure 22.2). Consider a time interval dt during which the left side of the primary solenoid in Figure 22.2 moves from $x = -a$ to $x = -a + v\,dt$ and the right side moves from $x = +a$ to $x = +a + v\,dt$. The change in current during this time can be thought of as turning off the surface current $\pm K$ at the old positions ($x = \pm a$) and turning on $\pm K$ at the new positions ($x = \pm a + v\,dt$). All four of these current changes (and therefore their superposition) can be exactly analysed using the wave pictured in Figure 22.2b. Assume there are no other current changes (i.e., the solenoid didn't move during any other time interval). Assume the vertical parts of the stationary loop are at $x = 0$ (inside the solenoid) and $x = -2a$ (outside the solenoid).

a. Compute and plot the electric fields $E(x = 0)$ and $E(x = -2a)$ as functions of time. Verify that the EMF has no contribution from the horizontal parts of the loop, and compute and plot the EMF as a function of time.

b. Now superpose the EMF's produced by many such "jumps" of length $v\,dt$ by the solenoid during successive time intervals, to get the effect of a smooth motion at the uniform velocity v. Is the result consistent with Faraday's law?   ◄

SECTION 22.5   INDUCTANCE

In the previous section we wrote down a general method for calculating the EMF in a "secondary" loop of wire (say 2 in Figure 22.13) due to a change in the current I_1 in another ("primary") loop (1 in the figure). It requires calculating the magneto-

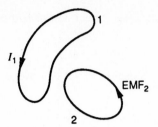

Figure 22.13   Two arbitrarily shaped loops of wire, with current I_1 in one.

static field (call it B_1) due to the current I_1, and then the flux through loop 2. It is not always easy to calculate the field, but we can say something about its dependence on I_1: it is **linear**. This follows indirectly from Maxwell's equations, or directly from formulas such as Eq. 20.50 giving the magnetostatic field B_1 as I_1 times an integral over the loop. Then the flux Φ_2 through loop 2, being an integral of B_1 over the area enclosed by this loop, is also proportional to I_1:

$$\Phi_2 = M_{21} I_1 \tag{22.13}$$

The proportionality constant M_{21} is called the **mutual inductance** of the two loops. Its units are **henries**; one henry is a volt–second per ampere. Then Faraday's law gives the EMF

$$\mathrm{EMF} = -M_{21} \frac{dI_1}{dt} \tag{22.14}$$

We have already essentially calculated M_{21} for the two solenoids of Example 22.1; comparing Eq. 22.14 and Eq. 22.7 gives

$$M_{21} = \frac{\pi \mu_0 N_p N_s b^2}{L} \tag{22.15}$$

for the case of a short solenoid of radius b, coaxial with a long solenoid of length L. In most other cases it is not so easy, but there are a few for which M_{21} can be exactly calculated.

▶   **Problem 22.12**

A current I_1 flows through the lower loop in Figure 22.14.

a. Compute the magnetic field at a point a distance z above the center of the lower loop. (You may use the result of a previous problem.)

b. Assuming $a \ll b$ so the field is essentially constant in the small upper loop, compute the flux Φ_2 through it, and thence the mutual inductance M_{21}.

c. Now reverse the roles of the loops: assume a current I_2 in the upper one, and find the flux Φ_1 in the lower one. (Treat the upper loop as a point dipole; you may obtain Φ_1 by integrating over **any** surface bounded by loop 1.) Show that $M_{12} = M_{21}$.   ◀

▶   **Problem 22.13**

Compute the mutual inductance M_{21} of the circuit shown in Figure 22.15a. Assume $b \gg a$, so you can treat the large square as a long straight wire in computing the field near the small square.   ◀

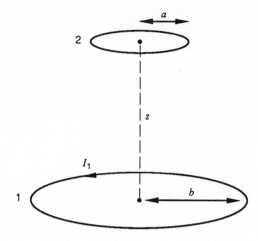

Figure 22.14   Two coaxial circular loops in parallel planes.

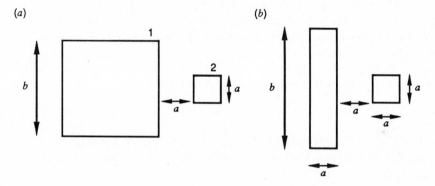

Figure 22.15   (a) Two square loops of dissimilar size, (b) a long rectangular loop and a small square one.

▶   **Problem 22.14**

Compute the mutual inductance of the two loops shown in Figure 22.15b, using the approximation $b \gg a$.   ◀

It is possible to derive a formal expression for the mutual inductance between an arbitrary pair of loops, such as those shown in Figure 22.16. Equation 20.31 gives the flux Φ_2 through loop 2 as a path integral of the vector potential $\mathbf{A}_1$ due to the current I_1 in loop 1, and $\mathbf{A}_1$ can in turn be written as a path integral over loop 1 (Eq. 20.40, with $I\,d\mathbf{l}$ for $\mathbf{j}\,d\tau$). Thus,

$$\Phi_2 = \int \mathbf{A}_1 \cdot d\mathbf{l}_2 = \frac{\mu_0}{4\pi} \int \left(\int \frac{I_1\,d\mathbf{l}_1}{s} \right) \cdot d\mathbf{l}_2 = \frac{\mu_0 I_1}{4\pi} \int\int \frac{d\mathbf{l}_1 \cdot d\mathbf{l}_2}{s} \quad (22.16)$$

where s is the distance between the two length elements $d\mathbf{l}_1$ and $d\mathbf{l}_2$ on the two

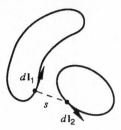

Figure 22.16 The two loops of Figure 22.13, showing path length elements $d\mathbf{l}$ along each and the distance s between the elements.

loops. From Eq. 22.13 we can write the mutual inductance as

$$M_{12} = \frac{\mu_0}{4\pi} \int \int \frac{d\mathbf{l}_1 \cdot d\mathbf{l}_2}{s} \tag{22.17}$$

Equation 22.17 is called the **Neumann formula**. It treats the two loops symmetrically ($d\mathbf{l}_1 \cdot d\mathbf{l}_2 = d\mathbf{l}_2 \cdot d\mathbf{l}_1$), making it clear that $M_{21} = M_{12}$ for any two loops. (Recall that we found this to be true in the particular case of Problem 22.12.) This fact is sometimes useful; for example it would be very difficult to calculate M_{12} directly for the squares in Figure 22.15, but it is easy to get it from M_{21}.

The special case of the mutual inductance in which loop 1 and loop 2 are the same is called the **self-inductance** L; it is defined by

$$\Phi = LI \tag{22.18}$$

where I is the current through the loop (no subscript is needed because there's only one) and Φ the flux. Evidently the EMF in the loop due to a change in its own current is then

$$\text{EMF} = -L \frac{dI}{dt} \tag{22.19}$$

A case in which we can approximately calculate the self-inductance is that of a long cylindrical solenoid with N turns and length l. Except near the ends, the magnetostatic field inside is $\mu_0 K$, where $K = NI/l$ is the linear current density. If the radius is b, the flux per turn is $\pi b^2 \mu_0 K$; the total flux is N times this,

$$\Phi = \frac{N\pi b^2 \mu_0 NI}{l} \tag{22.20}$$

so the self-inductance (Eq. 22.18) is

$$L = \frac{\pi b^2 \mu_0 N^2}{l} \tag{22.21}$$

(Note that this could also be obtained from Eq. 22.15 for two solenoids, by setting $N_p = N_s = N$.)

One must be **very** careful in dealing with the concept of self-inductance. We have managed to calculate it for a surface (2D) current density. The simplest case,

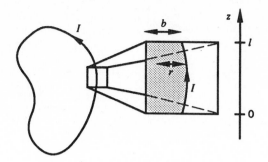

Figure 22.17 A loop of wire with an expanded view of a small segment.

however, would seem to be that of a 1D current density, a wire loop (this is the context in which the mutual inductance was defined). Unfortunately, the self-inductance always diverges for a 1D current density. To see this, consider the loop of wire shown in Figure 22.17. If the shape is not simple, it might be hard to compute the field everywhere on a surface bounded by the loop. However, we can approximately calculate it for a region near the wire, such as that shown in the expanded view in Figure 22.17. The wire in this vicinity resembles a piece of long straight wire, and for points near the wire (when b in the figure is small) the field is dominated by the effect of the nearby wire (because of the s^{-2} factor in the Biot–Savart law). Thus we can calculate the contribution to the flux from the shaded region using the field $B = \mu_0 I / 2\pi r$ of a long straight wire:

$$\Phi = \int \mathbf{B} \cdot d\mathbf{a} = \int_0^l dz \int_0^b dr\, B$$

$$= \frac{\mu_0 lI}{2\pi} \int_0^b r^{-1}\, dr = \frac{\mu_0 lI}{2\pi} [\ln b - \ln 0] = \infty$$

The flux (and therefore the contribution Φ/I to the self-inductance) diverges. This occurs because B diverges very near the wire. In a real physical wire of finite radius, of course, B does not diverge, and the flux is finite (see Problem 22.15).

▶ **Problem 22.15**

Consider a racetrack-shaped loop, made of cylindrical wire of radius R, having two long parallel straight sections, as in Figure 22.18. Ignore the regions near the top and bottom where the wire bends around, i.e., assume that for the entire height h of the loop the field is that of two straight wires. Assume that the current density is uniform, i.e., $j = I/\pi R^2$.

a. Calculate the B field due to the left wire, at a point x (use Ampère's law).

b. Add a similar contribution from the right wire to obtain the total field.

c. Compute the total flux $\Phi \equiv h \int_0^b B(x)\, dx$ through the loop, and hence the self-inductance $L = \Phi/I$. What happens to L in the limit $R \to 0$?

d. This flux will give the EMF along a path at the center of the wire. Now calculate the flux through a path at the inside edge of the wire ($x = R$, $x = b - R$). The flux is not exactly the same for all paths through the wire, so the self-inductance is not

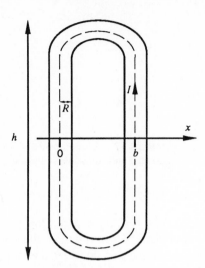

Figure 22.18 A wire loop whose self-inductance can be calculated.

completely well-defined for a finite-radius wire. Show that the fractional difference between these two fluxes is small if $R \ll b$, so this is not in practice a problem for thin wires.   ◀

Thus, it appears that a 2D surface current in a solenoid is the only case for which the concept of self-inductance, as we have defined it, makes perfect sense. If the current is more singular (1D) the self-inductance diverges, and if it is less singular (3D) the EMF depends on an arbitrary choice of path through the conductor. The latter problem can be avoided even in systems in which the total dimensions of the current distribution are not small, by assuming the current to be divided among many thin wires as illustrated in Problem 22.16.

▶ **Problem 22.16**

Compute the self-inductance of a long thick solenoid of length l with N turns of thin wire, extending from radius a to radius b. In computing the magnetostatic field, assume the current density **j** is azimuthal and constant between $r = a$ and $r = b$ (i.e., ignore the discreteness of the wires). However, in computing the flux and the EMF you must treat each turn of wire individually, because each has a different flux (assume each is essentially a circle, and they are connected in series so the EMF's add). Verify that your result reduces to our previous formula when $a = b$.   ◀

▶ **Problem 22.17**

Consider a solenoid (in an electric motor, for example) with a self-inductance $L = 1$ h, in which flows a current $I = 10$ a. It is reduced during an interval $dt = 1$ ms $= 10^{-3}$ s to zero (by unplugging it, for example). Calculate the EMF across the solenoid during this interval (assuming the current changes linearly). (This EMF causes the spark when the plug is pulled.)   ◀

■ **EXAMPLE 22.2 AN *RL* CIRCUIT**   Consider a circuit with an inductor (i.e., an arrangement with a substantial self-inductance, such as a solenoid with many turns), a resistor, and a battery, connected in series (so that each charge carrier must go through each of the three

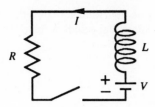

Figure 22.19 A series *RL* circuit.

components in turn), as in Figure 22.19. The switch is closed at time $t = 0$ and we would like to know the current as a function of time. The EMF of the battery is V, a constant. The total EMF in the circuit (regarding counterclockwise EMF's as positive) is then $V - L\,dI/dt$, and Ohm's law is

$$V - L\,\frac{dI(t)}{dt} = I(t)R$$

where $I(t)$ is the (counterclockwise) current in the resistor. If we eliminate the constant V by changing variables to $I'(t) \equiv I(t) - V/R$, we get a differential equation we have encountered before, with the exponential solution $I'(t) = A\exp(-\,t/\tau)$ where $\tau \equiv L/R$ (the time constant) and A is a constant. Our initial condition $I'(0) = -\,V/R$ determines A, and gives

$$I(t) = VR^{-1}(1 - e^{-\,t/\tau})$$

The rise of the current is shown in Figure 22.20. After a long time it approaches the value V/R that the resistor would have without the inductor; the effect of the inductor is to add "inertia" to the circuit, slowing down the change in the current.

▶   **Problem 22.18**

A current I_0 is flowing at time $t = 0$ in the circuit of Figure 22.21, with the switch closed, so the capacitor charge Q is zero. The switch is suddenly opened. Use the fact

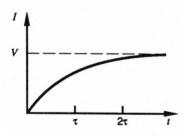

Figure 22.20 The current in an *RL* circuit, as a function of time.

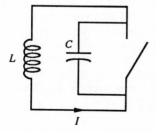

Figure 22.21 An *LC* circuit.

that the EMF's $-L\,dI(t)/dt$ and $Q(t)/C$ across the inductor and the capacitor respectively must match to write a second-order differential equation for $Q(t)$. Solve it, and give $Q(t)$ in terms of L, C, and I_0.   ◄

SECTION 22.6   ENERGY IN INDUCTORS

Consider an inductor through which a current is driven, such as that in Figure 22.19. The battery must do work, transporting charge outside the inductor from one terminal of the inductor to the other. The work is done against a potential difference (EMF) of $L\,dI/dt$, so the rate of work is

$$\frac{dW}{dt} = I(t)L\,\frac{dI(t)}{dt} = \tfrac{1}{2}L\,\frac{d}{dt}\,I^2(t)$$

If the total work and the current both start out at zero, the work required to increase the current to a value I is exactly

$$W = \tfrac{1}{2}LI^2 \tag{22.22}$$

independently of the exact form of the function $I(t)$.

This work represents energy that is flowing into the inductor from the outside; it must exist in some form inside the inductor. One possible form is the kinetic energy of the moving charge carriers; however, as you will see in Problem 22.19, this is totally negligible. The only other physical change that has occurred in the inductor is an increase in the magnetic field, so the extra energy must be magnetic field energy. We can easily verify this for the special case of a long solenoid, in which we know the field is $B = \mu_0 K = \mu_0 NI/l$, where N/l is the number of turns per unit length. If the cross-sectional area of the solenoid is A, the volume inside is Al, and the magnetic field energy is

$$U_B = \frac{B^2}{2\mu_0}Al = \frac{\pi b^2 \mu_0 N^2 I^2}{2l} = \tfrac{1}{2}LI^2$$

(using Eq. 22.21 for the inductance L). This agrees exactly with the work (Eq. 22.22) done in turning on the current and the field.

▶   **Problem 22.19**

Consider a solenoid of radius R whose winding forms a layer of thickness $d \ll R$, so that when the current per unit length is K, the current density $j \approx K/d$. The velocity v of the charge carriers (assumed to be electrons) is related to j by $j = nev$, where n is the number density of electrons and e is the electronic charge. Take n to be about one mole per cubic centimeter, so $n \approx 6 \times 10^{29}$ m^{-3}. Denote the electron mass by m, and show that the ratio of electron kinetic energy to magnetic field energy is $(2/Rd)(m/\mu_0 ne^2) = (2/Rd)(3 \times 10^{-17}\ \text{m}^2)$. Assuming that $R/d = 10$, how small would R need to be for this ratio to be unity?   ◄

▶   **Problem 22.20**

   a. Compute the magnetic field inside a toroidal solenoid of rectangular cross section whose inner and outer cylindrical surfaces have radii a and b respectively. Assume the height along the cylinder axis is h and there are N turns, each carrying current I. Do not assume $b - a \ll a$.

   b. Compute the self-inductance. Verify that it agrees with Eq. 22.21 in the limit $b - a \ll a$.

   c. Compute the magnetic field energy by integrating $B^2/2\mu_0$, and verify that it agrees with Eq. 22.22.   ◀

▶   **Problem 22.21**

   Compute the energies of the capacitor and the inductor of Problem 22.18 as functions of time. Verify that their sum is constant.   ◀

CHAPTER TWENTY-THREE

MAGNETIC MATERIALS

SECTION 23.1   BOUND CHARGES REVISITED: BOUND CURRENTS

In discussing dielectric materials (Chapter 18), we considered the possibility that an externally imposed electric field may shift bound charges within atoms of a material, polarizing it. This of course implies the existence of a bound current (a current of bound charge) for a short time, while the bound charge is shifting. However, this is a greatly oversimplified view: actually, there are **always** bound currents. The bound charges are continually in motion within the molecules; furthermore the molecules themselves are moving in random ways. The charge density is therefore continually fluctuating. The only way we can sensibly talk about a "static" charge density in a dielectric is by using cells large enough that the fluctuations become relatively unimportant (we noted a similar situation with Ohm's-law currents in imperfect conductors). Quantitatively, it turns out that the root-mean-square fluctuation of the charge density about its average value, expressed as a fraction of the average itself, is proportional to $N^{-1/2}$ where N is the average number of charges in a cell. A similar result holds for currents: if the face f is large compared to atomic dimensions, the random fluctuations of the current about its average value can be neglected. The statistical-mechanical study of such fluctuations is beyond the scope of this text, so we will deal only with scales that are large enough so that statistical fluctuations of this sort are negligible.

In this large-scale sense, currents and densities of bound charge are well-defined quantities. In a dielectric, the bound current flows only momentarily when an electric field is first turned on. It was possible for us to ignore it in electrostatics, which is concerned with what happens after a long time. In magnetostatics, when we impose a static **magnetic** field, it turns out that the bound current doesn't vanish as $t \to \infty$, even if we eliminate random fluctuations by using a large cell size dr. Magnetic fields can produce **systematic** (non-random) changes in the bound current; we will explore two possible mechanisms for this in the next sections. Our interest in the bound current arises from the fact that it produces additional magnetic fields that extend outside the material in question and can be measured and used for practical purposes, such as turning electric motors.

SECTION 23.2   PARAMAGNETISM

There are basically two effects that a magnetic field has on bound currents (i.e., on electrons circulating in atoms). One is that the plane of the orbit will tend to be

tilted by the field; we will see later that this tends to produce a magnetic moment **parallel** to the B field, and is called **paramagnetism**. The other is that the electron will be speeded up or slowed down by the field; this will produce a magnetic moment **opposed** to the B field, a phenomenon called **diamagnetism** ("dia" meaning opposed, as in "diametrically opposed").

To isolate the effects of tilting the orbit from those of changing the speed of the electron, let us suppose that the electron is constrained to move along a rigidly specified trajectory, which is only free to tilt. This is essentially equivalent to modeling the electron current by the current in a rigid wire loop; the fact that the charge is not really spread evenly around the loop but rather moves around in one lump does not affect the large-scale current or the fields produced by it. (Think of spreading the charge around the loop by dividing it into pieces; the motion of each shifted piece differs from that of the original charge only by a constant time delay, which doesn't affect the average current in a static situation.) So let us examine the behavior of a rigid loop in a magnetic field B, as shown in Figure 23.1. There is a magnetic force $I\mathbf{l} \times \mathbf{B}$ along each side of the square ($\mathbf{l}$ is a vector along the side). The forces on the front and back sides ($x = \pm \frac{1}{2}a$) pull outward on the loop, but cancel out of the net force and produce no torque. The forces on the right and left sides are also equal and opposite ($\pm Ia\mathbf{B}$, in the y-direction), and cancel out of the net force. However, each contributes a torque about the x-axis, with moment arm $\frac{1}{2}a \sin\theta$. So the net force on the loop is zero, but the torque is

$$N = Ia^2 B \sin\theta \tag{23.1}$$

In the present case, this can be written in the form

$$\mathbf{N} = \mathbf{m} \times \mathbf{B} \tag{23.2}$$

in terms of the magnetic moment $\mathbf{m} = Ia^2\hat{\mathbf{n}}$. (The orientation of $\hat{\mathbf{n}}$ is that of the field produced by the loop, which can be obtained by the right-hand rule.) It is not too difficult to show from the general formula (Eq. 20.64) for the magnetic moment of an arbitrary current distribution that Eq. 23.2 gives the correct torque in general, and that there is no net force on such a distribution (in a uniform field).

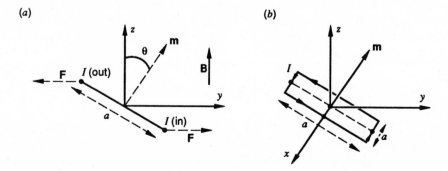

Figure 23.1 An $a \times a$ square loop tilted by an angle θ with respect to the magnetic field, which points along the z-direction. (a) View along the x-axis, which is the tilt axis. (b) Oblique view.

Note that the **direction** of the torque is such as to orient the magnetic moment parallel to the magnetic field **B**. Although there will be other torques acting on the electron orbit (exerted for example by neighboring atoms) which will cause it to point in other directions much of the time, there will be a systematic tendency for the moment to be parallel to **B** (i.e., for the plane of the orbit to be normal to **B**).

There are two important differences between this simple classical mechanical model and a real atom, which instead obeys quantum mechanics. In addition to the "orbital" magnetic moments we have been discussing, an electron has a magnetic moment due to its spin around its own axis. This spin motion can be modeled as a current loop in much the same way as the orbital motion, however, and the magnetic field exerts a torque on the electron due to its spin just as it does on the orbit. The spin magnetic moment will therefore tend also to point along **B**. The other quantum mechanical effect is that electrons are constrained to occupy definite atomic orbitals, which come in pairs having opposite directions of orbital and spin magnetic moments. When the orbitals for both directions have been filled with electrons, the magnetic moments cancel; there is no net current flowing around our loop. Such an atom (helium, for example) cannot exhibit paramagnetism. Only atoms or molecules having "unpaired electrons" exhibit paramagnetism; the sodium atom (with one unpaired electron outside a filled shell) is an example. It turns out that when there **are** unpaired electrons the paramagnetic effect usually dominates the opposing diamagnetic effect we discuss below. However, when there are no unpaired electrons diamagnetism dominates.

SECTION 23.3   DIAMAGNETISM

In addition to tilting the orbit (paramagnetism), a magnetic field can speed up or slow down the orbital motion of an electron in an atom or molecule. The rigid current-loop model we used above suppressed this effect by requiring the current I to remain constant. Of course, if the electron obeyed the laws of classical physics it would slow down even without any applied fields, gradually radiating its energy away and spiraling toward the nucleus. A complete treatment of the quantum mechanical effects that prevent this is beyond the scope of this text. However, we can mimic some of these effects by imagining the electron to be like a bead sliding along a rigid hoop that keeps it at a fixed radius, as in Figure 23.2. To study the effects of speeding up and slowing down without the complications of axis tilting we considered before, let us imagine the axis of rotation to be fixed and vertical. Neither of our two models is very realistic by itself, but between them they include all the important physical effects.

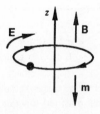

Figure 23.2   A model of an atom, in which an electron is constrained to stay at a fixed distance from the z-axis; **m** is the magnetic moment.

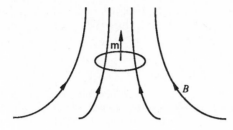

Figure 23.3 A current-carrying loop in an inhomogeneous magnetic field (B increases upward).

What happens when we turn on a B field in the z direction? This increases the flux through the hoop, and by Faraday's flux law there must be an EMF around the hoop. That is, there is an induced electric field parallel to the hoop. Lenz's law tells us the direction of the field: it must drive a current that creates a magnetic field inside the hoop **opposite** to the one we turned on. The magnetic moment of this additional current in the hoop is defined to point in the same direction as the field it creates inside the hoop, so the additional magnetic moment is opposite to the **B** field we turned on. This is the phenomenon of diamagnetism, and it occurs in all atoms and molecules. In those without unpaired electrons, it is the dominant effect, and the net magnetic moment is opposite to the applied magnetic field.

▶   **Problem 23.1**

Consider a magnetic dipole in an inhomogeneous magnetic field, such as the current loop in Figure 23.3. Assume $B_z(r, \phi, z) = B_0 + az$, $B_r(r, \phi, z) = -ar/2$, $B_\phi = 0$, where B_0 and a are constants.

a. Show that this is a possible magnetic field, i.e., $\nabla \cdot \mathbf{B} = 0$ and $\nabla \times \mathbf{B} = 0$.

b. Compute the force on a magnetic dipole at the origin with magnetic moment $\mathbf{m} = m\hat{\mathbf{z}}$. Model it as a circular loop of radius R normal to the z axis. (You must first calculate the current in the loop.)

c. The B field increases as you approach the pole of a magnet, so you can think of Figure 23.3 as depicting the field just below the south pole of a magnet. Will a paramagnetic material be attracted or repelled by the magnet? What if it was near the north pole? How about a diamagnetic material?

d. (Harder) Suppose the magnetic dipole had an **arbitrary** current distribution $\mathbf{j}(\mathbf{r})$, which is nonzero in a small region near the origin, and whose magnetic moment is $m\hat{\mathbf{z}}$. Show that the force is the same as in part (**b**).   ◀

SECTION 23.4   THE MAGNETIZATION FIELD **M**

We have seen that magnetic fields can cause non-random bound currents to flow, producing nonzero magnetic moments in atoms and molecules. These bound currents are important because they produce additional magnetic fields outside the material. We would like to be able to predict these additional fields quantitatively, by finding formulas giving the bound current in terms of the applied magnetic field. It turns out to be easiest to do this indirectly, by expressing the current in terms of a new vector field called the **magnetization**, which is related more simply than the

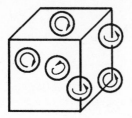

Figure 23.4 A material composed of molecules, each with a bound current, shown with one cell of a cubic lattice.

current itself to the applied magnetic field. To define the magnetization, think of a model material composed of separate molecules, each of which has an internal current distribution that may be influenced by the B field but that cannot leave the molecule (Figure 23.4). There are two equivalent ways to define the magnetization, in terms of the total magnetic dipole moment in a cell and in terms of the current circulating around an edge. We will use the latter because it makes it much easier to relate the magnetization to the bound current. To motivate the use of edges in this way, let us note that a moving charge bound within a molecule near the center of a face, as in Figure 23.5a, **cannot** affect the current through the face, on the average. At one instant it may produce an upward current, but the next instant it must come back down. Therefore, to study the effects of atomic currents, we need only consider the currents that encircle edges (as in Figure 23.5b), and thus pass through any particular face in only one direction. Let us then define an edge vector field **M**, the **magnetization**, as the current per unit length around an edge e:

$$M(e) = \frac{\text{(total bound current around } e)}{dl(e)} \tag{23.3}$$

The significance of M is that it determines the discrete current through each face due to the atomic bound currents. In this context, the total bound current is called the **magnetization current** and is denoted by j_M to distinguish it from any other current that may be present. As we have seen, only atoms at the edges of a face f contribute to $j_M(f)$ (Figure 23.5c). Those at edge e contribute a current $M(e)\,dl(e)$, so the total is

$$j_M(f) = da(f)^{-1} \sum_e M(e)\,dl(e) = \text{curl } M(f) \tag{23.4}$$

by the definition of the curl as the path integral per unit area.

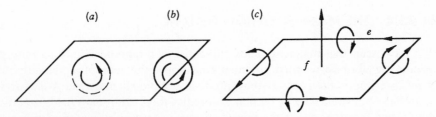

Figure 23.5 Molecules whose bound current crosses a face near (a) its center, (b) an edge; (c) the four edges contributing to the net bound current across a face f.

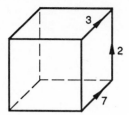

Figure 23.6   The discrete magnetization field for Problem 23.2.

Suppose we know the magnetization field in a material (we will see later how to calculate it). We can then use Eq. 23.4 to calculate the magnetization current j_M, and then use the methods of magnetostatics to find the magnetic field produced by j_M. It may appear that this can't possibly give the correct field of the bound current; the magnetization vector takes into account only a few of the magnetic dipoles (namely those near edges in the discrete lattice). Doesn't an atom near the center of a face or a cell contribute just as much to the magnetic field far away? The answer is that we are, in effect, estimating the effects of all the atoms by using a sample consisting of those near edges: because the cells are large, this sample contains many atoms and gives a good estimate.

▶   **Problem 23.2**

   a. Compute the magnetization current j_M in the discrete system depicted in Figure 23.6, wherever it is nonzero. The numbers are the nonzero values $M(e)$ of the magnetization field (in $\mu a/m$), and $dr = 0.1$ m.

   b. Compute the divergence of j_M for the cell shown. What can you say about it in general?   ◀

SECTION 23.5   MAGNETIC DIPOLE MOMENT PER UNIT VOLUME

The magnetization **M** is often defined as the magnetic dipole moment per unit volume. We have defined it in terms of currents around edges because Eq. 23.4 follows much more easily that way. However, we can easily show that the two definitions are equivalent. Consider a system of identical magnetic dipoles uniformly distributed with number density (number per unit volume) n in the vicinity of an edge e, as in Figure 23.7. Assume they are current loops with area A and current I, so each has dipole moment IA; the magnetic moment per unit volume is then nIA. Take the loops to be in a plane perpendicular to e. (The result can be proved without making any of these restrictive assumptions, but it is messier.) Then the current in a loop encircles the edge e only if the center of the loop is in the cylinder shown, which has cross-sectional area A. The average number of such loops is nAL, producing a total current $nALI$ encircling e. The magnetization (current per unit length) is then $M = nAI$, exactly the magnetic dipole moment per unit volume. This result does not mean that the dipole moment per unit volume in any **particular** volume is always exactly equal to the magnetization $M(e)$ at any **particular** edge e,

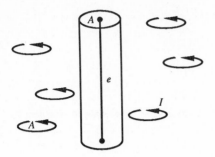

Figure 23.7 An edge e in a region having a uniform density of identical dipoles.

but that on the **average** they are equal, when the density of dipoles is slowly varying in space. The same is true of the two definitions of polarization (dipole moment per unit volume and charge shifted per unit area) which were discussed in Chapter 18.

SECTION 23.6   CONTINUUM PROBLEMS: SURFACE CURRENTS AND THE 10^{-10} METER RELAY

Sometimes the magnetization M in a material can be well approximated by a smooth vector function $\mathbf{M(r)}$. We would like to calculate the magnetization current $\mathbf{j}_M$ of such a configuration. This is much harder than the corresponding discrete calculation, just as it was in the case of dielectrics, because of complications at boundaries between materials. We must use a different smooth function for $\mathbf{M}$ in each material. In the vacuum region, of course, that function must be $\mathbf{M} = 0$. In the interior of each material we can use the continuum limit of Eq. 23.4,

$$\mathbf{j}_M(\mathbf{r}) = \nabla \times \mathbf{M(r)} \tag{23.5}$$

to calculate the function $\mathbf{j}_M(\mathbf{r})$ (which is also different in different materials). However, at a boundary (say between a material and a vacuum) there is a difficulty, which is illustrated in Figure 23.8. Consider the simplest magnetization field $\mathbf{M}$, which points in the tangential $\hat{\mathbf{t}}$ direction and is uniform in the material. Thus, each

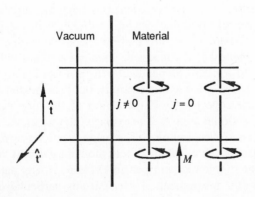

Figure 23.8 The boundary of a magnetic material, with a discrete lattice whose cells are centered on the boundary.

vertical edge within the material in the figure has the same current $M\,dr$ circulating around it. Each face parallel to the paper in the interior of the material has $j(f) = 0$, because the currents at its right and left edges cancel. However, the current density at a face straddling the interface is nonzero (in fact $M\,dr/dr^2$, pointing at the viewer in Figure 23.8). The problem is that this $j_M(f)$ diverges as $dr \to 0$; we cannot describe this current by a continuous function $\mathbf{j}_M(\mathbf{r})$. This problem is exactly analogous to the one faced in electrostatics in the presence of surface charges. We had to characterize the continuum charge density by a continuous function $\sigma(\mathbf{r})$, the surface charge density, in addition to the volume density $\rho(\mathbf{r})$ (see Section 19.2 and Appendix C). We can do something analogous here: the current near an interface of this type is characterized by two vector functions, the volume current density $\mathbf{j}(\mathbf{r})$ and a surface current density (current per unit length) $\mathbf{K}(\mathbf{r})$. (We can add M subscripts if they are magnetization currents.) These determine a discrete current $j(f)$ by

$$j(f)\,da(f) = j_p(\mathbf{r})\,da(f) + K_{p'}(\mathbf{r})\,dl \qquad (23.6)$$

where j_p is the component normal to the face f (Figure 23.9) of the continuous vector function $\mathbf{j}$, evaluated at the face center $\mathbf{r}$, and $da(f)$ is the area of f. In the second term, dl is the length of the intersection of f with the interface (zero if the interface doesn't touch f) and $K_{p'}$ is the component of the vector function $\mathbf{K}$ normal to $d\mathbf{l}$. If the face f is normal to the interface, $\hat{\mathbf{p}}$ and $\hat{\mathbf{p}}'$ will be the same; we will assume this in what follows, and write $\hat{\mathbf{p}}$ for both. We will be mostly concerned with the surface current density $\mathbf{K}$, which can be recovered from $j(f)$ in the $dr \to 0$ limit:

$$K_p(\mathbf{r}) = \lim_{dr \to 0} \frac{j(f)\,da(f)}{dl} \qquad (23.7)$$

(Note that the continuum $\mathbf{j}(\mathbf{r})$ drops out in this limit, because $da/dl \propto dr \to 0$. A more general treatment of continuum currents is found in Appendix C.2.) For the uniform-M case of Figure 23.8, we can calculate $\mathbf{K}(\mathbf{r})$ from Eq. 23.7 using $j(f) = M/dr$ at the interface:

$$K = \lim\left(\frac{M}{dr}\right) dr = M \qquad (23.8)$$

The direction of $\mathbf{K}$ is toward the viewer, perpendicular to $\mathbf{M}$. $\mathbf{K}(\mathbf{r})$ is a two-dimen-

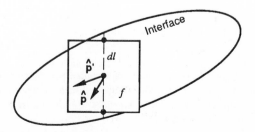

Figure 23.9 A face f intersecting a magnetic-material interface along a line of length dl. The unit vector normal to f is $\hat{\mathbf{p}}$, and $\hat{\mathbf{p}}'$ is a unit vector tangent to the interface and perpendicular to $d\mathbf{l}$.

sional vector field, in that it is defined only for **r** on the interface and has only components tangential to the interface.

It may seem odd that the current density inside our material having uniform M is exactly zero. After all, the mere statement that $M \neq 0$ says there are currents everywhere, circulating around edges. This oddity is a result of our need to consider faces much larger than atomic sizes to suppress fluctuations; the current through a face of roughly atomic dimensions would be nonzero, but fluctuate from face-to-face and from time-to-time in such a way as to average to zero. The magnetic field produced far away (outside the material) is what we can measure, and this is the same whether we use large or small faces. It is determined in this case entirely by the surface current K_M, with no contribution from the interior.

It may also seem odd that we have a surface current flowing for large distances along the surface of a nonconducting material, each of whose electrons is forbidden to leave its molecule. The problem lies in our view of a current as analogous to a marathon race, in which each electron moves along the entire current path. The magnetization surface current is actually more like a relay race, in which each charge moves only a short distance, but the effect is the same.

Let us now calculate $\mathbf{K}_M$ for a general interface between materials 1 and 2 (Figure 23.10), given continuum fields $\mathbf{M}_1$ and $\mathbf{M}_2$, which are not necessarily constant. Its component toward the viewer (along $\hat{\mathbf{p}}$) is given by Eqs. 23.4 and 23.7:

$$K_{Mp} = \lim_{dr \to 0} \left[M(e_1) + M(e_2) + M(e_3) + M(e_4) \right] \tag{23.9}$$

Two of the limits on the right-hand side are easy: e_3 stays in region 2, so $\lim M(e_3) = M_{2t}(\mathbf{r})$. Similarly, e_1 stays in region 1: $\lim M(e_1) = -M_{1t}(\mathbf{r})$ (the component of $\mathbf{M}_1$ along e_1, which is along $-\hat{\mathbf{t}}$). As in the case of dielectrics (Section 19.2), we may as well set $M(e_2)$ equal to some average of M_1 and M_2, say $\frac{1}{2}[M_1(e_2) + M_2(e_2)]$, for edges straddling the surface. The exact way we choose to average them doesn't matter, because as $dr \to 0$ the function M_{1n} is the same at the two normal edges e_2 and e_4, so $M_1(e_2) \to -M_1(e_4)$ and these edges cancel out of Eq. 23.9. We are left with

$$K_{Mp}(\mathbf{r}) = M_{2t}(\mathbf{r}) - M_{1t}(\mathbf{r}) \tag{23.10}$$

Thus the discontinuity in each tangential component of **M** determines the **other** tangential component of $\mathbf{K}_M$. We can summarize this in the form

$$\mathbf{K}_M = \hat{\mathbf{n}} \times (\mathbf{M}_2 - \mathbf{M}_1) \tag{23.11}$$

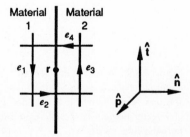

Figure 23.10 An interface between two different magnetic materials, with a lattice whose cells are centered on the interface. The t and p directions are tangent to the interface, and n is normal to it.

which you can verify has all the right components by expanding $\mathbf{M} = M_t\hat{\mathbf{t}} + M_p\hat{\mathbf{p}}$, etc. As a check on the general equation, we can verify that it gives the correct answer in the specific case of a uniform magnetization, Figure 23.8. There $\mathbf{M}_1 = 0$ and $\mathbf{M}_2$ is along $\hat{\mathbf{t}}$, and Eq. 23.11 gives $\mathbf{K}_M = M_2\hat{\mathbf{n}} \times \hat{\mathbf{t}} = M_2\hat{\mathbf{p}}$, in agreement with Eq. 23.8.

We now have the tools necessary to compute the magnetization current in a continuum problem from the magnetization $\mathbf{M}$: Eqs. 23.5 and 23.11.

■ **EXAMPLE 23.1 THE CURRENT IN A UNIFORMLY MAGNETIZED SPHERE**   Compute the magnetization current of a sphere with uniform magnetization $\mathbf{M}$ along the z-axis. The volume current density (Eq. 23.5) vanishes, and the surface current (Eq. 23.11) is, in spherical coordinates,

$$\mathbf{K}_M = \hat{\mathbf{n}} \times (0 - \mathbf{M}) = +M\sin\theta\,\hat{\boldsymbol{\phi}}$$

where $\hat{\mathbf{n}}$ is a radial unit vector.

▶   **Problem 23.3**

A magnetic material occupies a cube $0 < x < a$, $0 < y < a$, $0 < z < a$. The continuum magnetization field is $\mathbf{M}(x, y, z) = Cx\hat{\mathbf{y}}$. Outside the cube is a vacuum.

a. Compute the magnetization current everywhere (within the cube and at all six surfaces).

b. Compute the total magnetization current passing the plane $z = \frac{1}{2}a$, in the $+z$-direction. Do you get the answer you expect physically?   ◀

▶   **Problem 23.4**

Show that the magnetization current of the uniformly magnetized sphere in Example 23.1 is the same as the free current of a sphere with a uniform surface charge density σ, rotating about the z-axis. Compute the necessary angular frequency ω in terms of M and σ.   ◀

Usually the magnetization $\mathbf{M}$ is created by an externally imposed magnetic field $\mathbf{B}$; we will discuss how to calculate $\mathbf{M}$ from $\mathbf{B}$ in Section 23.8. However, there are some materials (so-called "hard" ferromagnets) in which $\mathbf{M}$ is essentially "frozen in," and is more or less independent of fields imposed externally. Such materials may often be found stuck to refrigerator doors. Isolated in space, such a magnetized material sets up its own magnetic field $\mathbf{B}$, which we can calculate if we know $\mathbf{M}$. This is because $\mathbf{M}$ determines the magnetization current $\mathbf{j}_M$ (which is the **only** current in this case), and we know how to get $\mathbf{B}$ from $\mathbf{j}$. This can be done exactly for a long cylindrical magnet such as that in Figure 23.11a, with a uniform axial magnetization M (a "bar magnet"). The surface current is $K = M$ (Eq. 23.8), and is just like the current in a cylindrical solenoid. We know how to compute the magnetostatic field of this current from Ampère's law—it is $B = \mu_0 K = \mu_0 M$ inside the solenoid (Eq. 20.10) and zero outside. In a real bar magnet of finite length L, as in a solenoid of finite length, the field isn't quite zero outside; the field lines look roughly like Figure 23.11b. If the radius $R \ll L$, the field is nearly constant $(= \mu_0 M)$ inside away from the ends and much smaller but nonzero outside. Very far away, it will be a magnetic dipole field. In the opposite limit $R \gg L$ the surface

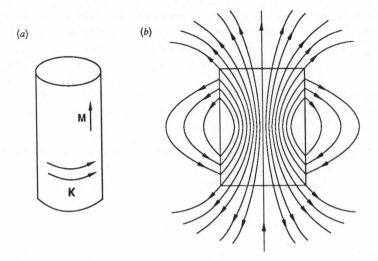

Figure 23.11 (*a*) A bar magnet. (*b*) Magnetic field lines near a bar magnet (or a solenoid).

current looks like that of a wire loop, so the field looks like the field of a loop (as shown in Figure 20.38*b*).

▶   **Problem 23.5**

A long circular cylinder of radius R has an azimuthal magnetization $\mathbf{M} = kr^2\hat{\phi}$, where k is a constant, r and ϕ are the usual cylindrical coordinates, and $\hat{\phi}$ is the azimuthal unit vector. Find the magnetic field due to $\mathbf{M}$, for points inside and outside the cylinder.

◀

▶   **Problem 23.6**

Find the magnetic field at the center of a nickel-shaped ferromagnetic cylinder of height t, radius $R \gg t$, and uniform axial magnetization (normal to the plane of the nickel) M.

◀

SECTION 23.7   THE H FIELD

We now have complete prescriptions for calculating the bound current in terms of the magnetization field $\mathbf{M}$, in both discrete and continuum problems. This is of no use whatsoever, of course, unless we can calculate $\mathbf{M}$, which usually depends on the magnetic field $\mathbf{B}$. Recall the analogous situation in electrostatics; the polarization $\mathbf{P}$ determined the bound charge, but was of little use unless it could be calculated from $\mathbf{E}$, for example in a linear material by $\mathbf{P} = X\epsilon_0\mathbf{E}$. Then we could eliminate the bound charge from Gauss's law for $\mathbf{E}$, leaving $\nabla \cdot (\mathbf{P}(\mathbf{E}) + \epsilon_0\mathbf{E}) = \rho_{\text{free}}$. Rather than solve directly for $\mathbf{E}$, it was convenient to solve for the auxiliary field $\mathbf{D} \equiv \mathbf{P} + \epsilon_0\mathbf{E}$, whose divergence is simply related to the (presumably known) free charge. Then $\mathbf{D}$, $\mathbf{E}$, and $\mathbf{P}$ are all proportional to each other.

Our salvation in magnetostatics is accomplished in much the same way. In many materials $\mathbf{M}$ is linearly related to $\mathbf{B}$ by a similar equation, and we can manipulate Ampère's law for **B** into an equation for an auxiliary field that involves only the **free** current (also presumed known). To see this, subtract Eq. 23.4 for $\mathbf{j}_M$ from Ampère's law (Eq. 20.4) for $\mathbf{j}$ to obtain an equation for the free current $\mathbf{j}_f$:

$$\nabla \times (\mathbf{B}/\mu_0 - \mathbf{M}) = \mathbf{j} - \mathbf{j}_M = \mathbf{j}_f \qquad (23.12)$$

The free current is that not due to the bound atomic currents and includes, for example, Ohm's-law currents in wires. Since $\mathbf{j}_f$ is experimentally measurable and the total $\mathbf{j}$ is not, this form of Ampère's law is often more useful than the original form relating **B** to $\mathbf{j}$. In symmetrical situations, such as the example below, it can be directly solved: it gives an auxiliary field **H**

$$\mathbf{H} = \frac{\mathbf{B}}{\mu_0} - \mathbf{M} \qquad (23.13)$$

In terms of **H**, Eq. 23.12 is

$$\nabla \times \mathbf{H} = \mathbf{j}_{\text{free}} \qquad \text{(Ampère's law for }\mathbf{H}\text{)} \qquad (23.14)$$

> The reader should be warned that the terminology for **B** and **H** in the literature is not well-standardized. Older texts tend to refer to **H** as the "magnetic field intensity" and **B** as the "magnetic induction," while more recent ones call the more fundamental field **B** the "magnetic field" and **H** just "**H**". We adopt the latter convention here.

The integral form of Ampère's law for **H** is

$$\int \mathbf{H} \cdot d\mathbf{l} = I_{\text{free}} \qquad (23.15)$$

where I_{free} is the total free current passing through the curve around which we are integrating **H**. As before, this allows us to solve certain highly symmetrical problems very easily.

■ **EXAMPLE 23.2 A FILLED SOLENOID**   Consider a long cylindrical solenoid filled with a magnetic material. We computed the field for the vacuum-filled case in Section 20.2; here we can again assume the magnetic field **B** and therefore **H** (which is proportional to it) is axial and vanishes outside the solenoid. Using an amperian loop running axially just inside and just outside the solenoid (see Figure 20.6), Eq. 23.15 becomes

$$HL = I_{\text{free}} = LK_f \qquad (23.16)$$

where K_f is the free surface current per unit length (the number of wires per unit length times the current in each). Thus we have determined H:

$$H = K_f \qquad (23.17)$$

Note that we didn't find the magnetic field **B** in the above example. In fact, we cannot do this until we write down a second relation among **B**, **M**, and **H**, which we will do in the next section. I have showed the Ampère's-law solution for **H** here only to emphasize that it is independent of the relation between **M** and **B**—it is **not** restricted to linear media. This is fortunate, because many important media (e.g., ferromagnets) are nonlinear.

■ **EXAMPLE 23.3**   We can re-calculate the magnetic field in the bar magnet of Figure 23.11 in a very simple way using the **H**-field concept. Use an amperian loop like the one in Example 23.2 in Ampère's law for **H**. But this time there is **no** free current, so the path integral of **H** is zero. Assuming as usual that the fields vanish far from the cylinder, so **H** = 0 outside, we conclude that **H** vanishes inside as well! We know **M**, so Eq. 23.13 defining **H** can be solved for the magnetic field: $B = \mu_0 M$, in agreement with our earlier answer.

▶   **Problem 23.7**

Consider an infinitely long cylindrical piece of hard ferromagnetic material, which has an axial magnetization $M = kr^2$, for some constant k.

    a. Calculate the magnetic field without using the **H**-field concept: First find the magnetization current everywhere (surface as well as volume), then calculate the magnetic field due to this current (you may use Ampère's law for **B**).

    b. Use Ampère's law for **H** to calculate **H** (assume **H** = 0 outside the cylinder), and then calculate **B**. Do you get the same answer you got for (**a**)? Which way is easier? ◀

▶   **Problem 23.8**

Clearly there is a parallel between **M** (the magnetic dipole moment per unit volume) and **P** (the electric dipole moment per unit volume). In a dielectric, $-\nabla \cdot \mathbf{P}$ gives the bound charge density; we can define a corresponding "magnetic pole density" by

$$\rho_M(\mathbf{r}) \equiv -\nabla \cdot \mathbf{M}(\mathbf{r}) \tag{23.18}$$

Consider a system in infinite space (not in a box), with no free current (the only currents are the bound currents in hard magnetic materials). You are given the magnetization field **M**.

    a. Of **H** and **B**, which is the correct analog of **E** in this context? To answer this, write equations for the divergence and curl of each. Which are like the equations for **E** in electrostatics?

    b. Use this analogy to sketch the **H**-field lines of a long, thin bar magnet and a short, fat (coin-shaped) bar magnet. To what electrostatic systems are they analogous? ◀

SECTION 23.8   LINEAR AND NONLINEAR MEDIA

Our eventual objective is to be able to calculate the fields produced externally by magnetic materials. Our motivation for introducing **M** was the promise that it (and therefore the bound current $\mathbf{j}_M$ and the fields produced by it) could be expressed simply in terms of **B**. There is a large class of materials in which **M** is just proportional to **B**; these are called linear magnetic media. In such media of course $\mathbf{H} = \mathbf{B}/\mu_0 - \mathbf{M}$ is also proportional to **B** and to **M**. In electrostatics, in which **P** was similarly proportional to **E**, we defined the dielectric susceptibility as the proportionality constant relating **P** to **E**. We could analogously define a magnetic susceptibility as the ratio of **M** to **B**. However, there is no way to experimentally determine **B** inside a material subjected to a magnetic field, as we could determine **E** inside a dielectric (in a parallel-plate capacitor, for example, $E = V/d$). But in symmetric cases (such as in a long solenoid) we can determine **H**, from Ampère's law (Eq. 23.15). Therefore, the measurable quantity is the ratio of **M** to **H**, denoted by χ_M and called the **magnetic susceptibility**

$$\mathbf{M} = \chi_M \mathbf{H} \tag{23.19}$$

Recalling that the molecular magnetic moments (and hence the magnetic moment per unit volume **M**) are parallel to the imposed field in a paramagnetic material, we see that the magnetic susceptibility χ_M is positive in such a material; we may regard $\chi_M > 0$ as the definition of a paramagnet. Accordingly, a diamagnetic material is one in which $\chi_M < 0$. Table 23.1 gives some typical values of χ_M; it is a dimensionless number.

Table 23.1 *Magnetic Susceptibilities of Selected Materials at Atmospheric Pressure and 20°C (Calculated from Data in Handbook of Chemistry and Physics, 63rd edition, CRC Press, Inc.)*

Material	χ_M
Bismuth	-16.5×10^{-5}
Gold	-3.4×10^{-5}
Copper	-1.0×10^{-5}
Water	-0.9×10^{-5}
CO_2	-1.2×10^{-8}
H_2	-0.2×10^{-8}
O_2	$+200 \times 10^{-8}$
Sodium (Na)	$+0.8 \times 10^{-5}$
NaCl	-1.4×10^{-5}
Aluminum	$+2.1 \times 10^{-5}$
Gadolinium	$+0.48$

Note that the gases (CO_2, H_2) whose molecules have only paired electrons are weakly diamagnetic, whereas O_2, which has unpaired electrons, is paramagnetic and has a much larger susceptibility. It is clear from the table that in most materials χ_M is very small, so the bound currents can be ignored for most purposes; these materials can be treated like vacuum in magnetostatics problems.

A material in which Eq. 23.19 holds is called a **linear**, **isotropic** material. The relation between **B** and **H** (Eq. 23.13) becomes

$$\mathbf{B} = \mu_0(\mathbf{H} + \mathbf{M}) = \mu_0(1 + \chi_M)\mathbf{H} \tag{23.20}$$

so if we define

$$\mu \equiv \mu_0(1 + \chi_M) \tag{23.21}$$

(called the **magnetic permeability**) we have

$$\mathbf{B} = \mu \mathbf{H} \tag{23.22}$$

Since $\chi_M = 0$ in a vacuum, $\mu = \mu_0$; μ_0 is the permeability of a vacuum. We can describe a magnetic material by giving either its susceptibility or its permeability; each determines the other through Eq. 23.21.

We can now finish the calculation of the fields in a long solenoid we began in Section 23.7. We got $H = K_f$, the free surface current. If the material inside is linear, with permeability μ, then

$$B = \mu H = \mu K_f$$

If the material is paramagnetic ($\mu > \mu_0$), this is larger than the field in a vacuum-filled solenoid. The magnetization is

$$M = \chi_M H = \chi_M K_f$$

From M we can get the bound surface current (Eq. 23.8); it is $K_b = M = \chi_M K_f$. If the material is diamagnetic ($\chi_M < 0$), the surface current cancels part of the free current and produces a weaker field than that in a vacuum-filled solenoid. If it is paramagnetic, the surface current **reinforces** the free current, producing a stronger B. The same is true in a ferromagnet, although there the relations are not linear; the detailed relation between M and H will be discussed in Section 23.12. As a crude approximation, however, we may think of a ferromagnetic material as a linear material having $M \cong \chi_M H$; the distinguishing feature of a ferromagnet is that $M \gg H$, so $\chi_M \gg 1$ (it can be of the order of several hundred).

We can now take the formula $u_B = B^2/2\mu_0$ that we obtained in Chapter 13 for the magnetic energy in a vacuum, and generalize it for magnetic materials. The easiest way to do this is by the work done when we turn on the current in a solenoid (Section 22.6). The rate of work done by the solenoid's EMF is the product of the total current and the EMF, so the rate of work done **on** the solenoid by the power source is the negative of this,

$$\frac{dW}{dt} = -K_f L \left(\frac{-d\Phi}{dt} \right) = K_f L A \frac{dB}{dt} = VH \frac{dB}{dt}$$

where $V = AL$ is the volume of the solenoid. The simplest way to state this result is that the work done per unit volume to change B by dB is

$$\frac{\text{work}}{\text{volume}} = H \, dB \tag{23.23}$$

Integrating this while the current is being turned on gives the total work per unit volume needed to build up a field, which we may identify with the energy density of the field:

$$u_B = \int_0^B \left(\frac{B'}{\mu} \right) dB' = \frac{B^2}{2\mu} = BH/2 \tag{23.24}$$

It must be emphasized that Eq. 23.24 is only valid in linear media, while Eq. 23.23 is much more general and holds even in ferromagnetic media.

▶ **Problem 23.9**

Compute **H**, **M**, and **B** everywhere in a toroidal solenoid (such as that in Figure 20.12 above) filled with a magnetic material of permeability μ. The wire winds around N times, and has current I. (Assume the fields are azimuthal, as in Problem 20.4.) ◀

▶ **Problem 23.10**

By subtracting the energy $B^2/2\mu_0$ of the B field from Eq. 23.24, derive a concise formula for the extra energy density due to the magnetization itself. ◀

SECTION 23.9 MAGNETOSTATIC PROBLEMS IN LINEAR MATERIALS

We now have enough information to solve any magnetostatic problem involving linear magnetic materials, of the type shown in Figure 23.12. It can involve arbitrary free currents and an arbitrary configuration of materials of different permeabilities μ. (Actually, it is not necessary here to restrict ourselves to linear materials; we can solve the problem as long as **B** is some known function of **H**.) To set up the problem discretely, choose which faces are inside the superconducting box (i.e., have no supercurrent through them). Allow $B(e) \neq 0$ at their edges. Assign a $\mu(e)$ to each edge so $B(e) = \mu(e)H(e)$; $\mu(e)$ is μ_1 if e is in region 1, μ_2 if region 2, and could be some average on the boundary. Assuming we know $j_{\text{free}}(f)$ for each face f, we just have to solve

$$\text{curl } H = j_{\text{free}} \tag{23.25}$$

at each nonsuperconducting face, and

$$\text{div}(\mu H) = 0 \tag{23.26}$$

at each of the vertices of these faces, simultaneously for H. No explicit mention need be made of the magnetization.

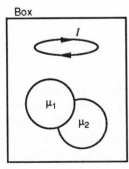

Box

Figure 23.12 An arbitrary system containing magnetic materials and free currents, enclosed in a superconducting box.

(a)   (b)

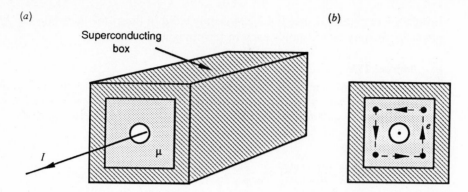

Figure 23.13   A long superconducting duct filled with a magnetic material, with a wire down its center. (a) Oblique view. (b) Cross section, showing discrete lattice edges.

■ **EXAMPLE 23.4**   Consider a long bar of magnetic material of magnetic permeability μ with a long straight wire through a hole in it, all surrounded by a superconducting box (Figure 23.13). We will discretize this quasi-2D system in such a way that in the cross section in Figure 23.13b there is only one nonsuperconducting face and there are four nonsuperconducting edges (dotted lines). Each edge is in the magnetic material, and hence has $\mu(e) = \mu$. Due to the 90° rotation symmetry, the edges are all equivalent, and Ampère's law (Eq. 23.25) becomes

$$dr^{-1} \sum_e H(e) = 4 dr^{-1} H(e) = \frac{I}{dr^2}$$

$$H(e) = \frac{I}{4\,dr} \qquad\qquad (23.27)$$

[The divergence equation (Eq. 23.26) is trivially satisfied.] From this we can compute the magnetic field

$$B(e) = \mu H(e) = \frac{\mu I}{4\,dr} \qquad\qquad (23.28)$$

If $\mu \gg \mu_0$ (i.e., if the material is ferromagnetic), the field in this example is much larger than the B field $\mu_0 I/4\,dr$ that would exist in a vacuum. This is an example of a general property of ferromagnetic materials: they allow us to produce (for a given free current) much larger magnetic fields than could be produced without them. The extra magnetic field is produced by the bound current, which flows on the inner surface of the magnetic material parallel to the free current I.

■ **EXAMPLE 23.5**   In the previous example, we could not easily take advantage of the larger field, or even measure it: it exists only in the magnetic material. We can get at the field by cutting a channel out of the bar, as shown in cross section in Figure 23.14, leaving a sort of pseudo-2D horseshoe magnet. Then three of the edges still have $\mu(e) = \mu$. The fourth is mostly in vacuum, and we will assign $\mu(e) = \mu_0$ to it. Each vertex has exactly one

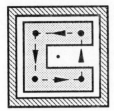

Figure 23.14   A discretization of a horseshoe magnet.

nonsuperconducting edge going out and one going in, so the divergence equation div$B = 0$ implies that the four values of $B(e)$ are the same. The values of $H(e)$ are then different, being $B(e)/\mu_0$ on the right and $B(e)/\mu$ on the other three edges.

Ampère's law gives

$$dr^{-1}\left[\frac{3B(e)}{\mu} + \frac{B(e)}{\mu_0}\right] = \frac{I}{dr^2}$$

$$B(e) = \frac{I}{[3\mu^{-1} + \mu_0^{-1}]\,dr}$$

This is smaller than Eq. 23.28, but still much larger (by a factor of 4, if $\mu \gg \mu_0$) than the value $\mu_0 I/4\,dr$ we could achieve without the magnetic material.

You might think it would be fairer to assign an average like $\frac{1}{2}(\mu_0 + \mu) \approx \frac{1}{2}\mu$ to the edge that is partly in vacuum. However, from the result you can see that it is really the **reciprocals** of the μ's that get averaged. Thus, a better average would be $[\frac{1}{2}(\mu_0^{-1} + \mu^{-1})]^{-1} \approx 2\mu_0$. Our guess was very close to this for $\mu \gg \mu_0$.

▶ **Problem 23.11**

Figure 23.15 shows a system designed to illustrate the tendency of magnetic materials to concentrate or channel a magnetic field. It shows a square slab of magnetic material with a square hole in it (i.e., topologically a torus), with a current-carrying wire wrapped around the top arm. The whole system is enclosed in a superconducting box, so there are

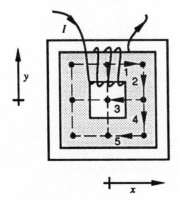

Figure 23.15   A magnetic circuit.

four nonsuperconducting faces and twelve edges. The symmetries $H_y(x, y) = -H_y(-x, y)$ and $H_x(x, y) = +H_x(-x, y)$ imply that the only inequivalent nonzero fields are on the edges labeled 1, 2, 3, 4, and 5. There is a current $2I$ through each of the upper two faces, directed away from the viewer; it comes back out through the faces above these, which are not shown because they are partly in the superconductor, so we don't know their total current. Assign $\mu(e) = \mu$ to all edges except e_3, and $\mu(e_3) = \mu_0$.

a. Write Ampère's law for the two inequivalent nonsuperconducting faces, and div $B = 0$ for the inequivalent vertices.

b. Solve for B_1, B_2, B_3, B_4, and B_5 in terms of μ, μ_0, I, and dr.

c. Compare B_5 and B_3 (the fields in and out of the magnetic material) for $\mu/\mu_0 = 1$ and 10^3. Show that in the latter case most of the magnetic flux follows the magnetic material, rather than taking a shortcut through the vacuum.   ◄

SECTION 23.10   CONTINUUM BOUNDARY VALUE PROBLEMS IN MAGNETOSTATICS

In a continuum treatment of a system such as Figure 23.12, containing linear magnetic materials, we are given the free current density $\mathbf{j}_{\text{free}}$ and the permeabilities μ_1 and μ_2. We want to solve for smooth functions $\mathbf{H}_1(\mathbf{r})$ and $\mathbf{H}_2(\mathbf{r})$, which give the fields in the two regions (by $\mathbf{B}_1(\mathbf{r}) = \mu_1\mathbf{H}_1(\mathbf{r})$, and so on). They are related to the discrete fields in the way described in Chapter 6, just as in the continuum theory of electrostatics. In each region the continuum limits of Eqs. 23.25 and 23.26 are $\nabla \times \mathbf{H} = \mathbf{j}_{\text{free}}$, $\nabla \cdot \mathbf{H} = 0$. (In regions with no free current, $\mathbf{H}$ can be expressed as the gradient of a magnetic scalar potential and we get Laplace's equation, which can be solved by methods similar to those we used in electrostatics.) At interfaces such as that between regions 1 and 2 in Figure 23.16a, the functions $\mathbf{H}_1$ and $\mathbf{H}_2$ must satisfy boundary conditions that we will derive from Eqs. 23.25 and 23.26. Applying

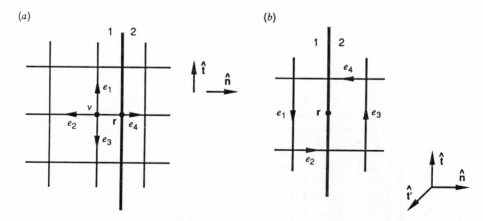

Figure 23.16 A boundary between magnetic materials 1 and 2, with a discrete lattice whose cells are centered on the interface, showing the edges needed to derive a boundary condition on (a) the normal component of *B* and (b) the tangential component of *H*.

$[\text{div } B](v) = 0$ at the vertex v in Figure 23.16a gives

$$dr^{-1}\big[B(e_1) + B(e_2) + B(e_3) + B(e_4)\big] = 0 \tag{23.29}$$

The discrete fields at e_1, e_2, and e_3 are obtained by evaluating B_1 at their centers $\mathbf{r}_1$, $\mathbf{r}_2$, and $\mathbf{r}_3$, which are inside region 1. At a boundary edge like e_4, we let $B(e_4) = \frac{1}{2}[B_{1n}(r_4) + B_{2n}(r_4)]$; we used a similar average in the electrostatic case. In the limit $dr \to 0$, $B(e_1) \to B_{1t}(\mathbf{r})$, $B(e_3) \to -B_{1t}(\mathbf{r})$ (so these two cancel) and $B(e_2) \to -B_{1n}(\mathbf{r})$. The divergence equation ends up as

$$B_{1n}(\mathbf{r}) = B_{2n}(\mathbf{r}) \tag{23.30}$$

To get a boundary condition on the tangential components, evaluate Eq. 23.25 at a face straddling the interface as in Figure 23.17:

$$dr^{-1}\big[H(e_1) + H(e_2) + H(e_3) + H(e_4)\big] = j_{\text{free}}(f) \tag{23.31}$$

Allowing for a surface free current $K_f(\mathbf{r})$ (for example a solenoid winding), the right-hand side can be expressed in terms of the continuum current by Eq. 23.6 as $\mathbf{j}_{\text{free}}(\mathbf{r}) + \mathbf{K}_{\text{free}}(\mathbf{r})/dr$. Multiplying by dr and letting $dr \to 0$, $H(e_2) = \frac{1}{2}[H_{1n}(\mathbf{r}_2) + H_{2n}(\mathbf{r}_2)] \to \frac{1}{2}[H_{1n}(\mathbf{r}) + H_{2n}(\mathbf{r})]$, which cancels with $H(e_4)$, leaving $H(e_3) = H_{2t}(\mathbf{r}_3) \to H_{2t}(\mathbf{r})$ and $H(e_1) = -H_{1t}(\mathbf{r}_1) \to -H_{1t}(\mathbf{r})$:

$$H_{2t}(\mathbf{r}) - H_{1t}(\mathbf{r}) = \lim\big[j_{\text{free},\,t'}(\mathbf{r})\, dr + K_{\text{free},\,t'}(\mathbf{r})\big]$$

$$= K_{\text{free},\,t'}(\mathbf{r}) \tag{23.32}$$

That is, the discontinuity in H_t across the boundary is $K_{\text{free},\,t'}$. We could also have reversed the roles of $\hat{\mathbf{t}}$ and $\hat{\mathbf{t}}'$, giving $H_{2t'} - H_{1t'} = -K_{\text{free},\,t}$. These can be combined into a vector equation

$$\mathbf{H}_{2(\text{tang})}(\mathbf{r}) - \mathbf{H}_{1(\text{tang})}(\mathbf{r}) = \mathbf{K}_{\text{free}}(\mathbf{r}) \times \hat{\mathbf{n}} \tag{23.33}$$

where $\mathbf{H}_{1(\text{tang})}$ is the 2D projection of $\mathbf{H}_1$ onto the plane of the interface (i.e., it has no normal component). The boundary conditions (Eqs. 23.30 and 23.33) allow one to solve (in principle, at least) any continuum boundary value problem involving static magnetic fields in materials. As in electrostatics, there are convenient mnemonics for remembering these boundary conditions; Eq. 23.30 is $\int \mathbf{B} \cdot d\mathbf{a} = 0$ for a very thin Gaussian surface sandwiching the boundary, as in Figure 23.17a, if we consider only the flat surface of the sandwich and ignore the sides. In addition, Eq. 23.32 is the integral form of Ampère's law for a rectangular loop such as that in Figure 23.17b, if we consider only the sides parallel to the interface: $\int \mathbf{H} \cdot d\mathbf{l}$ is $H_{2t} L - H_{1t} L$, and $I = K_{\text{free},\,t'} L$.

In applying these boundary conditions to a general problem, we would express the field $\mathbf{H}$ in each region as a linear combination of a complete set of solutions of $\nabla \times \mathbf{H} = 0$, $\nabla \cdot \mathbf{H} = 0$; it is most convenient to do this in terms of a magnetic potential. We will content ourselves here with simpler cases in which the fields can be used directly.

(a)   (b)

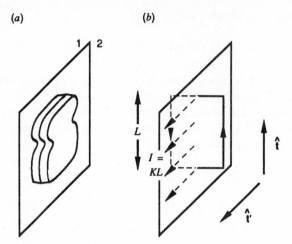

Figure 23.17   (a) A thin Gaussian surface whose large flat sides are parallel to an interface and whose nonflat sides have negligible area. (b) An amperian loop for obtaining the boundary condition on the tangential H.

■ **EXAMPLE 23.6 A LONG SOLENOID**   Consider the long cylindrical solenoid filled with magnetic material of permeability μ for which we have already found the fields by Ampère's law (Figure 23.11). We will assume $H = 0$ outside the solenoid (call this region 1). Let $\hat{t}$ be in the axial direction, so the surface free current density K_f is in the $\hat{t}'$ direction (azimuthal). Then, H inside the solenoid (region 2) is given by Eq. 23.32 as

$$H_{2t} = K_f$$

which agrees with our earlier result (Eq. 23.17).

■ **EXAMPLE 23.7 AN ELECTROMAGNET**   A more interesting but approximate simple application is to an electromagnet with a narrow gap d between its poles, as in Figure 23.18. The electromagnet is made by cutting a slice of thickness d out of a toroidal piece of magnetic material of permeability μ and winding N turns of wire around it. We assume the ring radius $R \gg a \gg d$, where a is the cross-sectional diameter of the material. Then the fields are nearly constant within the material (as in an infinite straight solenoid) and also within the gap. We will assume they are purely azimuthal, as shown in the enlargement of the gap, Figure 23.18 b. Then the continuity of the normal component of **B** (Eq. 23.30) implies $B_{material} = B_{gap}$. We will simply call them B. Then H is much smaller in the material:

$$H_{gap} = B/\mu_0$$

$$H_{material} = B/\mu \tag{23.34}$$

In this approximation, we can calculate the value of B from Ampère's law

$$\int \mathbf{H} \cdot d\mathbf{l} = H_m(2\pi R - d) + H_g d = \frac{B(2\pi R - d)}{\mu} + \frac{Bd}{\mu_0} = NI$$

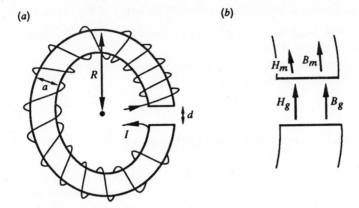

Figure 23.18 (*a*) A narrow-gap electromagnet. (*b*) An enlargement of the region near the poles, showing **H** and **B** inside the magnet and between its poles.

or

$$B = \frac{NI}{d\mu_0^{-1} + (2\pi R - d)\mu^{-1}}$$

If the material is ferromagnetic, with a very high effective permeability ($\mu/\mu_0 \gg 2\pi R/d$), then

$$B \cong \frac{NI\mu_0}{d}$$

which is much larger than the field $NI\mu_0/2\pi R$ we would have without the magnetic material. This is why magnets designed to produce high fields usually contain ferromagnetic materials and have the geometry shown in Figure 23.18.

▶ **Problem 23.12**

A large piece of magnetic material has a uniform magnetic field $\mathbf{B}_0$ (and therefore uniform **M** and **H**) except near a small cavity. Use the boundary conditions (Eqs. 23.30

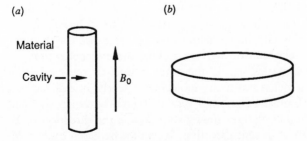

Figure 23.19 An infinite piece of magnetic material with uniform magnetic field, with cylindrical cavities having (*a*) length $\gg$ radius, and (*b*) length $\ll$ radius.

and 23.33) to determine the magnetic field near the center of the cavity, if the cavity is

a. needle-shaped, along $\mathbf{B}_0$ (Figure 23.19*a*). Assume the field in the cavity is uniform far from the ends of the needle.

b. coin-shaped, normal to $\mathbf{B}_0$ (Figure 23.19*b*). Assume the field in the cavity is uniform far from the edge of the coin.   ◄

SECTION 23.11 ELECTRODYNAMICS IN MATERIAL MEDIA

The evolution of the electric and magnetic fields in material media is governed by Maxwell's equations, just as it is anywhere else. However, Maxwell's equations involve the **total** current density **j**, including the bound current, which is not easily controlled or even measured in a magnetic medium. In a practical problem (such as the propagation of an electromagnetic wave through a material, as in the Chapter 24) we would much prefer a set of equations that refers only to the **free** current (and the free charge). Two of Maxwell's equations do not involve sources, and need not be changed:

$$\nabla \cdot \mathbf{B} = 0 \tag{23.35}$$

$$\frac{d\mathbf{B}}{dt} = -\nabla \times \mathbf{E} \tag{23.36}$$

We have actually written down material versions of the other two:

$$\nabla \cdot \mathbf{D} = \rho_{\text{free}} \qquad \text{(Gauss' law for } \mathbf{D}) \tag{23.37}$$

$$\nabla \times \mathbf{H} = \mathbf{j}_{\text{free}} \qquad \text{(Ampère's law for } \mathbf{H}) \tag{23.38}$$

However, these were derived only in **statics**. Can we generalize them to dynamics? This requires defining **H**, and therefore **M**, for nonstatic magnetic fields. This is tricky; there does not seem to be a completely unambiguous way to do it. The problem is that **M** was defined assuming a time-independent current distribution. This is true whether **M** is defined as the current around an edge or as the dipole moment per unit volume. So we must pick some dynamic definition of **M**, and this is necessarily somewhat arbitrary. The difficulty is that "the current around the edge *e*" is ambiguous; we could try to compute it by determining the current during a given time interval through one of the four faces touching *e* (see Figure 23.20) due to charges in molecules touching *e*. However, we would get different answers for the four faces. One solution would be to define $M(e)\,dl(e)$ to be the **average** of these answers. Whatever we choose for **M**, we define the magnetization current by $\mathbf{j}_M = \text{curl}\,\mathbf{M}$. In magnetostatics, $\mathbf{j}_M$ is the total bound current, all of which we imagine to be circulating inside stationary molecules. Clearly, in a dielectric polarized by an electric field there must be another type of bound current: when the electric field is turned on, the bound charges obviously move in the direction of **E**, so there is a non-circulatory current in that direction. It cannot be given by curl **M**—if **E** is uniform, for example, there is no net current circulating around any edge, and therefore no **M**. This extra bound current is called the **polarization current** $\mathbf{j}_P$. The

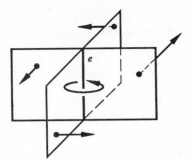

Figure 23.20 The four faces touching an edge *e*.

most consistent way to define it is simply as whatever bound current isn't accounted for by the magnetization:

$$j_P(f, t) = j_b(f, t) - \text{curl } M(f, t) \qquad (23.39)$$

The total bound current $j_b(f, t)$ is defined as the bound charge per unit area per unit time crossing the face f between $t - \frac{1}{2} dt$ and $t + \frac{1}{2} dt$.

Gauss' law (Eq. 23.37) in a time-dependent situation involves

$$D(f, t) = \epsilon_0 E(f, t) + P(f, t)$$

so we need to define $P(f, t)$. In statics, $P(f)$ was the total charge that had shifted across the face f; this can be obtained by adding successive bound currents:

$$P(f) = \sum_t j_b(f, t) \, dt \qquad (23.40)$$

When $M \neq 0$, a charge may be circulating around an edge of f and crossing f repeatedly, so the P so defined would be continually increasing. To avoid this, we should omit the magnetization current from Eq. 23.40, that is, use j_P instead of j_b. We can get P at a particular time t by adding only the charges $j_P \, dt$ that cross **before** that time:

$$P(f, t) = \sum_{t' < t} j_P(f, t') \, dt \qquad (23.41)$$

(here t is an integer multiple of dt, t' a half-integer multiple). (We assume that the fields that polarize and magnetize the material were turned on at some time in the past, before which P, j_b, and M were all zero.) We can show from the continuity equation for bound charge (div $j_b = -d\rho_b/dt$) and Eq. 23.41 that the usual relation holds between ρ_b and P:

$$\rho_b(c, t) = -\text{div } P(c, t) \qquad (23.42)$$

It is then easy to see that the derivation we gave in electrostatics for Eq. 23.37 (Gauss' law for **D**) is actually perfectly general.

The only Maxwell equation that has as yet no nonstatic version for materials is Ampère's law, Eq. 23.38. Starting from the general Maxwell equation

$$\frac{d\mathbf{E}}{dt} = c^2 \nabla \times \mathbf{B} - \frac{\mathbf{j}}{\epsilon_0}$$

let us replace $\mathbf{j}$ by $\mathbf{j}_{\text{free}} + \mathbf{j}_P + \nabla \times \mathbf{M}$ (using Eq. 23.39), multiply by ϵ_0, and rearrange:

$$\epsilon_0 \frac{d\mathbf{E}}{dt} + \mathbf{j}_P = c^2 \epsilon_0 \nabla \times \mathbf{B} - \nabla \times \mathbf{M} - \mathbf{j}_{\text{free}}$$

However, $\mathbf{j}_P = d\mathbf{P}/dt$ (from Eq. 23.41) so

$$\frac{d\mathbf{D}}{dt} = \nabla \times \mathbf{H} - \mathbf{j}_{\text{free}} \qquad (23.43)$$

by the definitions of $\mathbf{D}$ and $\mathbf{H}$. This is now a completely general version of Ampère's law for $\mathbf{H}$.

We have now derived a complete set of Maxwell equations for material media, in which only free charges and currents appear: Eqs. 23.35 to 23.37 and Eq. 23.43.

These equations are exactly true in any medium, in spite of our arbitrary choice of $\mathbf{M}$. Changes in the definition of $\mathbf{M}$ can cause magnetization current to disappear and be reincarnated as polarization current, changing what we call $\mathbf{D}$ and $\mathbf{H}$, but always in such a way as to affect the two sides of Eq. 23.43 identically. Another source of arbitrariness that we have not so far considered lies in the distinction between free and bound charge. In a semiconductor, for example, an electron that is bound at low temperatures may be quite free at room temperature, or when irradiated by light. A change in the definition of bound charge causes a reshuffling between $\nabla \cdot \mathbf{D}$ and ρ_{free} in Gauss' law (Eq. 23.37) and between $d\mathbf{D}/dt$ and $\mathbf{j}_{\text{free}}$ in Eq. 23.43, but leaves both equations exactly true.

In spite of their exactness, the free-charge Maxwell equations must be used with some caution in nonstatic situations because they require knowing how $\mathbf{H}$ is related to $\mathbf{B}$ and $\mathbf{D}$ to $\mathbf{E}$; these relations are generally different for time-varying fields than for static ones. This gives rise, for example, to dispersion of electromagnetic waves in media.

▶   **Problem 23.13**

Repeat the arguments of Section 13.3 leading to the discrete energy-flux field (defined at linkages between faces f and edges e) $S = E(f)B(e)/\mu_0$, using the material Maxwell equations and the electric energy density $u_E = \frac{1}{2}ED$ (Section 19.4). Show that the correct generalization is $S = E(f)H(e)$. The continuum limit (the Poynting vector in a linear material) is then

$$\mathbf{S}(\mathbf{r}, t) = \mathbf{E}(\mathbf{r}, t) \times \mathbf{H}(\mathbf{r}, t) \qquad (23.44)$$

◀

SECTION 23.12 FERROMAGNETIC MATERIALS

Phenomenologically, a ferromagnet is a material in which a small H field produces a magnetization thousands or millions of times larger than in a paramagnet (Figure 23.21). It is noteworthy that the maximum magnetization obtainable (the "saturated magnetization") is not greatly different in the two cases. The saturation magnetization is achieved when all of the molecules have their magnetic dipole moments (i.e., the axes of the orbital or spin motions of their unpaired electrons) pointing in the same direction, and its magnitude is the product of the moment of a single molecule and the density of molecules. Thus the remarkable properties of ferromagnets are **not** due to their molecules having much larger dipole moments than the molecules in a paramagnet. Rather, the difference lies in the fact that it requires astronomically large H (or B) fields to completely align the moments in a paramagnet. The B field exerts an aligning torque on each moment, which must compete with torques due to random thermal motions which tend to disalign the moments. The torques due to easily achievable H fields (say 10^4 a/m) are very small compared to these random torques, so they manage to achieve very little alignment.

Why does it take so little H field to align the moments in a ferromagnet? This question was first answered by the French physicist Pierre Weiss around 1900. The answer is that the H field doesn't **have** to align them: they are **already** aligned! A ferromagnetic material (Figure 23.22) is composed of regions called **domains**, within each of which the magnetic moments are essentially all parallel. The domain boundaries can be made visible by sprinkling small magnetic particles on the cut surface of the material (much as you make the field lines of a permanent magnet visible with iron filings) or by chemical etching. The moments align in this way because there is an interaction (called the "exchange" interaction) between nearest neighbors that causes them to tend to be parallel to each other.

The origin of the exchange interaction involves some complicated quantum mechanics, and is not yet completely quantitatively understood. Basically, it arises from the Pauli exclusion principle, one form of which says that two electrons with the same spin direction can't be in the same place at the same time. If two electrons

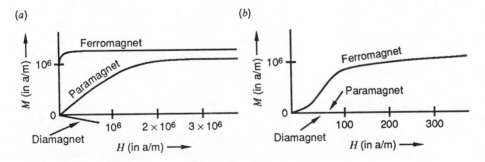

Figure 23.21 Magnetization curves for a ferromagnet, a strongly paramagnetic material, and a diamagnetic material: (a) with the same scale for M and H, (b) with the H scale greatly expanded to show the rise in ferromagnetic magnetization (the "paramagnet" and "diamagnet" curves are indistinguishable from the axis).

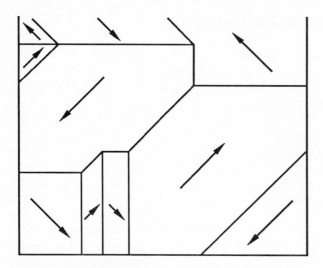

Figure 23.22 A sketch of a cross section of a ferromagnetic material, showing the magnetic domains (the arrow indicates the direction of magnetization in each).

in orbitals on neighboring atoms are antiparallel, they will sometimes be very close together (because the orbitals overlap somewhat), with a resultingly large Coulomb's-law repulsion, giving this antiparallel state a high energy. If the spins are parallel, the exclusion principle will keep them apart, and the Coulomb repulsion will be smaller; this state has lower energy. The two-atom system will drop to the state of lower energy, so the spins will be parallel. This energy difference is often large compared to the energy of random thermal motions (which is of the order of Boltzmann's constant times the temperature), so the latter does not destroy the alignment. It will do so at a high enough temperature, of course; every ferromagnetic material has a temperature, called the Curie temperature, above which it is merely paramagnetic. Iron, for example, has a Curie temperature of 770°C.

From the point of view of the exchange interaction, the lowest-energy state of the whole material is one in which **all** the moments are parallel. However, we have seen that a magnetized material produces a B field, which has energy density $B^2/2\mu_0$. By splitting into domains with different magnetization directions, so that the large-scale average magnetization vanishes, this magnetic-field energy can be decreased at relatively little cost in exchange energy (only the spins near domain boundaries are not parallel to their neighbors, and these are a small fraction of the spins). A quantitative version of this argument can be used to estimate the domain size giving the lowest total energy; it varies widely among ferromagnetic materials, but 1 μm is a typical value. On an atomic scale, this is huge, but on a macroscopic scale it is small enough that the domains escaped detection until Weiss' work.

How does an external H field affect these domains? It puts a torque on each dipole; one might think it would simply cause each dipole to rotate, so that the domain's magnetization vector would rotate. This can happen in materials without a regular crystal structure ("amorphous" materials) but most ferromagnetic materials are crystalline, and the magnetization vector prefers to point in certain discrete

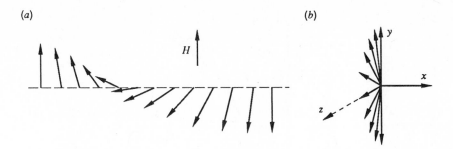

Figure 23.23   (*a*) One row of atomic magnetic moments in a Bloch wall between two magnetic domains; the moments all have approximately the same magnitudes, but are at different angles in the *yz* plane (*b*).

directions (this can be seen in Figure 23.22). So the external field (which is not strong enough to flip a spin to a direction opposite to its neighbors) has no effect in the interior of a domain. As one crosses a domain boundary, the spin direction changes gradually from one discrete direction to another as in Figure 23.23 (this is called a **Bloch wall**). You can see that a small rotation of each spin in the Bloch wall in response to a small torque can have the net effect of moving the wall one atom to the right, so that one more atom is pointing along the external field **H**, and one fewer opposite to it. Thus the domains magnetized along the external field grow, and those magnetized in other directions shrink. The average magnetization is then no longer zero, but points along the external **H** field. You can see from Figure 23.21*b* that the growth of M does not occur in any simple way (for example, proportionately to H). The magnetization curve of a ferromagnet is extremely hard to predict theoretically, because the domain wall motions are very sensitive to trace impurities and minor imperfections in the crystal lattice (dislocations) that depend on the details of how the sample of the material was prepared. Figure 23.21*b* was drawn for a sample of annealed iron (i.e., one which has been heated to give the imperfections a chance to heal). If you bent it back and forth a few times to introduce dislocations (a process known as "cold-working"), it would take a much larger H field to magnetize it—the steeply rising part of the curve would shift to the right, off the edge of the graph. It would become magnetically "harder." It would also become harder to bend: the same dislocations that impede domain boundary motion also impede plastic deformation. This association between difficulty of changing M ("magnetic hardness") and difficulty of plastic deformation ("mechanical hardness") is the reason for the use of the word "hard" in the magnetic context.

One effect of this sluggishness of the domain wall motion is that the magnetization does not entirely disappear when the external field is turned off. Thus, magnetically hard materials can be turned into "permanent magnets" by magnetizing them in a strong H field. Some of their domain walls get permanently stuck, preventing the return to zero magnetization when the field is turned off and they are placed on refrigerator doors. This means that Figure 23.21 does not tell the whole story: the curve there describes the growth of M only when it is initially zero. If after raising H to some value H_0, we then decrease it to $-H_0$, and then increase it again, we will trace out a curve like that shown in Figure 23.24. If we then

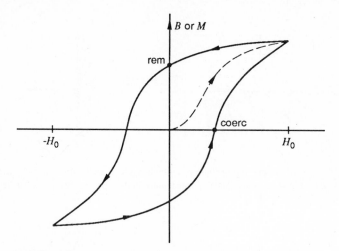

Figure 23.24 A hysteresis curve (solid line), reached from the $M = H = 0$ initial state by the dashed curve. Since $H \ll M$, $B \approx \mu_0 M$ and the vertical axis can be labeled either B or M. The value of B at the point labeled "rem" is called the **remanence**, and the value of H at the point labeled "coerc" is called the **coercivity**.

repeatedly cycle between $+H_0$ and $-H_0$, we will retrace the closed curve in the figure. This is called a **hysteresis loop**, after the Greek word for "lag": the return of the magnetization to zero lags behind the return of H to zero. The behavior is irreversible, in the sense that we can never follow the same curve in both directions, but only in the direction indicated by the arrows. There is an infinite number of hysteresis curves similar to this. There is, in fact, one for each choice of H_0. The curve shown assumes that H is changed slowly; if it is changed quickly enough so that the domain walls can't keep up, the curve will be fatter than that in the figure.

▶ **Problem 23.14**

   a. Estimate the susceptibility of the paramagnet in Figure 23.21 (the curve is not linear, but the susceptibility is defined in terms of the tangent at $H = 0$).

   b. It is obvious from Figure 23.21b that the concept of susceptibility is not very useful for a ferromagnet; the curve is not approximately linear, even near $H = 0$. Nonetheless, make an estimate for χ that might be useful in a calculation in which H ranges from 0 to 100 a/m by fitting a straight line through the origin to this part of the curve. ◀

▶ **Problem 23.15**

   From Eq. 23.23, it is apparent that the energy per unit volume required to take a sample along any curve in the B–H plane is $\int H \, dB$, which is the area between the hysteresis curve and the B-axis. The energy required to take a sample around a hysteresis loop is thus exactly the area of the loop. (The initial and final fields are the same, so the energy clearly does not go into field energy. In fact, it is turned into heat by the friction that

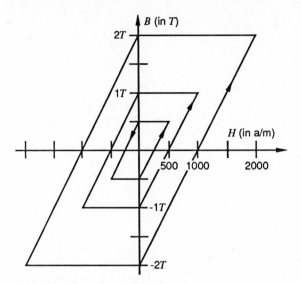

Figure 23.25 Hysteresis curves for a fictitious ferromagnetic material, for various values of H_0.

resists domain wall motion.) Consider a material with the hysteresis loops shown in Figure 23.25.

a. Compute the energy dissipated to heat, when the material is moved around the largest hysteresis loop shown.

b. Compute the power dissipated in a 1 cm^3 volume when the material is subjected to an AC field $H(t) = H_0 \cos \omega t$ with $\omega = 377$ s^{-1} and $H_0 = 10^3$ a/m. ◄

PART E

Electromagnetic Radiation

CHAPTER TWENTY-FOUR

SINUSOIDAL ELECTROMAGNETIC WAVES

SECTION 24.1 POLARIZATION AND EXPONENTIAL NOTATION

We have already looked at sinusoidal waves on a string (Chapter 2), which are solutions of coupled differential equations for the strain and velocity fields. The equations for pseudo-1D transverse electromagnetic fields (Eqs. 11.2 and 11.4) in the absence of current,

$$\frac{dE_y(x, t)}{dt} = -c^2 \frac{dB_z(x, t)}{dx} \tag{24.1}$$

$$\frac{dB_z(x, t)}{dt} = -\frac{dE_y(x, t)}{dx} \tag{24.2}$$

are exactly the same, so they also have sinusoidal solutions. Since there are two transverse directions (y and z, for a wave traveling in the x-direction), we can have waves whose **E** fields point in either of these directions (the equations for the z case are Eqs. 11.8 and 11.9). (Actually, **E** can point in any direction in the yz plane, but we will see later that it is sufficient to consider these two, because any wave can be expressed as a linear combination of them.) We have written the differential equations for the wave whose **E** field points in the y-direction; this is called the y-polarized wave.

The easiest way to find a solution to these equations is to recall that any choice of $E_y(x, t)$ (or any choice of stress in the string of Section 2.6) that moves at the velocity $\pm c$ can be made into a solution of the Maxwell equations. It will move with the velocity $+c$ if it is chosen to be a function of $x - ct$; let us choose $E_y(x, t) = E_0 \cos[k(x - ct) + \delta]$. This is conventionally written

$$E_y(x, t) = E_0 \cos(kx - \omega t + \delta) \tag{24.3}$$

where E_0, k, ω, and δ are constants called the **amplitude, wavenumber, angular frequency,** and **phase lag** respectively, and ω is defined by

$$\omega \equiv ck \tag{24.4}$$

The angular frequency has units radians/sec, or just s^{-1}. One sometimes gives frequencies (denoted by f) in units of cycles/sec, or hertz (Hz): $f = \omega/2\pi$.

▶ **Problem 24.1**

a. Show that if we define

$$\lambda \equiv 2\pi/k, \qquad T \equiv 2\pi/\omega \tag{24.5}$$

then λ is the **wavelength** [in the sense that if $E_y(x, t)$ has some value, say its peak value E_0, then $E_y(x + \lambda, t)$ has the same value] and T the **period** [in the sense that $E_y(x, t) = E_y(x, t + T)$].

b. Show that the wave in Eq. 24.3 moves with velocity c, in the sense that $E_y(x + ct, t) = E_y(x, 0)$: the wave at time t looks identical to the wave at time zero, but displaced by ct.   ◀

The Maxwell equations (Eqs. 24.1 and 24.2) for E_y involve only the z-component of **B**. The simplest way to find the correct sinusoidal form for B_z is to use complex exponential notation; because the algebra later in this chapter would become unwieldy without it, we may as well switch now. This means we will use complex numbers like $2 + 3i$. You can do algebra with complex numbers by treating i just like a real variable, using two facts:

$$i^2 = -1 \tag{24.6}$$

$$e^{i\theta} = \cos\theta + i\sin\theta \qquad \text{(Euler's formula)} \tag{24.7}$$

Given a complex number $a + bi$, where a and b are real numbers, we call a the **real part**, denoted $\text{Re}(a + bi)$, and b the **imaginary part**, denoted $\text{Im}(a + bi)$. Complex numbers are often depicted in the **complex plane**, in which the real part is plotted horizontally and the imaginary part vertically (Figure 24.1). The **magnitude** of $a + bi$, denoted $|a + bi|$, is a non-negative real number defined by

$$|a + bi|^2 = a^2 + b^2 \tag{24.8}$$

and is the distance from the origin in the complex plane. From Eq. 24.7 we obtain

$$\cos\theta = \text{Re } e^{i\theta}, \qquad \sin\theta = \text{Im } e^{i\theta} \tag{24.9}$$

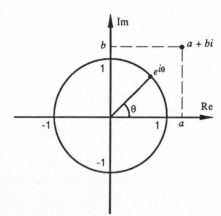

Figure 24.1 The complex plane, showing the complex numbers $a + bi$ and $e^{i\theta}$.

so we can rewrite Eq. (24.3) as

$$E_y(x, t) = E_0 \, \mathrm{Re} \, e^{i(kx - \omega t + \delta)} = \mathrm{Re} \, E_0 e^{i\delta} e^{ikx - i\omega t}$$

$$= \mathrm{Re} \, \tilde{E} e^{ikx - i\omega t} \tag{24.10}$$

(noting that the real number E_0 commutes with the Re operation, but complex numbers do not) where

$$\tilde{E} \equiv E_0 e^{i\delta} \tag{24.11}$$

is the **complex amplitude** of the sinusoidal field E_y (we will always denote complex numbers by characters with ~ over them). Any complex number can be written in the form of Eq. 24.11; δ is called the **phase** of the complex number $\tilde{E}$, and you can easily show that E_0 is its magnitude. The complex amplitude of a wave describes both the amplitude (in the usual sense) and the phase of the wave.

Evidently any sinusoidal function can be written in the form of Eq. 24.10; let us guess that

$$B_z(x, t) = \mathrm{Re} \, \tilde{B} e^{ikx - i\omega t} \tag{24.12}$$

where $\tilde{B}$ is an undetermined complex amplitude. Then the Maxwell equation (Eq. 24.1) becomes

$$\mathrm{Re} \, \tilde{E} \frac{d}{dt} e^{ikx - i\omega t} = -\mathrm{Re} \, c^2 \tilde{B} \frac{d}{dx} e^{ikx - i\omega t}$$

$$\mathrm{Re} \, \tilde{E}(-i\omega) e^{ikx - i\omega t} = -\mathrm{Re} \, c^2 \tilde{B}(ik) e^{ikx - i\omega t} \tag{24.13}$$

$$\mathrm{Re}[-i\omega \tilde{E} + ic^2 k \tilde{B}] e^{ikx - i\omega t} = 0$$

We now introduce a useful fact that we will need frequently in this chapter: if $\tilde{A}$ is a complex number,

$$\mathrm{Re} \, \tilde{A} e^{i\omega t} = 0 \qquad \text{for all } t$$

$$\text{if and only if} \tag{24.14}$$

$$\tilde{A} = 0$$

where the latter zero is the complex zero, $0 + 0i$. In the present instance, Eq. 24.14 implies

$$[-i\omega \tilde{E} + ic^2 k \tilde{B}] e^{ikx} = 0$$

But $e^{ikx} \neq 0$ (in fact, e^{ikx} always has magnitude 1, as you can see from Figure 24.1 or Eqs. 24.8 and 24.9), so we can divide it out. Using Eq. 24.4, we get

$$\tilde{B} = \frac{\tilde{E}}{c} \tag{24.15}$$

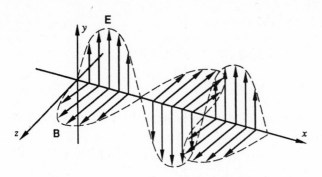

Figure 24.2 A snapshot showing the **E** and **B** fields at a fixed time t, as a function of x (they are independent of y and z; only those at $y = z = 0$ are drawn). We have chosen $\delta = 0$, $t = 0$; the wave moves to the right without changing form.

▶ **Problem 24.2**

Show that Eq. 24.2 is also satisfied by the $\tilde{B}$ given in Eq. 24.15. (This differential equation is redundant because we assumed the wave velocity was c; if we had not, we could have deduced it by demanding that Eq. 24.2 be satisfied by the solution of Eq. 24.1.)   ◀

Note that the ratio of $\tilde{B}$ to $\tilde{E}$ is **real**; that is, the two complex amplitudes have the same phase. This means that the physical fields E_y and B_z have the same phase, that is, they reach a maximum at the same position and time, as shown in Figure 24.2.

▶ **Problem 24.3**

Write equations similar to Eqs. 24.10, 24.12, and 24.15 for a wave moving to the left. Show that it can be obtained from the right-moving wave by replacing k by $-k$ and $\tilde{B}$ by $-\tilde{B}$. This is easy to remember because it simply corresponds to rotating Figure 24.2 by 180° about the y-axis.   ◀

▶ **Problem 24.4**

Prove the "useful fact" (Eq. 24.14).   ◀

SECTION 24.2   WAVES IN LINEAR MATERIALS

A general (dielectrically and magnetically) linear material is characterized by a dielectric permittivity ϵ and a magnetic permeability μ that may differ from the vacuum values ϵ_0 and μ_0. For most purposes μ can be assumed to be μ_0; very few ferromagnetic materials are transparent, and paramagnetic and diamagnetic susceptibilities are usually very small. However, because it is very little trouble, we will allow an arbitrary μ. The fields are governed by the material Maxwell equations 23.36 and 23.43,

$$\frac{d\mathbf{B}}{dt} = -\nabla \times \mathbf{E} \tag{24.16}$$

$$\frac{d\mathbf{D}}{dt} = \nabla \times \mathbf{H} \tag{24.17}$$

(our assumption of no current in the vacuum case must now become an assumption of no **free** current, because there will in general be bound currents due to polarization and magnetization). Equation 24.17 becomes, in our linear material,

$$\frac{d\mathbf{E}}{dt} = (\mu\epsilon)^{-1}\nabla \times \mathbf{B} \tag{24.18}$$

Thus the Maxwell equations are identical to those in a vacuum, except that c^2 is replaced by $(\mu\epsilon)^{-1}$. The solutions are therefore the same, with c replaced by

$$v \equiv (\mu\epsilon)^{-1/2} \tag{24.19}$$

Note that this reduces to c if $\mu = \mu_0$ and $\epsilon = \epsilon_0$; it differs from c by a factor

$$n \equiv \frac{c}{v} = \left(\frac{\mu}{\mu_0}\right)^{1/2}\left(\frac{\epsilon}{\epsilon_0}\right)^{1/2} \tag{24.20}$$

called the **index of refraction** for reasons that will be apparent in Section 24.5. If $\mu = \mu_0$, this is

$$n = K^{1/2} \tag{24.21}$$

where K is the dielectric constant ϵ/ϵ_0. Explicitly, E and B are given by Eqs. 24.10 and 24.12 with

$$\omega = vk \tag{24.22}$$

and

$$\tilde{B} = \frac{\tilde{E}}{v} \tag{24.23}$$

We can also calculate the energy density and the energy flux (the Poynting vector) in this wave. From Figure 24.2 it is apparent that the energy density is highly non-uniform in space; it is zero at the nodes (where $E = B = 0$) and largest at the maxima of E and B. It is also not constant in time: at any given point, it is largest when a maximum is passing. Quantitatively, using Eq. 19.53 we get the electric energy density

$$u_E(x, t) = \tfrac{1}{2}\epsilon E(x, t)^2$$

$$= \tfrac{1}{2}\epsilon[\text{Re}\,\tilde{E}e^{ikx - i\omega t}]^2 = \tfrac{1}{2}\epsilon E_0^2 \cos^2(kx - \omega t + \delta) \tag{24.24}$$

The magnetic energy density turns out to be the same: Eq. 23.24 gives

$$u_B(x, t) = \frac{B(x, t)^2}{2\mu} = \left(\frac{E}{v}\right)^2 / 2\mu = \frac{\mu\epsilon E^2}{2\mu} = u_E \tag{24.25}$$

(using Eqs. 24.19 and 24.23), so the total energy density is just double this,

$$u(x, t) = \epsilon E_0^2 \cos^2(kx - \omega t + \delta) \qquad (24.26)$$

Usually, we are not interested in the variations of u on the time scale T or the space scale λ; if we average u over any integer number of wavelengths in space or periods in time, $\cos^2$ averages to $\frac{1}{2}$.

▶   **Problem 24.5**

Show that the square of the cosine averages to $\frac{1}{2}$: $\lambda^{-1}\!\int_0^\lambda \cos^2(kx + \delta) = \frac{1}{2}$. [A good mnemonic for this is that the graph of the function $\cos^2$ can be turned into the constant $\frac{1}{2}$ by slicing off the peaks and filling the valleys with them.]   ◀

Thus the average energy is

$$\bar{u} = \tfrac{1}{2}\epsilon E_0^2 \qquad (24.27)$$

(we indicate averaged quantities by putting a bar over them).

To get the Poynting vector, we must use the result of Problem 23.13,

$$S(x, t) = E_y(x, t)H_z(x, t) = \frac{E_0^2}{\mu v}\cos^2(kx - \omega t + \delta) \qquad (24.28)$$

$$= v\epsilon E_0^2 \cos^2(kx - \omega t + \delta)$$

This is exactly v times the energy density u, as you might expect because the energy is traveling at speed v. The average energy flux is

$$\bar{S} = \tfrac{1}{2}v\epsilon E_0^2 = v\bar{u} \qquad (24.29)$$

SECTION 24.3   REFLECTION AND TRANSMISSION OF WAVES

Consider an electromagnetic pulse moving toward an interface between two materials, as in Figure 24.3a. If you compute the evolution of this pulse using the Maxwell equations, it will, of course, propagate in the usual simple way (without changing

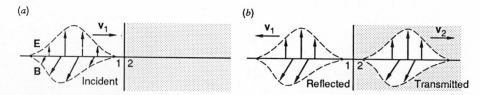

Figure 24.3   (a) A pulse moving toward an interface. (b) The system at a later time, showing the reflected and transmitted pulses.

shape) while it is in material 1, at a velocity v_1 determined by that material's ϵ and μ. However, when it reaches the interface, the value of ϵ changes, and the evolution is not so simple. But you will find that eventually the fields at the interface die out, and the system looks like Figure 24.3b: there are now **two** pulses. One, the **transmitted** pulse, is traveling to the right in material 2; the other, the **reflected** pulse, is traveling to the left in material 1. We would like to calculate the properties of these pulses; in particular, we would like to calculate the fraction of the energy that is reflected, and the fraction transmitted.

It turns out that this calculation is easiest if we replace the pulse in Figure 24.3a (the **incident** pulse) by a sinusoidal wave. This case is hard to depict and to visualize, because a sinusoidal wave extends by definition infinitely far to the left and right, and looks the same at all times (at least if we only look at intervals of the period T): we can't say exactly when it hits the interface, and the reflected wave is mixed up with the incident wave at all times. However, we can write equations for the waves, and distinguish the incident from the reflected wave by the fact that one has e^{ikx} and the other e^{-ikx} (see Problem 24.3). The incident wave is

$$E_I(x, t) = \text{Re}\, \tilde{E}_I \exp(ik_1 x - i\omega t) \tag{24.30}$$

(we will leave the y and z subscripts off E and B respectively in this section) and the reflected wave is

$$E_R(x, t) = \text{Re}\, \tilde{E}_R \exp(-ik_1 x - i\omega t) \tag{24.31}$$

[We will assume the reflected and transmitted waves have the same frequency ω as the incident wave. You can verify this either mathematically (see Problem 24.11) or intuitively (by observing that if $\omega/2\pi$ wave crests hit the interface per second, this same number must be reflected and transmitted).] Neither of these two formulas gives the actual field anywhere, because the incident and reflected waves are always superposed. The actual field to the left of the interface is

$$E(x, t) = \text{Re}\big[\tilde{E}_I \exp(ik_1 x) + \tilde{E}_R \exp(-ik_1 x)\big] e^{-i\omega t} \quad (x < 0) \tag{24.32}$$

If we assume the interface is at $x = 0$, this formula gives the field for $x < 0$. For $x > 0$, only the transmitted wave is present:

$$E(x, t) = \text{Re}\, \tilde{E}_T \exp(ik_2 x) e^{-i\omega t} \quad (x > 0) \tag{24.33}$$

where $k_2 = \omega/v_2$ is different from k_1.

We would like to calculate $\tilde{E}_R$ and $\tilde{E}_T$ from $\tilde{E}_I$. To do this we must use the continuum boundary conditions on **E** and **D** we derived in Section 19.2, and those on **B** and **H** from Section 23.10. Obviously, the conditions on normal components are irrelevant, because those components are zero here. The conditions on tangential components were that E and H must be continuous across the interface (we assume no free surface current here). The continuity of E means (substituting $x = 0$ into Eqs. 24.32 and 24.33)

$$\text{Re}\big[\tilde{E}_I + \tilde{E}_R\big] e^{-i\omega t} = \text{Re}\, \tilde{E}_T e^{-i\omega t} \tag{24.34}$$

at all times t. Moving $\tilde{E}_T$ to the left-hand side and using the useful fact (Eq. 24.14), we get

$$\tilde{E}_I + \tilde{E}_R = \tilde{E}_T \tag{24.35}$$

as one condition on $\tilde{E}_I$ and $\tilde{E}_R$. We need two complex equations to determine two complex unknowns, however (they are equivalent to four real equations for four real unknowns, the real and imaginary parts). To get the other equation, we must use the continuity of H. This requires knowing the magnetic fields in each wave. The complex amplitude of H_I can be obtained from Eq. 24.23:

$$\tilde{H}_I = \frac{\tilde{E}_I}{v_1 \mu_1} = \frac{n_1 \tilde{E}_I}{\mu_1 c} \tag{24.36}$$

Similarly,

$$\tilde{H}_T = \frac{n_2 \tilde{E}_T}{\mu_2 c} \tag{24.37}$$

The B field of the reflected wave has the opposite sign (Problem 24.3), so

$$\tilde{H}_R = -\frac{n_1 \tilde{E}_R}{\mu_1 c} \tag{24.38}$$

Continuity of H then requires

$$\tilde{H}_I + \tilde{H}_R = \tilde{H}_T \tag{24.39}$$

by an argument such as that for Eq. 24.35. We can write this in terms of the E's using Eqs. 24.36 to 24.38 and multiplying by c:

$$n_1 \mu_1^{-1} \tilde{E}_I - n_1 \mu_1^{-1} \tilde{E}_R = n_2 \mu_2^{-1} \tilde{E}_T \tag{24.40}$$

It is now convenient to assume $\mu_1 = \mu_2$, so we can cancel the μ's. It is straightforward to solve Eqs. 24.35 and 24.40 for $\tilde{E}_R$ and $\tilde{E}_T$ (the algebra is the same as though they were real):

$$\tilde{E}_R = \frac{n_1 - n_2}{n_1 + n_2} \tilde{E}_I \tag{24.41}$$

$$\tilde{E}_T = \frac{2n_1}{n_1 + n_2} \tilde{E}_I \tag{24.42}$$

These equations give the complex amplitudes (i.e., the amplitudes and phases) of the outgoing waves. Note that the phase of the transmitted wave is the same as that of the incident wave, as is that of the reflected wave if $n_1 > n_2$ (as it would be if the wave were going from glass with $n_1 \approx 1.5$ into air with $n_2 \approx 1$, for example). If $n_2 > n_1$ (e.g., air to glass) the sign changes, that is, the phase changes by $180°$.

Often we are not interested in the reflected amplitude so much as in the energy flux. The fluxes of the incident and transmitted waves can be computed from Eq. 24.29:

$$\bar{S}_I = \tfrac{1}{2} v_1 \epsilon_1 |\tilde{E}_I|^2 = \tfrac{1}{2} n_1 \mu_1^{-1} c^{-1} |\tilde{E}_I|^2 \tag{24.43}$$

$$\bar{S}_T = \tfrac{1}{2} n_2 \mu_2^{-1} c^{-1} |\tilde{E}_T|^2 \tag{24.44}$$

The calculation for the reflected wave is similar, but H has the opposite sign:

$$\bar{S}_R = -\tfrac{1}{2} n_1 \mu_1^{-1} c^{-1} |\tilde{E}_R|^2 \tag{24.45}$$

Thus the fraction of the incident energy transmitted (the **transmission coefficient**) is (assuming again $\mu_1 = \mu_2$)

$$T \equiv \frac{\bar{S}_T}{\bar{S}_I} = \frac{n_2}{n_1} \left| \frac{\tilde{E}_T}{\tilde{E}_I} \right|^2 = \frac{4 n_1 n_2}{(n_1 + n_2)^2} \tag{24.46}$$

The fraction reflected is the **reflection coefficient**

$$R \equiv \frac{|\bar{S}_R|}{\bar{S}_I} = \left| \frac{\tilde{E}_R}{\tilde{E}_I} \right|^2 = \frac{(n_1 - n_2)^2}{(n_1 + n_2)^2} \tag{24.47}$$

▶ **Problem 24.6**

Calculate the reflection coefficient in air from a piece of glass with $n = 1.5$.   ◀

▶ **Problem 24.7**

Verify that Eqs. 24.41, 24.42, 24.46, and 24.47 reduce to what you would expect when $n_1 = n_2$.   ◀

▶ **Problem 24.8**

Verify that $R + T = 1$ (conservation of energy).   ◀

▶ **Problem 24.9**

Derive equations like Eqs. 24.46 and 24.47 for the case where $\mu_1 \neq \mu_2$.   ◀

▶ **Problem 24.10**

We are being inconsistent in calculating the incident and reflected fluxes separately; neither $\tilde{E}_I$ nor $\tilde{E}_R$ by itself gives the actual electric field anywhere. To get the actual energy flux for $x < 0$ we must use the actual fields, given by Eq. 24.32 and a corresponding equation for H.

a. Show that the answer given by our incorrect calculation ($\bar{S}_I + \bar{S}_R$, Eqs. 24.43 and 24.45) happens to be correct.

b. Show that we would not have been so lucky if the waves had been going in the **same** instead of opposite directions.   ◀

▶   **Problem 24.11**

Suppose $\tilde{E}_I$, $\tilde{E}_R$, and $\tilde{E}_T$ are nonzero complex numbers, ω_I, ω_R, and ω_T are real numbers, and

$$\text{Re } \tilde{E}_I \exp(-i\omega_I t) + \text{Re } \tilde{E}_R \exp(-i\omega_T t) = \text{Re } \tilde{E}_T \exp(-i\omega_R t)$$

for all t. Show that $\omega_I = \omega_R = \omega_T$.   ◀

SECTION 24.4   COMPLEX VECTOR AMPLITUDES

We will now consider a plane wave that is not necessarily moving in the x-direction. Suppose the wave is moving in the direction of the unit vector $\hat{\mathbf{k}}$ in Figure 24.4. We have drawn a coordinate system with coordinates x', y', and z' (the "primed coordinate system") such that the wave is moving in the x'-direction, so its electric field can be written (Eq. 24.10)

$$E_{y'}(x', t) = \text{Re } \tilde{E}_{y'} e^{ikx' - i\omega t} \tag{24.48}$$

where $\tilde{E}_{y'}$ is a complex amplitude. To eliminate x' in favor of the unprimed coordinates, it is easiest to observe that $kx' = kx' + 0y' + 0z' = k\hat{\mathbf{k}} \cdot \mathbf{r}$, where $\mathbf{r}$ is the field point having coordinates (x', y', z') in the primed system and (x, y, z) in the unprimed system, and $\hat{\mathbf{k}}$ has coordinates $(1, 0, 0)$ in the primed system. Thus,

$$\mathbf{E}(\mathbf{r}, t) = \hat{\mathbf{y}}' \text{Re } \tilde{E}_{y'} e^{i\mathbf{k} \cdot \mathbf{r} - i\omega t} \tag{24.49}$$

where $\mathbf{k} \equiv k\hat{\mathbf{k}}$ is called the **wavevector** of the sinusoidal wave; the dot product can be evaluated in either coordinate system. An arbitrary wave can be written as a linear combination of the y'-polarized wave we have written down, and another polarized along z', with complex amplitude $\tilde{E}_{z'}$. The total electric field is thus

$$\mathbf{E}(\mathbf{r}, t) = \hat{\mathbf{y}}' \text{Re } \tilde{E}_{y'} e^{i\mathbf{k} \cdot \mathbf{r} - i\omega t} + \hat{\mathbf{z}}' \text{Re } \tilde{E}_{z'} e^{i\mathbf{k} \cdot \mathbf{r} - i\omega t} \tag{24.50}$$

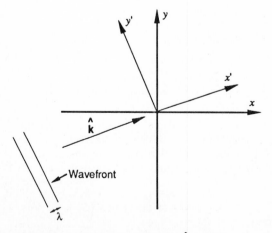

Figure 24.4   An arbitrary direction $\hat{\mathbf{k}}$, showing the primed coordinate system in which $\hat{\mathbf{k}}$ is along the x'-axis. For simplicity, we have drawn $\hat{\mathbf{k}}$ in the xy-plane, but our results do not depend on this.

Equation 24.50 is the most general sinusoidal wave we can write down. It is still a plane wave; a surface ("wavefront") along which it has a given value is described by $\mathbf{k} \cdot \mathbf{r} = $ constant, which is the equation of a plane.

To simplify our notation, we will introduce the concept of a **complex vector**, which is simply an ordered triple of complex numbers, just as an ordinary ("real", in this context) vector is an ordered triple of real numbers. Thus the **complex vector amplitude**

$$\tilde{\mathbf{E}} \equiv \hat{\mathbf{y}}' \tilde{E}_{y'} + \hat{\mathbf{z}}' \tilde{E}_{z'} \tag{24.51}$$

[which is the triple $(0, \tilde{E}_{y'}, \tilde{E}_{z'})$ in the primed coordinate system] is a complex vector. This allows us to write Eq. 24.50 in the coordinate-free form

$$\mathbf{E}(\mathbf{r}, t) = \operatorname{Re} \tilde{\mathbf{E}} e^{i\mathbf{k} \cdot \mathbf{r} - i\omega t} \tag{24.52}$$

[The real part of a complex vector $\tilde{\mathbf{A}}$ is a real vector defined by taking the real parts of its components, i.e., $\operatorname{Re}(\tilde{A}_x, \tilde{A}_y, \tilde{A}_z) \equiv (\operatorname{Re} \tilde{A}_x, \operatorname{Re} \tilde{A}_y, \operatorname{Re} \tilde{A}_z)$. You can show that this allows us to move the real unit vectors $\hat{\mathbf{y}}'$ and $\hat{\mathbf{z}}'$ past the Re symbol.] There is an immense amount of complication buried in Eq. 24.52—for example, the y' and z' components need not have the same phase—most of which we will not have to deal with here.

You will note that Eq. 24.51 for $\tilde{\mathbf{E}}$ is not the most general complex vector. It lacks a component along $\mathbf{k}$, that is, it is **transverse**. This can be shown formally from Gauss' law:

$$\nabla \cdot \operatorname{Re} \tilde{\mathbf{E}} e^{i\mathbf{k} \cdot \mathbf{r} - i\omega t} = 0$$

$$\frac{d}{dx} \operatorname{Re} \tilde{E}_x \exp(ik_x x + ik_y y + ik_z z - i\omega t) + \frac{d}{dy} \cdots = 0 \tag{24.53}$$

$$\operatorname{Re} \tilde{E}_x (ik_x) \exp(ik_x x + ik_y y + ik_z z - i\omega t) + \cdots = 0$$

Since this is true for all t,

$$ik_x \tilde{E}_x \exp(i\mathbf{k} \cdot \mathbf{r}) + ik_y \tilde{E}_y \exp(i\mathbf{k} \cdot \mathbf{r}) + \cdots = 0$$

$$\mathbf{k} \cdot \tilde{\mathbf{E}} = 0 \tag{24.54}$$

The real part of this equation states that the real part of $\tilde{\mathbf{E}}$ is perpendicular to $\mathbf{k}$, and the imaginary part says that the imaginary part of $\tilde{\mathbf{E}}$ is perpendicular to $\mathbf{k}$.

Armed with the concept of complex vector amplitude, it is easy to write down the electric and magnetic fields of the most general plane wave. Assume a form for $\mathbf{B}$ similar to Eq. 24.52:

$$\mathbf{B}(\mathbf{r}, t) = \operatorname{Re} \tilde{\mathbf{B}} e^{i\mathbf{k} \cdot \mathbf{r} - i\omega t} \tag{24.55}$$

The first Maxwell equation (Eq. 24.16) then implies

$$\operatorname{Re} \tilde{\mathbf{B}}(-i\omega) e^{i\mathbf{k} \cdot \mathbf{r} - i\omega t} = -\operatorname{Re}(i\mathbf{k}) \times \tilde{\mathbf{E}} e^{i\mathbf{k} \cdot \mathbf{r} - i\omega t} \tag{24.56}$$

Here, as in Eq. 24.53, each component of ∇ brings out a factor of the corresponding component of $i\mathbf{k}$; in effect, we can replace ∇ by $i\mathbf{k}$. The cross product of a real vector and a complex vector is a complex vector, defined in the obvious way by taking the cross products of the real and imaginary parts separately. Using the by now familiar fact in Eq. 24.14, we get

$$-i\omega\tilde{\mathbf{B}} = -i\mathbf{k} \times \tilde{\mathbf{E}} \tag{24.57}$$

which gives the magnetic field amplitude in terms of the electric field amplitude:

$$\tilde{\mathbf{B}} = \omega^{-1}\mathbf{k} \times \tilde{\mathbf{E}} \tag{24.58}$$

Note that this magnetic field is also transverse, as required by $\nabla \cdot \mathbf{B} = 0$.

The field must also satisfy the other Maxwell equation (Eq. 24.17), which leads similarly to

$$-i\omega\epsilon\tilde{\mathbf{E}} = i\mu^{-1}\mathbf{k} \times \tilde{\mathbf{B}} \tag{24.59}$$

or (using Eq. 24.58)

$$-\epsilon\mu\omega^2\tilde{\mathbf{E}} = \mathbf{k} \times (\mathbf{k} \times \tilde{\mathbf{E}}) = \mathbf{k}(\mathbf{k} \cdot \tilde{\mathbf{E}}) - k^2\tilde{\mathbf{E}} \tag{24.60}$$

using the BAC–CAB rule (Problem 20.17). This still looks rather complicated; however, note that $\mathbf{k} \cdot \tilde{\mathbf{E}}$ vanishes by Gauss' law (Eq. 24.54). Thus, (assuming $\tilde{\mathbf{E}} \neq 0$, i.e., at least one of the six real numbers involved is nonzero)

$$\epsilon\mu\omega^2 = k^2$$
$$\omega = vk \tag{24.61}$$

where $v = (\epsilon\mu)^{-1/2}$ is the velocity of light in our material and k is the magnitude of the vector $\mathbf{k}$.

We can calculate the Poynting vector of this most general wave, using Eq. 23.44:

$$\mathbf{S}(\mathbf{r}, t) = \mathbf{E}(\mathbf{r}, t) \times \mathbf{H}(\mathbf{r}, t)$$
$$= \mathrm{Re}(\tilde{\mathbf{E}}e^{i\mathbf{k}\cdot\mathbf{r}-i\omega t}) \times \mathrm{Re}(\tilde{\mathbf{H}}e^{i\mathbf{k}\cdot\mathbf{r}-i\omega t}) \tag{24.62}$$

Each term in each component of this vector cross product is a product of the form

$$\mathrm{Re}(\tilde{A}e^{-i\omega t})\,\mathrm{Re}(\tilde{B}e^{-i\omega t}) \tag{24.63}$$

A useful formula for the time average of such a quantity is proved in Problem 24.13: it is

$$\tfrac{1}{2}\,\mathrm{Re}(\tilde{A}^*\tilde{B}) \tag{24.64}$$

where $\tilde{A}^* \equiv \mathrm{Re}\,\tilde{A} - i\,\mathrm{Im}\,\tilde{A}$ is called the **complex conjugate** of $\tilde{\mathbf{A}}$. Thus the time average of Eq. 24.62 is

$$\bar{\mathbf{S}}(\mathbf{r}, t) = \tfrac{1}{2} \operatorname{Re}(\tilde{\mathbf{E}}^* \times \tilde{\mathbf{H}})$$

$$= \tfrac{1}{2}\omega^{-1}\mu^{-1} \operatorname{Re}\left[\tilde{\mathbf{E}}^* \times (\mathbf{k} \times \tilde{\mathbf{E}})\right] \tag{24.65}$$

$$= \tfrac{1}{2}v^{-1}\mu^{-1} \operatorname{Re}(\tilde{\mathbf{E}}^* \cdot \tilde{\mathbf{E}})\hat{\mathbf{k}}$$

where we have used Eqs. 24.58 and 24.61, and the BAC–CAB rule. The Re in Eq. 24.65 is unnecessary, because the dot product of a complex vector with its complex conjugate is always real; we will denote it by $|\tilde{\mathbf{E}}|^2$.

To summarize the results of this section, the most general sinusoidal plane wave in a material is given by Eqs. 24.52, 24.55, and 24.58, with an arbitrary complex vector amplitude $\tilde{\mathbf{E}}$, which is transverse (i.e., satisfies Eq. 24.54). Its energy flux vector is along the wavevector, and is given by Eq. 24.65.

▶   **Problem 24.12**

Write the x-component of Eq. 24.16 explicitly in terms of components, substituting Eqs. 24.52 and 24.55 for **E** and **B**, perform the differentiation, and show that the result is the x-component of Eq. 24.56.   ◀

▶   **Problem 24.13**

Show that for any two complex numbers $\tilde{A}$ and $\tilde{B}$, the time average (over an integer number of periods) of Eq. 24.63 is Eq. 24.64. Hint: $(e^{i\theta})^* = e^{-i\theta}$, and the complex conjugate of a product is the product of the conjugates. Use $\operatorname{Re} \tilde{C} = \tfrac{1}{2}(\tilde{C} + \tilde{C}^*)$ three times.   ◀

▶   **Problem 24.14**

a. Consider a wave propagating along the x-axis with complex vector amplitude $\tilde{\mathbf{E}} = E_0\hat{\mathbf{y}} + iE_0\hat{\mathbf{z}}$, where E_0 is a real constant. Calculate the electric and magnetic fields at the origin, as functions of time. Show that $\mathbf{E}(t)$ can be written in the form $E_0\hat{\mathbf{u}}(t)$, where $\hat{\mathbf{u}}(t)$ is a unit vector that rotates with frequency ω about the x-axis; this wave is said to have **circular polarization**.

b. Show that if $\hat{\mathbf{E}} = E_1\hat{\mathbf{y}} + iE_2\hat{\mathbf{z}}$ (E_1 and E_2 are real constants), then the electric field vector at the origin traces out an ellipse (the equation of an ellipse is $ay^2 + bz^2 = c$, for positive constants a, b, and c). This is called **elliptical polarization**; both linear and circular polarization are special cases of elliptical polarization.

c. Show that for **any** (transverse) complex vector amplitude $\tilde{\mathbf{E}}$, you can find a coordinate system with axes y' and z' and a phase angle ϕ such that $\tilde{\mathbf{E}} = (E_1\hat{\mathbf{y}}' + iE_2\hat{\mathbf{z}}')e^{i\phi}$. That is, elliptical polarization is the most general kind. (Hint: Consider the function $\tilde{r}(\theta) \equiv \tilde{E}_{y'}/\tilde{E}_{z'}$, where θ is the angle by which you have rotated the coordinate system about the x-axis. You need to show that $\tilde{r}(\theta)$ crosses the imaginary axis; if it is to the right when $\theta = 0$, where is it when $\theta = \pi/2$?)   ◀

SECTION 24.5   REFLECTION AND REFRACTION AT OBLIQUE INCIDENCE

Consider a plane wave incident on an interface, with its wavevector in an arbitrary direction, and choose a coordinate system whose x-axis is normal to the interface, as in Figure 24.5. Denote the complex amplitude, wavevector, and frequency of the

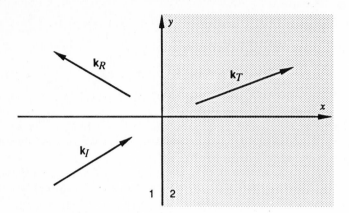

Figure 24.5 A plane wave incident on a dielectric interface, showing the wavevectors of the incident, reflected, and transmitted waves.

incident wave by $\tilde{\mathbf{E}}_I$, $\mathbf{k}_I$, and ω. In analogy with Section 24.3, we might suppose that there would be a reflected and a transmitted wave, with wavevectors $\mathbf{k}_R$ and $\mathbf{k}_T$. Their frequencies must be the same as the incident frequency, just as in Section 24.3. This determines the magnitudes of their wavevectors through Eq. 24.61:

$$k_R = k_I = \frac{\omega}{v_1}, \qquad k_T = \frac{\omega}{v_2} \tag{24.66}$$

We must deduce their directions and amplitudes from the boundary conditions. The latter are more complicated than in the case of normal incidence, because the normal components of the fields no longer vanish.

Let us start with the condition requiring the continuity of the transverse electric field at $x = 0$:

$$\operatorname{Re} \tilde{\mathbf{E}}_I^t \exp(i\mathbf{k}_I^t \cdot \mathbf{r} - i\omega t) + \operatorname{Re} \tilde{\mathbf{E}}_R^t \exp(i\mathbf{k}_R^t \cdot \mathbf{r} - i\omega t) = \operatorname{Re} \tilde{\mathbf{E}}_T^t \exp(i\mathbf{k}_T^t \cdot \mathbf{r} - i\omega t) \tag{24.67}$$

where $\mathbf{k}_I^t$ is the tangential part $(0, k_{Iy}, k_{Iz})$ of the incident wavevector (k_{Ix} is not needed here because $x = 0$), $\tilde{\mathbf{E}}_I^t$ is the tangential part of the complex vector amplitude of the incoming wave (i.e., the x-component has been removed), and similarly for the reflected and transmitted waves. Recall that we used the fact that an equation such as this was true for all t to conclude that the three frequencies must be the same (Problem 24.11). Since $k_{Iy}y$ appears in exactly the same way as $-\omega_I t$, we can use the same argument to conclude that the three $\mathbf{k}$'s have the same y-component. We can do the same for the z-component (but **not** the x-component, because Eq. 24.67 is not true for all x), leading to

$$\mathbf{k}_I^t = \mathbf{k}_R^t = \mathbf{k}_T^t \tag{24.68}$$

that is, the tangential part of the wavevector is conserved in the process of reflection or transmission.

> Note that we used **only** the fact that we are dealing with sinusoidal plane waves in deriving Eq. 24.68; we have not yet invoked the Maxwell equations. Tangential wavevector conservation is a property of all reflection and transmission processes, for example refraction of sound waves at an air–water interface or reflection of water surface waves at the side of a swimming pool. In a form that is slightly modified to account for the inhomogeneity of the reflecting surface, it is also applicable to Bragg reflection of x-rays and diffraction of electron beams at metal surfaces.

Let us now choose the y-axis along $\mathbf{k}_I^t$; as a result of Eq. 24.68 this means that all three wavevectors are in the xy plane: the incident, reflected, and transmitted waves are always coplanar. Thus all the vectors in Figure 24.5 are in the plane of the paper, which is referred to as the **plane of incidence**. From Eqs. 24.66 and 24.68 we can immediately deduce the x-component of $\mathbf{k}_R$:

$$k_{Rx}^2 = k_R^2 - |\mathbf{k}_R^t|^2 = k_I^2 - |\mathbf{k}_I^t|^2 = k_{Ix}^2$$

(24.69)

$$k_{Rx} = \pm k_{Ix}$$

It is clear that $\mathbf{k}_R \neq \mathbf{k}_I$, so we must choose $k_{Rx} = -k_{Ix}$: the reflected wavevector is identical to the incident one except that its component normal to the interface is reversed, as shown in Figure 24.6. The two triangles on the left are congruent, hence

$$\theta_R = \theta_I$$

(24.70)

which is called the **law of reflection**: the **angle of reflection** is equal to the **angle of incidence**.

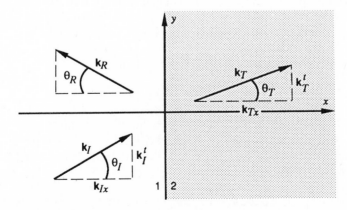

Figure 24.6   The normal (x) and tangential components of the incident, reflected, and transmitted wavevectors.

The **angle of refraction** can also be determined from Eqs. 24.66 and 24.68:

$$\sin \theta_T = \frac{|k_T^t|}{k_T} = \frac{k_{Iy}}{(v_1 k_I/v_2)} = \frac{v_2}{v_1} \sin \theta_I$$
(24.71)
$$n_2 \sin \theta_T = n_1 \sin \theta_I$$

which is, of course, **Snell's law** of refraction.

Let us now return to the boundary conditions on the tangential electric field (Eq. 24.67), which simplifies through the use of Eqs. 24.14 and 24.68 to

$$\tilde{\mathbf{E}}_I^t + \tilde{\mathbf{E}}_R^t = \tilde{\mathbf{E}}_T^t$$
(24.72)

The corresponding equation for the tangential H field is obtained from Eq. 24.58:

$$\mu_1^{-1}(\mathbf{k}_I \times \tilde{\mathbf{E}}_I)^t + \mu_1^{-1}(\mathbf{k}_R \times \tilde{\mathbf{E}}_R)^t = \mu_2^{-1}(\mathbf{k}_T \times \tilde{\mathbf{E}}_T)^t$$
(24.73)

The continuity of the normal component of D (Eq. 19.12), in the absence of free surface charge, becomes

$$\epsilon_1 \tilde{E}_{Ix} + \epsilon_1 \tilde{E}_{Rx} = \epsilon_2 \tilde{E}_{Tx}$$
(24.74)

and that of the normal component of B (Eq. 23.30) is

$$(\mathbf{k}_I \times \tilde{\mathbf{E}}_I)_x + (\mathbf{k}_R \times \tilde{\mathbf{E}}_R)_x = (\mathbf{k}_T \times \tilde{\mathbf{E}}_T)_x$$
(24.75)

These six complex linear equations determine the six complex components of the vectors $\tilde{\mathbf{E}}_R$ and $\tilde{\mathbf{E}}_{T'}$ in terms of $\tilde{\mathbf{E}}_I$. To solve them, it is easiest to decompose $\tilde{\mathbf{E}}_I$ into components, solve the problem separately for each component, and superpose the resulting reflected and transmitted waves at the end. Since $\tilde{\mathbf{E}}_I$ is transverse, it has components only along the z' (out-of-paper) and y' directions in Figure 24.4. We will solve the problem in detail for the latter case, and leave the former as Problem 24.18.

Consider an incident wave polarized parallel to the plane of incidence, so that $\tilde{\mathbf{E}}_I$ has no z-component (it is in the plane of the paper in Figure 24.7). Then $\tilde{\mathbf{B}}_I$ (Eq. 24.58) is perpendicular to the plane of incidence, so it has no normal (x) component; the nonzero components of $\tilde{\mathbf{E}}_I$ and $\tilde{\mathbf{B}}_I$ are (from Figure 24.7)

$$\tilde{E}_{Iy} = \tilde{E}_I \cos \theta_I, \qquad \tilde{E}_{Ix} = -\tilde{E}_I \sin \theta_I$$
$$\omega \tilde{B}_{Iz} = (\mathbf{k}_I \times \tilde{\mathbf{E}}_I)_z = k_I \tilde{E}_I$$
(24.76)

The symmetry of this problem under reflection in the plane of the paper suggests that the reflected and transmitted electric fields should also be in the plane of the paper (see Problem 24.19). Thus, Eqs. 24.76 hold for the transmitted fields as well. For the reflected wave, the signs are different:

$$\tilde{E}_{Ry} = \tilde{E}_R \cos \theta_R, \qquad \tilde{E}_{Rx} = +\tilde{E}_R \sin \theta_R$$
$$\omega \tilde{B}_{Rz} = (\mathbf{k}_R \times \tilde{\mathbf{E}}_R)_z = -k_R \tilde{E}_R$$
(24.77)

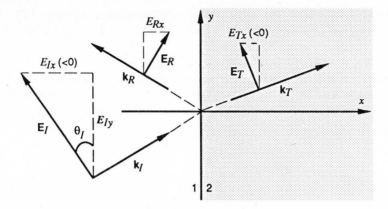

Figure 24.7   The electric field amplitudes for the case of polarization parallel to the plane of incidence. We show only the real parts of the complex vectors; a diagram showing the imaginary parts would look the same.

The tangential parts of the electric field in Eq. 24.72 are now just y-components:

$$\tilde{E}_I \cos \theta_I + \tilde{E}_R \cos \theta_I = \tilde{E}_T \cos \theta_T \qquad (24.78)$$

and Eq. 24.73 becomes (using Eq. 24.61 and multiplying by c/ω)

$$\mu_1^{-1} n_1 \tilde{E}_I - \mu_1^{-1} n_1 \tilde{E}_R = \mu_2^{-1} n_2 \tilde{E}_T \qquad (24.79)$$

Introducing the notation

$$\alpha \equiv \frac{\cos \theta_T}{\cos \theta_I}, \qquad \beta \equiv \frac{n_2 \mu_1}{n_1 \mu_2} \qquad (24.80)$$

(α and β can be expressed as functions of θ_I, n_1, and n_2 using Snell's law), Eqs. 24.78 and 24.79 are

$$\tilde{E}_I + \tilde{E}_R = \alpha \tilde{E}_T \qquad (24.81)$$

$$\tilde{E}_I - \tilde{E}_R = \beta \tilde{E}_T \qquad (24.82)$$

They can be easily solved:

$$\tilde{E}_R = \frac{\alpha - \beta}{\alpha + \beta} \tilde{E}_I \qquad (24.83)$$

$$\tilde{E}_T = \frac{2}{\alpha + \beta} \tilde{E}_I \qquad (24.84)$$

These are the **Fresnel equations** for the case of polarization parallel to the plane of incidence. (We never used the second pair of boundary conditions: Eq. 24.75 vanishes in this case, and Eq. 24.74 is redundant, as you can verify by substituting our solution into it and using Snell's law.)

The transmission coefficient is defined as the ratio of the transmitted to the incident energy per unit area of the interface. We must be a little careful in using the Poynting vector $\overline{S}$ (Eq. 24.65) for this; $\overline{S}$ was the energy per unit area **normal to the wavevector**. The area in the plane of the interface differs from this by a factor of $\cos \theta$; thus,

$$T \equiv \frac{\overline{S}_T \cos \theta_T}{\overline{S}_I \cos \theta_I}$$

$$= \frac{\cos \theta_T}{\cos \theta_I} \frac{v_1 \mu_1}{(v_2 \mu_2)} \left(\frac{2}{\alpha + \beta} \right)^2 \tag{24.85}$$

$$= \frac{4 \alpha \beta}{(\alpha + \beta)^2}$$

The reflection coefficient is easier, because the cosine factors cancel:

$$R = \frac{(\alpha - \beta)^2}{(\alpha + \beta)^2} \tag{24.86}$$

The reflection and transmission coefficients are plotted in Figure 24.8. You can

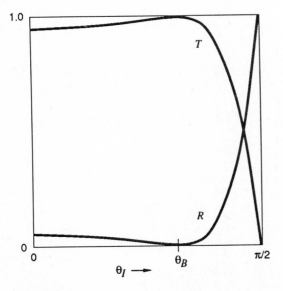

Figure 24.8 The reflection and transmission coefficients as a function of angle of incidence, for the case $n_2/n_1 = 1.5$, $\mu_1 = \mu_2$ (e.g., reflection from glass).

easily see from Eqs. 24.85 and 24.86 that $R + T = 1$. Note that there is an angle (called **Brewster's angle**, θ_B) at which there is no reflected wave. For the case in Figure 24.8, $\theta_B = 56°$. If the incident light is unpolarized (i.e., is a mixture of light polarized perpendicular and parallel to the plane of incidence) only the light perpendicular to the plane of incidence will be reflected: the reflected light is then completely linearly polarized. This phenomenon is exploited by light-polarizing sunglasses, which reduce "glare" (light reflected from horizontal surfaces, which is mostly horizontally polarized if the angle of incidence is near Brewster's angle) by allowing only vertically polarized light to pass.

▶   **Problem 24.15**

Show that Eqs. 24.83 to 24.86 reduce to the results of Section 24.3 when $\theta_I = 0$.   ◀

▶   **Problem 24.16**

Consider reflection at Brewster's angle with $\mu_1 = \mu_2$.

a. Show that $\theta_I + \theta_T = \frac{1}{2}\pi$. Hint: Use Snell's law to show $\sin 2\theta_I = \sin 2\theta_T$.
b. Show that $\theta_B = \tan^{-1}(n_2/n_1)$.   ◀

▶   **Problem 24.17**

Show that for $n_2 < n_1$ and incident polarization parallel to the plane of incidence, as we increase θ_I there is an angle θ_c beyond which Snell's law can no longer be solved for θ_T. Show that $R \to 1$ as $\theta_I \to \theta_c$. For $\theta_I > \theta_c$, there is no transmitted wave, and $R = 1$; this is the phenomenon of **total internal reflection**. Sketch R as a function of θ_I for the case $n_2/n_1 = 2/3$.   ◀

▶   **Problem 24.18**

Derive the Fresnel equations and the reflection and transmission coefficients for an incident wave polarized **perpendicular** to the plane of incidence. As always, you should check your result in any known limit. Also check that $R + T = 1$. Does R ever vanish? Compute $\tilde{E}_R/\tilde{E}_I$, $\tilde{E}_T/\tilde{E}_I$, R, and T for $\theta_I = 0$, $\frac{1}{6}\pi$, $\frac{1}{3}\pi$, and $\frac{1}{2}\pi$ in the case $n_2/n_1 = 1.5$, $\mu_1 = \mu_2$. Sketch each as a function of θ_I.   ◀

▶   **Problem 24.19**

Show that the reflected and transmitted electric fields are in the plane of Figure 24.7 if the incident electric field is.   ◀

SECTION 24.6   WAVES IN CONDUCTORS

In this chapter we have until now assumed our material to be an insulating dielectric: the free-current term was left out of the Maxwell equation for $d\mathbf{D}/dt$ (Eq. 24.17). In an imperfect conductor, there is a free current given by Ohm's law (Eq. 21.10)

$$\mathbf{j}_f(\mathbf{r}, t) = \sigma(\mathbf{r})\mathbf{E}(\mathbf{r}, t) \tag{24.87}$$

so that the more general Maxwell equation (Eq. 23.43) is

$$\frac{d\mathbf{D}}{dt} = \nabla \times \mathbf{H} - \mathbf{j}_f = \nabla \times \mathbf{H} - \sigma\mathbf{E} \tag{24.88}$$

The derivation of the most general sinusoidal plane wave then proceeds almost as in Section 24.4. However, we will see that if we insist that ω and $\mathbf{k}$ be real there is no solution in general. So we will allow them to be a complex scalar $\tilde{\omega}$ and a complex vector $\tilde{\mathbf{k}}$. The fields of the most general wave are then similar to Eqs. 24.52 and 24.55:

$$\mathbf{E}(\mathbf{r}, t) = \operatorname{Re} \tilde{\mathbf{E}} \exp(i\tilde{\mathbf{k}} \cdot \mathbf{r} - i\tilde{\omega}t) \tag{24.89}$$

$$\mathbf{B}(\mathbf{r}, t) = \operatorname{Re} \tilde{\mathbf{B}} \exp(i\tilde{\mathbf{k}} \cdot \mathbf{r} - i\tilde{\omega}t) \tag{24.90}$$

Substituting these into the Maxwell equations 24.16 and 24.88, we find that the $d\mathbf{B}/dt$ equation is formally (i.e., except for the fact that $\mathbf{k}$ and ω must now be interpreted as being complex) unchanged. It leads again to Eq. 24.58 expressing the magnetic field amplitude in terms of the electric field amplitude,

$$\tilde{\mathbf{B}} = \tilde{\omega}^{-1}\tilde{\mathbf{k}} \times \tilde{\mathbf{E}} \tag{24.91}$$

Recall that in the $d\mathbf{D}/dt$ equation, we needed to use Gauss' law (and the fact that the free charge density vanished) to prove that $\tilde{\mathbf{E}}$ is transverse, so a $\mathbf{k} \cdot \tilde{\mathbf{E}}$ term dropped out. It is no longer obvious that we can do this; the free charge density is not necessarily zero in a conductor. (In fact, because the real and imaginary parts of $\tilde{\mathbf{k}}$ may point in different directions, it isn't even obvious what "transverse" means; we will interpret it as meaning $\tilde{\mathbf{k}} \cdot \tilde{\mathbf{E}} = 0$.) So we must return to Eq. 24.88; substituting Eqs. 24.89 and 24.90 for $\mathbf{E}$ and $\mathbf{B}$ gives

$$-i\tilde{\omega}\epsilon\tilde{\mathbf{E}} = i\mu^{-1}\tilde{\mathbf{k}} \times \tilde{\mathbf{B}} - \sigma\tilde{\mathbf{E}} \tag{24.92}$$

(using the fact in Eq. 24.14, which is still true for complex $\tilde{\omega}$). This equation provides us with an alternate proof that $\tilde{\mathbf{E}}$ is transverse—it can be solved to give $\tilde{\mathbf{E}}$ as a constant times $\tilde{\mathbf{k}} \times \tilde{\mathbf{B}}$, which is transverse. Thus we can proceed to substitute Eq. 24.91 into Eq. 24.92 and use $\tilde{\mathbf{k}} \cdot \tilde{\mathbf{E}} = 0$, obtaining

$$-\epsilon\mu\tilde{\omega}^2\tilde{\mathbf{E}} = -\tilde{k}^2\tilde{\mathbf{E}} + i\sigma\mu\tilde{\omega}\tilde{\mathbf{E}} \tag{24.93}$$

where $\tilde{k}^2 \equiv \tilde{k}_x^2 + \tilde{k}_y^2 + \tilde{k}_z^2$ is complex. This gives a version of Eq. 24.61 with an extra term involving σ:

$$\epsilon\mu\tilde{\omega}^2 = \tilde{k}^2 - i\sigma\mu\tilde{\omega} \tag{24.94}$$

The first thing we notice about this relation between $\tilde{k}$ and $\tilde{\omega}$ is that it still has an i in it; it is not possible in general to make both $\tilde{\omega}$ and $\tilde{k}$ real. This complicates matters considerably. It is possible to make either $\tilde{\omega}$ **or** $\tilde{k}$ real, but the other must be complex. In discussing reflection from an interface, in which $\tilde{\omega}$ must be the same on the material and vacuum sides, we clearly should make $\tilde{\omega}$ real, so $\tilde{\mathbf{k}}$ must be allowed to be complex. In general the complex vector $\tilde{\mathbf{k}}$ has six real components. To avoid unnecessary complication we will restrict ourselves to cases in which the real and imaginary parts point in the same direction:

$$\tilde{\mathbf{k}} = \tilde{k}\hat{\mathbf{k}} \tag{24.95}$$

where $\tilde{k}$ is a complex number and $\hat{k}$ is a **real** unit vector. Then Eq. 24.94 becomes

$$\tilde{k} = \pm(\epsilon\mu)^{\frac{1}{2}}\tilde{\omega}[1 + i\sigma/\epsilon\tilde{\omega}]^{\frac{1}{2}} \qquad (24.96)$$

Taking the square root of a complex number is easy when it is in the "polar" form $Ae^{i\theta}$,

$$[Ae^{i\theta}]^{\frac{1}{2}} = \pm A^{\frac{1}{2}}e^{\frac{1}{2}i\theta} \qquad (24.97)$$

and any complex number $a + bi$ can be put in this form:

$$A^2 = a^2 + b^2, \qquad \theta = \tan^{-1}\left(\frac{b}{a}\right) \qquad (24.98)$$

The factor in the square brackets in Eq. 24.96 and its square root are shown in Figure 24.9.

No generality is lost by choosing the x-axis along $\hat{k}$. Taking a wave linearly polarized along the y-direction so $\tilde{\mathbf{E}} = \tilde{E}\hat{y}$, we get

$$E_y(x, t) = \text{Re}\,\tilde{E}\exp(i\tilde{k}\cdot\mathbf{r} - i\tilde{\omega}t) = e^{-k''x}\,\text{Re}\,\tilde{E}e^{ik'x-i\omega t}$$
$$\qquad (24.99)$$
$$= \tilde{E}e^{-k''x}\cos(k'x - \omega t)$$

Here we have used the common notations k' and k'' for the real and imaginary parts of $\tilde{k}$ respectively, and we have taken $\tilde{E}$ to be real. Equation 24.99 describes a spacially exponentially damped (or growing, if $k'' < 0$) cosine, as shown in Figure 24.10. Because it diverges as $x \to -\infty$, it is not a suitable solution of Maxwell's equations for a conducting material that extends throughout all of space. However, real conducting materials have boundaries, and we will see that it is possible to use the right-hand part of the wave in Figure 24.10 to fit boundary conditions. The e-folding distance d of this wave is

$$d \equiv \frac{1}{k''} \qquad (24.100)$$

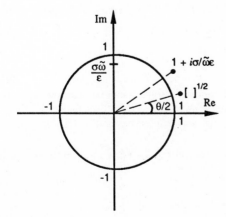

Figure 24.9 The complex plane, showing $1 + i\sigma/\tilde{\omega}\epsilon$ and its square root.

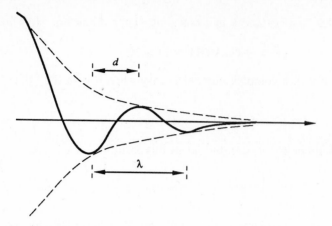

Figure 24.10 The electric field (Eq. 24.99) of a sinusoidal plane wave in a conductor. The exponential envelope $\pm e^{-k''x}$ is indicated by dashed lines.

and is called the **skin depth**; it is a measure of how far the field extends into a conductor from a boundary. The wavelength is defined by

$$\lambda \equiv \frac{2\pi}{k'} \tag{24.101}$$

The magnetic field can be computed from Eq. 24.91:

$$B_z(x, t) = \omega^{-1}|\tilde{k}|\tilde{E}e^{-k''x}\cos\left(k'x - \omega t + \tfrac{1}{2}\theta\right) \tag{24.102}$$

where $\tfrac{1}{2}\theta$ is the phase angle of $\tilde{k}$ (Figure 24.9). Thus B is **out of phase** with E—it lags behind in time.

■ **EXAMPLE 24.1 A POOR CONDUCTOR**   Skin depth and wavelength are, in general, very complicated if we try to express them directly in terms of the conductivity. However, we can simplify them if the imaginary term in Eq. 24.96 is very small. This limit, in which

$$\sigma \ll \epsilon\omega \tag{24.103}$$

is called the limit of a **poor conductor**.

It appears that this is also the "high frequency" limit: it can be achieved by increasing ω, for **any** σ. But we must be careful here; as we pointed out in Chapter 21, Ohm's law assumes that the fields are slowly varying, and it must be modified at high frequency.

In this limit the imaginary term in the square root in Eq. 24.96 is small, and we can use the binomial theorem

$$[1 + x]^{1/2} = 1 + \tfrac{1}{2}x + O(x^2)$$

to give

$$\tilde{k} \approx (\epsilon\mu)^{1/2}\omega\left[1 + \tfrac{1}{2}\frac{i\sigma}{\epsilon\omega}\right] \qquad (24.104)$$

so

$$k' \approx (\epsilon\mu)^{1/2}\omega + O(\sigma^2) \qquad (24.105)$$

(the conductivity doesn't affect the wavelength to lowest order in σ). The imaginary part is

$$k'' \approx \tfrac{1}{2}\sigma\left(\frac{\mu}{\epsilon}\right)^{1/2} \qquad (24.106)$$

so the skin depth

$$d \approx \frac{2(\epsilon/\mu)^{1/2}}{\sigma} \qquad (24.107)$$

is independent of frequency in this limit.

Consider the special case of a glass container in a microwave oven, with $\sigma = 10^{-10}$ mho/m. If the frequency $f = \omega/2\pi = 30$ GHz(gigahertz) $= 3 \times 10^{10}$ Hz, $\epsilon = 2\epsilon_0$, and $\mu = \mu_0$, then $\sigma/\epsilon\omega = 3 \times 10^{-11}$. This is a "poor conductor." Then Eq. 24.107 gives $d = 7 \times 10^7$ m — the lid of the container can be quite thick without blocking out the microwave energy. If you did this calculation for pure water, using its DC conductivity of 2×10^4 mho/m, you would find $d \approx 10^3$ m; it also would be transparent to the microwaves. This is wrong, of course: the problem is that the conductivity of water at microwave frequencies is many orders of magnitude larger than its DC conductivity. The actual skin depth is of order 10^{-1} or 10^{-2} m.

We now want to generalize our treatment of reflection and refraction to an interface at which one material (the one on the right, in Figure 24.11) is conducting. To simplify the algebra we will restrict ourselves to the case of normal incidence, so $\mathbf{k}_I$ (which is real because material 1 is nonconducting) is along the x-axis in Figure 24.11. Conservation of the tangential wavevector (Section 24.5) implies that the reflected and transmitted wavevectors $\mathbf{k}_R$ and $\mathbf{k}_T$ must be also. The same considerations as before determine $\mathbf{k}_R = -\mathbf{k}_I = -k_I\hat{\mathbf{x}}$. However, the transmitted wavevector $\tilde{\mathbf{k}}_T = \tilde{k}_T\hat{\mathbf{x}}$ is complex, and given by Eq. 24.96; choice of the $+$ sign is imposed by the requirement that E_y (Eq. 24.99) not diverge as $x \to \infty$. To write the boundary conditions, we must write expressions for all three fields. We will assume the incident wave is linearly polarized in the y-direction, so each field is given by an expression similar to Eq. 24.99, with different amplitudes $\tilde{E}_I$, $\tilde{E}_R$, and $\tilde{E}_T$. The fields are all transverse (tangent to the interface, in this case of normal incidence) so that the boundary conditions on the normal components of the fields are not needed. The continuity of the tangential E field gives exactly Eq. 24.72, and that of the tangential H field Eq. 24.73. These lead again to Eqs. 24.78 and 24.79 for $\tilde{\mathbf{E}}_R$ and $\tilde{\mathbf{E}}_T$, where the index of refraction for region 2 is defined by

$$\tilde{n}_2 \equiv \frac{c\tilde{k}_2}{\omega} \qquad (24.108)$$

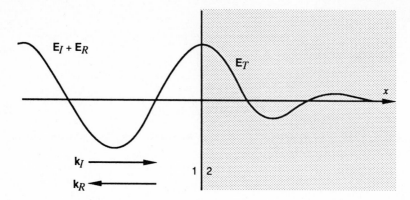

Figure 24.11 An interface between a nonconducting material (1, left) and a conductor (2, right) showing the electric field on both sides.

and is complex because $\tilde{k}_2$ is. At normal incidence, Eq. 24.80 gives $\alpha = 1$, but

$$\tilde{\beta} \equiv \frac{\tilde{n}_2 \mu_1}{n_1 \mu_2} \tag{24.109}$$

is complex. The formal solution is still given by Eqs. 24.83 and 24.84:

$$\tilde{E}_R = \frac{1 - \tilde{\beta}}{1 + \tilde{\beta}} \tilde{E}_I \tag{24.110}$$

$$\tilde{E}_T = \frac{2}{1 + \tilde{\beta}} \tilde{E}_I \tag{24.111}$$

But now the Fresnel coefficients multiplying $\tilde{E}_I$ are **complex**: the reflected and transmitted waves are out of phase with the incident one.

The reflection coefficient can be calculated just as in Section 24.5, and is the squared magnitude of the complex Fresnel coefficient:

$$R = \left| \frac{1 - \tilde{\beta}}{1 + \tilde{\beta}} \right|^2 \tag{24.112}$$

There is no transmission coefficient in this case—because the fields die out as we go into the conducting material, the Poynting vector does too: no energy is transmitted all the way through. The energy is **absorbed** (turned into heat) in the conductor (see Problem 24.20). We can define an **absorption coefficient** as the ratio of absorbed to incident energy per unit area of the interface; this is formally given by Eq. 24.85, and turns out to be $1 - R$, as you would expect from conservation of energy.

The reflection coefficient can be expressed directly in terms of the conductivity, permeability, and permissivity using Eqs. 24.96, 24.108, and 24.109, but it is usually easier to get it by calculating $\tilde{\beta}$ first. R is fairly simple in the limit of a poor conductor (Example 24.1), and in the opposite limit of a **good conductor**.

■ **EXAMPLE 24.2 A GOOD CONDUCTOR**   Consider the limit

$$\sigma \gg \epsilon\omega \tag{24.113}$$

(here $\sigma = \sigma_2$ and $\epsilon = \epsilon_2$; we assume medium 1 is a vacuum). Then we can ignore the left-hand side of Eq. 24.94, giving

$$\tilde{k} = (i\sigma\mu\omega)^{1/2} = e^{i\pi/4}(\sigma\mu\omega)^{1/2} \tag{24.114}$$

because $i = e^{i\pi/2}$. The real and imaginary parts are then the same $[e^{i\pi/4} = 2^{-1/2}(1+i)]$:

$$k' = k'' = (\sigma\mu\omega/2)^{1/2} \tag{24.115}$$

Assuming $\mu_2 = \mu_0$, Eq. 24.96 gives

$$\tilde{\beta} = \tilde{n} = \frac{c\tilde{k}}{\omega} = \left(\frac{\sigma}{\epsilon_0\omega}\right)^{1/2} e^{i\pi/4} \tag{24.116}$$

Thus, $\tilde{\beta}$ is large in our limit $\sigma \gg \epsilon\omega$ (assuming ϵ/ϵ_0 is of order 1). The reflection coefficient (Eq. 24.112) becomes

$$R = \left|(1 - \tilde{\beta}^{-1})^2(1 + \tilde{\beta}^{-1})^{-2}\right| \approx \left|1 - 4\tilde{\beta}^{-1}\right|$$

$$\approx 1 - 4\,\mathrm{Re}\,\tilde{\beta}^{-1} = 1 - \left(\frac{8\omega\epsilon_0}{\sigma}\right)^{1/2} \tag{24.117}$$

where we have used the binomial theorem to expand in powers of the small number $\tilde{\beta}^{-1}$.

For the case of silver ($\sigma \approx 6 \times 10^7$ mho/m) at the frequency of visible light ($\lambda = 500$ nm in vacuum, so $\omega = ck = 2\pi c/\lambda = 4 \times 10^{15}$ s^{-1}), this gives $R = 0.93$. The skin depth is $d = 1/k'' = 3$ nm, which is much less than a wavelength ($\lambda = 2\pi/k' = 20$ nm)—the wave does not look sinusoidal in the silver at all, but is almost exponential. This is a general property of good conductors: $d/\lambda = 1/2\pi$, so the wave has almost decayed away before it completes a single cycle. At the microwave frequency of Example 24.1, ω is 10^5 times smaller, so $R \approx 0.9998$. Thus, very little energy is absorbed by the walls of a microwave oven if they are made of a high-conductivity material: the microwaves just bounce around inside until they hit some food. The skin depth for these microwaves is $10^{2.5}$ times larger than at optical frequencies, but this is still only 1 μm.

▶   **Problem 24.20**

   a. Compute the time-averaged Poynting vector for the wave in a conductor given by Eqs. 24.99 and 24.91.

   b. Compute the rate $r(x, t) \equiv \mathbf{j} \cdot \mathbf{E}$ at which work is done on the free charges in a unit volume (called the **Joule heating rate**), and its time average $\bar{r}(x)$. Compute the total Joule heating rate (per unit area in the yz-plane) to the right of the plane $x = a$, $\int_a^\infty \bar{r}(x)\, dx$.

   c. Are the answers to parts (**a**) and (**b**) consistent with the law of conservation of energy?

◀

▶   **Problem 24.21**

The conductivity of sea water is about 5 mho/m, and $\epsilon = 80\epsilon_0$. Using the good-conductor (low-frequency) approximation, it is apparent that the skin depth gets very large at low frequencies. Compute the highest frequency at which it would be possible to communicate with a submarine at a depth of 20 m (assume that this requires a skin depth $d \geq 10$ m). Is your good-conductor assumption valid at this frequency? What vacuum wavelength does this correspond to? Transmitters at this sort of frequency have been considered for communication with nuclear submarines. The problem is that antennas are very inefficient if they are much smaller than a wavelength—how much would an antenna of length $\frac{1}{2}\lambda$ cost at \$1000/m?   ◀

THE CONTINUUM LIMIT

SECTION C.1 SINGULAR CHARGE DENSITIES IN THE CONTINUUM LIMIT: POINT, LINE, AND SURFACE CHARGES

The most fundamental definition of a continuum charge distribution is that it is a prescription for calculating how much charge is in each particular region of space—that is, a prescription for determining a discrete charge distribution for any lattice with any spacing dr.

In solving Poisson's equation in continuum electrostatics problems (Section 16.7), we have assumed that the continuum charge distribution is described by a continuous function $\rho(\mathbf{r})$, which can be evaluated at cell centers to obtain the corresponding discrete distribution. (We omit here the subscript we have used previously to indicate that ρ is not discrete; this is implied by the use of the continuous argument $\mathbf{r}$.) However, in other continuum problems we have encountered charge distributions that cannot be described in this way: point, line, and surface charges. These are "singular" distributions, meaning that there are special places where the density ρ diverges as $dr \to 0$. We have described these by using instead of ρ a charge q, linear charge density λ, and a surface charge density σ respectively. In each case, there was a procedure for determining the corresponding discrete density. In the case of a surface charge, for example, the discrete density was determined by the fact that the charge $\rho(c)\,d\tau$ in a cell containing an area da of the surface is $\sigma\,da$.

We have not yet discussed the possibility that more than one type of charge density (point, linear, surface, or volume) is present at once. In dielectrics this is a distinct possibility. For example, a piece of dielectric may have a volume charge density and also a bound surface charge such as that on a dielectric in a capacitor (Section 18.3), which cannot be described by a volume density [$\rho(c)$ diverges as $dr \to 0$]. The purpose of the present section is to define what we mean by such a "hybrid" continuum distribution.

The most general continuum charge distribution (Figure C.1) is described by

a. the position and magnitude of each point charge q.

b. the location and linear charge density $\lambda(\mathbf{r})$ of each (possibly curved) line charge; $\lambda(\mathbf{r})$ is a continuous function, but only its values on the curve are relevant.

c. the location [specified, for example, by the equation $f(x, y, z) = 0$ of the surface] and surface charge density $\sigma(\mathbf{r})$ of each surface charge.

d. the volume charge density $\rho(\mathbf{r})$.

529

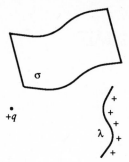

Figure C.1 A continuum charge distribution containing one of each of the possible types of charge density: point, line, and surface charges.

To give a precise meaning to this continuum distribution, we must give a prescription for determining the discrete density for a particular cell lattice. In general this can be taken to be

$$\rho(c)\, d\tau(c) = \rho(\mathbf{r})\, d\tau + \sigma(\mathbf{r})\, da + \lambda(\mathbf{r})\, dl + Q(c) \tag{C.1}$$

for the cell c shown in Figure C.2. Here da is the area of the surface contained in the cell (which is, of course, zero if the cell does not touch the surface), dl is the length of the line charge contained in the cell (also zero if they don't touch), and $Q(c)$ is the sum of the point charges in the cell (in Figure C.1, it is q if the charge is in c, 0 if not). The continuous function $\rho(\mathbf{r})$ can be evaluated at the cell center $\mathbf{r}$, and σ and λ at the centers of da and dl respectively. (You may object that this does violence to cases in which the functions vary rapidly over the cell. It is true that it would be somewhat more accurate to use average values of λ, σ, and ρ over the length, area, or volume in the cell. However, evaluating them at the center is much easier—it is what's called a "quick and dirty" expedient—and in the continuum limit is doesn't matter.)

We also need a way to reverse this process, that is, to get the continuous functions describing the continuum charge distribution from the discrete density. This is easiest for the most singular distribution, the point charge: The charge q at a point $\mathbf{r}$ is given by

$$q = \lim_{dr \to 0} \rho(c)\, d\tau(c) \tag{C.2}$$

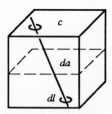

Figure C.2 A cell c containing fragments of a surface charge, a line charge, and possibly point charges (not shown).

(where c is the cell containing $\mathbf{r}$) because when we use Eq. C.1 for $\rho(c)\,d\tau(c)$ all terms but the point charge term $Q(c) = q$ vanish as $dr \to 0$. If we then subtract off this most-singular part, we can extract the next-most-singular part, the linear charge density, from what's left:

$$\lambda(\mathbf{r}) = \lim_{dr \to 0} \frac{[\rho(c)\,d\tau(c) - Q(c)]}{dl} \tag{C.3}$$

This follows, because when we substitute Eq. C.1 for ρ, the first two terms have factors $d\tau/dl$ and da/dl, which vanish as $dr \to 0$. (Normally we use coordinate axes so dl and da are along axes; for simplicity we can also assume cartesian coordinates so $dl = dr$, $da = dr^2$, $d\tau = dr^3$.) You must be careful to apply Eq. C.3 **only** for $\mathbf{r}$ on the line charge; otherwise the denominator dl vanishes for small enough dr. Continuing in the same vein, we get

$$\sigma(\mathbf{r}) = \lim_{dr \to 0} \frac{[\rho(c)\,d\tau(c) - \lambda(\mathbf{r})\,dl - Q(c)]}{da} \tag{C.4}$$

and

$$\rho(\mathbf{r}) = \lim_{dr \to 0} \frac{[\rho(c)\,d\tau(c) - \sigma(\mathbf{r})\,da - \lambda(\mathbf{r})\,dl - Q(c)]}{d\tau} \tag{C.5}$$

Thus we can calculate all the continuous functions that characterize a continuum charge distribution, from the discrete density in the limit $dr \to 0$.

■ **EXAMPLE C.1** Consider the continuum charge distribution shown in Figure C.3. It has a point charge q_0 at the origin, a uniform (**r**-independent) line charge with $\lambda(\mathbf{r}) = \lambda_0$ along the x-axis, a uniform surface charge $\sigma(\mathbf{r}) = \sigma_0$ on the xz plane, and a uniform volume charge $\rho(\mathbf{r}) = \rho_0$ everywhere. For the cubic discrete lattice whose $z = 0$ cross section is shown in Figure C.3b, the discrete charge densities in the cells labeled 1, 2, and 3 are

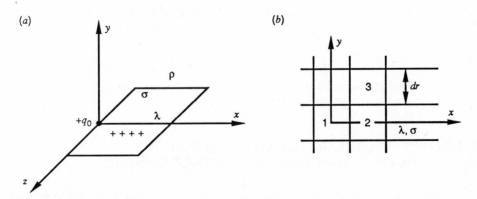

Figure C.3 (a) A continuum charge distribution, (b) a discrete lattice for this distribution, showing the cell labeling.

given by Eq. C.1 as

$$\rho\left(c_1 = (0,0,0)\right) = \rho_0 + \frac{\sigma_0}{dr} + \frac{\lambda_0}{dr^2} + \frac{q}{dr^3}$$

$$\rho\left(c_2 = (dr,0,0)\right) = \rho_0 + \frac{\sigma_0}{dr} + \frac{\lambda_0}{dr^2} \qquad\qquad (C.6)$$

$$\rho\left(c_3 = (dr,dr,0)\right) = \rho_0$$

At any cell c on the surface charge but not the line charge (i.e., whose center has $y = 0$ but $x > 0$ and $z > 0$) we have

$$\rho(c) = \rho_0 + \frac{\sigma_0}{dr} \qquad\qquad (C.7)$$

▶ **Problem C.1**

   a. Evaluate Eqs. C.2 to C.5 for q and the continuous functions ρ, σ, and λ at $\mathbf{r} = 0$, starting with the discrete density in Eq. C.6. Verify that you get back what we started with.

   b. Do the same at $\mathbf{r} = (x, 0, z)$ for $x > 0$, $z > 0$, from Eq. C.7.   ◀

▶ **Problem C.2**

   a. Compute the discrete density, at cells centered at $(0,0,0)$, $(x,0,0)$, $(x,y,0)$, and $(x,0,z)$ [x, y, z represent positive numbers], for the continuum distribution specified by $q = q_0$, $\lambda(x,0,0) = Ax^2$ (note that λ need be defined only on the line), $\sigma(x,0,z) = Bz$, $\rho(x,y,z) = \rho_0 + Cy$. The point, line and plane charges are situated as in Figure C.3.

   b. Substitute the answer to part (**a**) into Eqs. C.2 to C.5 and verify that we get $\lambda(\mathbf{r})$, $\sigma(\mathbf{r})$, and $\rho(\mathbf{r})$ back for $\mathbf{r} = 0$, and for $\mathbf{r} = (x,0,0)$.   ◀

▶ **Problem C.3**

   Repeat Problem C.2, but in Eq. C.1 use the **average** values of $\rho(\mathbf{r})$, $\sigma(\mathbf{r})$, $\lambda(\mathbf{r})$ over the volume, area, or length in the cell instead of the values at the center. (This will change the answer only for λ. Why?) Do you still get the same answer back in (**b**)?   ◀

▶ **Problem C.4**

   A continuum distribution gives the discrete density (for each choice of dr)

$$\rho\left(c \text{ at } (0, y, 0)\right) = A + Cy + \frac{B}{dr^2}$$

$$\rho\left(c \text{ at } (x, y, z)\right) = A + Cy \qquad \text{when } x \neq 0 \text{ or } z \neq 0$$

   Find the locations of the singularities (point, line, or surface charges) and the continuous functions that describe them.   ◀

SECTION C.2   SINGULAR CURRENT DENSITIES IN THE CONTINUUM LIMIT: THIN-WIRE AND SURFACE CURRENTS

The most general continuum current includes currents in infinitely thin wires, surface currents (which could be magnetization currents as in Section 23.6 or free currents in thin solenoid windings as in Section 20.2), and volume currents.

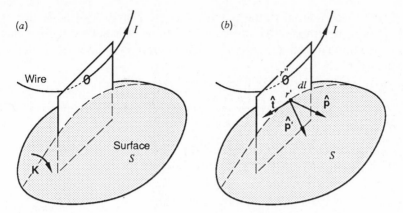

Figure C.4   (*a*) A continuum current distribution having all three kinds of current density: thin-wire, surface, and (not shown) volume. (*b*) The unit vector $\hat{\mathbf{p}}$ normal to the face f, and two unit vectors tangent to the surface S: $\hat{\mathbf{p}}'$ (normal to dl) and $\hat{\mathbf{t}}$ (along dl).

A wire current is specified by giving the 1D path of the current and its value $I(\mathbf{r})$ at each point. (If I changes from point to point, a charge $\lambda(\mathbf{r})$ will build up—in a static situation I is uniform along the wire.) A surface current is specified by giving the equation of the surface and the surface current density $\mathbf{K}(\mathbf{r})$.

To determine the discrete current $j(f)$ for the square face f shown in Figure C.4, we use an analog of Eq. C.1:

$$j(f)\, da(f) = j_p(\mathbf{r})\, da(f) + K_{p'}(\mathbf{r}')\, dl + I(\mathbf{r}'') \tag{C.8}$$

Here $da(f)$ is the area of the face f and j_p is the component of the continuum function $\mathbf{j}(\mathbf{r})$ along the unit vector $\hat{\mathbf{p}}$ normal to f (Figure C.4b) evaluated at the center of the face. The length of the intersection of the surface S and the face is dl, and $K_{p'}(\mathbf{r}')$ is the component of the continuum function $\mathbf{K}$ normal to the curve dl, evaluated at the center $\mathbf{r}'$ of dl. Finally, $I(\mathbf{r}'')$ is the current in the wire, evaluated at the point where it intersects the face.

Equation C.8 defines a general continuum current distribution, in that it determines a discrete current density $j(f)$ for each cell size dr. As in the case of the charge distribution of the previous section, we can recover the continuum functions from the discrete density by taking limits:

$$I(\mathbf{r}) = \lim_{dr \to 0} j(f)\, da(f) \tag{C.9}$$

where f is the face intersecting the wire closest to $\mathbf{r}$. You may verify this by substituting Eq. C.8 for $j(f)$; the terms involving $j_p(\mathbf{r})$ and $K_{p'}(\mathbf{r}')$ vanish in the limit. Subtracting the most-singular I term, we can get $\mathbf{K}(\mathbf{r}')$: for each of the three axis directions p of our discrete lattice

$$K_{p'}(\mathbf{r}') = \lim_{dr \to 0} \frac{[j(f)\, da(f) - I(f)]}{dl} \tag{C.10}$$

where f is the face normal to the p-axis closest to $\mathbf{r}'$ and $I(f)$ is the wire current through f. ($\mathbf{K}$ is determined by two components, so it is sufficient to evaluate this for two of the three directions, each of which determines a different p'.) Finally, for each direction p,

$$j_p(\mathbf{r}) = \lim_{dr \to 0} \frac{\left[j(f)\, da(f) - I(f) - K_{p'}(\mathbf{r}')\, dl \right]}{da(f)} \tag{C.11}$$

where f is the face closest to $\mathbf{r}$ and dl is the length of the intersection of f with the surface S (Figure C.4b)—it is zero if f does not cut the face.

PROGRAM GUIDES

The microcomputer programs that accompany this text are distributed through the instructor, so that updates for new computers can be distributed rapidly. The inside back cover of this book has been left blank so that the disk sleeve can be glued to it if desired. A guide for the TURBO Pascal (3.01A) version of the program disk is given in this Appendix, along with modifications necessary for the AppleSoft Basic version. (Apple Pascal versions exist for the programs through POP; these are very similar to the TURBO version.) Whichever version you have, you should first read the file named "WHATSNEW" on your disk for any additional modifications to the user's guide, and mark these modifications in the guide below.

SECTION G.1 USER'S GUIDE TO IBM PC TURBO PASCAL DEMONSTRATION PROGRAMS (VERSION 1.0)

These programs will run on any PC, PCjr, AT, XT, or compatible; some of them require 128 KB of memory. [Until a TURBO 4.0 version is completed, they will require a color graphics adaptor (CGA)]. To run the first program, turn on the PC with a DOS disk in it (label side up, oval cutout first). After it boots and you have pressed the "enter" key twice for the date and time, it will give the A⟩ prompt. Insert the program disk, type "mass" (no quotes), and push the "enter" key. You may eventually want to copy some or all of the programs onto a disk formatted with DOS on it so you can start it up with only one disk.

1. MASS

This program simulates the motion of a mass attached to a spring obeying Hooke's law. The displacement and velocity of the mass are updated according to Eqs. 1.4 and 1.5 of the book. The program asks to be told what time interval dt and spring constant per unit mass (k/m) to use. Reasonable values are 0.03 and 1.0. It also asks for an initial displacement and velocity; try 0.0 and 1.0 at first. It then asks how long (i.e., for how much real time) you want to simulate the motion; to start, try 10. It prints out the displacement and velocity at the end of this time. During the simulation it displays the mass as a solid square "icon," moving between rigid walls at $x = 1$ and $x = -1$.

It can be used to simulate a pendulum in real time; this is particularly effective if the real pendulum swings just above the moving pixel. Choosing parameters for this is described in Example 1.1 of the main text.

2. MASSES

The MASSES program simulates the motion of a chain of masses connected by springs. It will ask you whether you want to simulate longitudinal or transverse motion; if you haven't come to the latter in the text you should select "l" (this is the default, which means it is used if you just press "enter"). It also asks whether you want periodic boundary conditions; again you can just press "enter" if you haven't come to this. If you don't ask for periodic boundary conditions, it uses zero-displacement BC's: the first and last masses are attached to rigid walls. It asks how you want to enter the initial displacements and velocities: if you are starting with the default (longitudinal) display, type "h" to enter them by hand (the options for sinusoidal and pulsed initial conditions generally show up better with the transverse display). The program lists default values for the time interval dt, the chain length N, the spring constant per unit mass k/m, and the total time to be simulated. One can use these values, or enter new ones. The initial conditions are specified by giving the mass label r, followed by its displacement and velocity. When you are using a longitudinal display, the displacements and velocities are in units of the mass separation, so a displacement of 1.0 moves a mass to its neighbor's equilibrium position. When you are using the transverse display, they are in units of pixels (the screen height is 200 pixels). When $r > N$ is entered, the simulation begins, assuming unmentioned masses have zero displacement and velocity.

For example, entering $r = 1$, $x = 0$, $v = 1$, and $r = 999$ gives the initial conditions of Section 1.2 of the main text (the leftmost mass banged with a hammer).

3. TVS

The functions attributed in the text to the generic program TVS are carried out by the actual program MASSES (described above) on the PC disk. Press "t" when it asks you whether you want a transverse or a longitudinal display. The longitudinal display gets very crowded if there are more than six masses or so; a mass can't move very far before its icon merges with that of its neighbor. The transverse display is less crowded because it uses the whole screen, instead of just one line of it. You can think of it as just a transverse representation of a motion that is really longitudinal, or as a picture of a real transverse motion. (MASSES and TVS, as well as several other generic programs, were combined on the PC disk to save space; a TurboPascal .COM file takes about 15 kB even if the actual program is very short, because many utility procedures are included even if they are not called.)

You can construct pulses and sinusoidal waves with MASSES (answer p or s when it asks how you want to specify the initial conditions). In the sinusoidal case, the program asks for the number of half-waves you want (M, as in Eq. 2.3 of the text). This must be an integer for the motion to stay sinusoidal, but nonintegers are allowed (for example, $M = 2.5465$ gives Figure 2.1). MASSES prints the amplitude A in the initial condition $x(r, 0) = A \sin(Qr)$. The pulses generated by TVS are of the Gaussian form $x(r, 0) = A \exp[4(r - r_0)^2 / W^2]$ where the width W (try 2 to 6) and amplitude A are solicited by the program. Three options for choosing the velocity are allowed: zero velocity (yielding a rightward and a leftward pulse, as described in Section 2.3), a pure rightward pulse (Section 2.6), or a pure leftward one. As many pulses as desired can be superimposed; when you have enough and want to look at the evolution, type 0 for the pulse width. You will note that these pulses are reversed in phase upon reflection from the ends of the system because of the fixed-end boundary condition.

4. RING

As with TVS, the functions of the generic program RING are performed by MASSES, by answering "y" to the question about periodic BC's (these are described in Section 2.4 of the text). The system is then periodic with (spatial) period N. The important effect of this difference is that pulses are not reflected from the boundaries, but appear to jump to the other end of the chain. Note also that if you choose initial velocities whose sum (the center-of-mass velocity of the system) is nonzero, the masses will eventually drift off the screen. In constructing sinusoidal initial conditions, you need an integer number of **full** waves to avoid a sharp corner at the boundary. The program constructs traveling waves if it is using periodic BC's and standing waves if not.

5. MAX1

This program solves the discrete 1D Maxwell equations, which are equivalent to the equations for stress and velocity in a ring of masses (Eqs. 2.28 and 2.29). These are in turn equivalent to the equations solved by MASSES; the new feature of MAX1 is that it displays the stress and velocity as well as the displacement. The displacement is displayed on the top half of the screen, which looks just like the display of MASSES with its transverse option. The bottom half of the screen displays the stress (by a curly line whose length is proportional to $|s|$, pointing upward from the baseline if $s > 0$, downward if $s < 0$) and the velocity (by a similar, but straight, line). The stress determines the displacement field only up to an additive constant; this constant is chosen to center the masses in the top half of the screen.

6. POP

This program computes the evolution of two interacting populations. If you want to use the linear equations of Section 4.2 of the text, answer "n" to the question about nonlinear equations. First try the default choices: non-interacting populations ($V_W = N_H = 0$) with growth rates $R_H = 0.01$, $R_W = -0.02$, initial populations $H = 10$, $W = 10$, and $dt = 1$, $t_{max} = 50$. The hare population grows exponentially, and the wolf population declines exponentially. You can reduce the initial net growth rate dH/dt of the hares to zero by using voracious wolves with $V_W = 0.01 = R_H/W$. Note that the wolves still die off, so eventually the hare population begins to increase. Adding a hare nutritiousness $N_H = -R_W W/H = 0.02$ prevents the wolf population from declining: this is a steady state. It is interesting to examine effects of wolf overpopulation: try $V_W = 0.005$, $N_H = 0.005$ with $H, W, T_{max} = 3, 6, 250$ (the wolf population declines rapidly due to insufficient food but then recovers), or $3, 15, 400$ (it declines faster and barely recovers), or $3, 18, 260$ (the wolves eat the hares into extinction, and then become extinct themselves).

7. NONLIN

You can use POP to solve nonlinear equations (see Section 4.2) by answering "y" to the appropriate question. To see how the nonlinearity prevents the wolf populations from going negative when there is an initial overpopulation, try $R_H = 0.01$, $R_W = -0.02$, $V_W = 0.0015$, $N_H = 0.005$, with initial $H = 3$, $W = 19$, using $dt = 2$, $t_{max} = 1000$.

8. MIG

MIG simulates 1D diffusion, using Eqs. 4.13 and 4.14. It performs the functions of the generic programs MIG1 and MIG2 discussed in the text. The only difference is whether the next-nearest neighbor as well as the nearest-neighbor cell to a face is allowed to contribute to the migration there. To get the simple equations of Section 4.3, answer "n" to the question about k' and MIG2. MIG allows you to set up initial conditions and save them on files for later use. A few such data files are on the distribution disk, for use in the problems. The cell populations are displayed as vertical bars, whose length is proportional to the population.

9. SOURCE

The function of generic program SOURCE is also performed by MIG; answer "y" to the question about sources and sinks. The program allows you to set the initial density field starting from a uniform density instead of zero. You can get just a source by setting the sink rate to zero.

10. MAXWELL

MAXWELL simulates the evolution of electric and magnetic fields using Maxwell's equations, in the form of Eqs. 9.31 and 9.32 of the text. The first time you use MAXWELL you should ask for the demonstration of the field displays. MAXWELL is designed for fields that are created by pseudo-two-dimensional currents parallel to the plane of the screen, which is taken to be the xy-plane. As shown in Section 10.2, such currents have only x and y components, and are independent of z (the coordinate perpendicular to the screen); the nonzero components are $j_x(x, y)$ and $j_y(x, y)$. The allowed electric fields have the same symmetry, and the magnetic field has a component only perpendicular to the screen: $B_z(x, y)$.

The screen display is a cross section of the 3D cell lattice (Figure G.1b), described in Section 10.2, in a plane passing through the centers of the cells. The cells appear as squares, separated by lines that are cross sections of the square faces of the cells (only the faces normal to the x and y directions appear). The corners of the squares are the cross sections of edges of the 3D lattice. As in the main text, the electric field is defined at faces (i.e., the sides of the squares in Figure G.1a) and the magnetic field at edges (i.e., the corners of the squares). The electric field is depicted by a bar normal to the face (Figure G.1c) that starts at the center of the face and points in the direction of the field. The length of the bar is proportional to the magnitude of the field. The magnetic field B is depicted by a roughly square cloud of dots at the cell corner, whose color (or brightness on a monochrome monitor) indicates the sign; a positive B points **away** from the viewer. The number of dots is proportional to the magnitude of the B field.

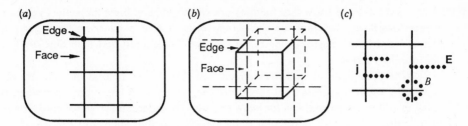

Figure G.1   (a) A schematic view of the screen display of program MAXWELL, which is a cross section of a 3D lattice (b). (c) The actual appearance of the screen.

MAXWELL allows you to choose arbitrary initial E and B fields (or even to change them in mid-simulation, although this is not recommended). You can impose an AC or DC current $J(f, t) = J_{amp}(f) \cos(\omega t)$, with arbitrary amplitude field $J_{amp}(f)$. In order that the novice not need to worry about setting the fields (or the options and other complications described below), the initial conditions and options can be read from a file, of which several (e.g., TACHMAN, CAPACITR, SOLENOID) are included in a subdirectory on the distribution disk.

Even if the options and initial conditions are read from a file, the program allows you to change them. (Be careful if you change the system size: this jumbles the initial fields, and you must re-set them everywhere.) Even if you do not change the options, the program allows you to ascertain the values of the fields (in SI units) or change them. This is done by moving a cursor (shaped vaguely like a voltmeter) by pressing the arrow keys. The cursor moves in increments of $\frac{1}{2}dr$, so it can be put over a corner, a cell center, or a face center. In each case the value of the field at the cursor position is displayed on the screen. You may change the E, B, ρ, or J fields by pressing e, b, r, or j and entering a number. You can read fields outside the screen window, but of course you can't see the cursor then — you have to track it by its x and y coordinates, which are always displayed. If you want to display the field in a region outside the present window, you can move the window. Pressing "c" centers the window on the present cursor position. Another option, "t," toggles the current readout on the meter between the amplitude J_{amp} (this is the default, because you normally want to set this at the beginning) and the actual current $J_{amp}\cos(\omega t)$. The graphics display always gives the actual current; while you are setting them at $t = 0$ the two are of course the same.

To get the program to run at a reasonable speed, all the fields are stored as integers (floating point operations are much slower). As a consequence, the program must tread a rather narrow line between arithmetic overflow and excessive roundoff error. If the program suddenly stops while the fields are being updated, chances are that an overflow has occurred; try it again with smaller values for the fields. The three-digit field readout should not be taken to reflect the accuracy of the values; the uncertainty in the current is $1/16$ pixel, and that of the other fields is usually somewhat smaller. There is a difference between screen overflow (which only affects the graphics display) and arithmetic overflow (which causes the numbers to be wrong). Most arithmetic overflows will cause a warning to be printed and the program to stop; this is not absolutely guaranteed, however, because in the interests of speed overflow checks are only made on the arithmetic operations most likely to overflow. Screen overflows cause a "hole" to appear in the affected field display;

if you don't need to know that field (or are willing to read it off the meter) you can ignore screen overflows. To clean up the screen after a screen overflow (and get rid of the "screen overflow" message), press "h" for the help screen and then any key to re-write the graphics screen.

When you input or edit the options, you first choose dt and the cell size dr. These determine a dimensionless light speed $c\,dt/dr$, which is an important parameter. Because integer arithmetic is used, this (or more precisely its square) must be rational. The program automatically looks for a rational approximation to the value you choose. If you have trouble with overflow, it may be because the numerator of this quantity is large; try changing dt or dr a little to make it smaller.

The next thing you choose is the size of the square display window (as a multiple of dr), which can be larger (if you want to see the whole system) or smaller (if you want to see only a part of the system, with higher resolution) than the system, whose size you choose later. The next options are the boundary conditions; you can choose periodic boundary conditions in the x and/or y directions (the alternative is a $B = 0$ boundary condition).

There are several options that are designed to make it possible to simulate the evolution of static fields. This is not trivial in a finite system — when a current flows to create a charge distribution, electromagnetic waves are generated that carry their energy away to infinity in an infinite system. In a finite system they bounce back and forth forever (recall that the discrete Maxwell equations conserve energy). The longer-wavelength waves can be absorbed by a conducting boundary; the program allows you to choose its width and conductivity, with suitable defaults. Unfortunately, short waves of length $2\,dr$ are not efficiently absorbed, and some other mechanism is needed to damp them. The best way to damp these is to introduce an extra term into the Maxwell equations, allowing diffusion of the E or B field: if D is the diffusivity of B, for example, the Maxwell equation for dB_z/dt gets an extra term $D(d^2B_z/dx^2 + d^2B_z/dy^2)$. (No d^2/dz^2 term is needed; this might be called a **transverse** diffusivity.) If the diffusivity of B is nonzero, B will eventually diffuse to the boundary and disappear; thus you would use this in a system evolving to an electrostatic field, in which B would disappear anyway in an infinite system. Similarly, a diffusivity of E causes E to diffuse away and disappear, and would be used in computing a magnetostatic field. (Don't make them both nonzero: then **both** of the fields will disappear.)

Introducing these diffusivities may seem to be an extreme thing to do, because it modifies the updating equations everywhere (not just at the boundary). However, if properly done it is quite innocuous, in two senses: (1) It has no effect in the continuum limit, and (2) even for finite dr, it has no effect on the static fields. To show (1), note that in taking the continuum limit we can let D become arbitrarily small, say $D = A\,dr$ for some constant A. It can be shown that this is still adequate to damp the short waves (which keep getting shorter as $dr \to 0$) but has no effect on a wave of fixed, finite wavelength. To show (2) for the case of electrostatics, note that the static fields are determined by Maxwell's equations with B set to zero, in which the diffusivity of B is irrelevant. In the case of magnetostatics, a transverse diffusivity of E causes Gauss' law to be violated, but after $E \to 0$ you are still left with the correct magnetostatic field.

To convince yourself that these heroic measures are necessary to achieve convergence to static fields in a reasonable time, choose one of the static-field-creating data files (e.g., DIPOLE or SOLENOID) and edit the options to turn off the absorber and/or the diffusivity.

After setting the field-absorbing options, you will be asked to set the frequency $f = \omega/2\pi$, in gigahertz, of the current ($f = 0$ gives DC, of course). Then you set the duration of the current (use a large number like 999 if you want it to stay on during the whole simulation, dt if you want a transient DC current to create a charge and then disappear, or anything in between). Finally you set t_{max}, the total time to simulate.

During the simulation, after each update of the E or B field (i.e., at every multiple of $\frac{1}{2}dt$) you are allowed to choose whether you want to step the fields forward $\frac{1}{2}dt$ at a time (type s) or continue nonstop until you reach the total time you specified previously. While you are stepping you can use the meter to read the fields after each update. At the end you can use the meter again to read the fields; when you want to quit reading them press "q" and you have the option of continuing the simulation by entering a new t_{max}, or stopping it by pressing "enter."

SECTION G.2   MODIFICATIONS TO GUIDE FOR APPLESOFT BASIC DEMONSTRATION PROGRAMS (VERSION 1.0)

To run these programs, turn on your Apple-compatible microcomputer with the DOS disk in it. After it has finished booting, insert the program disk (label side up, oval cutout first) and type LOAD "MASS" (or other program name) and "RUN."

1. MASS

Use 0.1 instead of 0.03 for dt; the Apple is somewhat slower. To improve speed, the program uses low-resolution graphics (which has large pixels, half the size of a character) — the icon is a single pixel. Note that very large or small numbers are printed in scientific notation: $2.5E - 5$ means 2.5×10^{-5}.

2. MASSES

The AppleSoft version of MASSES simulates only longitudinal motion, with zero-displacement boundary conditions walls.

3. TVS

TVS is similar to MASSES, but simulates only **transverse** motions, with zero-displacement boundary conditions. Displacements are specified in pixels, instead of in units of the mass separation as in MASSES.

4. RING

RING simulates transverse motions of a ring of masses. It is identical to TVS except that it uses periodic boundary conditions and constructs traveling waves instead of standing waves when the "s" option is used.

5. MAX1

The AppleSoft version displays the stress and the velocity by differently colored bars; it does not display the displacement.

6. POP

The AppleSoft version uses only **linear** equations (version NONLIN is a separate program). You can alter the total time TM of the simulation by changing a BASIC line. For the wolf overpopulation simulations, replace the values given in the PC guide by $H, W, TM = 5, 10, 300; 5, 23, 400;$ and $5, 30, 260.$

7. NONLIN

This is similar to POP, but for nonlinear equations. Use initial $H = 3$, $W = 38$ in the suggested simulation.

8. MIG1

This program computes the cell populations for a system of migrating objects, using the equations of Section 4.3. You must enter the coefficient K in these equations. The AppleSoft version does not use initial-condition files, but gives you a choice of three options for setting the initial populations: (1) by hand, in a manner like that described for the displacements in MASSES, (2) as a bell-shaped curve (like the pulse option in TVS), or (3) as an almost-linear (zig-zag) density (this is for demonstrating that such a density is converted rapidly into a linear one).

9. MIG2

This is just like MIG1 except you are also allowed to enter the coefficient K' describing the effect of the next-nearest neighbor cell population on the migration at a face.

10. SOURCE

This is another modification of MIG1, which allows you to enter a source rate as well as a population for each cell. They must be entered by hand (you don't have options 2 and 3 of MIG1) but you can specify a "background population," which is used for the cells you don't mention.

11. MAXWELL AND MAXDEMO

The demonstration of the AppleSoft field displays is in a separate program, MAXDEMO, which you should run before you use MAXWELL for the first time.

The screen display is similar to that described in the PC guide, but since it uses low-resolution graphics the AppleSoft B-field display is a block of large square pixels. The color indicates the sign of B — on a black and white screen, bright means it points out toward you and the dimmer color means it points away. The current is not displayed.

The AppleSoft version does not allow AC currents or read initial conditions from files. It will ask whether you want to input initial electric and magnetic fields and/or currents. These are set by moving the cursor to the appropriate face or edge. The I, J, K, and M keys move the cursor up, left, right, and down (as suggested by their arrangement on the keyboard); when you get to someplace where you want to put a Field or a Current, type F or C and then input the numerical value. The sign convention is that a positive field or current on a face normal to a given axis points in the direction in which the cursor was last moved along that axis (ignoring intervening moves along the perpendicular axis). Unlike the PC version, a magnetic field is positive if it points out of the screen. The program will ask whether you want the current to be transient (meaning it is nonzero only at time $\frac{1}{2}dt$: type P, for Pulse) or steady (type S). You must specify whether you are looking for a static electric field (B is then allowed to diffuse, as described in the PC guide) or a static magnetic field (so E should diffuse). It will ask what total time you want to simulate. At the end it will display the fields numerically. Usually all the calculated fields will not fit on the screen at once, so it displays them in slices, starting from the left. The field at each face or edge is printed at a screen position that corresponds to the position of the face or edge. Face orientations are to the right and **downward** (not upward). Cell centers (at which there is no field defined) are indicated by the letter C; this allows you to tell which numbers are at faces and which are at edges. The numbers are truncated (not rounded) to one digit after the decimal point, so you can only trust them to ± 0.1 (unlike the PC version, all numbers are floating point in AppleSoft).

BIBLIOGRAPHY

- The following is a list of introductory electricity and magnetism texts.

Corson, D. R., and Lorrain, P., *Electromagnetic Fields and Waves*, 2nd Ed., Freeman, New York, 1970.

Griffiths, David J., *Introduction to Electrodynamics*, Prentice-Hall, 1981.

Nayfeh, M. H. and Brussel, M. K., *Electricity and Magnetism*, Wiley, 1985.

Purcell, E. M., *Electricity and Magnetism*, 2nd Ed., McGraw-Hill, 1970.

Reitz, J. R., Milford, F. J., and Christy, R. W., *Foundations of Electromagnetic Theory*, 3rd Ed., Addison-Wesley, 1979.

Wangsness, Roald J., *Electromagnetic Fields*, 2nd Ed., Wiley, 1986.

- A standard graduate-level text is

Jackson, J. D., *Classical Electrodynamics*, 2nd Ed., Wiley, 1976.

- A number of special topics have had to be omitted or abridged in the present book for lack of space; some of these are listed below, with references in which discussions of them may be found at a level accessible to readers of this book.

Antenna theory: Nayfeh and Brussel.

Circuit theory: Purcell; Nayfeh and Brussel; Reitz, Milford and Christy.

Dispersion (frequency dependence of the dielectric constant): Reitz, Milford and Christy; Griffiths.

Ferroelectricity: Reitz, Milford and Christy.

Interference in reflection from thin films: Reitz, Milford and Christy.

Mathematical methods (continuum): Boas, *Mathematical Methods in the Physical Sciences*, 2nd Ed., Wiley, New York, 1983.

Maxwell's equations: The authoritative source is Maxwell, J. Clerk, *Electricity and Magnetism*, originally published 1891, available in Dover paperback. A reprint of his original paper, *A Dynamical Theory of the Electromagnetic Field*, has been edited by Thomas F. Torrance (Columbia University Press).

Numerical analysis (more advanced than in this book): Koonin, Steven E., *Computational Physics*, Benjamin/Cummings, Menlo Park, CA, 1986. Another book on general computational physics is Gould, Harvey and Tobochnik, Jay, *Computational and Simulation Physics*, Addison-Wesley, Reading, MA, 1987

Plasmas: Reitz, Milford and Christy.

Radiation from moving charges: Griffiths; Reitz, Milford and Christy.

Relativity: Griffiths; Nayfeh and Brussel.

Statistical theories of magnetism (Langevin theory of paramagnetism, Weiss theory of ferromagnetism): Reitz, Milford and Christy; Nayfeh and Brussel.

Superconductivity: Reitz, Milford and Christy.

Units other than SI or MKS (Gaussian, esu, cgs): Wangsness; Nayfeh and Brussel; Jackson.

Waveguides and resonant cavities: Reitz, Milford and Christy; Nayfeh and Brussel (see also the instructor's manual for the present text).

SYMBOL INDEX

$\equiv$: identity; "is defined as"
$\int$: continuum integral, 50
   discrete integral, 44
∂: boundary symbol, 99
∇^2: Laplacian, 284
∇: "Del," 117, 118
α: proportionality symbol, 157
∞: infinity

χ_M: magnetic susceptibility, 483
δ: Kronecker delta function, 28
   phase, 503
ϵ: set-inclusion symbol, 94
ϵ: permittivity, 360
ϵ_0: permittivity of space, 173
Φ: magnetic flux, 421, 455
ϕ: azimuthal angle, 128, 129
   phase, 4
γ: type of ionizing radiation, 208
Λ: gauge function, 421
λ: charge per unit length, 235
   mass per unit length, 51
   wavelength, 504
μ: magnetic permeability, 224, 400, 484
   mass per unit area of "trampoline," 155
μc: microcoulomb $(10^{-6}c)$, 177
μT: millitesla $(10^{-3}T)$
θ: colatitude in spherical coordinates, 129
ρ: electric charge density, 169
   population density, 65
   resistivity, 443
σ: conductivity, 439
   surface charge density, 311
τ: relaxation time, 445
   time constant, 467
Ω: Ohm, 443
ω: angular frequency, 4, 202, 503

Subject Index

Problem Index

Example Index

KEY EQUATIONS OF ELECTRODYNAMICS

Continuity equation	$d\rho/dt = -\operatorname{div} j$ (Eq. 9.6)
Ampère's law	$dE/dt = c^2 \operatorname{curl} B - j/\epsilon_0$ (Eq. 9.31) or $dD/dt = \operatorname{curl} H - j_{\text{free}}$ (Eq. 23.43)
Faraday's law	$dB/dt = -\operatorname{curl} E$ (Eq. 9.32)
Gauss' law	$\operatorname{div} E = \rho/\epsilon_0$ (Eq. 14.22) or $\operatorname{div} D = \rho_{\text{free}}$ (Eq. 18.9) $\operatorname{div} B = 0$ (Eq. 14.24)

Maxwell's Equations

Lorentz force	$\mathbf{F} = q\mathbf{v} \times \mathbf{B}$ (Eq. 12.1), $\mathbf{F} = I\mathbf{l} \times \mathbf{B}$ (Eq. 12.11)
Energy density	$\frac{1}{2}\epsilon_0 E^2 + \frac{1}{2}B^2/\mu_0$ (Eqs. 13.17 and 13.36)
Poynting vector	$\mathbf{S} = \mathbf{E} \times \mathbf{H}$ (Eqs. 13.48 and 23.44)
Coulomb's law	$E = q/4\pi\epsilon_0 r^2$ (Eq. 14.29), $V = q/4\pi\epsilon_0 r$ (Eq. 15.4)
Potential	$E = -\operatorname{grad} V$ (Eq. 15.1)
PE of test charge	$PE = qV$ (Eq. 15.12)
Poisson's equation	$\nabla^2 V = -\rho/\epsilon_0$ (Eq. 16.1) or $\dfrac{1}{dr^2}\Sigma[V(c') - V(c)] = -\dfrac{\rho(c)}{\epsilon_0}$ (Eq. 16.4)
Capacitance	$Q = CV$ (Eq. 16.42)
Polarization charge	$\rho_b = -\nabla \cdot P$ (Eq. 18.3)
D field	$D = \epsilon_0 E + P$ (Eq. 18.8)
Electric Susceptibility	$P = \epsilon_0 \chi E$ (Eq. 18.10)
Permittivity	$D = \epsilon E$ (Eq. 18.12), $K = \epsilon/\epsilon_0$ (Eq. 18.14)
Surface charges	$D_n = \sigma_{\text{free}}$ (Eq. 19.12), $P_n = \sigma_{\text{bound}}$ (Eq. 19.16)
Magnetic vector potential	$\nabla \times \mathbf{A} = \mathbf{B}$ (Eq. 20.24), $\mathbf{A} = (\mu_0/4\pi)\int \mathbf{J}/s$ (Eq. 20.40)
Biot–Savart law	$\mathbf{B} = (\mu_0 I/4\pi)\int d\mathbf{l} \times \hat{\mathbf{s}}/s^2$ (Eq. 20.50)
Magnetic dipole moment	$m = IA$ (Eq. 20.66)
Ohm's law	$J = \sigma E$ (Eq. 21.1), $V = IR$ (Eq. 21.6)
EMF	$EMF = \int \mathbf{E} \cdot d\mathbf{l}$ (Eq. 21.18)
Faraday's flux law	$EMF = -d\Phi/dt$ (Eq. 22.6)
Inductance	$EMF = -M\, dI/dt$ (Eq. 22.14)
Magnetization current	$\mathbf{j}_m = \nabla \times \mathbf{M}$ (Eq. 23.4)
H field	$H = B/\mu_0 - M$ (Eq. 23.13)
Magnetic Susceptibility	$M = \chi_M H$ (Eq. 23.19)
Permeability	$B = \mu H$ (Eq. 23.22)

Fundamental Constants

Name	Symbol	Value	Section
Permittivity of space	ϵ_0	$8.85 \times 10^{-12}\ c^2/m^2\ N$	9.3
Speed of light	c	$3.0 \times 10^8\ m/s$	9.3
Permeability of space	μ_0	$4\pi \times 10^{-7}\ N/a^2$	13.4
Elementary charge	e	$1.6 \times 10^{-19}\ c$	
Electron mass	m_e	$9.1 \times 10^{-31}\ kg$	
Proton mass	m_p	$1.7 \times 10^{-27}\ kg$	

The Art of
Floral Arranging

Learning from the Master Florists at FlowerSchool New York

Eileen W. Johnson

Photographs by Brie Williams

Gibbs Smith, Publisher
TO ENRICH AND INSPIRE HUMANKIND

Salt Lake City | Charleston | Santa Fe | Santa Barbara

First Edition
11 10 09 08 07   5 4 3 2 1

Text © 2007 Eileen Johnson
Photographs © 2007 Brie Williams

All rights reserved. No part of this book.may be reproduced by any means
whatsoever without written permission from the publisher, except brief portions
quoted for purpose of review.

Published by
Gibbs Smith, Publisher
P.O. Box 667
Layton, Utah 84041

Orders: 1.800.835.4993
www.gibbs-smith.com

Designed by Mary Ellen Thompsen and Sheryl Dickert Smith
Produced by Mary Ellen Thompsen
Printed and bound in China

Library of Congress Cataloging-in-Publication Data

Johnson, Eileen W.
   The art of floral arranging : learning from the master florists at
Flowerschool New York / Eileen Johnson ; photographs by Brie Williams. —
1st ed.
       p. cm.
   ISBN-13: 978-1-4236-0103-6
   ISBN-10: 1-4236-0103-3
   1. Flower arrangement. 2. Florists—New York (State)—New York. 3.
Flowerschool New York. I. Williams, Brie. II. Title.

SB449.J56 2008
745.92—dc22
                                        2007009196

To my mother,

who taught me about

grace and beauty

Acknowledgments

I would like to thank all the florists who are in this book for their participation.

Being a teacher requires that one is willing to share with others the knowledge and talent acquired through a lifetime of work. They have all been truly generous in this.

Each of them has taught me something I did not know before and each of them has impressed me with their creative vision and dedication to their métier.

Contents

Introduction

I have always enjoyed entertaining, cooking for friends, welcoming people into my home, setting a table with fresh flowers.

In the fall of 2002, while thinking of ideas for having my own business, I happened to come across a small article in the New York Times about a school in Paris entitled "Fly to Paris and Learn the Art of French Floral Arrangement." I had no idea that such schools existed and I was intrigued. I loved flowers, and I bought them on a regular basis; however, I never learned how to put them together with any real panache. Was this a skill that could be learned in a classroom the way you could learn to cook? And as much as I loved an excuse to visit Paris, didn't we have great florists in New York?

I had the germination of an idea—why not have a school in New York where we could highlight the work of some of our great florists? We had cooking schools with famous chefs teaching in them, why not a flower school?

I soon found myself going to Paris and taking some classes under the tutelage of Christian Tortu, one of the most renowned florists in Europe. From the first class, I was hooked—I had to bring the concept to New York.

Back in New York, the first florist I spoke with was Michael George, a florist well known for his modern, cutting edge style. He was quite taken with the idea. He had heard that the great choreographer and dancer George Balanchine set up the School of American Ballet because he felt that the most important thing an artist can do is teach, and thus ensure that his legacy would be preserved. Michael wanted to share his gifts with others.

We both decided that if I set up a school, it should only have the best designers we could find. The first teacher we brought on after Michael was Chris Giftos, who had been doing the flowers at the Metropolitan Museum of Art for thirty-three years and was planning to retire.

After Chris, I asked his protégé, Remco Van Vliet, if he would teach for the school. He was followed by his brother and partner, Cas Trap. Subsequently, Charles Masson, Meredith Waga, and Felipe Sastre signed on to teach. I had a truly wonderful team.

What I loved about the school from the start was the enthusiasm of the teachers and the students when they were able to finish their creations and take them home. There was—and still is—the sense of "I did that?" when the arrangements would leave the school. The other thing I have loved about the school is that there is no set dogma. If you have six different teachers, you will get six different points of view on everything from the conditioning of flowers to the style of arranging. It is up to the students to choose the philosophy or method to which they wish to subscribe. And that makes our classes all the more lively and varied!

If we have accomplished one thing in setting up FlowerSchool New York, it is to make beautiful flowers less intimidating and more accessible to those people who love them and fall under their spell as I have.

CHAPTER I

Cas Trap and the Flower Market

I t is still dark when the flowers are loaded onto the truck that will drive around to a
number of high-end florists in Manhattan. The first light of dawn is just beginning to
show in the sky. It is 5:30 in the morning on what turns out to be a glorious September day.
Twenty-eighth Street is coming to life as the market opens up to the first buyers of the day.

This is where it all starts—with the flowers, the selection, the abundance. Even if you were not limited by budget, you would have to make a choice. Having more flowers and more money to spend on them doesn't guarantee that you will have a better bouquet. What matters is having the eye to select, the ability to choose and to eliminate that which does not add to the composition.

Each one of our florists at FlowerSchool New York shops at this market on a regular basis. And Cas Trap, as the manager of the Dutch Flower Line, one of the major importers and wholesalers on the block, is there to help them—suggesting new hybrid roses, an interesting foliage imported from South Africa, peonies from New Zealand while New York is deep in the doldrums of winter, precious lilies of the valley, exotic parrot tulips from Holland, and luscious calla lilies from Ecuador. All of these flowers arrive in boxes from Kennedy

Airport where they have made the journey from field to cut product in as little as thirty hours.

Cas, a third generation florist from Holland, knows more about flowers than practically anyone in the market. He grew up making bouquets for his father's flower shop—sometimes as many as one hundred a day on a busy Saturday—and each one made in his hand. The customers would come in the shop with their own vases and buy a bouquet to put in them.

For us at FlowerSchool New York, Cas re-creates the Dutch Bouquet of his youth. This bouquet is similar to a Biedermeier Bouquet. According to some history books, it was created in Switzerland in the late 1800s. A Dutch bouquet can be made as large as your grip can get. Cas remembers having the honor of making a hand-tied bouquet for the Queen of the Netherlands, Queen Juliana, when she visited his hometown. The bouquet was quite large—large enough to go with the Queen's rather considerable frame.

FIRST STEP, THE SELECTION OF FLOWERS

The flowers we chose for this bouquet were

10 stems of viburnum berries

20 stems of leucadendrum (safari sunset)

20 stems of rudbeckia

20 stems of mambo spray roses

20 stems of dahlias

For Cas's bouquet we wanted some bright colors. We also wanted to use different textures. Since it was a lovely autumn day, we started with fall hues. It is important to decide on a tone and stick with it—be it blues, pinks, or reds. We decided on orange, which is the national color of Holland and consequently Cas's favorite color. In the orange family we chose one very beautiful flower—in this instance, it was a mambo spray rose that caught his eye. The rest of the flowers were chosen with regard to how they would go with that rose. We ended up with a selection of yellows, oranges, browns, and reds—colors that work naturally together.

To create the Dutch Bouquet

STEP 1

What you will need is a good pair of garden clippers and a rubber band.

STEP 2

You take all the flowers out of their wrappers and clean them of all their leaves up to about two inches below the flowers. Take most of the petals off the viburnum berries. Clean the roses of their thorns as well as their leaves. Separate the flowers into different groups on a clean worktable.

STEP 3

First, you need a base—some woody stemmed flowers. Here, Cas starts with the heaviest stemmed flowers—the viburnum berries. He makes a circle by bringing his thumb and index finger together and puts each stem through it, placing the stems in a circle holding the stems loosely close to the flowers. Then he twists the bouquet around in his hand and adds more of the woody stems, always making sure that his grip is not too tight.

STEP 4

He then makes a concentric circle with the next group of flowers—this time the leucadendron flowers—all the while turning the bouquet around with his other hand.

STEP 5

He adds the other flowers in the front and at a diagonal—all in the same direction and in this order—the rudbeckia, the spray roses, and lastly the dahlias. The stems should look like this.

STEP 6

Then he holds them with one hand as loosely as possible and takes the clippers in the other hand, cutting the stems evenly about 6 inches from the flowers. For this bouquet we chose a vase that was about 8 inches high.

STEP 7

Then with the rubber band stretched out, he wraps the band around the stems. (You might want to have another person help you with this.)

STEP 8

Cut the stems evenly across the bottoms.

STEP 9

The real trick to the Dutch Bouquet is to have it standing on its stems alone without holding it. This is not the easiest thing to accomplish so you might have to do it several times before you get it right.

STEP 10

Try the flowers in your vase—the stems should hit the bottom of the vase and the flowers and the greens should be just slightly above the level of the top of the vase.

STEP 11

At this point, you can cut the rubber band and let the flowers loosen up in the water or keep the rubber band on for a tighter composition.

THE COMPLETED BOUQUET

If you have made this bouquet correctly, all the
stems will be in the vase at a diagonal and each
one of the flowers will have its own place in the
bouquet. As you can see, the stems at the bottom of
the vase all face in a diagonal direction. The flowers
are balanced on all sides of the arrangement. Although the
Dutch Bouquet has a casual look, it is not haphazard. Texture is
aided by the rudbeckia and the viburnum berries, a unity of palette is
achieved with the different oranges of the roses playing against the yellow
of the dahlias, and the bits of green in the leaves are nicely counterbalanced by
the stems in the vase.

CHAPTER II

Michael George,
a Modernist Approach
to Design

Michael George's style has been imitated but rarely mastered by other florists. It is with great generosity that Michael shares his two signature bouquets.

TULIPS

Michael George's technique is clean, simple, and monochromatic. It is a technique that he has refined over the past forty years of being a florist. If Michael had not become a florist, he probably would have become an architect. However, he decided to follow in the footsteps of his father, one of New York's first society florists. Michael calls his style "Graphic Minimalism" and indeed it is graphic, but hardly minimal. When he uses flowers, he uses them in great abundance, sometimes putting as many as two hundred tulips in a centerpiece. The minimalism comes in his monochromatic colors and his attention to the detail of the stems. For many of Michael's arrangements, the stem is as important as the flower. He always prefers to use glass vases that highlight the sculptural flow from the stem to the flower.

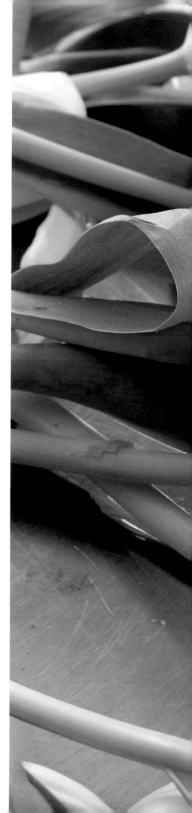

TO CREATE THE TULIP BOUQUET

Here Michael has used a rectangular vase, six inches by six inches by eight inches, and has massed together sixty tulips. We will start off with a smaller vase and forty tulips.

STEP 1

Use a rectangular vase that is approximately six inches by four inches by three inches. You will need at least forty tulips or four bunches in order to create this very sculptural design.

STEP 2

First, cut one-half inch off the bottom of the stems, and place them in a bucket of cold water leaving the wrapping paper on. The flowers need to be hydrated immediately as they have often been without water for long periods of time. Keeping the wrapping paper on helps make the tulips stand upright.

STEP 3

After the tulips have spent at least one hour in the water, take them out of the bucket, unwrap them, and lay them out on a clean worktable.

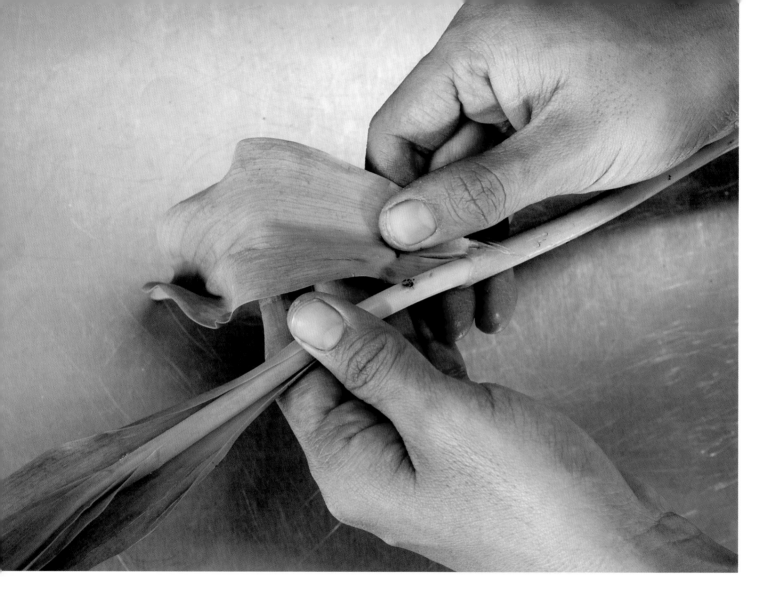

STEP 4

Clean the tulips by carefully pulling off all of the lower leaves and most of the leaves near the head, letting the small leaves just below the head stay attached. Tulips can often be quite dirty so you might want to dip them again in a bucket of clean cold water.

STEP 5

After you have cleaned all the tulips, lay them flat on the worktable and make a pile of tulips about eight tulips wide. When you place the tulips on the pile, put the stems that are leaning toward the left on the left side, the stems leaning toward the right on the right side, and the straight stems in the middle.

STEP 6

Make sure that all the tulips are the same length by measuring the tops with the palm of your hand.

STEP 7

Hold onto the tulips with one hand and then slide a knife under them so that you are able to pick them up with both hands in a bunch.

STEP 8

Gently place a see-through rubber band around the stems to hold them tightly. The rubber band should be about four to five inches below the heads. You may want to have someone else hold the tulips while you do this. Make sure that the rubber band is not twisted.

STEP 9

Lay the tulips back on the table with the flowers hanging off the side of the table so they don't crush. Then with the knife make a clean cut several inches up the stems.

STEP 10

Then pick up the tulips and hold the stems in both hands. Twist the stems one way with your hands on the base and the other way with your hands near the flowers.

STEP 11

Carefully check to see that the tulips are all the same height and symmetrical, with none of the flowers sticking their heads up above the others. If they are out of alignment, you can pull or push them gently up or down by the stems. The tulips should all be facing outwards.

STEP 12

Test the length of the bouquet by putting it in the vase. You may need to clean the vase after this since tulips may still be quite sandy. The tulip stems should be at a maximum of two inches above the vase.

Here, you will probably have to make two cuts. Make the first cut conservatively so you do not accidentally make the tulips too short. Make the second cut so that the tulips are the length that looks right. Michael always suggests that you make three cuts—the first is clearly too long, the second is close, but not quite as short as you will go in the final cut. Then, when you are sure you have the correct length, make the final cut. If you cut them too short on either of the first two go-rounds you will have no other recourse than to work with a smaller vase.

STEP 13

Place the tulips in the vase with water and primp them gently so that they form a round dome. The primping is important here: you must make sure that the tulips are all the same height even if it means pulling some up or down with the rubber band still attached. Be careful when doing this that the delicate stems do not break!

Note: Remember tulips are among the few flowers that grow after they are cut so you may want to make this arrangement the day you are planning to use it.

ROSES

Some of Michael's most loyal clients are fashion editors. An assistant editor knows she has arrived when she receives Michael George's distinctive dome of roses at her desk as a gift from a designer whom she has written about.

What the fashion editor probably doesn't realize is the attention to detail that goes into Michael's roses before they are even put in the vase.

Many people (especially professional florists) are shocked when they hear that Michael places the ends of the stems of fresh roses into boiling water—yet that is exactly what he does when he conditions them.

Roses from South America are sold in packs of twenty-five at the wholesale market so you may want to do this arrangement with someone else and share an order of three or four bunches so that you can each do a small arrangement. Generally you should buy at least ten percent more flowers than you need because in packing flowers and working with them, there is often breakage.

Prepare a clean glass vase about five inches by five inches by five inches and fill it with fresh cold water. You will also need some brown paper packaging (about 36" wide), a stapler, a small glass bowl (two inches deep and at least four inches round), clippers, and a florist's knife.

 ## TO CREATE THE ROSE BOUQUET

STEP 1

First, unwrap the roses from the packaging and clip about one inch from the bottom of the stems of the roses and put them into a large bucket of cold water to hydrate them.

STEP 2

After at least an hour (or even overnight) soaking in water, the roses are ready to be conditioned. Take them out of the bucket and the wrapping paper and lay them down on a clean work surface.

STEP 3

Taking one rose at a time, stand the rose up with the flower facing you. Gently slide your knife down the stem and cut off the thorns and most of the leaves starting at about three inches below the flower.

STEP 4

Lay the cleaned roses in a small pile on the table in front of you so that the heads of the roses are lined up neatly together.

STEP 5

Fold the brown paper packaging into a triangle and wrap the flowers with the paper making sure that the bottom of the triangle is clasping the stems. Staple the brown paper together so that the roses are comfortably packed together.

STEP 6

Using the clippers, clip the bottoms of the stems evenly about two inches or as long as the shortest stem.

STEP 7

Boil water to fill up the small round bowl and pour it into the bowl. Stand the stems up in the bowl and hold them for three minutes. The roses can be placed next to a wall if they are evenly cut at the bottoms.

STEP 8

Take the roses out of the hot water and immediately plunge them into a bucket filled with clear cold water. You can do this up to twenty-four hours ahead of time if you have a cool place for the roses.

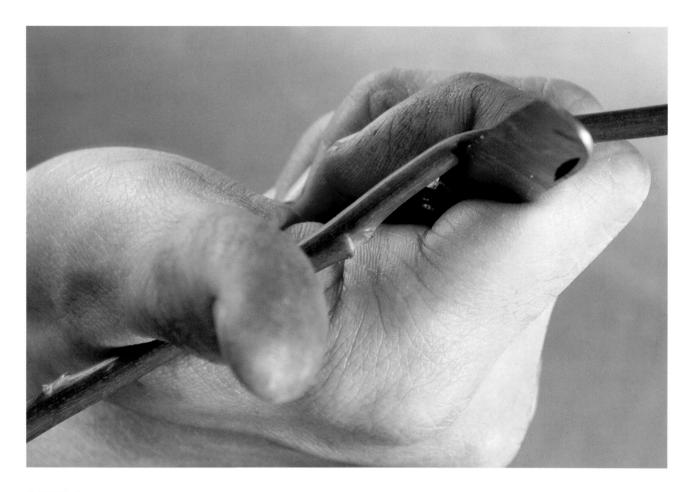

STEP 9

Take your florist's knife and hold the handle between your thumb and your index finger with the other three fingers securing the knife. Pull the knife across the stem of each rose diagonally and away from your body, making sure that your thumb is not in the way. Use the lower part of the knife so that you have more control. Don't put pressure on your thumb. Roses are cut at a diagonal so that they have the maximum ability to drink water. This is the way you cut roses.

STEP 10

Place one rose diagonally in the corner of the vase and cut the stem so that the flower sits on the top edge of the vase. Cut another rose and place it in the opposite corner at the same height as the first one. Then add another rose and place the stem under the opposing stem. Do the next corner, cutting to measure each rose so that it is the same height as the others. Keep going around the vase adding stems clockwise under the stems that are already in the water until the vase is full.

THE COMPLETED ROSE BOUQUET

Ideally, you should have some neat stems crisscrossing at the bottom of the vase, and the roses on top should form a flat dome with none higher or lower than the others. If you have cut a few of the stems too short, you can tuck them into the middle layers, which require slightly shorter stems than the outer layers.

Depending on your taste, as well as how you have cut the roses, the dome will either be rounded or flat. Flat on the top is more graphic and architectural; rounded is somewhat more romantic.

CHAPTER III

Chris Giftos, a House in Connecticut, Country Flowers

Chris Giftos is a New York florist extraordinaire. After thirty-three years of arranging flowers for the Metropolitan Museum of Art, he retired and is now busier than ever lecturing around the country and doing events for those lucky enough to get him.

Chris, more than anyone else, is responsible for the grand flowers that have been gracing the entrance or "great hall" of the Metropolitan Museum of Art. The flowers are a bequest of Lila Acheson Wallace; Chris was the person who executed them so brilliantly these many years. His legacy has changed the way we perceive flowers in public places throughout much of the country. Over the past three decades he has planned many of New York's most glamorous parties, working with such fashionable luminaries as Pat Buckley, Oscar de la Renta, and Anna Wintour.

Chris has been arranging flowers since he first started delivering them at the age of fifteen. He was lucky enough to have his first job at a shop on Madison Avenue in New York that catered to the many wealthy and social residents of Park and Fifth Avenues, including Kennedys, Vanderbilts, and Astors.

We are privileged to have Chris spend some of his precious time teaching for us at FlowerSchool New York.

He invited us up to his Connecticut farmhouse one early fall day so that we could photograph and record his personal taste in flowers.

FLOWER SELECTION

When we purchased the flowers from Dutch Flower Line, we were guided by Cas, who assured us that Chris always likes to use kale, sunflowers, and leucadendron in early autumn. The rest of the flowers were chosen to work well with those first three.

Dill Flowers (Anethem)

Kale (Brassica)

Cockscomb (Celosia)

Sunflowers (Helianthus)

St. John's Wort

(Hypericum)

Leucadendron Goldstrike

Loosestrife (Yellow

Lysmachia)

Rosemary (Rosmarinus)

Sedum (Sedum)

TO CREATE THE CENTER TABLE BOUQUET

First we went to Chris's workshop where he often does flowers for his home.

STEP 1

Chris clipped a piece of chicken wire to fit the vase that he had selected for us. This chicken wire is used to establish a base for the flowers so that they stay in place where you put them.

STEP 2

The first flowers that he put in the arrangement were the sedum and the cockscomb, both of which have thick woody stems. Using his eye to determine the measurement, he cut the stems to a level where they would rise several inches above the vase.

STEP 3

He put the sunflowers in next, interspaced around the sedum and some of the cockscomb. The sunflowers that were bent rather than straight were placed so that they hung over the sides of the vase.

STEP 4

He added the hypericum, more cockscomb, the dill, and the rosemary so he had a good balance of colors with the yellows, reds, greens, and burgundies. Chris works so quickly and intuitively we had to slow him down so that we could capture his movements with a camera. He doesn't seem to labor over his ingredients; instead, like a talented and experienced chef, he appears to almost throw

things together and then produces something quite extraordinary. His arrangement looks effortless, however many years have gone into creating such an unfussed look.

STEP 5

The bouquet for the center of the table was no taller than the distance from his elbow to his fist—one of the most important rules for centerpieces.

 ## TO CREATE THE SIDE TABLE BOUQUET

Using the same flowers, Chris created another arrangement for the side table, this time keeping the stems taller. The two bouquets are similar without being the same. Basically they reference each other and create a unity of design in the dining room, complementing the old, original beams from the eighteenth century and the rustic country furniture. Here is a photo of Chris's dining table with one of the arrangements that he made for us.

 TO CREATE THE GARDEN BOUQUET

It was such a lovely fall day that we decided to take advantage of the gentle breeze and work outside for the next arrangement. Chris decided to make it in his classic crisscross style instead of using chicken wire. This crisscross style is the foundation for many of the designs that our florists create.

We chose a clear round glass container 12 inches high and 8 inches in diameter. By using a glass container, we could see the stems that would hold the flowers.

STEP 1

Chris first cleaned off all the leaves from the stems that would be below the waterline. He laid the flowers out in groups. We decided to use all the flowers that we used in the dining room, but we added kale and loosestrife (yellow lysmachia) to the mixture.

STEP 2

Chris started with the heaviest and thickest stems, in this case the sedums. He created a matrix to hold the other flowers in place by placing the sedum stems at four equidistant points in the vase. If this were a rectangular vase, he would place the sedum stems in the four corners. Each time he adds a stem, it goes under the preceding stem.

STEP 3

He next added the hypericum berries in between the sedum stems.

STEP 4

He cleaned off the leaves of the cockscomb and added two stems at a somewhat higher level, and then put two stems at a lower level so the bright red had a place of prominence among the other colors.

STEP 5

He next added the kale, fluffing it up a little before he inserted it in the vase. Then he put the sunflowers at a similar height to the cockcombs, at opposite sides of the vase.

STEP 6

Lastly he placed more sedum, the dill, and the loosestrife with their more delicate stems at strategic points in the arrangement. Chris has created a bouquet that looks as if the flowers were just picked from his garden.

THE COMPLETED GARDEN BOUQUET

He, like many of our other instructors at FlowerSchool New York, believes that flowers should be appropriate for the season and the setting. He created grand flowers at the Metropolitan Museum of Art, which was a grand setting, while here he has given us an equally exquisite look perfect for a New England afternoon in autumn.

CHAPTER IV

Charles Masson, La Grenouille, Artful Flowers

As soon as you walk into Charles Masson's restaurant, La Grenouille, you are aware that you are in the presence of someone who loves flowers. His father, who opened the restaurant on December 19, 1962, in "a quiet night in the midst of a snow-storm," passed on this passion to Charles. It all starts out with love, according to Charles Masson. If you don't love the flowers, they will not be happy.

When Charles took a hiatus from the restaurant and had his own flower business, he would go into client's homes and could almost always tell by the state of the flowers the emotional state of the family.

73

Fresh flowers always would thrive in an atmosphere that was filled with emotional warmth and unity. Bickering couples or unhappy children would make even the freshest flowers wilt within a few days.

As proprietor of La Grenouille, one of the most beautiful and renowned French restaurants in Manhattan, Charles makes beautiful flowers, like great food, an integral part of his daily existence. And he has taken his obsession with them both to great heights in his restaurant. Charles has many things to say about flowers that could be as easily applied to food—or, for that matter, to getting yourself dressed in the morning. And even though his creations may appear elaborate, they are based upon the simplest, quality ingredients. Everything he does evokes nature— there are no clashing colors, no ingredients that do not work in

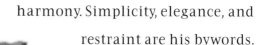

harmony. Simplicity, elegance, and restraint are his bywords. Although the flowers that Charles has in his restaurant are admittedly spectacular, they work on an intimate level as well. Looking at them for the first time you are impressed by their grandness—the overall impression is quite spectacular as you walk into the restaurant. As you look at his large bar flowers, you start to see more details: berries at the lip of the vase perhaps, ranunculus bunched next to the roses, delphiniums scattered at the top. None of this is haphazard. The theme is color, texture, and nature.

Charles's bar flowers set the theme for the flowers that are on each table in the restaurant. Throughout the year, he offers a

seasonal signature drink—this one is champagne with a reduction of fresh peaches, currants, raspberries, and blueberries. The season is summer when all of those fruits and flowers are at their freshest. The colors, ranging from a pale green to a soft pink to a more vibrant pink to blues and purples, are all the colors that he has used both for the dramatic bar displays as well as for the small bouquets at each of the tables in the restaurant.

TIPS FOR CREATING BOUQUETS

TIP 1

First of all, Charles does not believe in having too many ingredients in a dish or too many types of flowers in a bouquet. He even suggests in his classes that the students start out with only two types of flowers and see how they come out. I have seen him make an arrangement out of flowers with broken stems that would have been discarded by most florists.

TIP 2

He doesn't like contrast; instead he looks for harmony of color and texture. Unlike Michael George, Charles does not like symmetry—he does not see it in his vision of nature and he does not see it in flowers that he puts together.

TIP 3

His advice to us is to start with the construction. The larger bouquets are done first, and they are based on the same grid or crisscross that we have used before. Here Charles references a bird's nest—a perfect construction that nature has generously given us to copy. The larger stems (usually branches) form a base for all the other stems to follow. The thickest are placed in the vase first. The branches hold all the other flowers and berries so there is no need for chicken wire in the glass vases.

TIP 4

Charles is almost fanatical about keeping the water in the vases clean—the water is changed in each vase every single morning. He does this in the very large vases by using a siphon. You can replicate this by putting a garden hose into a vase of water and then draining the water into a large container that is on a lower platform (or on the floor).

The smaller vases are washed and cleaned every morning by members of his staff. This attention to detail ensures that the flowers last the full six days from Monday morning to Saturday night—with some still fresh enough for Charles to take home for his Sunday dinner!

TIP 5

One of the tricks of the trade that Charles shows us is how to make it look as if there are more flowers in the vase than there actually are. Here he places a vase with sunflowers in a mirrored corner so that when we look at the flowers from each angle, they appear to have double the number than are actually there. He accomplishes the same effect gracefully in the ladies room with a single lily stem placed in a tall glass vase at a strategic mirrored corner. You may notice in the photograph that it has the effect of three vases or twelve lilies from the single stem.

TO CREATE A SIMPLE DESIGN

At the end of many of our classes, Charles shows us how to make the simplest of designs by using an empty wine bottle and one beautiful, open rose. Here the wine bottle is clear glass, and the stem has a graceful arch that is about eight inches higher than the vase. It is this same sense of discipline, style, and lightness of touch that he has given to his more important and grander compositions. Like many of the great works of art that he talks about in his classes—those by Matisse, Manet, Van Gogh, and others—his compositions reveal their depth, their lack of artifice, and their true beauty to those who take the time to study their many aspects.

CHAPTER V

Remco Van Vliet, Grand Flowers at the Met

Before Remco turned thirty, he held one of the most prestigious positions designing flowers in America—the Great Hall of the Metropolitan Museum of Art and all the events held at the museum.

Every Monday when the museum is closed to outside visitors, Remco puts together flowers for the Great Hall that will have to last for the week. There are several staff members at the museum who will refresh the flowers during the week; however, the basic arrangement must be strong enough to look good for seven days. It is a formidable job, which he accomplishes brilliantly within a short few hours each week.

In addition to doing the flowers at the Metropolitan Museum of Art, Remco regularly designs flowers for the Museum of Modern Art's events, the New York Philharmonic, and a growing group of private clients.

He was handpicked by Chris Giftos to be his successor after having worked with him for several years. Remco started young—he is a third generation florist who grew up in Holland doing the flowers for his father's flower shop. His style is lush and modern at the same time, and his classes at FlowerSchool New York are among our most lively. Remco has a depth of knowledge about flowers that we rarely see in other florists. Like many of his well-trained Dutch counterparts, he can give the Latin names for the flowers as well as their common names in English, French, and of course Dutch. Although he can purchase flowers from all over the world at any time, Remco has a great respect for seasonality in flowers.

Once a year, in spring, we have a class in

which Remco shows us how to work with large branches. He makes cutting large branches like cherry look easy—and even with the best clippers, believe me it is not!

He always does the flowers in the Great Hall around a seasonal theme. The season here is late summer, almost early fall. In this photograph, we have pictured hydrangeas, berries, and weeping willows—just some of the flowers that he will use in his grand composition.

THE SELECTION OF FLOWERS

To create these magnificent arrangements Remco used only five elements altogether:

> Long-stemmed hydrangeas
>
> Magnolia leaves
>
> Bittersweet
>
> Weeping willows
>
> Salmon queen lilies

TO CREATE THE LARGE FORMAL URN ARRANGEMENT

STEP 1

Using one of the large urns that are placed in four spaces within the formal entrance of the Great Hall, Remco prepares a smaller container within a larger vase and uses a grid of chicken wire to hold the stems in place.

STEP 2

He then puts the weeping willows into the larger of the two containers on the two sides, cascading them over the front.

STEP 3

He next places the magnolia leaves in the larger of the two containers. The leaves fill out the spaces left by the weeping willows and form a grid in the chicken wire that will hold the rest of the flowers.

STEP 4

Stem by stem, he next adds the very long hydrangeas, cleaning off the leaves before he puts them in.

STEP 5

He adds the bittersweet berries and intersperses them with the lilies, which you may notice are still tightly closed. These lilies will open up later in the week and change the look of the arrangement making it fuller and more dramatic as you can see in the picture above. This photograph was taken on the sixth day after the arrangement was made.

THESE FRESH FLOWERS ARE
THE CONTINUING GIFT OF
LILA ACHESON WALLACE

TO CREATE THE LARGE FORMAL VASE ARRANGEMENT

Here Remco starts to prepare for the arrangement that will dominate the entire entranceway—this is the large vase on the information desk. Unlike the urns, which back the wall, this arrangement will be viewed from all sides, so many more flowers and branches are needed.

One of the innovations that Remco brought with him to the Met was a base for an urn that rotates like a lazy Susan so that he does not have to keep moving the ladder as he adds more branches.

STEP 1

He starts with the magnolia leaves, which form the base of the composition. Standing on the ladder, Remco cleans the leaves off the bottom of the branches with his bare hands. He does this casually and quickly as if he has been doing this for most of his life—which he has. I would not suggest that you do this if you are making such a large arrangement for the first time. Instead you should carefully prepare all the branches and flowers so they can be tucked into the arrangement without taking a break to clean them.

STEP 2

The next addition is the curly willows, which he rather generously cascades over the entire arrangement. Even though he can rotate the arrangement, he still moves the ladder several times since this display will be seen from a 360 degree angle throughout the room.

STEP 3

He adds the hydrangea at strategic points on the top and the bottom of the arrangement.

STEP 4

He then puts in the bittersweet berries and the lilies, which at this point are so closed that they are barely perceptible.

What Remco has created is a vision of what Lila Acheson Wallace intended when she made her bequest for flowers: she wanted to have people feel that they were being welcomed as if they were entering the home of someone who was thoughtful enough to set out flowers.

INFORMATION

Remco Van Vliet outside the museum after he has finished putting together the four side arrangements and the information desk arrangement in the Great Hall of the Met.

CHAPTER VI

Meredith Waga of Belle Fleur, Romantic Flowers

Meredith Waga designs the most romantic flowers I have ever seen. Her look is feminine, yet modern. There is nothing old-fashioned about a Meredith Waga arrangement. Her classes for us at FlowerSchool New York usually feature a mother-daughter theme, and they often attract three generations from the same family—mother, daughter, and grandmother. Her compositions are always perfectly balanced with a touch of whimsy added so that they have a quality of lightness that is much sought after in the fashion business.

Besides her gifts and events business, Meredith does many, many weddings. On the day of each wedding, she insists on giving the bridal bouquet to the bride herself so that she can savor the moment when the bride first sees the flowers that she will carry down the aisle.

Meredith had been working for Calvin Klein when she decided that she wanted to work with flowers. We find that many of our students come to us with a fashion background. When they see the wonderful arrangements for the fashion shows and the many beautiful flowers that are given to editors as thanks for their coverage they are drawn to floral design. Fashion and flowers—like food and flowers—often work comfortably together. Just as fashions in dress change so do fashions in flowers. One of the most popular classes at FlowerSchool is the "Flowers and Fashion" class. We have done classes featuring the "black bouquet" as well as the currently popular monochromatic white bouquet.

For FlowerSchool New York, Meredith makes an all-white arrangement that would be suitable as a centerpiece for a wedding, a graduation party, or a festive dinner. She puts all the flowers that she will use in her arrangement on a showcase sideboard in her office.

 ## IN ORDER TO MAKE AN ARRANGEMENT LIKE MEREDITH'S, YOU WILL NEED:

One brick of florist's foam soaked and submerged in water before you start

A round glass bowl about twelve inches in diameter and 8 inches deep

A sharp pair of scissors

Two large calathea leaves

10 delphiniums

10 hypericum berries

5 hyacinths

One orchid stem

10 calla lilies

10 ornithogalum

5 hydrangeas

10 lisianthus

10 snowball viburnum

10 roses

several grass stems

All of the flowers should be the same color. If you cannot find white flowers, pale pinks would be a good choice for this bouquet.

TO CREATE
THE ROMANTIC
CENTERPIECE

STEP 1

First, Meredith lines the
bowl with the two large
Calathea leaves.

STEP 2

She then cuts the florist's foam so that it fits comfortably inside the bowl that is lined with the leaves. Then she places it in the bowl.

STEP 3

Using a sharp pair of scissors, Meredith carefully cuts the Calathea leaves so that they are at the same level as the bowl.

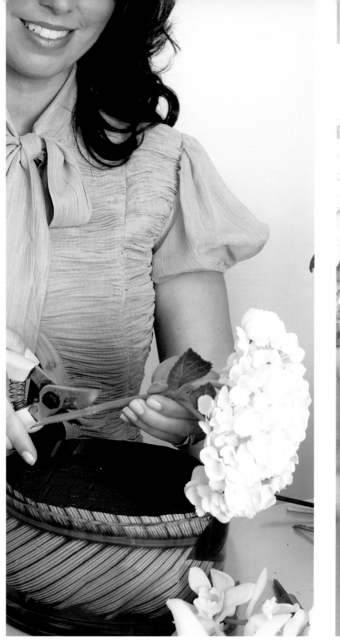

STEP 4

She then cuts the hydrangea and pushes it into the florist's foam at an angle so that it cascades slightly over the side of the bowl.

STEP 5

Next, she puts the roses in, sticking them around and alongside the hydrangeas.

STEP 6

Using water tubes for the orchids, she inserts the tubes into the foam right up against the roses. So far, Meredith has put all the flowers into the centerpiece that have thicker stems, or, like the orchids, are in water tubes. This follows the principle that we have seen before of putting the hardier, larger-stemmed flowers into an arrangement first.

STEP 7

The next step is to add the other more delicate flowers—the calla lilies, snowball viburnum, lisianthus, hyacinths, delphinium, hypericum berries, and ornithogalum. You might notice that she has several buds of the lisianthus peeking out over the other flowers. These buds, along with their flowers, add a touch of asymmetry to the arrangement.

STEP 8

Having put as many flowers as is humanly possible into the centerpiece, Meredith adds a final crowning touch—a stem of grass with each end held by a small green wooden stake. She then splits the grass stem down the middle, making it look like a decorative handle for a basket made of flowers.

THE COMPLETED CENTERPIECE

Her centerpiece is complete. You may notice that the arrangement is not symmetrical; however, it is balanced. There are hydrangea flowers anchoring the arrangement on each side, with roses between them. Interspersed are the more delicate flowers. Looking more closely, you may notice that the grass on the right has been balanced by the lisianthus bud that peeks over the other flowers. It is this attention to detail that has made Meredith Waga one of our most popular teachers as well as one of the most sought-after florists among Manhattan sophisticates.

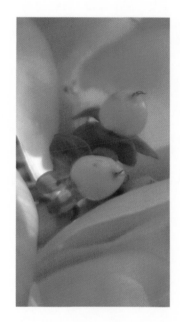

CHAPTER VII

Felipe Sastre,
an Exotic Bridal Bouquet

What could be more fitting for our last chapter than to end with a bridal bouquet? Like the bride in a fashion show, the bridal bouquet is appearing last; only this time the bridal bouquet is not white and traditional, but rather a spicy yet subtle mixture of colors and textures. Felipe Sastre, who teaches our intensive introductory and advanced classes at FlowerSchool New York, likes to work with the tropical materials that he grew up surrounded by. Raised in Veracruz, Mexico, by a Cuban father and a Mexican mother, his look is informed by a south-of-the-border color sense yet infused with the sophistication of his training under the tutelage of Michael George.

Flowers speak to Felipe and the language they speak is exotic, tropical, and often with bright colors. When Felipe was a young child, he loved to help his mother in her garden. When he grew up, he went to sea, becoming a merchant marine who traveled all over the world before he found his true métier. Helping out in a friend's shop during Valentine's Day, he felt that he was truly in "paradise." It was his first time in a flower shop, and he immediately knew what he wanted to do with the rest of his life.

Here, Felipe shows us how to handwire orchids and leaves to make an unusual bridal bouquet.

TO MAKE THIS BRIDAL BOUQUET, YOU WILL NEED:

Cymbidium orchids (one large stem)

Hypericum berries (green) one bunch

Spider orchids (one bunch)

One stalk of variegated leaves such as dracena marginata

One roll of 1/2-inch floral tape

Medium gauge floral wire

A wire cutter

Brown grosgrain ribbon

TO CREATE THE BRIDAL BOUQUET

STEP 1

First Felipe prepares all the elements that will go into the bouquet. He cuts the blossoms of the cymbidium orchids so that there is a one-inch stem left on the bloom.

STEP 2

This is the wire that he will use. He bends the top of the wire so that it looks like the outside of a paperclip.

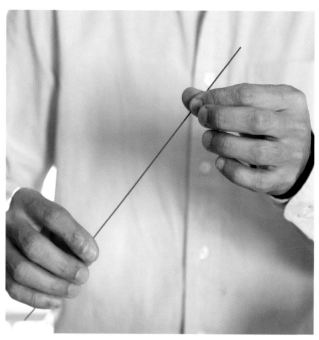

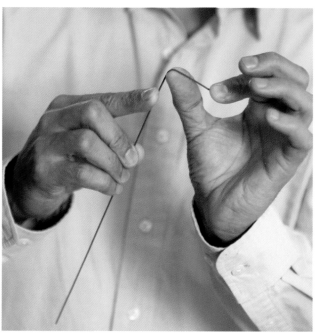

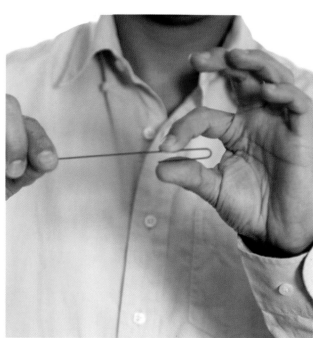

STEP 3

He pushes the straight part of the wire gently through the center of the orchid and catches it on the part of the wire that is bent. When you look at the orchid, you will only see the tip of the wire deep inside the flower.

STEP 4

He then takes the floral tape and cuts it the length of the wire, while at the same time holding the delicate stem of the orchid.

STEP 5

Then he attaches the floral tape to the folded wire; he wraps it carefully so that the wire is wrapped neatly around the stem. He continues wrapping the floral tape around the wire so that the wire is completely hidden by the tape.

STEP 6

He wraps all the cymbidium orchids with the floral tape, sets them aside, and then wraps the spider orchids the same way. Here he is holding the end of the tape and wire in his left hand while spinning the tape in his right.

STEP 7

He then cleans all of the larger leaves off the hypericum berries, keeping only the smallest leaves next to the berries at the top.

STEP 8

He cleans six variegated leaves off the stalk. Once again he twists the wire so that the end looks like the outside of a paper clip. Then he puts the larger end of the leaf through the wire and holds the end as if he were about to make a ribbon. He wraps the "paperclip" part with the floral tape and continues covering the rest of the wire with the tape.

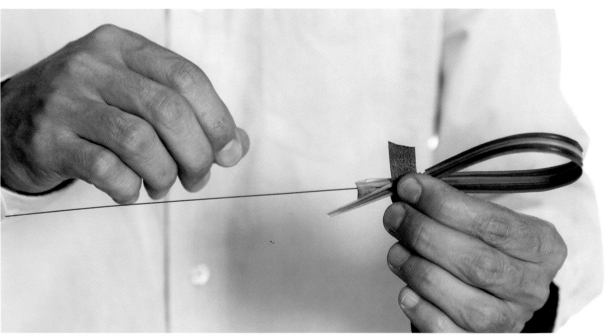

STEP 9

Here are all of the elements of the bouquet neatly separated into containers. Felipe is now ready to put the bouquet together. The first step for assembly is similar to that used by Cas Trap in the Dutch Bouquet (see chapter 1) as he holds a cymbidium orchid loosely in his hand and forms a triangle with two other cymbidiums. He then adds stem after stem of hypericum berries forming a concentric circle. His next concentric circle will be the spider orchid, followed by the variegated leaves. The last circle will be the balance of the cymbidium orchids interspersed with the variegated leaves.

STEP 10

Holding the bouquet in his left hand, he uses the clippers to cut the bouquet so that it is about eight inches long on the bottom. He wraps the stems together with the florist tape all the way down to the bottom of the wires. Once the wires have been covered by the tape, he can wrap the ribbon, turning the bouquet with his right hand so that it is evenly covered from the top to the bottom.

STEP 11

He finishes the bridal bouquet by putting 2 pearl-headed pins diagonally through the ribbon close to the spot where everything branches out. This way the ribbon is securely fastened and should not unravel as the bride proceeds down the aisle.

THE
COMPLETED
BRIDAL
BOUQUET

Resources

WHITE FLOWERS

Agapanthus

Anemones

Anthuriums

Asters

Cattelya Orchids

Chrysanthemums

Cosmos

Cymbidium Orchids

Delphiniums

Dendrobium Orchids

Dianthus (carnations)

Digitalis

Euphorbia

Freesia

Gardenias

Gerberas

Gladiolas

Gypsophilia (baby's breath)

Hydrangeas

Irises

Ornithogalums

Phalaenopsis Orchids

Phlox

Protea

Roses

Stephanotis

Sweet Peas

Tulips

Viburnums

Zinnias

YELLOW FLOWERS

Acacia (mimosa)

Alstroemeria

Chryanthemums

Cymbidium Orchids

Dianthus (carnations)

Digitalis

Eremurus

Euphorbia

Forsythia

Freesia

Gerbera

Gladiolas

Heliconias

Irises

Narcissus

Roses

Tulips

Zinnias

ORANGE FLOWERS

Asclepias

Chrysanthemums

Dianthus (carnations)

Euphorbia

Freesia

Gerbera

Gladiolus

Lilies

Marigolds

Poppies

Roses

GREEN FLOWERS

Amaranthus

Bupleurum

Chrysanthemum

Cymbidium Orchids

Dendrobium Orchids

Dianthus (carnations)

Euphorbia

Gladiolus

Roses

Viburnums

Zinnias

BLUE FLOWERS

Agapanthus
Delphiniums
Hydrangeas
Lilies
Nigellas

VIOLET AND PURPLE FLOWERS

Alliums
Alstroemerias
Anemones
Anthuriums
Asters
Catteleya Orchids
Chrysanthemums
Dendrobiums
Dianthus (carnations)
Freesia
Gladiolus
Hydrangeas
Iris
Lavender
Lilacs
Phlox
Roses
Salvia
Statice
Sweet Peas
Trachelium
Tulips
Verbena

RED AND PINK FLOWERS

Alstroemeria
Amaranthus
Anemones
Anthuriums
Celosia
Chrysanthemums
Cosmos
Cymbidium Orchids
Dianthus (carnations)
Gerbera
Ginger
Gladiolus
Heliconia
Hypericum
Leucadendron
Lilies
Nerines
Nigella
Phlox
Protea
Roses
Sedums
Sweet Peas
Tulips
Zinnias

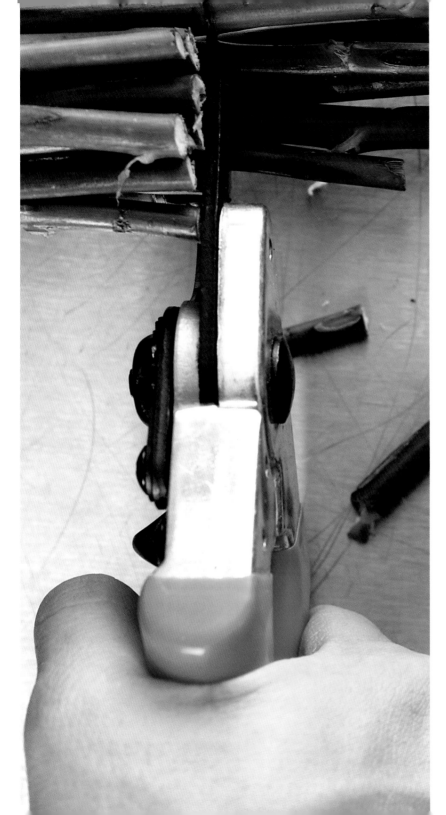

EQUIPMENT

Floral shears by Falco are particularly good for cutting branches such as cherry. We use Falco #2 at FlowerSchool.

The simplest Swiss Army knife is always good for cleaning roses and cutting woody stems.

Good clippers and scissors from such companies as Smith & Hawkin or other garden suppliers are essential.

Florist foam or Oasis can usually be purchased at your local florist or garden center.

TIPS FROM THE INSTRUCTORS AT FLOWERSCHOOL NEW YORK

When purchasing flowers, look at the foliage—make sure the leaves are fresh. That will give you the best clue to see whether the flowers are old or not.

When in doubt, go monochromatic! Mixing several types of the same color flowers always works well in a centerpiece or a bouquet. Massive amounts of the same color and type of flower in a vase, such as 40 irises or 30 white roses, always makes a dramatic statement.

As soon as you get flowers, cut them and put them in clean, cold water. Flowers need to drink! There are different philosophies about conditioning them; however, everyone agrees that after they have been out of water for a long time, they need hydration immediately.

When you arrange flowers, make sure that there are never any leaves below the waterline of the vase. These leaves can aid the formation of bacteria, which will cloud the water and eventually kill the flowers.

Make sure that your vase is clean. Change the water every other day and cut the stems so that the flowers are always drinking fresh water.

Keep cut flowers away from direct sunlight and heat. Never put them on a radiator.

BUYING FRESH FLOWERS

Buying from a quality local florist is one of the best ways to guarantee the freshness of your flowers.

Whole Foods has a good selection of fresh flowers, many of them from organic producers.

Farmers Markets are a good source of flowers that are locally grown and in season.

WEBSITES

Some wholesale markets such as the San Francisco Flower Mart are open to the public during restricted hours. Their website is: *www.sfflmart.com*

Here are some additional websites that may be helpful:

FLOWER SITES

About Flowers (SAF)
www.aboutflowers.com

Flowers. Alive with Possibilities.
www.flowerpossibilities.com

TRADE ORGANIZATIONS

American Institute of Floral Designers (AIFD)
www.aifd.org

California Cut Flower Commission
www.ccfc.org

California State Floral Association
www.calstatefloral.com

International Cut Flower Growers Association
www.rosesinc.org

Kee Kitayama Research Foundation
www.kkrf.org

Master Florists Association
www.masterfloristsassn.org

NORCAL - California Association of Flower Growers and Shippers
www.norcalflowers.org

Society of American Florists
www.safnow.org

Wholesale Florist & Floral Supplier Association
www.wffsa.org

TRADE PUBLICATIONS

Florists' Review
www.floristsreview.com

Flowers &
www.flowersandmagazine.com

ADDITIONAL WHOLESALE MARKETS

The Boston Flower Exchange
www.thebostonflowerexchange.com

Los Angeles Flower District Association
www.laflowerdistrict.com

San Diego International Floral Trade Center
www.floraltradecenter.com

OTHER

San Francisco Design Center
www.sfdesigncenter.com

San Francisco Wholesale Produce Market
www.sfproduce.org

Alamance Community College
Library
P.O. Box 8000
Graham, NC 27253